CLINICAL BIOCHEMISTRY: TECHNIQUES AND INSTRUMENTATION

CLINICAL BIOCHEMISTRY: TECHNIQUES AND INSTRUMENTATION

A Practical Course

John S. Varcoe

Royal Melbourne Institute of Technology, Australia

World Scientific
Singapore • New Jersey • London • Hong Kong

Published by

World Scientific Publishing Co. Pte. Ltd.

P O Box 128, Farrer Road, Singapore 912805

USA office: Suite 1B, 1060 Main Street, River Edge, NJ 07661

UK office: 57 Shelton Street, Covent Garden, London WC2H 9HE

British Library Cataloguing-in-Publication Data
A catalogue record for this book is available from the British Library.

ISBN 981-02-4550-5
ISBN 981-02-4556-4 (pbk)

Printed in Singapore.

CONTENTS

Chapter 2. Fluorimetry.

Chapter 4. Atomic Emission and Absorption.   4-1

Chapter 5. Ion Selective Electrodes.   5-1

Chapter 6.   Oxygen and Carbon Dioxide Electrodes.

Chapter 7.   Chromatography.

Chapter 8.   Thin Layer Chromatography and Extraction Techniques.

Chapter 14. Coulometry, Osmometry and Refractometry.   14-1

PREFACE

Clinical biochemistry (clinical chemistry) is an analytical and interpretive science. The analytical part involves the determination of the level of chemical components in body fluids and tissues. The interpretive part examines these results and uses them in the diagnosis of disease, the screening for susceptibility to specific diseases and for monitoring the progress of treatment. The analysis can be for the presence of abnormal components, normal components in abnormal amounts, and for monitoring the level of therapeutic drugs.

The analysis requires the use of a variety of techniques and instruments. This book is designed to cover the major techniques and analytical instruments used in clinical biochemistry. Hands-on experience greatly enhances the understanding of analytical techniques and instrumentation. Hence, the book is also designed to be used as a laboratory manual.

Each chapter is based on a specific technique, or techniques, with associated instrumentation. These are discussed in some detail, leading into the practical exercises. The first exercises in most chapters are a general introduction to the technique, leading on to those with a clinical bias. Where applicable, the clinical practical exercises are associated with a case history and/or the discussion of the relevance of the assay to diagnosis and prognosis and to monitoring recovery. The practical exercises are set out in an easy to follow step-by-step fashion with details of reagent preparation etc. Each chapter concludes with a selection of appropriate references.

The book is based on material taught in the degree and masters courses in medical laboratory science and clinical toxicology at the Royal Melbourne Institute of Technology (Australia). As with most teaching institutions, access to the latest instrumentation is limited due to a lack of available funds. Hence, some exercises use instruments that have been around for a while. These have an advantage over the latest instruments in that they require a greater understanding of how the instrument works, rather than pressing a few keys on a keyboard. This helps to dispel the "black-box" mentality that some students have towards analytical instruments. We were fortunate in that the major Melbourne hospitals and private pathology laboratories gave us their old instruments when they updated. Hence, we were not too out of date. Where specific instruments are used in the exercises, the analytical procedure should be able to be modified, fairly easily, to suit other instruments.

Most analytical work carried out in a clinical biochemistry (clinical chemistry) laboratory is now undertaken using automated instruments. There are a whole variety of these on the market and in service. These instruments are designed for easy use and can be run by people with little training. This book is designed to educate the users of these instruments about the analytical principles that lie beneath the surface of these pieces of equipment.

The book is mainly designed for students and staff in medical science courses, clinical biochemistry (clinical chemistry) hospital and private pathology laboratory staff. The book will also be of use in a general biochemistry course, as most of the contents are of general interest. Some of the material would be of use in certificate and diploma courses in medical technology. Analytical chemistry students will also find this book of value.

ACKNOWLEDGMENTS

I acknowledge the help and advice received from my colleagues at the Royal Melbourne Institute of Technology (RMIT) in relation to the contents of this book. I would especially like to thank They Ng who read the manuscript and made valuable suggestions for improvements. I also extend my thanks to David Propert who made some good suggestions regarding the Molecular Diagnostics chapter. Thanks also go to Brian Stevens for sharing his extensive knowledge of atomic absorption techniques. I should also like thank friends and acquaintances in the Victorian hospitals, private pathology laboratories, as well as instrument and reagent supply companies for their help during my time at RMIT and with some aspects of writing this book. Special thanks go to David Casley for the use of some of the notes he gave to students, for whom he was a demonstrator. Parts of Appendices 3 and 4 are based on these notes.

Appendix 7 (Units of Enzyme Activity) is based on Chapter 4 (Units of Enzymic Activity), p 26-27 of Enzyme Nomenclature - Recommendations (1972) of the International Union of Pure and Applied Chemistry and the International Union of Biochemistry, Elsevier Scientific Publishing Company.

I should like to thank the following companies for permission to use material from their commercial kits:

Amersham Pharmacia Biotech	:	Cyclic AMP kit.
ANSYS, Inc.	:	TOXI-LAB kit.
Dade Behring	:	Syva EMIT®-TOX kit.
Ortho-Clinical Diagnostics	:	Amerlex Digoxin kit.
Zeneca Diagnostics	:	CF (4) m-PCR kit.

This book has developed from practical courses in clinical biochemistry and toxicology, taught at RMIT, over a period of 27 years. I am most grateful to past students who have offered comments for the improvement of exercises included in this book.

As most of the exercises were developed for courses run at RMIT, possible copyright problems have been resolved with the RMIT Copyright Office.

ABBREVIATIONS

This list of abbreviations, used in the book, is mainly non-standard abbreviations. Standard English and mathematical abbreviations and symbols are not listed. Standard chemical symbols and abbreviations for the chemical elements are also not listed. Abbreviations for standard measurement units, including subdivisions and multiples, are not included. Standard journal abbreviations are also not listed. Three letter abbreviations for amino acids are not included, but the single letter abbreviations are included. Abbreviations used by instrument manufacturers on the controls of their instrument, that appear in the text, are not covered in the list below. Greek letters, and their use in this text, can be found in Appendix 5.

$\perp$	:	Perpendicular
//	:	Parallel
*	:	Electron anti-bonding orbital (Eg. π*)
*	:	Excited state of an atom or molecule
*	:	Labelled molecule (Eg. Ag*)
[]	:	Concentration (Eg. [Na] = Sodium concentration)
A	:	Absorbance
A	:	Adenine
A	:	Adult
A	:	Alanine
A	:	Analyte
A	:	Angelman
a	:	Ion activity
A^-	:	Dissociated acid
A/G	:	Albumin/Globulin ratio
aa	:	Amino acid
AACC	:	American Association for Clinical Chemistry
Ab	:	Antibody
Abs	:	Absorbance
AC	:	Adenyl Cyclase
AC	:	Alternating Current
Ag	:	Antigen
AGE	:	Advanced Glycated End product
AlAT	:	Alanine Aminotransferase
Alk Phos	:	Alkaline Phosphatase
AMP	:	Adenosine Monophosphate
AP	:	Alkaline Phosphatase
AR	:	Analar (reagent quality)
ARMS	:	Amplification Refractory Mutation System
ASO(s)	:	Allele Specific Oligonucleotide(s)
ATP	:	Adenosine Triphosphate
B	:	Bound label
B	:	Brain
B	:	Non-protonated base
B Pt	:	Boiling Point
b	:	Bases (Eg. kb = kilobases (10^3 bases) and Mb = megabases (10^6 bases))
β^+	:	Reduced production of β globin
Bg	:	Background
BH^+	:	Protonated base
Bis (bis)	:	Methylene bisacrylamide

B_O	:	Bound label of the zero standard
β^o	:	Absence of β globin
bp	:	Base pairs
Bq	:	Becquerel (disintegrations per second) (MBq = 10^6 Bq)
BRCA	:	Breast cancer genes
BSA	:	Bovine Serum Albumin
BSTFA	:	N,O-Bis (trimethylsilyl)-trifluoroacetamide
C	:	Child
C	:	Cross-linker (Bis percentage of total acrylamide)
C	:	Cytosine
C or c	:	Concentration
c-AMP	:	Cyclic AMP
C3	:	Complement component
C4	:	Complement component
Cal	:	Calibrate
Cat	:	Catalogue
cDNA	:	Complementary DNA
CFTR	:	Cystic Fibrosis Transmembrane Conductance Regulator
CK	:	Creatine Kinase
C_O	:	Bq of zero standard
CoA	:	Coenzyme A
CoA-SH	:	Free coenzyme A (free -SH group)
COHb	:	Carboxyhaemoglobin
Conc	:	Concentration
cPDE	:	Cyclic nucleotide phosphodiesterase
cpm (c/m)	:	Counts per minute
CRT	:	Cathode Ray Tube
C_s	:	Concentration of the standard
Cs	:	Control sample
CSL	:	Commonwealth Serum Laboratories (Australia)
CTP	:	Cytidine Triphosphate
C_u	:	Concentration of the unknown
C_x	:	Bq of standards or unknown
D	:	Aspartic acid
D	:	D lines of sodium (2 emission lines in the yellow, 589.0 and 589.6 nm)
d	:	Deoxy (Eg. dATP = deoxyadenosine triphosphate)
d	:	Electron sub-shell
DC	:	Direct Current
dd	:	Dideoxy (Eg. ddATP = dideoxyadenosine triphosphate)
DELFIA	:	Delayed Enhanced Lanthanide Fluorescence Immunoassay
DNA	:	Deoxyribonucleic Acid
dNTPs	:	Deoxyribonucleotide triphosphates
dp	:	Particle diameter
DPC	:	Diagnostic Products Corporation
dps (d/s)	:	Disintegrations per second
E	:	Electric field strength (volts/cm)
E	:	Electric potential
E	:	Electronic (energy)
E	:	Enzyme
E	:	Evidence
E	:	Glutamic acid
e^-	:	Electron

EC	:	Enzyme Commission
ECD	:	Electron Capture Detector
Ed	:	Edition
Ed(s)	:	Editor(s)
EDTA	:	Ethylenediaminetetraacetic Acid
EEL	:	Evans Electroselenium Limited
EI	:	Enzyme Inhibitor complex
EIS	:	Enzyme Inhibitor Substrate complex
ELISA	:	Enzyme Linked Immunoadsorbent Assay
EMIT	:	Enzyme Multiplied Immunoassay Technique
Enz	:	Enzyme
ES	:	Enzyme-Substrate complex
ESR	:	External Standard Ratio
et al	:	And others
eV	:	Electron volt (the energy imparted to an electron by a potential of one volt) ($keV = 10^3$ eV and $MeV = 10^6$ eV)
F	:	Faraday's constant
F	:	Father
F	:	Filter (spectrophotometer diagrams)
F	:	Fluorescent molecule
F	:	Phenylalanine
FET	:	Field Effect Transducer
FID	:	Flame Ionisation Detector
FISH	:	Fluorescence *In Situ* Hybridisation
G	:	Glycine
G	:	Grating (diffraction)
G	:	GTP regulatory site
G	:	Guanine
g	:	Centrifugal force (in relation to gravity)
G-6-P	:	Glucose-6-phosphate
G-6-PD	:	Glucose-6-phosphate Dehydrogenase
GC	:	Gas Chromatography
GC-MS	:	Gas Chromatography - Mass Spectrometry
GLC	:	Gas Liquid Chromatography
GSC	:	Gas Solid Chromatography
GTP	:	Guanosine Triphosphate
H	:	Heart
H	:	Histidine
H	:	Hormone
H	:	Plate height (HETP)
h	:	half life
h	:	Planck's constant
h	:	Reduced plate height (number of packing particles per plate)
HA	:	Undissociated acid
HAT	:	Hypoxanthine, Aminopterin and Thymidine
Hb	:	Haemoglobin
HbA	:	Major adult form of haemoglobin
HbA_0	:	Major component of HbA
HbA_1	:	Modified HbA, containing more negative charge
HbA_{1C}	:	Major component of HbA_1 (glycated β-chains of Hb)
HbA_2	:	Minor adult haemoglobin
HbF	:	Foetal haemoglobin

HbS	:	Sickle cell haemoglobin
HD	:	Huntington's Disease
HDL	:	High Density Lipoprotein
h_{eff}	:	Effective reduced plate height
HETP	:	Height Equivalent to a Theoretical Plate
HGPRT	:	Hypoxanthine-Guanine Phosphoribosyl Transferase
HIV	:	Human Immunodeficiency Virus
HPLC	:	High Performance Liquid Chromatography
HPLC-MS	:	High Performance Liquid Chromatography - Mass Spectrometry
HV	:	High Voltage
I	:	Inhibitor
I	:	Intensity of transmitted light
I	:	Ionic strength
i	:	Angle of incidence
ICP	:	Inductively Coupled Plasma
ICP-MS	:	Inductively Coupled Plasma - Mass Spectrometry
ID	:	Internal Diameter
IEF	:	Isoelectric Focusing
Ig	:	Immunoglobulin
IgA	:	Secretory immunoglobulin
IgG	:	Major immunoglobulin
IgM	:	First immunoglobulin produced by plasma cells
IL	:	Instrumentation Laboratory
INT	:	Iodonitrotetrazolium violet.
I_o	:	Intensity of incident light
IR	:	Infra Red
IRMA	:	Immunoradiometric Assay
IS	:	Internal Standard
ISE	:	Ion Selective Electrode
IU	:	International Units
J	:	Joining (segment of Ig gene)
K	:	Absorption coefficient (light)
K	:	Affinity constant (antibodies)
K	:	Dissociation (ionisation) constant (acids and bases)
K	:	Distribution coefficient (chromatography)
K	:	Equilibrium constant (chemical reactions)
K	:	Inner electron shell
K	:	Lysine
K	:	Response factor (of a detector)
k	:	Rate constant
k'	:	Column capacity factor (the number of column volumes needed to elute a solute)
$\bar{k}'$	:	Average column capacity factor
K_m	:	Michaelis constant (substrate concentration at half maximum rate (velocity))
K_{m_I}	:	K_m with inhibitor present (apparent K_m)
L	:	Length
L	:	Lens (spectrophotometer diagrams)
LCR	:	Ligase Chain Reaction
LD	:	Lactate Dehydrogenase
LDL	:	Low Density Lipoprotein
LED	:	Light Emitting Diode

LFT(s)	:	Liver Function Test(s)
LKB	:	LKB-Produkter AB
Ln	:	Natural logarithm
M	:	Gram molecular weight/litre solution
M	:	Marker
M	:	Mirror (spectrophotometer diagrams)
M	:	Molecule
M	:	Mother
M	:	Motor
M	:	Muscle
M	:	Normal α_1-Antitrypsin gene or gene product
m-RNA	:	Messenger RNA
M.Wt.	:	Molecular Weight
max	:	Maximum
MDE	:	Mutation Detection Enhancement
MEEC	:	Micellar Electrokinetic Capillary Chromatography
MMA	:	Methylmalonic Acid
mol	:	Mole (gram molecular weight)
MRC	:	Medical Research Council
MS	:	Mass Spectrometry
MU	:	Methyl Umbelliferone
MU	:	Methylene Units
MUP	:	Methyl Umbelliferyl Phosphate
N	:	Efficiency (number of theoretical plates)
N	:	Normal
n	:	Layer in photodiode
n	:	Non-bonding electron orbitals
n	:	Number
NAD	:	Nicotinamide Adenine Dinucleotide
NADH	:	Reduced NAD
NADP	:	Nicotinamide Adenine Dinucleotide Phosphate
NADPH	:	Reduced NADP
NBT	:	Nitro Blue Tetrazolium
N_{eff}	:	Effective plate number
No	:	Number
NPD	:	Nitrogen Phosphorus Detector
NSB	:	Non-specific Binding
NTP(s)	:	Nucleotide Triphosphate(s)
o	:	Product (in ELISA diagrams)
o	:	Unlabelled molecule (Eg. Ag^o)
OD	:	Optical Density
Osm	:	Osmoles
Ox	:	Oxidised
P	:	Counts per second of the photopeak
P	:	Patient
P	:	Product
p	:	Electron sub-shell
p	:	Layer in photodiode
p	:	Partial pressure
p.p.m.	:	Parts per million
PAGE	:	Polyacrylamide Gel Electrophoresis
PCR	:	Polymerase Chain Reaction

PD	:	Potential Difference
PEG	:	Polyethylene Glycol
pH	:	Logarithm of the reciprocal of the H^+ concentration (log $1/[H^+]$, ie - log $[H^+]$).
PHA	:	Pulse Height Analyser
Pi	:	Inorganic phosphate
pI	:	Isoelectric point (pH at which a molecule has no net charge)
pK	:	Logarithm of the reciprocal of the dissociation (ionisation) constant (log $1/K$, ie - log K) (this is the pH at which there are equal concentrations of ionised and non-ionised forms of a molecule)
PKU	:	Phenylketonuria
PLOT	:	Porous Layer Open Tubular (GC columns)
PM	:	Photomultiplier
PMS	:	Phenazine Methosulphate
POPOP	:	1,4-di-(2-(5-phenyloxazolyl))-benzene
PPO	:	2,5-diphenyloxazole
Prep	:	Preparation
psi (p.s.i.)	:	Pounds per square inch
PTH	:	Parathyroid Hormone
PVC	:	Polyvinyl Chloride
PW	:	Prader-Willi
Q	:	Fluorescence intensity
q	:	Quencher molecule (fluorescence)
QC	:	Quality Control
R	:	Arginine
R	:	Gas Constant
R	:	Receptor
R	:	Reference
R	:	Relaxed form of haemoglobin
r	:	Angle of refraction
r	:	Radius
RCH	:	Royal Children's Hospital (Melbourne)
Red	:	Reduced
R_f	:	Relative migration of a solute to the solvent front
RFLP(s)	:	Restriction Fragment Length Polymorphism(s)
RI	:	Refractive Index
RIA	:	Radioimmunoassay
RI_s	;	Refractive Index of sample
RI_w	:	Refractive Index of water
RNA	:	Ribonucleic Acid
rpm	:	Revolutions per minute
R_s	:	Resolution
RT	:	Room Temperature
S	:	An α_1-Antitrypsin mutation
S	:	Counts per second of the sum peak
S	:	Serine
S	:	Sickle cell
S	:	Slit (spectrophotometer diagrams)
S	:	Substrate
S or s	:	Sample
s	:	Electron sub-shell
s	:	Saturation

S_{0x}	:	Ground state of solvent molecules
S_{0y}	:	Ground state of solute molecules
S_{1x}	:	Lowest excited singlet state of solvent molecules
S_{1y}	:	Lowest excited singlet state of solute molecules
Sat	:	Saturation
SCOT	:	Support-Coated Open Tubular (GC columns)
SD	:	Standard Deviation
SDS	:	Sodium Dodecyl Sulphate
SDS-PAGE	:	Sodium Dodecyl Sulphate - Polyacrylamide Gel Electrophoresis
SG	:	Specific Gravity
SIM	:	Selective Ion Monitoring (Mass Spectrometry)
Sin	:	Sine
S_{nx}	:	Solvent molecules excited energy levels
SPE	:	Serum Protein Electrophoresis
STR(s)	:	Short Tandem Repeat(s)
T	:	Taut form of haemoglobin
T	:	Temperature
T	:	Thymine
T	:	Total acrylamide (as a percentage of a gel)
T	:	Total counts per second
T	:	Transmittance
t'_R	:	Corrected retention time
TBE	:	Tris-Borate EDTA buffer
TC	:	Total Counts
TCA	:	Trichloroacetic Acid
TCD	:	Thermal Conductivity Detector
TDM	:	Therapeutic Drug Monitoring
Temp	:	Temperature
THFA	:	Tetrahydrofolic Acid
TLC	:	Thin Layer Chromatography
t_m	:	Retention time of a non-retained compound
TMCS	:	Trimethylchlorosilane
TMED	:	N,N,N',N'-Tetramethylethylenediamine
TPP	:	Thiamine Pyrophosphate
t_R	:	Retention time
Tris	:	Tris(hydroxymethyl)aminomethane
TSH	:	Thyroid Stimulating Hormone
TTP	:	Thymidine Triphosphate
U	:	Units
U	:	Unknown
U	:	Uracil
Us	:	Unknown sample
UTP	:	Uridine Triphosphate
UV	:	Ultra Violet
V	:	Vibrational (energy)
V	:	Victim
V or v	:	Velocity
v	:	Volume
V_m	:	Volume of mobile phase
V_{max}	:	Maximum velocity
V_{max_I}	:	Maximum velocity with inhibitor present
VNTR(s)	:	Variable Number of Tandem Repeat(s)

V_O	:	Void volume (column chromatography)
V_s	:	Volume of stationary phase
W	:	Tryptophan
W	:	Window (spectrophotometer diagrams)
W and w_b	:	Base peak width from tangents drawn at the inflection points
w	:	Water
WCOT	:	Wall-Coated Open Tubular (GC columns)
w_h	:	Peak width at half peak maximum height'
w_i	:	Peak width at inflection points
X	:	Stop codon
X	:	X axis (abscissa, horizontal axis)
x'	:	First McReynolds constant (used for GC column selection - other McReynolds constants are y', z', u' and s')
X^+	:	Ion-pair reagent
XX	:	Female (sex chromosomes)
XY	:	Male (sex chromosomes)
Y	:	Y axis (ordinate, vertical axis)
Y^-	:	Ion-pair reagent
Z	:	An α_1-Antitrypsin mutation
z	:	Ionic charge

CHAPTER 1

SPECTROPHOTOMETRY

PHOTOMETRIC ANALYSIS.

This method of analysis is one of the most useful assay techniques in biochemistry. Comparison is made between the amount of light absorbed by the unknown and that absorbed by a series of known standards.

Visual comparison can be made by comparing the intensity of a coloured solution with a series of known standards. This method was used in the Lovibond Comparator, where the unknown was compared with a series of coloured discs. A dilution method can also be used whereby the unknown is diluted to match the intensity of the standard. An example of this is the Icteric Index, used to give a rough idea of the serum bilirubin concentration. The Icteric Index is the number of times serum is diluted to give a colour corresponding to a 100 mg/l solution of potassium dichromate. Visual judgement has, however, been almost completely replaced by the use of photoelectric detectors.

If a beam of electromagnetic radiation is passed through a transparent layer of solid, liquid or gas some of that radiation may be selectively absorbed. If the emergent beam is then passed through a prism, dark bands will be seen corresponding to the wavelengths that have been absorbed. This is known as the absorption spectrum.

Atoms and molecules can exist in a number of electronic states or energy levels. Energy can be absorbed by these atoms or molecules, moving their electrons into a higher energy state. On returning to the ground state, energy is released in the form of heat and sometimes as light emission of a lower frequency than that of the excitation beam, in which case we have fluorescence.

The energy imparted by electromagnetic radiation is directly proportional to the frequency of the radiation and so the energy required to alter the electronic state of an atom or molecule corresponds to a specific frequency of radiation. It is, therefore, this frequency that is absorbed by the sample under consideration.

The portion of the electromagnetic spectrum of interest covers the ultraviolet (UV), visual and infrared (IR) region. The UV region is subdivided into near and far UV. Atmospheric oxygen absorption sets the limit of near UV and, hence, the highest frequency that can be conveniently used in UV spectrophotometry. The corresponding wavelengths (velocity of light ÷ the frequency) are given below:

Far ultraviolet		10 - 190 nm
Near ultraviolet		190 - 380 nm
Visible region	Violet	380 - 430 nm
	Blue	430 - 490 nm
	Green	490 - 575 nm
	Yellow	575 - 590 nm
	Orange	590 - 650 nm
	Red	650 - 780 nm
Infra-red		0.78 - 300 μm

Most clinical determinations are carried out in the visual range; however, UV is required for a number of assays. UV absorption spectra can be useful in drug identification and quantitation. IR is used mainly in the identification of organic radicals, which have specific absorption bands in the IR. IR is not often used in clinical biochemistry.

The internal energy of a molecule can be considered to be the sum of the energy of its electrons (electronic energy), the energy of vibrations between its constituent atoms (vibrational energy) and the energy associated with the rotation of the molecule (rotational energy).

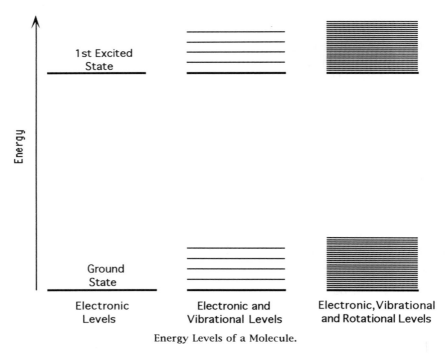

Energy Levels of a Molecule.

The quantum theory tells us that changes in any of these types of energy must be in finite steps. Changes in electronic energy levels involve the largest energy changes. Vibrational energy level changes involve a smaller energy change and rotational energy level changes are still smaller.

In complex molecules, the energy levels associated with the electrons are spread such that UV or visible light can impart enough energy to increase the electronic energy level of the molecule. Changes in the vibrational levels can be produced by the absorption of infrared radiation and changes in rotational energy can be produced by the absorption of infrared or microwave radiation.

The absorption of UV and visible light is associated mainly with the electrons in π bonding orbitals moving into the π anti-bonding configuration π^* and n (non-bonding) electrons moving to π^*. π bonding orbitals are associated with multiple bonds between two atoms. Eg.

$$\begin{array}{ccc} \diagdown & \diagup \\ C = C & \text{and} & -N=N- \\ \diagup & \diagdown \end{array}$$

n non-bonding orbitals are associated with atoms having lone pairs of electrons. Eg.

$$- \overset{|}{\underset{|}{C}} - \overset{\cdot\cdot}{\underset{\cdot\cdot}{Br}} : \quad \text{and} \quad \overset{\diagdown}{\underset{\diagup}{C}} = \overset{\cdot\cdot}{\underset{\cdot\cdot}{O}}$$

π and π^* orbitals are illustrated below (π orbitals are in the thin line and π^* are in the thick lines):

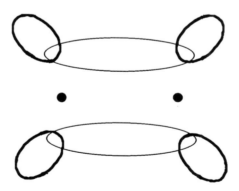

Where conjugated double bond exist, the electron orbitals extend over the conjugation site and this alters the π and π^* energies, as illustrated below:

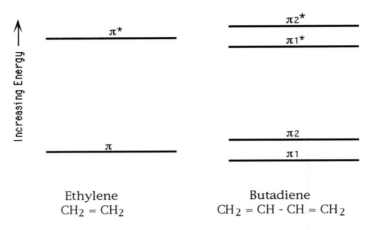

Ethylene
$CH_2 = CH_2$

Butadiene
$CH_2 = CH - CH = CH_2$

Hence, the energy required to move an electron from π_2 to π_1^* is less than that required to move from π to π^*. As a result, the absorbed light is of a longer wavelength. One also finds that the amount of absorbed light is increased. This is illustrated for the above compounds, in hexane:

	λ_{max}	ε
Ethylene	163 nm	15,000
Butadiene	217 nm	21,000

λ_{max} is the wavelength at which maximum absorption takes place. ε is the absorption of a 1 molar solution with a 10 mm path length.

Similar effects occur when groups containing n electrons are conjugated with a π electron group. These groups are known as chromophores.

About 6 conjugated double bonds are requires for the compound to absorb in the visual range and become coloured.

Transition metals can form coloured complexes with organic compounds. The colour is due to electron shifts in the metal atom, which are modified by the organic part of the molecule.

Other groups that do not significantly absorb light, when part of a molecule, may alter the absorption characteristics of that molecule. These groups are known as auxochromes and include groups such as amines and alcohols. These can either increase or decrease the λ_{max} and ε_{max}.

If electronic energy changes were the only energy changes that took place, the compound would show very narrow absorption lines corresponding to the wavelengths of light with just the right amount of energy to produce this change. However, one finds that changes in vibration energy levels and rotational energy levels also take place. These are added to the excited electronic energy of the molecule. Tightly packed molecules (as seen in solutions) exert influences on each other and effect the energy levels of adjacent molecules.

Therefore, a large number of energy levels can be reached in the excited molecules, each level corresponding to a specific wavelength of light being absorbed. There are so many of these absorbed wavelengths that the absorption spectrum of a solution exists as a broad peak rather than a number of sharp absorption lines.

The individual absorption lines can still be seen in loosely packed molecules seen in gases. This is illustrated by the UV absorption spectra of benzene between 225 and 275 nm, as benzene vapour and as a 1 in 1,500 solution of benzene in methanol (0.1 nm band width):

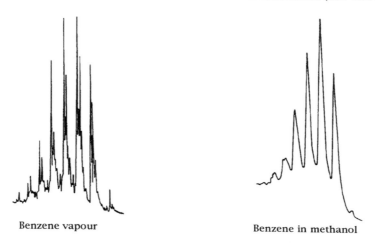

Benzene vapour   Benzene in methanol

In the spectrum of benzene vapour, the vibrational fine structure can be seen superimposed on the main electronic bands. This fine structure is lost in the benzene solution.

The Effect of Solvent.

Solvents can effect absorption spectra, as well as absorbing UV light of shorter wavelengths.

Conjugated dienes and aromatic hydrocarbons show little change in absorption spectra with different organic solvents. However, unsaturated carbonyl compounds, for example, have increased π to π^* orbitals (A blue shift) and decreased n to π^* orbitals (A red shift) with solvents less polar than water.

The wavelength at which a solvent absorbs 10% of the incident light (Abs $\approx$ 0.05) in a 10 mm cuvette is called the cut-off wavelength. Some examples (in nm) are:

Water & Acetonitrile	:	190.
Ethanol & Methanol	:	210.
Chloroform	:	245.
Benzene	:	280.
Acetone	:	330.

LAWS OF RADIANT ENERGY ABSORPTION.

For any particular absorbing system, where the incident radiation intensity and wavelength are constant, two factors affect the degree of absorption; these are: (i) The concentration of absorbing substance and (ii) The length of the energy path through the absorbing system.

Lambert's Law.

This states that the proportion of monochromatic light absorbed by a homogeneous medium is independent of the intensity of the incident light and that each layer of equal thickness of the medium absorbs an equal fraction of the radiant energy that traverses it.

Eg. If our absorbing medium absorbs 25% of the radiant energy, 75% emerges. If a similar piece of absorbing medium of equal thickness is then placed in the emerging light path this absorbs 25% of that light (ie. 25% of the 75% of the original radiant energy). Therefore, 56.25% of the original radiant energy emerges from the second absorbing medium.

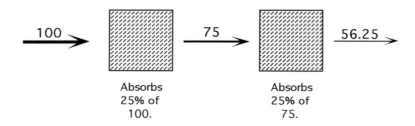

Beer's Law.

Beer applied this law to solutions. His law states that the fraction of light absorbed by a solute in a transparent solvent depends only on the quantity of solute in the light path.

Therefore, doubling the concentration of the solute is the same as having two cuvettes of the original concentration in the light path and Lambert's law applies as above. As the concentration of the solute increases, or the light path length through the solution increases, the proportion of light that emerges falls exponentially:

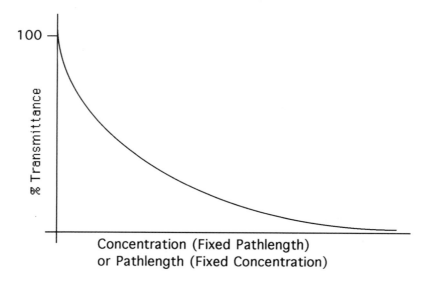

The relationship between the intensity of the emitted light and that of the incident light is given by the formula:

$$I \quad = \quad I_0 10^{-KCL}$$

Where $\quad$ I $\quad$ = $\quad$ The intensity of the transmitted light.

$\quad\quad\quad$ I_0 $\quad$ = $\quad$ The intensity of the incident light.

$\quad\quad\quad$ K $\quad$ = $\quad$ Absorption Coefficient (the absorbance of the solute at a known concentration for a light path of 10 mm). This is often the Molar Absorption Coefficient (ε).

$\quad\quad\quad$ C $\quad$ = $\quad$ Concentration of the solute.

$\quad\quad\quad$ L $\quad$ = $\quad$ Length of light path through the solution in 10 mm units.

Dividing both sides by I_0 we have:

$$\frac{I}{I_0} \quad = \quad 10^{-KCL}$$

$\frac{I}{I_0}$ is known as the transmittance (T) and is usually expressed as percentage transmittance, ie. 100 x T.

Take logarithms of both sides in the above equation:

$$Log_{10} \frac{I}{I_0} \quad = \quad -KCL$$

Reverse signs:

$$-Log_{10} \frac{I}{I_0} \quad = \quad KCL$$

$$ie. \quad Log_{10} \frac{I_0}{I} \quad = \quad KCL$$

$Log_{10} \frac{I_0}{I}$ is known as the absorbance (Abs or A).

It used to be known as the optical density (OD). This is now an obsolete term and should be avoided.

$$ie. \quad Abs = KCL$$

Beer-Lambert's Law.

The above relationship is summed up in this law, which states that the absorbance (Abs) of a solution is directly proportional to the concentration of the absorber and to the length of the path of radiant energy through that solution.

The law refers to the absorption of monochromatic light. "Monochromatic" light is a term applied to a narrow band of light. However, the dimensions of this band have not been officially defined. Thus, for the same solute, K remains constant, and using a 10 mm path length cuvette, L is also kept constant. Therefore, Abs is directly proportional to the concentration (C) of the solute.

Therefore, when Beer's law is obeyed we have:

$$\frac{Abs_u}{Abs_s} = \frac{C_u}{C_s}$$

Where u is the unknown and s the known standard.

Beer's law is not always obeyed. Unless it is known to be obeyed, over the concentration range under consideration, a standard curve must be plotted. The Abs of a series of standards is plotted against concentration:

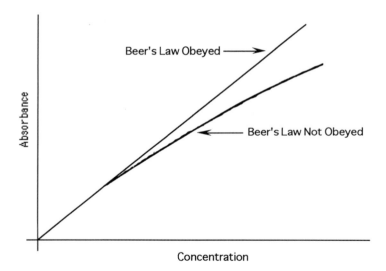

Interrelationship Between Absorbance and Transmittance.

This is given by the formula:

$$Abs = Log_{10} \frac{1}{T}$$

Also $$Abs = 2 - Log_{10} \%T$$

COLOUR AND COLORIMETERS.

The colour of a solution depends on the absorption wavelengths and is due to the unabsorbed light. White light can be divided into red, yellow and blue and so if the red is absorbed the resultant colour will be due to the yellow and blue, ie. green.

Hence, to obtain maximum absorption of a green solution, a red filter would be used. Likewise, a blue filter would be used for an orange (red + yellow) solution.

However, it is better to know the absorption characteristics of the solution of interest and to match this with a filter having transmission characteristics at the same wavelengths.

THE BLANK.

Strictly speaking, both transmittance and absorbance are concerned with the incident light; ie. that falling on the solution. It is impractical to measure this in absolute terms, so the light falling on the photo-detector, after it has passed through the reagent (allowing no colour development), is used as a reference.

The instrument is set to 100% transmittance (Zero **Abs**) using this "Reagent Blank". The blank is then replaced by the test solution and the emergent beam now represents the incident beam minus the light lost through absorbance.

INSTRUMENTATION.

The basic components of colorimeters and spectrophotometers are shown below:

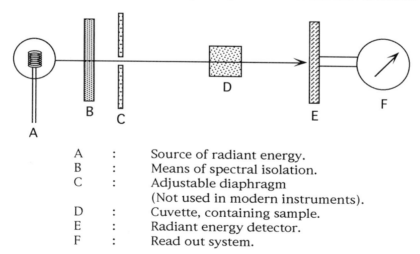

A	:	Source of radiant energy.
B	:	Means of spectral isolation.
C	:	Adjustable diaphragm
		(Not used in modern instruments).
D	:	Cuvette, containing sample.
E	:	Radiant energy detector.
F	:	Read out system.

Sources of Radiant Energy.

A tungsten filament lamp provides a continuous spectrum between 320 and 1,000 nm and was, therefore, the main light source for measurement in the visual range. These lamps are cheap and reliable. Even though light emission takes place below 320 nm, lower wavelengths are absorbed by the glass envelope. The end of the useful life of the lamp is often seen by the darkening of the glass envelope due to the deposition of tungsten, which has evaporated from the filament.

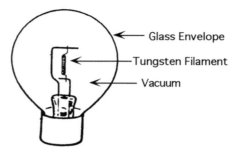

Tungsten Filament Lamp

Tungsten-halogen (quartz-iodine) lamps are now the most commonly used sources for the visible range and for part of the near UV. These lamps are smaller than the standard tungsten filament lamps and use quartz envelopes instead of glass, but still employ a tungsten filament. They are run at hotter temperatures than the standard tungsten filament lamps. The near UV and blue output increases with the temperature of the tungsten filament and so the radiation intensity is greater in the 300-400 nm range. The quartz envelope transmits UV light and also has the ability to withstand considerable thermal stress.

These lamps have a longer useful operating life, as they do not suffer from darkening of the envelope. This is prevented because the evaporating tungsten reacts with the halogen (iodine) to form a volatile halide. This halide decomposes, on contact with the hot filament, to redeposit the tungsten back on the filament.

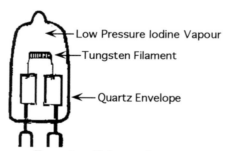

Tungsten-Halogen Lamp

In the UV region, deuterium (or, as in some older instruments, hydrogen) discharge tubes are used. These emit a continuous spectrum between about 190 and 400 nm. The deuterium lamp provides a greater intensity of light in the region around 200 nm than does the hydrogen lamp. Atmospheric absorption of light below 190 nm sets this wavelength as the lower limit of most ultra violet spectrophotometers.

The envelopes of these lamps are made of high grade silica to permit transmission of the UV light.

The cathode has to be heated before an arc is struck between the two electrodes. This heater current is switched off when the arc is struck. The lamp, therefore, requires a few seconds of cathode heating before the lamp will fire. In most modern instruments this sequence of events is automatic when the UV lamp is switched on.

One should be aware that these lamps require a high voltage input and that the UV emission is harmful to the eyes.

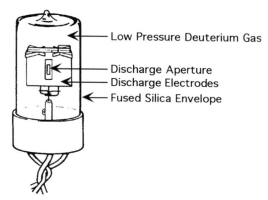

— Low Pressure Deuterium Gas

— Discharge Aperture
— Discharge Electrodes
— Fused Silica Envelope

Deuterium Discharge Lamp

Any fluctuation in the voltage applied to the lamps produces considerable fluctuation in the light output from the lamps. This results in current fluctuations produced by the photo-detector. For this reason a voltage stabiliser is fitted to most instruments.

Means of Spectral Isolation.

Two methods can be used for isolating the required spectral band. In the colorimeter (filter photometer), filters are used to filter out the unrequired wavelengths. This can be done by two types of filter, the glass and wratten filters or by interference filters.

The glass filter consists of one or more layers of coloured glass, whereas the wratten filter consists of a layer of coloured gelatine between clear glass plates. Interference filters rely on the principle of interference of light being reflected from 2 semi-transparent thin films separated by a spacer of x nm. Light of the desired wavelength (2 x) is transmitted. Other wavelengths are either reflected or absorbed in the filter due to the destructive interference of light waves. However, light of x and $\frac{1}{2}$ x nm will also be transmitted. These can be removed by a sharp cut-off filter included with the interference filter.

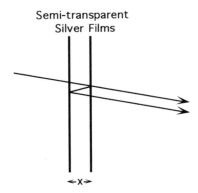

Semi-transparent
Silver Films

←x→

Most glass filters have a bandwidth of between 25 and 50 nm; whereas, bandwidths as low as 4 nm can be produced by interference filters. The bandwidth (also called half bandwidth) is the wavelength interval between points on the transmission curve that are equal to half the peak transmittance:

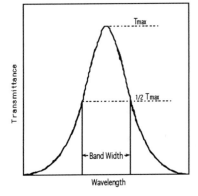

In the spectrophotometer, white light is dispersed by a grating (or prism) monochromator into a spectrum and the desired band is isolated by mechanical slits.

Older instruments used glass prisms for the visual range or quartz prisms to cover the UV and visible regions. Bandwidths of 0.5 nm, or less, can be obtained using prisms.

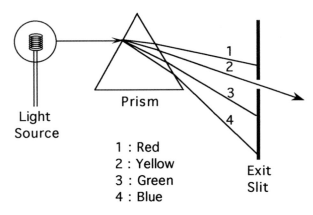

Light
Source

Prism

1 : Red
2 : Yellow
3 : Green
4 : Blue

Exit
Slit

1-12

Most prism instruments employed a mirror to reflect the light back through the prism. This doubles the dispersion of the light. The prism is rotated so that the desired wavelength leaves the monochromator via a fixed exit slit.

Unfortunately the wavelength dispersion by a prism is not linear. Dispersion is greatest in the ultra violet, less in the blue region, with small dispersion in the red; ie. the dispersion decreases as the wavelength increases. The wavelength control is, therefore, calibrated with a non-linear scale or calibrated with a linear scale but attached to the prism by an accurately calibrated cam .

Most modern spectrophotometers use a diffraction grating. These consist of a series of parallel grooves engraved on a reflecting surface. There are normally around 1,200 lines per mm. The commercial gratings are made from a master grating and the replica is coated with a reflective aluminium surface. Alternatively, gratings can be made by etching a holographic pattern.

The surface of each groove acts as a mirror. When a parallel beam of light is shone on the grating, light is reflected from the surface of each of the grooves. This reflected light is then subjected to interference. When the reflected light from one groove is separated from that reflected by the next groove by a distance equivalent to a whole wavelength, the reflected waves will be in phase and will thus be reflected undisturbed. When the separation is not a whole number of wavelengths the reflected light will interfere, as the reflected light is now out of phase. These waves cancel each other out and no light is propagated. By altering the angle of the incident light, the distance between the reflected segments is altered and a new wavelength will be reflected undisturbed.

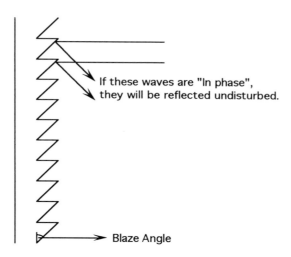

If these waves are "In phase", they will be reflected undisturbed.

Blaze Angle

If the wavelength (λ) is reflected undisturbed, then $\frac{\lambda}{2}$, $\frac{\lambda}{3}$ etc., will also be reflected by the same incident angle. This gives rise to 2nd and 3rd order spectra:

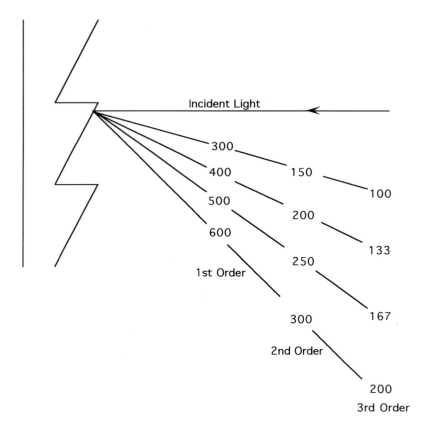

The unwanted components are removed by "Order Sorting" filters. These are usually sharp cut-off filters that remove light below a specific wavelength. Eg:

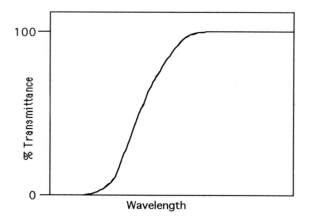

The blaze angle of a grating can be adjusted to give maximum efficiency for the reflection of a particular wavelength. The grating is then said to be blazed at that wavelength.

The grating has the advantage over a prism in that its angular dispersion is linear with wavelength. This means that a fixed slit width will allow the same bandwidth of light to pass for the whole spectral range of the instrument. The same slit width on a prism instrument will allow a very narrow bandwidth in the UV but a broad bandwidth in the red. Band widths of 0.1 nm, or less, can be obtained with grating instruments.

The light leaving the monochromator exit slit appears as a triangle, when light intensity is compared with wavelength:

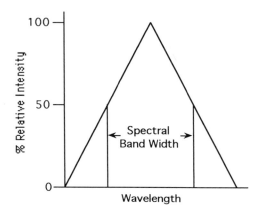

The disadvantages of the grating are that a considerable amount of radiation energy is wasted in the unwanted spectral orders and they are a greater source of stray light due to light scatter from the many surfaces of the grating.

Adjustable Diaphragm.

This is a mechanical means of regulating the intensity of the light that enters the cuvette and passes on to the photodetector. In some spectrophotometers, an optical wedge is used instead of a diaphragm. By this means the instrument can be adjusted to 100% transmittance (Zero Abs) for any wavelength and any blank.

Modern instruments have, however, replaced these mechanical systems with an amplification stage in the output from the photodetector. Rather than adjusting the amount of light reaching the photodetector to produce a specific output level, the output of the photodetector is amplified to reach this specific level.

Cuvettes.

For the range 320-1,000 nm, glass cuvettes can be used; however, below 320 nm quartz cuvettes must be used. Special ultraviolet grade silica cuvettes are required for work below 200 nm. Moulded polystyrene disposable cuvettes are available for use in the visible range and part of the near UV.

"Matched" glass cuvettes are selected to be within 0.5%T of each other at 365 nm and UV "matched" cuvettes to be within 1.5%T at 240 nm.

"Autocells" which allow semi-automatic filling and emptying of special cuvettes can be fitted to a number of instruments. This allows for increased speed in analysis. These cuvettes are made to the same high optical standards as the ordinary type of cuvette.

The cuvettes that are most used have a 10 mm path length. However, some instruments will take cuvettes of unusual path length.

Cuvettes should be handled only by the ground surface and the optical surface should be wiped with a cleaning tissue before each measurement. Cuvettes should not be filled higher than 15 mm from the top. This avoids spillage into the instrument.

Pasteur pipettes must not be used to fill or empty cuvettes as these can scratch the optical surfaces.

The contents of a cuvette can be mixed with a plastic cuvette stirrer (never a Pasteur pipette) or by placing parafilm on the top and inverting the cuvette a few times.

Radiant Energy Detectors.

(a)   Barrier Layer Cells.

This type of detector consists of a copper or iron plate (+) covered by a semi-conducting layer of cuprous oxide or selenium. A collector electrode (-), of a light transmitting metal such as gold, is laid on top. If the circuit is closed, illumination will produce a flow of electrons. The current flowing is proportional to the light intensity.

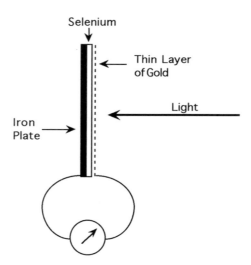

This type of cell exhibits "fatigue effects". On illumination, a rapid rise in current is produced to a value above the apparent equilibrium, which then slowly decreases.

The barrier layer cell is cheap and does not require a power supply. However it is not very sensitive and is limited to the 400-750 nm range. It is only used in colorimeters. These detectors are "photocells"; however, the term "photocell" has also been used for phototubes.

(b)   Vacuum Phototubes.

These photodetectors consist of a cathode coated in an alloy that emits electrons when exposed to light. The cathode is surrounded by a wire mesh anode and both electrodes are in a glass or silica evacuated tube.

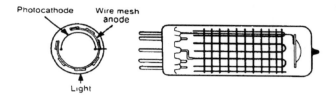

Light passing through the anode mesh causes the cathode to emit electrons. These electrons are captured by the anode and a current flows in the circuit. This output is then fed to an amplifier.

The cathodes are normally made sensitive to light of the wavelength range 190-650 nm or from 600-1,000 nm. As a result, instruments that use phototubes often have two phototubes.

Even in the dark, phototubes produce a small current. This is known as the "dark current" and correction must be made in instruments using phototubes. This dark current is temperature dependent. Phototubes show some fatigue on prolonged exposure to high levels of illumination. They are, therefore, protected by a shutter when not in use or when the cuvette housing is opened.

    (c)   **Photomultipliers.**

These operate in a similar way to the phototube, but include a number of amplification steps within the detector itself. Between the cathode and the anode, a number of additional electrodes are arranged. These are known as dynodes. Light passes through the silica window and strikes the cathode. As with the phototube, electrons are emitted from the cathode. These are then attracted to the first dynode, which is positive with respect to the cathode. These electrons on hitting the first dynode cause it to emit more electrons, which are then attracted to the second dynode, which is more positive than the first. This process is repeated by a number of dynodes so that a cascade effect is set up. The final shower of electrons is then collected by an anode and a large current flows in the circuit.

There are two main types of photomultiplier. The "Side-window" type has the dynodes set out so that electrons move in a complicated path within the photomultiplier. This type is compact, but is limited to about nine dynode stages. The "end-on" type has the dynodes in a parallel arrangement and can, therefore, accommodate more dynodes than the "side-window" type. The photocathode is much larger and, therefore, less sensitive to any variation in the optical path.

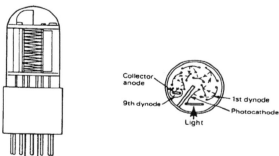

Side-window Photomultiplier

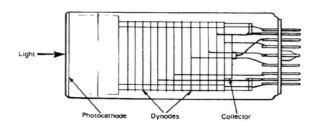

End-on Photomultiplier

Photomultipliers have a wider wavelength range than phototubes and so only one is needed for the full range of the instrument. They also have a very fast response time to changes in light intensity and are, therefore, used in double beam instruments.

This type of detector is not as prone to fatigue effects as photocells and phototubes. However, they can be burnt out by exposure to daylight. They require a high voltage supply to the dynodes. This is usually cut off when the cuvette compartment is opened. These detectors also have a "dark current".

(d) Photodiodes.

Photodiodes are a relatively new innovation in spectrophotometers. The construction of a photodiode is illustrated below:

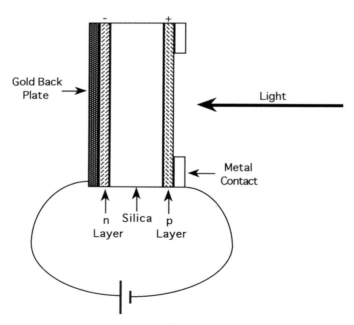

The detector consists of very high resistance silica into which trivalent atoms are incorporated at the top (the p layer) and pentavalent atoms at the bottom (the n layer). The interaction of the trivalent atoms with the silica lattice leaves a net positive charge on the p layer and the pentavalent atoms leave a net negative charge on the n layer.

The charge between these two layers is balanced by an external voltage, as illustrated above. Hence, in the dark, no current flows.

Photons, passing through the p layer, interact with the central silica layer and generate electrons. These electrons then flow from the detector. This output current is linear, with respect to incident light, over a very large range.

The response time is very rapid. Hence, these detectors have a wider response range and a faster response time than photomultipliers. Most, however, only respond to visible and near infra red light. However, they can be modified to respond in the ultraviolet range; as illustrated below:

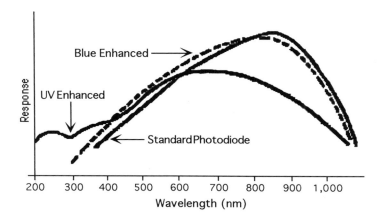

Photodiodes are much cheaper and smaller than photomultipliers. Hence, a battery of them can be used in diode array spectrophotometers. The optics of these instruments is illustrated below:

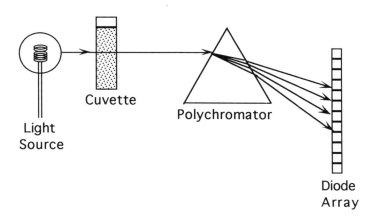

It will be noted that the sample is placed before the polychromator (usually a grating), rather than after the monochromator, as in a standard spectrophotometer.

Each diode has a condenser in parallel, as shown below. Light hitting the diode generates current that discharges the condenser. The condensers are recharged every 100 milliseconds. The amount of current needed to recharge the condenser is proportional to the amount of light hitting that diode in the 0.1 second.

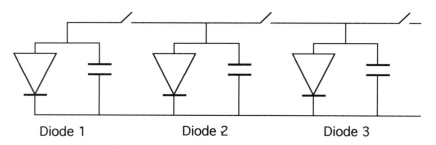

Diode 1   Diode 2   Diode 3

There are about 500 diodes in the instrument and each covers about 1 nm of the spectrum. These are scanned 10 times (1 second) and the averaged results used to generate a "Digitised" spectrum. The generated spectra are better than that illustrated below, due to the large number of sampling diodes:

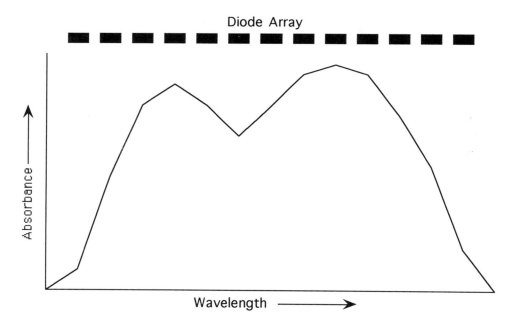

The advantages a diode array spectrophotometer are:

    1.   Rapid spectra generation (1 second).
    2.   Wavelength accuracy (no moving gratings or prisms).
    3.   Improved sensitivity (average of 10 scans per spectrum).
    4.   Less stray light (less components to scatter light).

Readout Systems.

(a)   Meters.

Moving coil meters are the most common form of meter readout; however, mirror galvanometers are used in some old colorimeters.

The simplest form of meter readout is to use the meter directly as an ammeter to monitor the output from the detector unit. The meter is then calibrated with a linear transmittance scale and/or a logarithmic absorbance scale.

A linear absorbance scale can be produced by using the exponential voltage decay of a capacitor or the logarithmic characteristics of a diode to convert the linear detector output into a logarithmic form for display on the meter. This linear absorbance response can then be amplified so that direct concentration readings can be made.

A galvanometer can be employed as a nullpoint indicator where the output from the detector unit is balanced by that from a potentiometer. The potentiometer is fitted with a linear transmittance scale and a logarithmic absorbance scale. This gives more accurate results than direct meter readings, as meters are seldom completely linear.

(b)   Digital Displays.

These are less likely to be misread than meters and are now common on new spectrophotometers. They are also more linear than meters and hence more accurate.

These can take the form of numerals in an evacuated tube. The appropriate numeral is illuminated by passing a heating current through it.

Another common type is the 7-bar light emitting diode. The ⌐¦ system has appropriate bars illuminated to give the numbers from 0-9.

A Typical Single Beam Spectrophotometer.

The Pye Unicam SP6 is an example of a single beam spectrophotometer.

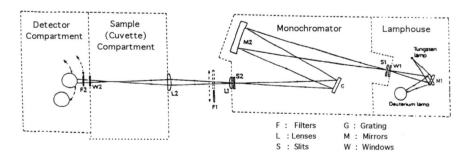

F : Filters   G : Grating
L : Lenses   M : Mirrors
S : Slits   W : Windows

The instrument is fitted with tungsten-halogen and deuterium lamps. The desired lamp is selected by a lever that moves the lamp mirror (M1) to the correct position (Tungsten-halogen: 325 - 1,000 nm, Deuterium: 195 - 325 nm).

The monochromator is a focus compensated version of the basic Monk- Gillieson arrangement (a plane grating operating in a convergent beam).

The grating is a holographic master with 600 grooves per mm and blazed at 240 nm. The grating and mirrors are in a sealed unit to protect them from dust and fumes. The spectral band width for the instrument is 8 nm.

A filter (F1) is automatically placed in the light beam between 270 and 410 nm to remove stray light.

The detectors are blue and red sensitive vacuum phototubes. The desired detector is selected with a lever on the left-hand front of the instrument. (The red sensitive one for 570 - 1 000 nm and the blue for 195 - 570 nm). A filter (F2) is permanently mounted in front of the red detector to absorb 2nd and 3rd order spectral emission.

A third photocell, which is blackened so that no light reaches it, is wired into the circuit with opposite polarity to the other two detectors. Thermally induced "dark current" is thus compensated in the other two photo-detectors.

The output from the phototube is amplified and displayed on a meter readout or as a digital display. The meter can be set to 10% T, 100% T, 0-1 Abs, 0-2 Abs or concentration.

The digital version of instrument can be switched to 100% T, 0-2 Abs or concentration. In the concentration mode, the absorbance can be multiplied by 0.2 up to 10, by altering the gain of the amplifier. 1.0 absorbance units can also be subtracted from absorbance readings on either instrument to give more accurate readings in the absorbance range of 1.0 - 2.5.

Double Beam Instruments.

Single beam instruments have the disadvantage that the output from the photodetector varies rapidly with a change in wavelength. This is due to variation in the source output, monochromator efficiency and detector sensitivity. Voltage fluctuations and temperature changes can also produce a change in output from the photodetector. These disadvantages can be overcome by using a double beam instrument.

The essentials of a double beam spectrophotometer are shown below:

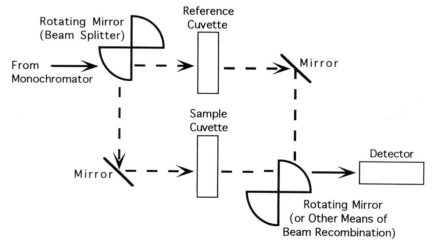

Light from the monochromator is split by a beam chopper into two alternating light pulses that pass alternatively through the reference (blank) and sample cuvettes. These two beams are then recombined before they reach the photodetector.

The signal from each light beam is resolved as square wave alternating currents. These are rectified to direct currents. The ratio of the sample current (I) to the reference current (I_0) is calculated to give transmission. After logarithmic conversion, the result can be displayed as absorbance. In effect, the absorbance of the blank is continuously subtracted from the absorbance of the sample.

Variations in source energy, monochromator efficiency or detector sensitivity effect both beams almost simultaneously, as the beam is split 50 or 100 times per second. As a result, the ratio of the two beams will remain unaltered. This means that the monochromator can be motor driven to produce an absorption spectrum without having to readjust the instrument with each change in wavelength.

Illustrated below is a single beam absorbance scan of 1% H_2SO_4, 200 mM cobalt ammonium sulphate in 1% H_2SO_4 and a double beam scan of the cobalt ammonium sulphate read against the 1% H_2SO_4, between 360 and 620 nm, in a Varian 635 spectrophotometer with a 1 nm spectral bandwidth:

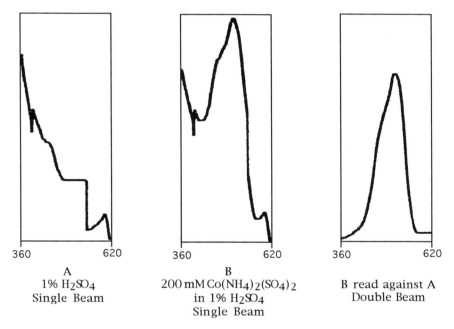

360 620	360 620	360 620
A	B	B read against A
1% H_2SO_4	200 mM $Co(NH_4)_2(SO_4)_2$	Double Beam
Single Beam	in 1% H_2SO_4	
	Single Beam	

A Typical Double Beam Spectrophotometer.

The Shimadzu Graphicord UV-240 is an example of a double beam spectrophotometer.

This instrument is fitted with tungsten-halogen and deuterium lamps. The lamp is selected automatically by inserting a mirror into the light path of the deuterium lamp, when the tungsten lamp is being used. The microprocessor normally changes lamps at 330 nm, but changeover can be selected anywhere in the region 294 - 364 nm.

The monochromator is of the Czerny-Turner mounting type, where two mirrors (M3 and M4) are located symmetrically so as to eliminate aberration. The grating is holographically blazed with 1,600 lines per mm. The spectral bandwidth can be set between 0.02 nm and 5 nm in 0.01 nm steps.

The light, leaving the monochromator, is alternately directed via the reference and sample cuvettes by the rotating mirror (M9), at 50 Hz. The non-absorbed light is then detected by a photomultiplier tube.

The photomultiplier output is fed through a pre-amplifier and separated into sample and reference signals. The reference signal is fed into a feedback system on the detector gain, so that there is a constant output from the reference signal. Thus the sample signal becomes proportional to the sample transmittance, irrespective of wavelength changes, etc.

The instrument operates over the wavelength range of 190 - 900 nm.

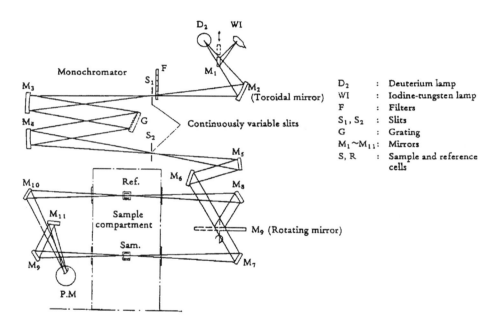

The recorder uses temperature sensitive paper and is microprocessor controlled. The microprocessor can undertake some simple calculations as well as recording the absorbance at up to 3 different wavelengths.

The wavelength accuracy is automatically checked, during the warm-up procedure, using the deuterium emission lines. The baseline is also automatically corrected during the warm-up period. This warm-up period takes 9 minutes.

Instrumental Photometric Error.

Making photometric measurements usually involves 3 steps:

1. Setting the instrument to 0% transmittance with no light reaching the photodetector.

2. Setting the instrument to 100% transmittance (Zero **Abs**) with light passing through the reagent blank.

3. Reading the absorbance or transmittance of the sample.

Errors can occur at any of these steps. On a statistical basis this will produce the largest errors in results when the sample has a high or low absorbance. Between the transmittance range 20-60% (**Abs** 0.22-0.7) this error will be at its minimum and increases outside this range.

Modern digital display instruments, with stabilised electronics, produce much smaller instrumental error than older ones. Reliable readings can be obtained between 0.01 and 2 absorbance units. The upper limit is set by thermal and shot noise of the instrument.

Thermal noise is due to heat induced motions in electron carriers. This is proportional to the square root of the absolute temperature.

Ultimately, shot noise reduces precision at high absorbance. Shot noise is due to the discontinuous nature of light energy and electrical current. For example, photons arrive at the cathode of a detector at random, even though the overall intensity of the light beam is constant. These variations become more obvious as the light intensity is reduced.

The precision profile is illustrated below:

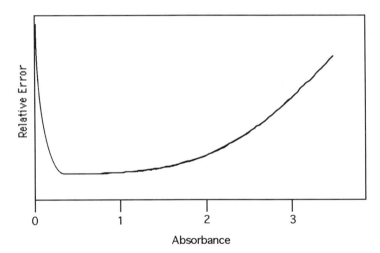

"Warm up" period.

After switching the instrument on, a finite time is required for the electrical circuits to stabilize. During this period the readout device will tend to drift and so no analytical measurements should be made. With most instruments this period lasts for about 10 minutes, after which the instrument should be quite stable.

Optimum Spectral BandWidth.

Three parameters define the shape of peaks in an absorption spectrum. These are (1) peak height (2) wavelength of maximum absorption (λ_{max}) and (3) the natural bandwidth.

The natural bandwidth is the width of the absorption band at half the peak height. These are illustrated below:

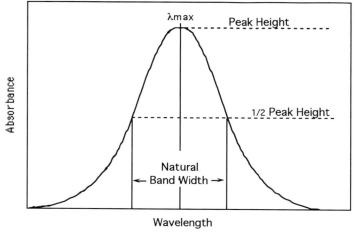

Coloured solutions usually have fairly wide natural bandwidths in the visual range. Correct absorbance values can be obtained, on these solutions, using fairly wide spectral bandwidths. However, many solutions that absorb in the UV (and some that absorb in the visual range) have fairly narrow natural bandwidths. In these cases, it is very important to use the right spectral bandwidth. This is illustrated below, where the narrow absorption bands (λ_{max}: 445.7, 453.4 and 460.0 nm) of a holmium oxide filter are scanned with 5.0, 1.0, 0.2 and 0.08 nm spectral band widths in a Shimadzu UV-240:

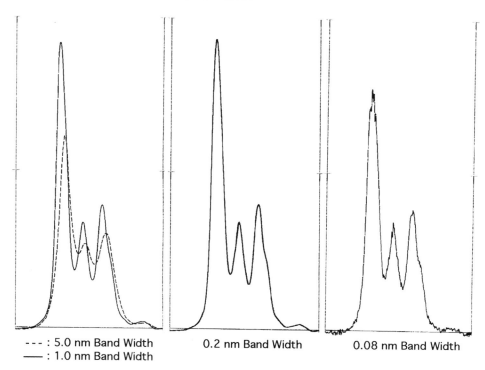

--- : 5.0 nm Band Width
—— : 1.0 nm Band Width   0.2 nm Band Width   0.08 nm Band Width

It will be noted that the peak heights are reduced with the spectral bandwidth of 5.0 nm, compared with 1.0 nm. This is due to increasing amounts of poorly absorbed light away from the λ_{max}. A bandwidth of 0.2 nm only gives a slight improvement over 1.0 nm, and there is too much instrumental noise when a bandwidth of 0.08 nm is used.

To obtain the correct spectrum, the spectral bandwidth should be $\frac{1}{10}$, or less, of the natural bandwidth of the peak. However, care must be taken with very narrow bandwidths so that instrumental noise does not become a problem.

The value of the λ_{max} is also shifted, when the 1 and 5 nm bandwidth scans are compared. This shift is partly associated with the scan speed at which the spectrum was obtained.

As mentioned above, scan speed is also very important in obtaining the correct spectrum. This is illustrated below, where the same absorption bands are scanned between 30 and 1,200 nm per minute in a Hitachi U-3000:

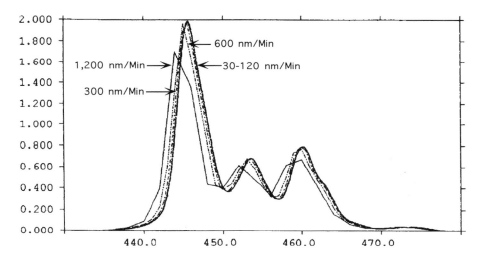

Note the shift of the λ_{maxs} to lower wavelengths with scans faster than 120 nm per minute (The instrument scans from the high to low wavelength). Also note that with the fastest scan (1,200 nm per minute) only a few data points are collected and an angular absorption spectrum is generated.

Stray Light.

Stray light is light, of wavelengths other than that which has been selected, which reaches the photodetector. The most likely source is the scattered light emerging from the monochromator. Scattering can occur from scratched or dirty optical surfaces and from the edges of the monochromator slits and baffles. However, with grating instruments, imperfect ruling of the grating leads to many surfaces where light scatter can take place. This is the major source of stray light with grating instruments.

Light of different spectral orders, from the grating, is also a source of stray light. This is usually removed by "Order Sorting" filters. In this case the shorter wavelengths of the second and third (etc.) order spectra are absorbed by the filters.

The major source of stray light in prism instruments is usually dust on the optical surfaces, causing light scatter.

Room light, entering the cuvette chamber or the photodetector compartment, is not a usual source of stray light. However, this can become a problem with damaged spectrophotometers.

Stray light can have an effect on spectral shape and on absorbance. False peaks can occur at the lower end of the spectrophotometer range and the absorption maxima wavelengths can shift when stray light is a problem. This is illustrated below, where a series of NaI solutions are scanned in the UV, using the Shimadzu UV-240:

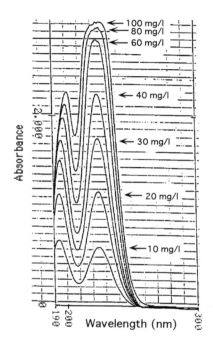

As can be seen, the response of the instrument, at 225 nm, tails off above about 50 mg/l. At 190 nm, the response tails off above 10 mg/l and the maximum absorbance reached is about 1.2. The peak at 195 - 200 nm is a false peak, as the real absorbance increases below 200 nm.

When there is no stray light, a plot of absorbance against concentration will give a straight line. With the presence of stray light, the Absorbance/Concentration plot is no longer linear at high concentration:

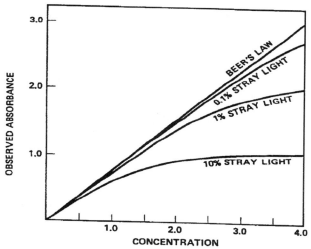

At low absorbance, the major light component reaching the photodetector is of the desired wavelength. At high absorbance, most of the desired wavelength has been absorbed and so most of the light reaching the detector is of the unabsorbed stray light wavelengths. Hence, the measured absorbance is less than the true absorbance:

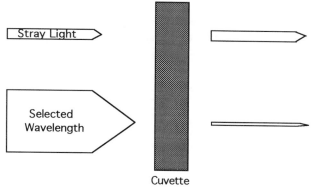

Stray light is expressed as a percentage transmission.

The effect of stray light is worse at the extremes of the wavelength range of spectrophotometers. At the UV end of the spectrum, the light output of the deuterium lamp is low compared with the stray light of a longer wavelength. At the red end of the spectrum, the detector response is poor compared with stray light of a shorter wavelength.

In the 325-400 nm region, there will be less stray light from a deuterium lamp than from a tungsten-halogen lamp. This is due to the high output of visual range light from the tungsten lamp that contributes to the stray light.

The American Society for Testing Materials has suggested that stray light figures for spectrophotometers should be based on the following, using a light path of 10 mm:

220 nm,	Deuterium lamp	:	10 g/l NaI
340 nm,	Tungsten-halogen lamp	:	50 g/l $NaNO_2$
370 nm,	Tungsten-halogen lamp	:	50 g/l $NaNO_2$

The NaI solution absorbs all the emitted light at 220 nm, but lets through light above 265 nm. The NaNO$_2$ absorbs all light at 340 and 370 nm, but lets through light above 400 nm.

Correction for Background Absorption.

Turbid solutions or complex samples containing plasma or homogenates have background absorbance spectra due to lipid or protein light absorption and/or light scatter. These backgrounds are superimposed upon the absorption spectra of the analytes of interests. A number of methods are available to reduce or eliminate this type of interference.

(a) Bichromatic Measurements.

Absorbance measurements are made at (or near) the λ_{max} of the analyte and at a wavelength at (or near) the base of the absorption peak. The latter is then subtracted from the former.

(b) The Allen Correction.

Absorbance measurements are made at the λ_{max} of the analyte and at 2 wavelengths equidistant from the absorption peak. The latter two are averaged then subtracted from the former.

(c) Difference Spectrophotometry.

Where the analyte has different spectra at different pHs, the sample can be split into two. The pH of each is adjusted appropriately and the absorption spectrum of one (sample) is recorded against the other (reference). This is illustrated below for salicylic acid, the active metabolite of aspirin:

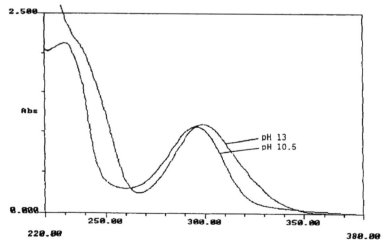

The absorption spectra of salicylate at pH 10.5 and 13.
The λ_{max} values of interest are: pH 10.5 : 296 nm and pH 13 : 299.

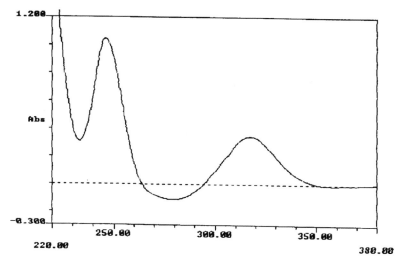

Difference spectra of salicylate (pH 13 read against pH 10.5).
The λ_{max} of interest are 246 and 317 nm.

The chemical structures of salicylate, at pH 10.5 and 13, are illustrated below:

pH 10.5   pH 13

(d) Derivative Spectrophotometry.

This plots the rate of change in absorbance per wavelength unit, against actual wavelength. This produces a 1st derivative spectrum. A 2nd derivative spectrum is the rate of change of the 1st derivative plotted against wavelength. 3rd, 4th and 5th derivatives can be produced, each producing a more complex spectrum than the previous.

The only derivative spectra of use are the 1st, 2nd and 4th. λ_{max} values of broad peaks can be obtained accurately from the 1st derivative. These correspond to zero crossing points (see below).

2nd derivative spectra produce sharp negative peaks corresponding to λ_{max} values; whereas, 4th derivative spectra produce even sharper, but positive, peaks corresponding to the λ_{max} values. Peak heights, generated in 2nd or 4th derivative spectra can be used in quantitative work.

Illustrated below are the zero order (normal), 1st, 2nd and 4th derivative spectra of a double absorption band from a didymium filter obtained on a Shimadzu UV-160, between 560 and 610 nm. This instrument has a fixed bandwidth of 3 nm. The absorption bands have λ_{max} values of 573 and 586 nm. The values obtained are about 1.5 nm lower.

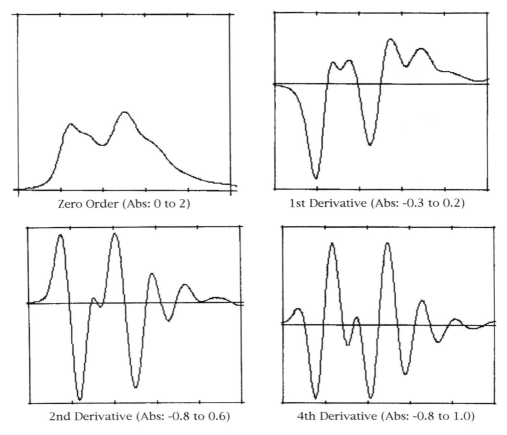

Zero Order (Abs: 0 to 2)

1st Derivative (Abs: -0.3 to 0.2)

2nd Derivative (Abs: -0.8 to 0.6)

4th Derivative (Abs: -0.8 to 1.0)

Problems with derivative spectra include instrumental noise, which will produce extra peaks to complicate the spectra. Instruments that can not store the initial spectrum generate derivative spectra as $\dfrac{dA}{dt}$ rather than $\dfrac{dA}{d\lambda}$. As a result, λ_{max} values depend on the wavelength scan rate, with the generated λ_{max} shifting from the true λ_{max} as the scan rate is increased.

Below is an example of the use of 2nd derivative spectra to identify the presence of amphetamine in a liver sample (Amphetamine has λ_{max} values of 251.5, 257 and 263 nm):

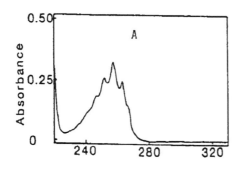

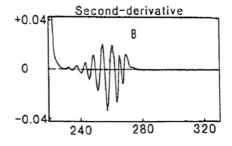

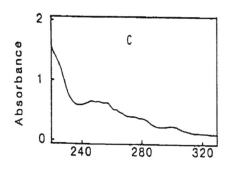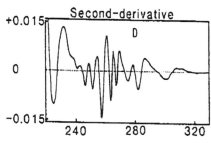

A : Zero order spectrum of amphetamine (pH 2).
B : 2nd derivative spectrum of amphetamine (pH 2).
C : Zero order spectrum of liver homogenate (pH 2).
D : 2nd derivative spectrum of liver homogenate (pH 2).

CHECKING INSTRUMENT PERFORMANCE.

Wavelength Alignment.

The absorbance of a solution is usually read at the wavelength corresponding to its maximum absorbance (λ_{max}). This gives the greatest sensitivity and precision to the method. λ_{max} is the top of an absorption peak. With a broad peak, the absorbance values at wavelengths close to λ_{max} will be almost the same as those at λ_{max}. However, with a narrow peak, the absorbance values will rapidly fall as the wavelength moves away from λ_{max}.

If one is using the molar absorption coefficient (ε) to calculate the concentration of a solution, and the instrument wavelengths are not correctly aligned, considerable errors can occur in the assay. (ε values are given for specific wavelengths).

λ_{max} values are useful in the identification of drugs and other xenobiotics. Hence, it is very important that the instrument is correctly aligned.

In checking the alignment, at least 2 wavelengths should be examined. An instrument could be correct at one wavelength and incorrect at a second, due to an error in the dispersion calibration.

The common ways to check wavelength accuracy are to use the emission lines of a discharge lamp or to use the absorption bands of specific filters.

Most UV instruments use deuterium (or hydrogen, in the older instruments) discharge lamps as their UV source. These have continuous emission in the UV, but have specific emission lines in the visual range. Two of the hydrogen emission lines are the 486.1 nm [D_2 : 486.0] (medium intensity) and 656.3 nm [D_2 : 656.1] (strong). These can be used to check the alignment, if the hydrogen or deuterium lamp (and single beam) can be selected for the visual range. Modern double beam instruments align themselves, by this method, during the warm up period.

Holmium oxide glass is the most commonly used filter for wavelength alignment checks. This had a number of sharp absorption bands. The ones at 360.9 nm and 536.2 nm are convenient for checking alignment.

Wavelength Repeatability.

This measures the ability of the spectrophotometer to return to the same wavelength, after being moved away from that wavelength. Poor repeatability involves "slack" in the mechanism that moves the grating or prism. On a good grating instrument the repeatability is about ± 0.05 nm. This may be in the order of ± 2 nm in a poor instrument. With the non-linear dispersion of a prism instrument, ± 0.1 nm in the UV can be associated with ± 5 nm in the red.

A means of measuring this is to measure the absorbance of a substance that has sharp absorption peaks and to make the measurements at a wavelength that is on the steep side of the peak. Any small variations in the wavelength setting will give large changes in the absorbance values.

Using the green food colouring (see photometric accuracy and linearity, below) 370 nm and 650 nm are suitable values for measuring wavelength repeatability. Note that the same cuvette must be used for these measurements, as variations in cuvette absorption will effect the results.

Spectral BandWidth.

The spectral bandwidth is the wavelength interval between points on the transmission curve that are half the peak height.

The 656.1 nm deuterium emission line can be used to check spectral bandwidths. This emission line is less than 0.001 nm wide. If this line is scanned, the detector output corresponds to the proportion of the emitted light that passes through the monochromator:

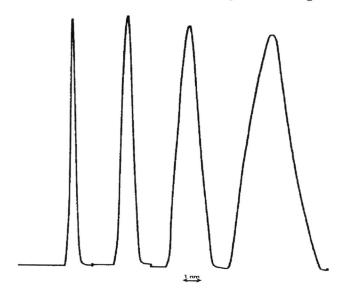

Detector response to the 656.1 nm deuterium emission line when scanned with the 0.2, 0.5, 1.0 and 2.0 nm band width slits in a Varian 635 spectrophotometer set in the transmission mode. The emission line was scanned between about 653 and 659 nm.

1-34

Stray Light.

Stray light can be checked with NaI and $NaNO_2$, as mentioned on page 1-29.

Photometric Accuracy and Linearity.

If one is using the molar absorption coefficient to calculate the concentration, then it is essential that the instrument displays the correct absorbance value, over a reasonable concentration range.

The American National Bureau of Standards gives the following absorbance values for a 36.7 mM solution of cobalt ammonium sulphate, in 1% (v/v) H_2SO_4:

Wavelength (nm)	Absorbance
400	0.012 5
450	0.077 3
500	0.163 5
510	0.174 2
550	0.077 5
600	0.013 7

Increasing concentrations of cobalt ammonium sulphate (up to about 400 mM), in 1% H_2SO_4, can also be used to check photometric linearity.

Potassium dichromate solutions (dilutions from 200 mg/l), in 10 mM H_2SO_4, can be used to check photometric linearity at 257 nm and 350 nm.

Another useful solution is a green food dye solution (5 parts of FDC yellow No 5 and 1 parts of FDC blue No 1 [25 g/l in H_2O + polypropyl glycol]). This is diluted 1 in 2 000 to give a stock solution. This stock is further diluted, and used neat, in the linearity check. Measurements are made at 257 nm, 410 nm and 630 nm (the λ_{maxs} of the dye). These values cover the most used wavelength range of the instrument.

These solutions are known to obey Beer's law.

EXERCISES

Where cuvettes are used, they should be thoroughly cleaned. This includes an ethanol wash, if they are to be used below 220 nm.

Always handle them by the ground (non-optical) surfaces and wipe the optical surfaces with soft paper tissues before putting them in the spectrophotometer.

Always fill cuvettes to about 15 mm from the top. Do not use Pasteur pipettes to fill, empty or stir cuvettes as they can scratch the optical surfaces.

EXERCISE 1-1

PHOTOMETRIC ACCURACY AND LINEARITY

It is advisable to check the photometric accuracy and linearity of spectrophotometers every now and again; especially when the instrument is getting older.

As mentioned above, cobalt ammonium sulphate solutions are one of the recommended reagents for doing this. A correction for mis-matched cuvettes will be made prior to the absorbance measurements.

REAGENTS AND EQUIPMENT:

1. 1% (v/v) Sulphuric Acid.
2. 200 mM Cobalt Ammonium sulphate in 1% Sulphuric Acid.
    Dissolve 39.46 g of dry $CoSO_4(NH_4)_2SO_4:6H_2O$ in 1% sulphuric acid and make to 500 ml in a volumetric flask.

Glass Cuvettes.
Soft Paper Tissues.
5 and 10 ml Bulb Pipettes.
Single Beam Spectrophotometer(s).
Double Beam Spectrophotometer(s).

PROCEDURE:

1. Using the 200 mM cobalt ammonium sulphate solution and 1% H_2SO_4, prepare the following solutions, in test tubes:

	A	B	C
ml of cobalt ammonium sulphate solution	5.0	10.0	15.0
ml of 1% H_2SO_4	15.0	10.0	5.0

2. Fill two glass cuvettes with 1% H_2SO_4 and place them in a single beam spectrophotometer.

3. Select one cuvette as the blank and the other as the sample.

4. For each of 400, 450, 500, 510, 550 and 600 nm, set the blank to read 0.100 absorbance units and read the absorbance of the sample.

5. Replace the sample 1% H_2SO_4 with solution A (Wash the cuvette, 2 or 3 times, with this solution before filling).

6. For each of the above wavelengths, reset the blank to 0.100 and read the absorbances of solution A.

7. Repeat the above with solutions B, C and the neat 200 mM cobalt ammonium sulphate.

8. Subtract the mis-match values (Step 4) from the other absorbance values, at their appropriate wavelength.

9. Fill two glass cuvettes with 1% H_2SO_4 and place them in a double beam spectrophotometer.

10. Set the instrument to scan between 360 and 620 nm, with an absorbance range of 0 - 1.0, a bandwidth of 2 or 5 nm and overlay scans.

11. Autozero the instrument at the start wavelength (600 nm) and run the scan.

12. Use the cursor to obtain the absorption values at the above wavelengths.

13. Repeat the scan, using the 4 cobalt ammonium sulphate solutions, but do **not** autozero the instrument before each scan. Obtain the absorbance values as in step 12.

14. Subtract the mis-match values (Step 12) from the others, at their appropriate wavelength.

For each instrument, plot corrected absorbance values against concentration, including each wavelength as a separate line on the graph.

If linearity is obeyed (Beer's Law), these lines should be straight.

If linearity is obeyed, use the graph line values to calculate the absorbance values of a 36.7 mM cobalt ammonium sulphate solution and compare these values with those quoted by the NBS.

If linearity is not obeyed, use the values from solution A (50 mM) to calculate the 36.7 mM values and compare these with the NBS values.

EXERCISE 1-2

SPECTRAL BAND WIDTH

The emission lines from a deuterium lamp in the visual region, or from metal cathode lamps, are very narrow and can be considered as fine lines (Bandwidths of less than 0.001 nm). If one of these lines is scanned by a spectrophotometer, the detector output corresponds to the proportion of the emitted light that passes through the monochromator at wavelengths at, or near, the emission wavelength.

The 656.1 nm deuterium line will be scanned with each of the available bandwidths on a suitable spectrophotometer.

The spectral bandwidths are then calculated as the wavelength interval between points on the transmission curve that are half the peak height (See page 1-14).

EQUIPMENT:

A scanning spectrophotometer that can be set to single beam and with a deuterium lamp that can be used in the visual region (Eg. The Varian 635).
A recorder connected to the spectrophotometer recorder output and set for 10 mV input.

PROCEDURE:

The procedure is written for the Varian 635. Instrument bandwidths may be different on other instruments.

1.   Select the UV lamp, %T and single beam on the 635.

2.   Select a bandwidth of 2.0 nm and locate the 656 nm emission line (Should be within 1 nm of the value on the indicator scale).

3.   Alter the "Wavelength" control, gently, to give maximum meter response (maximum transmission).

4.   Adjust the "Energy" control to give a meter reading of 0.9 (90%). The recorder input should be adjusted to give a value of about 90% (9 on the chart).

5.   Set the wavelength to 660 nm and adjust the recorder "Zero" to give a recorder pen reading of about 2% (0.2 on the chart paper).

6.   Start the chart at 10 cm/min.

7.   Then press the "Scan rate" of 10 nm/min on the 635.

8.   After the pen returns to the base line, stop the scan and then stop the chart.

9.   Repeat steps 2-8 for the 1.0 nm, the 0.5 nm and the 0.2 nm band widths.

The chart response should be similar to that on page 1-35.

For each scan, measure the peak height above the base line. Divide this value by two and mark this value on the trace. Measure the distance between these two points, on the emission trace, to the nearest mm.

$$1 \text{ mm} = 0.1 \text{ nm}$$

Compare these measured values with those stated on the instrument.

EXERCISE 1-3

SPECTRAL CHARACTERISTICS OF A FILTER

The bandwidth ("Half Band Width") and the nominal wavelength (Peak transmittance), of a filter, are important characteristics of a filter. This information is required when selecting the right filter for a filter instrument.

The technique of wavelength scanning with a single beam instrument will be employed in this exercise.

EQUIPMENT:

Single Beam Spectrophotometers.
Green Colorimeter Filters (EEL 624, or equivalent).

PROCEDURE:

1. Select a green filter and a single beam instrument.

2. Record the transmittance values (against air) of the filter, at 5 nm intervals, between 450 and 650 nm. Make measurements at 2 nm intervals, near the transmission peak, to obtain the peak value.

    Remember to set the instrument to 100% T, on air, for each wavelength, before reading the %T of the filter.

Plot these values against the wavelength and read off the nominal wavelength and bandwidth.

EXERCISE 1-4

WAVELENGTH CHECK

A holmium oxide filter will be used to check the wavelength accuracy of a double beam spectrophotometer.

EQUIPMENT:

Holmium Oxide Filter.
Double Beam Spectrophotometer.

PROCEDURE:

1. Set the instrument to scan between 300 and 700 nm, 0-2 absorbance units, on a scan speed of 100 nm/min or less, using a 1 nm band width.

2. Insert the holmium oxide filter in the sample holder and carry out a scan.

3. Obtain the peak absorption wavelengths.

4. Compare these peaks with:

333.8 nm	453.4 nm
360.8 nm	460.0 nm
418.5 nm	536.4 nm
445.7 nm	637.5 nm

EXERCISE 1-5

SPECTRAL BAND WIDTH AND RESOLUTION

The absorbance value measured by an instrument is the average value of absorbance for the light of all the wavelengths that leave the monochromator. The narrower the spectral band width, the closer the measured absorbance is to the true absorbance. As the bandwidth is increased, wavelengths that are absorbed to a lesser extent are averaged with the absorption peak wavelength; and so the overall absorption is reduced.

With increasing band width the ability of the instrument to resolve narrow absorption peaks is reduced and these peaks can even be lost, as they become incorporated into an overall broad absorption band.

Also, with a wide bandwidth and relatively narrow absorption bands, Beer-Lambert's law is no longer obeyed; and so standard curves have to be run with each assay.

To obtain true peak heights, the spectral bandwidth should be less than 1/10 of the natural bandwidth of the absorption peak. However, instrumental noise can be a problem with very narrow bandwidths.

These effects will be illustrated in this exercise by scanning a triple absorption peak of holmium oxide glass, using 0.05 (or 0.08) nm, 0.2 nm, 1.0 nm and 5.0 nm spectral bandwidth.

The effect of scan speed will also be examined.

EQUIPMENT:

Holmium Oxide Filter.
Double Beam Spectrophotometer.

PROCEDURE:

1.   Set up the instrument as follows:

Absorbance	:	0-2
Wavelength	:	430-480
Scan	:	Overlay
Scan speed	:	100 nm/min or less
Line type	:	Solid
Band width	:	5.0 nm

2.   Place the holmium oxide filter in the sample holder and close the lid.

3.   Autozero the instrument.

4.   Carry out the scan.

5.   After the scan, select a bandwidth of 1.0 nm and dashed line. Autozero the instrument and run the scan.

6. After this scan, select a bandwidth of 0.2 nm and dashed/dot. Autozero the instrument and run the scan.

7. Feed the chart paper and select a new overlay scan.

8. Select a bandwidth of 0.05 or 0.08 nm and a solid line. Autozero the instrument and run the scan. Select the bandwidth to illustrate moderate instrumental noise.

9. Feed the chart paper and select a new overlay scan.

10. Repeat steps 1-4, but use a 0.5 nm bandwidth.

11. Repeat the scan with a range of faster scan speeds, using a different line type for each scan.

Comment on the scans, including peak absorption and wavelength accuracy (The triple peak λ_{max} values are 445.7, 453.4 and 460.0 nm).

EXERCISE 1-6

CUVETTE CUT-OFF WAVELENGTHS

Knowing where cuvette material absorbs UV light is very important in selecting the correct type of cuvette for UV work.

UV absorption by glass, quartz and plastic cuvettes will be examined in this exercise, using a double beam spectrophotometer.

EQUIPMENT:

Glass, Quartz and Plastic (Polystyrene) Cuvettes.
Double Beam Spectrophotometer(s).
Soft Paper Tissues.

PROCEDURE:

1. Set the instrument for:

Transmittance	:	0-100%
Wavelength	:	Lowest-340
Scan	:	Overlay
Scan speed	:	300 - 500 nm/ min
Line type	:	Solid
Band width	:	2.0 nm

2. Fill the glass, quartz and plastic cuvettes with distilled water.

3. Place one of these cuvettes in the sample holder of a double beam spectrophotometer.

4. Auto zero the instrument and run the scan.

5.   Repeat the scan with the other 2 cuvettes, but use a different line type for each.

Note the wavelength at which the transmission has been reduced by 10% (ie. 90% T on the trace). The cuvette should not be used below this wavelength.

EXERCISE 1-7

THE ABSORPTION SPECTRUM OF BENZENE

Benzene has a number of electronic energy levels, which show up as absorption bands in the ultra violet. Superimposed upon this electronic spectrum are bands associated with changes in the vibrational energy of the molecules. These vibrational energy levels can be seen as fine line spectra when benzene vapour is examined. This vibrational "fine structure" is lost when a solution of benzene, in a transparent solvent such as ethanol, is scanned. In this case only the electronic energy levels are seen as broader peaks.

To pick up the fine spectra, a narrow spectral bandwidth has to be used in the spectrophotometer. One has to pick an optimal bandwidth because, if one uses a too narrow one, the background "noise" starts to dominate the spectra and the signal to "noise" ratio falls below an acceptable value. 0.1 nm is the bandwidth used in this exercise. This gives good resolution of the fine structure of the benzene absorption spectrum with only a small amount of background "noise". A double beam spectrophotometer will be used to scan benzene vapour and then benzene in ethanol, between 220 and 280 nm.

REAGENTS AND EQUIPMENT:

1.   AR Benzene.
2.   1/1,500 Benzene in Ethanol.
        Add 100 μl of AR benzene to 150 ml of AR ethanol and mix.

Quartz Cuvettes with Lids.
Double Beam Spectrophotometer.
Soft Paper Tissues.

PROCEDURE:

1.   Set the instrument for:

Absorbance	:	0-2
Wavelength	:	220-280 nm
Scan	:	Sequential
Scan speed	:	100 nm/min or less
Line type	:	Solid
Band width	:	0.1 nm

2.   Pipette 50 μl of Analar benzene into a quartz cuvette and place the lid on the cuvette.

3.   Place the cuvette in the sample cuvette holder on the instrument and close the lid. Autozero the instrument.

4.   Run the scan.

5.   Place a cuvette containing 1/1,500 benzene in ethanol in the sample cuvette holder and a cuvette containing ethanol in the reference cuvette holder. Close the lid and autozero the instrument.

6.   Run the scan.

Note the fine-line structure in the vapour spectrum, which is lost in the solution.

EXERCISE 1-8

THE ABSORPTION SPECTRA OF NAD AND NADH

NAD and NADP are the coenzymes involved in most biological oxidation-reduction reactions. These coenzymes are reduced to NADH and NADPH as the substrate is oxidised.

The reduced form of these coenzymes has an additional absorption band (λ_{max} of 340 nm) in the UV, not present in the oxidised form. The spectra of the 2 coenzymes are very similar.

Oxidation-reduction reactions can, therefore, be monitored by observing the change in absorption at 340 nm. Other enzyme catalysed reactions can be linked to oxidation-reduction reactions, which can then be monitored at 340 nm.

The absorption spectra of NAD and NADH will be examined in this exercise.

REAGENTS AND EQUIPMENT:

1.   50 µM NAD.
            Dissolve 7 mg of NAD in 200 ml of distilled water.
            Make up fresh and keep on ice.
2.   50 µM NADH.
            Dissolve 7 mg of NADH in 200 ml of distilled water.
            Make up fresh and keep on ice.

Quartz Cuvettes.
Double Beam Spectrophotometer.
Soft Paper Tissues.

PROCEDURE:

1.   Set the instrument for:

Absorbance	:	0-1
Wavelength	:	220-400 nm
Scan	:	Overlay
Scan speed	:	300-500 nm/min
Line type	:	Solid
Band width	:	2 nm

2.   Fill a quartz cuvette 50 µM NAD and place in the sample holder of a double beam spectrophotometer.

3. Fill a quartz cuvette with distilled water and place in the reference holder of the instrument.

4. Autozero the instrument and run the scan.

5. Obtain the λ_{max} values.

6. Replace the NAD cuvette with one containing 50 μM NADH, autozero the instrument and run the scan using a dashed line.

7. Obtain the λ_{max} values.

Compare the 2 spectra.

EXERCISE 1-9

STRAY LIGHT

Stray light is light, of wavelengths other than that which has been selected, which reaches the photodetector. The most likely source is the scattered light emerging from the monochromator. Scattering can occur from scratched or dirty optical surfaces and from the edges of the monochromator slits and baffles. However, with grating instruments, imperfect ruling of the grating leads to many surfaces where light scatter can take place. This is the major source of stray light with grating instruments.

Light of different spectral orders, from the grating, is also a source of stray light. This is usually removed by "Order Sorting" filters. In this case the shorter wavelengths of the second and third (etc.) order spectra are absorbed by the filters.

The major source of stray light in prism instruments is usually dust on the optical surfaces, causing light scatter.

Room light, entering the cuvette chamber or the photodetector compartment, is not a usual source of stray light. However, this can become a problem with damaged spectrophotometers.

Stray light can have an effect on spectral shape and on absorbance. False peaks can occur at the lower end of the spectrophotometer range and the absorption maxima wavelengths can shift when stray light is a problem.

When there is no stray light, a plot of absorbance against concentration will give a straight line. With the presence of stray light, the Absorbance/Concentration plot is no longer linear at high concentration (See page 1-29).

At low absorbance, the major light component reaching the photodetector is of the desired wavelength. At high absorbance, most of the desired wavelength has been absorbed and so most of the light reaching the detector is of the unabsorbed stray light wavelengths. Hence, the measured absorbance is less than the true absorbance:

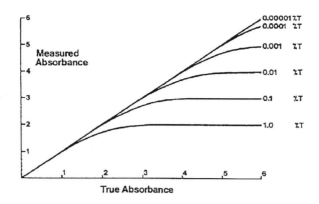

Stray light is expressed as a percentage transmission.

The effect of stray light is worse at the extremes of the wavelength range of spectrophotometers. At the UV end of the spectrum, the light output of the deuterium lamp is low compared with the stray light of a longer wavelength. At the red end of the spectrum, the detector response is poor compared with stray light of a shorter wavelength.

In the 325-400 nm region, there will be less stray light from a deuterium lamp than from a tungsten-halogen lamp. This is due to the high output of visual range light from the tungsten lamp that contributes to the stray light.

In this exercise, sodium iodide solutions with concentrations between 10 and 100 mg/l, will be scanned between 190 and 300 nm in a double beam spectrophotometer. This will illustrate the deviation from Beer's law and the presence of false absorption peaks in the far UV.

REAGENTS AND EQUIPMENT:

1.   100 mg/l Sodium Iodide.
                Dissolve 100 mg of NaI in 1 l of distilled water.

High Grade Quartz (Silica) Cuvettes.
Test Tubes.
10 ml Graduated Pipettes.
Double Beam Spectrophotometer capable of reaching 190 nm.
Soft Paper Tissues.

PROCEDURE:

1.   Set up the following, in test tubes, using the 100 mg/l NaI solution:

	1	2	3	4	5	6	7
ml of 100 mg/l NaI	1.0	2.0	3.0	4.0	6.0	8.0	≅ 10
ml of distilled water	9.0	8.0	7.0	6.0	4.0	2.0	-

2. Set the instrument for:

Absorbance	:	0-3or 4
Wavelength	:	190 -300 nm
Scan	:	Overlay
Scan speed	:	300-500 nm/min
Line type	:	Solid
Band width	:	2 nm

3. Place a quartz cuvette containing distilled water in the reference holder.

4. Place a quartz cuvette containing solution 1 in the sample holder and close the lid.

5. Autozero the instrument and run the scan.

6. After the scan, repeat the scan on each of the other 6 solutions, autozeroing before each scan.

Results similar to those on page 1-28 should be obtained.

(a) Plot a graph of absorbance against concentration for the 225 nm peak and **at** 190 nm, on the same piece of graph paper.

(b) Note the wavelength of the first absorption peak (lowest wavelength one) at each NaI concentration.

(c) From the absorbance values of the 100 mg/1 NaI solution, calculate the percentage transmission (stray light) at 190 and 225 nm.

EXERCISE 1-10

DERIVATIVE SPECTRA

Didymium filters will be used to demonstrate 1st, 2nd and 4th derivative spectra. These derivatives will be used to obtain accurate λ_{max} values for an absorption double peak (λ_{maxs} of 573 and 586 nm) of didymium glass.

EQUIPMENT:

Didymium Glass Filters.
Double Beam Spectrophotometer Capable of Generating Derivative Spectra.

PROCEDURE:

1. Set the instrument for:

Absorbance	:	0-2
Wavelength	:	550-620 nm
Scan	:	Overlay
Scan speed	:	100 nm/min

Line type	:	Solid
Band width	:	2 nm

2. Place a didymium filter in the sample holder of the spectrophotometer and autozero the instrument.

3. Run the scan to generate the zero order spectrum.

4. Obtain the 1st derivative and use the zero points to accurately read off the two λ_{max} values.

5. Select sequential scans and generate the 2nd derivative spectrum.

6. Obtain the trough values to give the λ_{maxs}.

7. Generate the 4th derivative spectrum.

8. Obtain the peak values to give the λ_{maxs}.

If the instrument uses $\frac{dA}{dt}$ to generate the derivative spectra, the λ_{max} values will not be accurate.

EXERCISE 1-11

THE DETERMINATION OF SERUM CALCIUM

This exercise is an example of a colorimetric method.

Calcium forms a complex with o-cresolphthalein complexone (phthalein purple), which is violet in alkaline solution. The absorbance of this complex is measured spectrophotometrically at 578 nm.

Other divalent metal ions also form complexes, and so 8-hydroxyquinoline is included in the reaction mixture to chelate magnesium etc. Urea is also included to diminish turbidity produced by lipaemic specimens and to enhance the complex formation. Ethanol is also in the reaction mixture to inhibit colour development in the blank.

Diethanolamine is used as the buffer (with acetic acid) to maintain the pH of the reaction mixture at 11.7. It also enhances the colour intensity.

As haemoglobin complexes also absorb at 578 nm, a correction has to be made for this. Hence, after the initial absorbance has been measured, the calcium-phthalein purple complex is dissociated by the addition of EDTA. The absorbance is then read again (this absorbance is due to the haemoglobin complexes only). Hence the difference in the two absorbances is due to the calcium present in the sample.

Bilirubin, in high levels, tends to produce decreased values and some drugs will yield elevated results.

Case History.

The plasma is from a 58 year-old female. 20 years ago, she had a partial thyroidectomy as treatment for hyperthyroidism. She has now been admitted to hospital for cataract removal. She is showing signs of increased neuro-muscular excitability. A request had been made for serum calcium and phosphate determinations to confirm a tentative diagnosis of hypoparathyroidism.

The parathyroids are small organs that sit on the back of the thyroid:

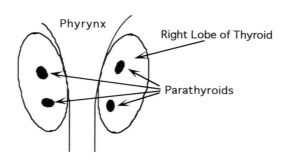

Their total weight is only 100 - 200 mg and they can easily be damaged during a partial thyroidectomy. They secrete parathormone (PTH) in response to a low extracellular calcium concentration. Parathormone is an 84 amino acid protein hormone.

It acts on the kidney to decrease phosphate reabsorption and to increase calcium reabsorption. It also acts on the kidney to increase the activation of Vitamin D (25-Hydroxycholecalciferiol → 1,25-Dihydroxycholecalciferol). This, in turn, increases calcium and phosphate absorption in the intestine. Parathormone also acts directly on bone, stimulating the osteoblasts to mobilise calcium. Hence, the plasma calcium level is increased and the phosphate reduced.

With hypoparathyroidism, the opposite occurs. The low extracellular calcium leads to increased neuromuscular excitability. This can be seen as abnormal reflexes and spasms. Finally tetany and death can occur if the level goes low enough. Cataracts and psychiatric conditions can be seen in conditions of prolonged hypocalcaemia.

REAGENTS AND EQUIPMENT:

1.   Diethanolamine Buffer (2 M, pH 11.7).
        Dissolve 300 g of urea in about 500 ml of distilled water. Add 210 g of diethanolamine and dissolve. Adjust the pH to 11.7 with glacial acetic acid. Make to 1 l with distilled water.
        Stable for 2 months.
        For reagent 3.
2.   Chromogen.
        Dissolve 64 mg of o-cresolphthalein complexone and 1.16 g of 8-hydroxyquinoline in 250 ml of ethanol. Add 2.5 ml of glacial acetic acid and mix. Add 300 g of urea and about 300 ml of distilled water to dissolve the urea. Make to 1 l with distilled water.
        Stable for 3 months.
        For reagent 3.
3.   Colour Reagent.
        Mix equal volumes of reagents 1 and 2.
        Use within 1 day.

4.   0.1 M TCA.
           Dissolve 16.3 g of TCA in 1 l of distilled water.
           For reagent 5.
5.   4.0 mM Calcium Standard.
           Dissolve 200 mg of dry AR $CaCO_3$ in reagent 4 and make to 500 ml in a volumetric flask,
           with reagent 4.
6.   150 mM EDTA.
           Dissolve 558 mg of $Na_2EDTA:2H_2O$ in 10 ml of distilled water.
7.   0.9% NaCl.
           Dissolve 450 mg of NaCl in 50 ml of distilled water.
           For reagent 8.
8.   Patient's Serum.
           50 ml of serum (or heparinised plasma) + 30 ml of 0.9% NaCl.
           Keep on ice.
9.   QC Control Serum.
           Reconstitute a QC sample with a known calcium value.
           Keep on ice.

Test Tubes.
10, 20 and 50 μl Pipettors.
Reagent dispenser for reagent 1, set to 5.0 ml.
Glass Cuvettes.
Spectrophotometers.
Soft Paper Tissues.

PROCEDURE:

1.   Set up the following tubes, using the standard calcium solution (4 mmol/l):

	A	B	C	D	E	F	U_1	U_2	QC_1	QC_2
ml of colour reagent	5.0	5.0	5.0	5.0	5.0	5.0	5.0	5.0	5.0	5.0
μl of Ca standard (4 mmol/l)	0	10	20	30	40	50	-	-	-	-
μl of H_2O	50	40	30	20	10	0	-	-	-	-
μl of unknown serum	-	-	-	-	-	-	50	50	-	-
μl of control serum	-	-	-	-	-	-	-	-	50	50
Equivalent serum calcium concentration (mmol/l)	0	0.8	1.6	2.4	3.2	4.0	-	-	-	-

2.   Mix the tubes and stand for 30 minutes.

3.   Read the absorbance of all tubes, against tube A, at 578 nm. Pour each solution back
      into the tube after the reading (Abs 1).

4.   Add 50 μl of 150 mmol/l EDTA to tubes A, U_1, U_2, QC_1 and QC_2 and mix.

5.   Read the absorbance of the unknown and control tubes against A, at 578 nm (Abs 2).

6. Calculate Abs 1 - Abs 2 for the controls and unknowns and take the average of each duplicate.

Plot a standard curve and read off the values of the unknown and control sera.

Reference Range: 2.1 - 2.6 mmol/l.

EXERCISE 1-12

THE DETERMINATION OF CARBON MONOXIDE IN BLOOD

Carbon monoxide is a colourless, odourless, tasteless and non-irritating gas, resulting from the incomplete combustion of organic matter. CO is the most abundant pollutant in the lower atmosphere. A large number of accidental and suicidal deaths occur each year from its inhalation.

Carbon monoxide is toxic because it can displace oxygen from haemoglobin, thus reducing the availability of oxygen to the tissues. The iron of haemoglobin has an affinity for CO, which is 220 times greater than for O_2. This means that 0.1% CO in air will result in a 50% carboxy-haemoglobinaemia. Hence, CO is potentially dangerous, even in very low concentrations.

The rate at which arterial blood approaches equilibrium with CO in the inspired air depends upon a number of factors. These include the diffusion capacity of the lungs and the alveolar ventilation, which depends upon the level of exercise of the subject.

Not only does CO have a greater affinity for haemoglobin than oxygen, but it also has an effect on the oxygen dissociation curve:

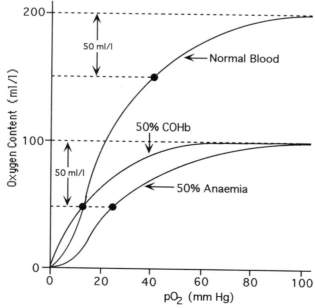

In a haemoglobin molecule that has bound two molecules of CO and two of O_2, then only two O_2 molecules are available for dissociation as the oxygen partial pressure is reduced. The CO tends to hold the molecule in the arterial (R) form and, hence, the molecule does not revert to the venous (T) form as the two oxygens are removed. As a result, a lower partial pressure of O_2 has to be reached before the molecule gives up its oxygen (ie. the curve is shifted to the left and becomes hyperbolic rather than sigmoidal). This is illustrated above, and compared with a case of 50% anaemia, as regards the release of 50 ml of oxygen from one litre of blood (at the points marked with a filled circle).

Factors that govern the toxicity of CO include the concentration of CO in the inspired air, the duration of exposure, the respiratory minute volume, the cardiac output, the oxygen demand of the tissues and the haemoglobin concentration of the blood. As a result, a poor correlation exists between the carboxyhaemoglobin content of blood and the signs and symptoms of poisoned patients.

Anaemic persons are more susceptible and people with a high metabolic rate tend to succumb earlier. Canaries were used to detect CO in mines because of their high metabolic rate. They die before toxic symptoms are noted in man. Children, due to their higher metabolic rate, tend to be more affected than adults.

The signs and symptoms of CO poisoning are characteristic of hypoxia. Even though there is poor correlation between the blood level of carboxyhaemoglobin and symptoms, an indication is given in the table below:

CORRELATION BETWEEN % COHb AND SIGNS AND SYMPTOMS OF CO POISONING

% of Blood Saturation	Signs and Symptoms
0-10	No symptoms.
10-20	Tightness across forehead; possibly slight headache, dilation of cutaneous blood vessels.
20-30	Headache; throbbing in temples.
30-40	Severe headache, weakness, dizziness, dimness of vision, nausea and vomiting, collapse.
40-50	Same as above with greater possibility of collapse and loss of consciousness; increased respiration and pulse.
50-60	Loss of consciousness, increased respiration and pulse; coma with intermittent convulsions. Cheyne-Stokes respiration.
60-70	Coma with intermittent convulsions, depressed cardiac function and respiration, possible death.
70-80	Weak pulse and slowed respiration; respiratory failure and death.

The foetus is particularly susceptible to the effects of CO, as the gas readily crosses the placenta. This can lead to brain damage.

Delayed toxicity can sometimes be seen. This is associated with an autoimmune response resulting from damage to the blood-brain barrier. This has a sudden onset after apparent recovery from poisoning. The symptoms consist of behavioural changes, confusion, disorientation, fever and neurological disturbances. The condition is usually progressive, with death resulting within a few weeks.

Treatment for CO poisoning is to displace the CO with O_2. The patient must be removed from exposure and given oxygen to breathe. Hyperbaric O_2 will displace the CO faster than O_2 at atmospheric pressure; however, care must be taken to avoid O_2 poisoning. Exchange transfusions have been used with moribund victims.

The major source of CO is from the burning of fossil fuels. Furnaces and cars are the major producers of CO, which result in CO poisoning. This is usually associated with poor ventilation. It is also noteworthy that most victims of fires die from acute CO poisoning rather than from burns.

Another source of exposure to CO is smoking. The average carboxyhaemoglobin level of heavy smokers (>40/day) is 5.9%.

The average concentration of CO in the atmosphere is 0.1 ml/m^3 (0.1 p.p.m.). The level can reach 100 ml/m^3 in heavy traffic and in underground garages. The toxicity associated with various atmospheric levels of CO is indicated in the table below:

CARBON MONOXIDE TOXICITY

% (v/v) in Air	Response
0.01	Allowable for an exposure of several hours.
0.04-0.05	Can be inhaled for 1 hour without appreciable effect.
0.06-0.07	Causing a just noticeable effect after 1 hour's exposure.
0.1-0.12	Causing unpleasant, but not dangerous symptoms after 1 hour's exposure.
0.15-0.20	Dangerous for exposure of 1 hour.
0.4 and above	Fatal in exposure of less than 1 hour.

An indication of high levels of CO in the blood is given by an abnormal red colour of the skin, mucous membranes and fingernails. This is due to the cherry-red colour of carboxyhaemoglobin.

CO in biological specimens can be determined by two methods:

    (a)   Release and measurement of the CO from carboxyhaemoglobin,
            and
    (b)   Measurement of carboxyhaemoglobin.

An example of B will be used in this exercise.

Oxyhaemoglobin is deoxygenated to haemoglobin by sodium dithionite; whereas, carboxyhaemoglobin is unaffected by dithionite. This fact, together with use of the characteristic absorption bands of these compounds, is used in this method.

The absorption peaks (in nm), in the visual range, are listed below:

	α	β
Hb	555	—
HbO$_2$	577	542
COHb	567	539

The wavelengths used in this method are 539 nm (the β band of COHb) and 555 nm (the Hb band). The absorbance of the diluted bloods at these wavelengths are measured, after treatment with dithionite, and the values of A_{539}/A_{555} ratios are calculated.

Case History.

A 25 year-old man was found unconscious in his garage. The car's engine was running and the garage doors were closed. On admission to hospital, a blood sample was taken for carbon monoxide determination.

REAGENTS AND EQUIPMENT:

1. Dilute Ammonia Solution.
    Make 4 ml of conc (SG 0.880) Ammonia to 1 l with distilled water.
    Place in a 25 ml dispenser.
2. Normal Blood.
    Keep on ice.
3. Patient's Blood.
    Add 10 ml of normal blood to a 250 ml conical flask. Tilt the flask slightly and very gently bubble CO through the blood from a Pasteur Pipette, for a few minutes. Make sure that the rate is slow enough for the blood not to froth too much.
    Add 10 ml of normal blood and decant into 50 ml reagent bottle and then mix.
    Keep on ice.
4. Sodium Dithionite (Sodium hydrosulphite).
    This tends to oxidise on storage, so it is a good idea to remove the top layer of crystals before use.
5. Watch Glass with 10 mg of Sodium Dithionite.
    An indication of how much to use in the assay.

Small Spatulas.
Oxygen Cylinder with Regulator and tube to a Pasteur pipette.
Carbon Monoxide Cylinder, adjacent to a fume cupboard, with Regulator and Bubble Flow Meter to a tube and Pasteur pipette.
150 ml Separating Funnels.
Wassermann Tubes.
Double Beam spectrophotometer, capable of measurements at 2 wavelengths.
Glass Cuvettes.
100 μl Pipettors and Tips.
Soft Paper Tissues.
Parafilm.

PROCEDURE:

Preparation of Standards:

1. Add 4 ml of normal blood to two 150 ml separating funnels.

2. Displace the air of one funnel with O_2 and the air of the second with CO.

3. Replace the stoppers and rotate the funnels to spread the blood over the surface of the funnels. Rotate gently for about 1-2 minutes.

4. Repeat steps 2 and 3, twice. This gives the 0 and 100% COHb standards.

5. Run about 1 ml of each blood into Wassermann tubes and cover with parafilm.

Assay:

1. Add 100 µl samples of the standards and the patient's blood, to 25 ml of dilute ammonia solution (This prevents protein precipitation) in test tubes and mix well.

2. Programme a double beam spectrophotometer for 2 wavelength absorption measurements and enter 539 and 555 nm. Select a 1 nm bandwidth.

    The instrument should calculate the A_{539}/A_{555} ratios.

    Autozero the instrument with nothing in the light paths and the lid closed.

3. In turn, add about 3 ml of each solution, from step 1, to a cuvette followed by about 10 mg of sodium dithionite and mix well by placing a piece of parafilm on the cuvette and inverting it several times.

4. Place the cuvette in the sample holder of the instrument, together with a dilute ammonia blank in the reference holder.

5. Press "Start". Wait for the values to appear on the screen before removing the cuvette.

6. After the last sample, obtain a copy of the results.

Calculate the A_{539}/A_{555} ratios, if the instrument has not done this.

Subtract the A_{539}/A_{555} ratio of the 0% COHb from the 100% COHb ratio. This value is equivalent to 100% saturation. Subtract the 0% COHb from the patient's ratio and calculate the % saturation.

Reference Values:

Non-smokers	:	0.5 - 1.5% saturation.
Smokers	:	4-5% saturation.
Heavy Smokers (above 40/day)	:	up to 8-9% saturation.

Major symptoms can be expected with values above 25%. Lethal levels range from 50% and upwards, although this can be lower in old or diseased persons. Levels of 60-70% will result in death if not treated.

EXERCISE 1-13

THE DETERMINATION OF PLASMA SALICYLATE

The chemical structures of the common salicylates are shown below:

Aspirin   Salicylic Acid   Methyl Salicylate

Salicylic acid is so irritating that it can only be used externally. However, less irritating derivatives can be used internally. The most common is aspirin, the ester of acetic acid. Methyl salicylate is the methyl ester of salicylic acid and is used externally in lineaments etc.

Aspirin is absorbed, as such, by the gastric mucosa; but is hydrolysed by the mucosal cells and, hence, enters the circulation as salicylic acid.

Aspirin is the most commonly used analgesic. It also has anti-pyretic, anti-inflammatory properties and is used as an anti-rheumatic agent. Accidental, or intentional, overdoses with salicylates are still very common in both children and adults. The symptoms of overdose are tinnitus, deafness, nausea and vomiting. More severe cases result in severe vomiting, hyperventilation, hyperthermia, convulsions, coma, acid-base and blood glucose disturbances. The clinical course may progress to pulmonary oedema, haemorrhage, acute renal failure, oliguria or death.

Treatment is by emesis or by gastric lavage if the patient is comatose. This is followed by the administration of activated charcoal to absorb the remaining drug in the intestine. Fluids are then give orally to combat dehydration that is a common finding with salicylate overdose.

The administration of bicarbonate, to make the urine alkaline, also speeds up the elimination of the drug from the body. This forced alkaline diuresis is recommended with adult salicylate levels above 500 mg/l and above 300 mg/l for children. The bicarbonate also helps to combat the metabolic acidosis seen with salicylate overdose.

There are a number of methods available for the determination of salicylate in the body fluids. One of the simplest methods is that of Trinder, where salicylates react with ferric salts to produce a violet colour. However, difference spectrophotometry of salicylates at pH 13 and 10.5 will be used in this exercise.

Salicylates are strong acids. They will be rendered non-ionised by the addition of 6 M HCl and extracted into chloroform. The salicylates are then ionised by the addition of an excess NaOH and extracted into the aqueous phase. pH 10.5 and 13 solutions are then prepared and the pH 13 solution scanned against the pH 10.5 (reference), between 225 and 380 nm.

Case History.

A 25 year-old woman presented at casualty saying that she had swallowed an unknown number of aspirin tablets about 2 hours previously. She was lethargic and gasping for breath. She complained of ringing in the ears. A Trinders spot test for salicylates in urine was positive (A purple colour with Hg and FeCl). Treatment for salicylate overdose was started and a blood sample collected for a quantitative plasma salicylate determination.

The patient's plasma and 3 standards will be treated by the method described above. This procedure removes most of the interfering material present in plasma.

REAGENTS AND EQUIPMENT:

1.   Chloroform.
2.   6 M HCl.
         Add 60 ml of Conc (11.6 M) HCl to 56 ml of distilled water and mix well.
3.   0.45 M NaOH.
         Dissolve 18 g of NaOH in 1 l of distilled water.
4.   160 g/l Ammonium Chloride.
5.   Salicylic Acid Standard (500 mg/l).
         Dissolve 290 mg of sodium salicylate in 500 ml of distilled water.
6.   Patient's Plasma.
         Dissolve 35 mg of sodium salicylate in 50 ml of plasma.
         Keep on ice.

25 x 150 ml Screw-capped Culture Tubes.
200 μl, 500 μl and 1.0 ml Pipettors with tips.
Wassermann Tubes.
Quartz Cuvettes.
Double Beam Spectrophotometer.
Soft Paper Tissues.
Centrifuge Tubes.
Bench Centrifuge.

PROCEDURE:

1.   To 4 25x150 mm screw-capped culture tubes, add 20 ml of chloroform.

2.   Add 1.0 ml of the patient's plasma to one tube and 0.2, 0.6 and 1.0 ml of the 500 mg/l salicylic acid standard to the remaining tubes.

3.   Add 0.8 and 0.4 ml of water to the 0.2 and 0.6 ml standards to make a total of 1.0 ml of aqueous solution in each tube.

4.   Add 1.0 ml of 6 M HCl to each tube, cap and shake vigorously for one minute.

5.   Stand for 1 minute and then remove the aqueous layers (top), by aspiration, and discard.

6.   Filter the chloroform layers into clean dry screw capped culture tubes.

7.   Add 8.0 ml of 0.45 M NaOH to each tube, cap and shake vigorously for 1 minute.

8.   When the layers separate, transfer the aqueous layers to centrifuge tubes and centrifuge for 5 minutes.

9.   Transfer two 3.0 ml aliquots of each aqueous layer, to Wassermann tubes.

10. To one of the two tubes of each set, add 0.5 ml of 160 g/l ammonium chloride and mix. This gives the pH 10.5 solution.

11. To the other tube in the set, add 0.5 ml of 0.45 M NaOH and mix. This is the pH 13 solution.

12. Set the instrument for:

Absorbance	:	-0.3 to 1.2
Wavelength	:	225-380 nm
Scan	:	Overlay
Scan speed	:	300-500 nm/min
Line type	:	Solid
Band width	:	2 nm

13. Place the pH 10.5 solution, in a quartz cuvette, in the reference light path and the pH 13 solution, in a quartz cuvette, in the sample light path. Autozero the instrument.

14. Record the difference spectrum, for each set. Autozero before each scan and use a dashed line for the patient's sample. After each scan, obtain the peak absorption values.

Plot the absorbance values of the standards at 246 nm (for low concentration) and 317 nm (for higher concentration) and read off the patient's value. Extrapolate the standard curve, if the patient's value is above the top standard.

Reference Values:

Therapeutic values range from 50 mg/l for mild analgesic levels to 150-300 mg/l for optimal anti-inflammatory effect in rheumatic patients.

Tinnitus occurs in the range 200-400 mg/l. Hyperventilation usually occurs with levels above 350 mg/l. Acidosis, and other signs of intoxication, are seen with levels above 450 mg/l.

The Done nomogram, below, is useful, in interpreting the result in relation to the time of the collection of sample after the ingestion of salicylate.

Done Nomogram for Salicylate Poisoning

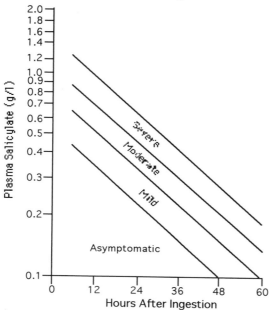

Cautions for use of chart:

(1). The patient has taken a single acute ingestion and is not suffering from chronic toxicity.

(2). Levels in the toxic range drawn before six hours should be treated.

(3). Levels in the nontoxic range drawn before six hours should be repeated to see if the level is increasing.

Paediatrics, (1960), 26, 805.

EXERCISE 1-14

THE DETERMINATION OF PLASMA PARACETAMOL

Phenacetin and its active metabolite, paracetamol (acetaminophen), are effective alternatives to aspirin, as analgesics and antipyretics. However, unlike aspirin, their anti-inflammatory activity is weak. They lack many of the side effects of aspirin and so both were commonly used. In normal doses, paracetamol is less toxic than phenacetin and so phenacetin has now been withdrawn from use in Australia.

Paracetamol, however, is very hepatotoxic in large doses. This is due to the production of a minor metabolite (Acetimidoquinone).

The majority of paracetamol is detoxified by conjugation with glucuronate or sulphate (Approx 94%). About 2% is excreted unchanged, which leaves 4% to be oxidised by the cytochrome P-450 system to acetimidoquinone. This is then detoxified by conjugation with glutathione.

However, the liver's glutathione tends to be used up when paracetamol is present in large amounts. This leaves the acetimidoquinone free to have its effect, which results in liver cell necrosis. Hepatic damage is seen when about 70% of the liver's glutathione is depleted.

This is produced by about 16 grams of paracetamol in adults and 4 grams in children. The actual value depends on normal absorption and metabolic activity.

However, liver damage is associated with higher plasma levels of paracetamol in children, than in adults. This is probably due to a lower rate of acetimidoquinone production in children, than in adults.

The chemical structure of the paracetamol is shown below, with some of its metabolites:

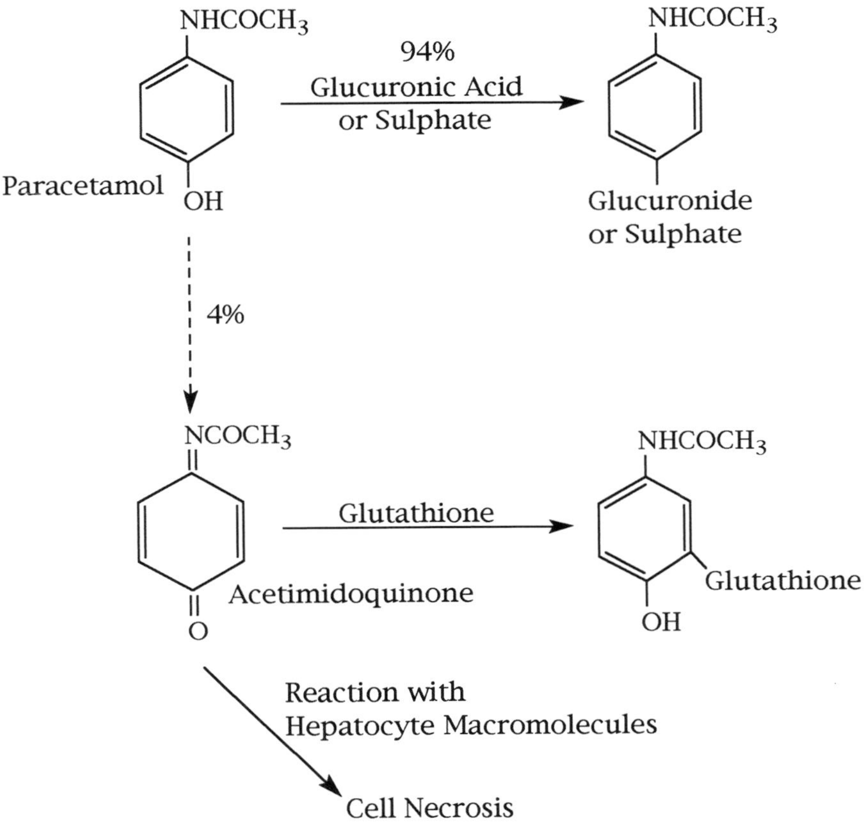

The paracetamol overdose patient presents with the following:

Stage I   -   2-24 hours.
Anorexia, nausea and vomiting.
A general feeling of malaise not unlike the common cold or flu.

Stage II   -   1-3 days.
Improvement.
The patient begins to feel better - may be hungry and want to get out of bed. At this time, the LFTs (alanine and aspartate aminotransferases, bilirubin, prothrombin time, etc) become abnormal. Right quadrant pain may occur.

Stage III   -   3-5 days.
Hepatic necrosis with peak abnormalities in liver function. Very high levels of LFT enzymes may be found, usually peaking at day 4.

Stage IV   -   7-8 days.
           Hepatic function returns to normal with general clinical improvement.

A small percentage of patients progress to hepatic encephalopathy and death. However, some survivors may have long-term liver damage.

The patient's plasma level of paracetamol is crucial in deciding on treatment. The plasma level peaks 3-4 hours after ingestion and then falls off exponentially. The relationship of plasma levels at specific times after ingestion and the probability of liver damage is shown as follows:

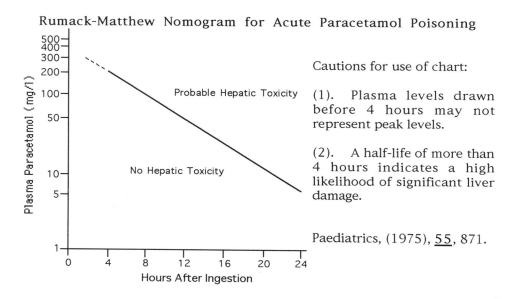

Rumack-Matthew Nomogram for Acute Paracetamol Poisoning

Cautions for use of chart:

(1). Plasma levels drawn before 4 hours may not represent peak levels.

(2). A half-life of more than 4 hours indicates a high likelihood of significant liver damage.

Paediatrics, (1975), 55, 871.

If the value falls above the line then treatment must be given. Rapid treatment is essential and should be started before the laboratory results are available, if there is any suggestion of paracetamol poisoning (Treatment is not effective after 16 hours post-ingestion).

The treatment is gastric lavage and administration of N-acetyl cysteine (preferably intravenously). This supplies -SH groups depleted by the loss of glutathione.

Even though phenacetin is converted to paracetamol, its other metabolic pathways are more significant when large doses are taken. Lethal doses of phenacetin do not produce liver damage. However, a metabolite causes the oxidation of haemoglobin to methaemoglobin, with cyanosis, respiratory depression and cardiac arrest resulting.

A number of methods are available for the determination of paracetamol. The most common ones are (1) hydrolysis to p-aminophenol and then coupling with o-cresol to form an indophenol blue, and (2) nitration to form a yellow product (The Glynn and Kendal Method).

However, deproteinised solutions will also be used in a direct UV absorption method in this exercise. The Allen correction will be used to reduce the interference caused by other UV absorbing material in the sample.

Paracetamol is first converted to its phenoxide, by the addition of sodium hydroxide. This ion has an absorption peak at 255 nm. The phenoxide ion has an absorption trough at 233 nm. This wavelength and 277 nm are the additional wavelengths used in the Allen calculation.

Case History.

On return from work, a man found his 23 year-old wife vomiting. She complained of nausea and dizziness. On admission to hospital, she admitted to swallowing a number of Paracetamol tablets as a parasuicidal gesture, because she was experiencing difficulties in her marriage. A blood sample has been collected for a plasma paracetamol determination.

REAGENTS AND EQUIPMENT:

1.   1/3 M H_2SO_4.
       100 ml of M H_2SO_4 + 200 ml of distilled water.
2.   10% Sodium Tungstate.
       Dissolve 10 g of $Na_2WO_4:2H_2O$ in distilled water and make to 100 ml.
3.   0.1 M NaOH.
       Dissolve 4.0 g of NaOH in 1 l of distilled water.
4.   Paracetamol Standard (500 mg/l).
       Dissolve 250 mg of paracetamol (p-Acetamidophenol) in 500 ml of distilled water.
5.   Patient's Plasma.
       Dissolve 20 mg of paracetamol in 50 ml of plasma.
       Keep on ice.

Eppendorf Centrifuge and Tubes.
Spectrophotometer, Preferably Double Beam.
Quartz Cuvettes.
Soft Paper Tissues.
100 and 200 μl Pipettors and Tips.
Test Tubes.
Wassermann Tubes.

PROCEDURE:

1.   Pipette 300 ml of 1/3 M H_2SO_4 and 300 ml of 10% sodium tungstate into 2 Eppendorf tubes and mix.

2.   Add 400 μl of the patient's plasma to each tube and mix well.

3.   Centrifuge for two minutes, at 14 000 g.

4.   Set up the following standards, in Wassermann tubes, using the standard paracetamol (500 mg/l) and mix well:

	A	B	C	D	E
μl of standard paracetamol (500 mg/l)	0	100	200	300	400
μl of H_2O	400	300	200	100	0
μl of 1/3 M H_2SO_4	300	300	300	300	300
μl of 10% sodium tungstate	300	300	300	300	300

5. Add 3.0 ml of 0.1 M NaOH to 7 test tubes labelled A, B, C, D, E and 2 of U (Unknown).

6. Add 100 µl of the solutions from step 4 and 100 µl of the supernatants from step 3 to their appropriate tubes and mix.

7. Read the absorbance of each solution, against tube A, at 233, 255 and 277 nm in a spectrophotometer.

Calculations:

Calculate the corrected absorbance from:

$$\text{Absorbance at 255 nm} - \left(\frac{\text{Abs at 233 nm} + \text{Abs at 277 nm}}{2} \right)$$

Plot a standard curve and read off the duplicate unknowns and average them. The patient's blood sample was collected about 6 hours after ingestion. Will antidote therapy be of use to this patient?

Reference Values:

Use the Rumack-Matthew nomogram on page 1-61 to assess the likelihood of hepatic damage. Treatment should be applied when there is a likelihood of liver damage.

GENERAL REFERENCES:

Frings, C. S. and Gauldie, J., (1989), Spectral techniques. Chapter 3 of Clinical Chemistry - Theory, Analysis and Correlation (2nd Edition). Eds. Kaplan, L. A. and Pesce, H. J., p 49-63, The C. V. Mosby Co.

Willard, H. H., Merritt, L. L., Dean, J. A. and Settle, F. A., (1981), Chapters 1,2 and 3 (p 1-104) of Instrumental Methods of Analysis (6th Edition), Wadsworth Publishing Co.

Dryer, J. R., (1965), Applications of Absorption Spectroscopy of Organic Compounds. Chapter 2, Ultraviolet Spectroscopy, p 4-20, Prentice-Hall Inc.

Kemp, W., (1975), Organic Spectroscopy. Chapter 4, Ultraviolet and Visible Spectroscopy, p 153-184.

Cooper, J. W., (1980), Spectroscopic Techniques for Organic Chemists. Chapter 8, Ultraviolet Spectroscopy, p 227-248, John Wiley and Sons.

Brown, S. B., (1980), An Introduction to Spectroscopy for Biochemists. Chapter 2, Ultraviolet and Visible Spectroscopy, p 14-69, Academic Press.

Human, M., (1985), Basic UV/Visible Spectrophotometry: A booklet published by LKB Biochrom.

Steward, J. E., (1977), Introduction to Ultraviolet and Visible Spectrophotometry: A booklet published by Pye Unicam Ltd.

SPECIFIC REFERENCES:

Cottrell, C. T., (1978), Derivative and Log Spectrophotometry: A booklet published by Pye Unicam Ltd.

Rand, R. N., (1969), Practical Spectrophotometric Standards, Clin. Chem., 15, 839-863.

Frings, C. S. and Broussard, L. A., (1979), Calibration and Monitoring of Spectrometers and Spectrophotometers, Clin. Chem., 25, 1013-1017.

Chamran, M. and Keiser, R., (1977), Maintaining Optimum Spectrophotometer Performance, A. Soc. Clin. Path., Technical Improvement Service, No. 27. (Available from Perkin-Elmer.)

West, M. A. and Kemp, D. R., (1976), Practical Standards for UV Absorption and Fluorescence Spectrophotometry, Int. Lab., May/June, 27-39. Reprinted by Activon Scientific Products.

Optimum Parameters for Spectrophotometry, (1973). A booklet published by Varian.

Plasma Calcium.

Lorentz, K., (1982), Clin. Chim. Acta, 126, 327-334.

Blood Carboxyhaemoglobin.

Tietz, N. W. and Fierech, E. A., (1973), Ann. Clin. Lab. Sc. 3, 26.

Plasma Salicylates.

Williams, L. A., Linn, R. A. and Zak, B., (1959), J. Lab. Clin. Med., 53, 156 - 162.

CHAPTER 2

FLUORIMETRY

FLUORESCENCE.

Many compounds absorb light in the UV and visual range and have absorption bands specific for the compound under consideration. The frequency of the light absorbed is proportional to the energy required to move the molecule from its ground state to an excited state.

Most compounds, in solution at room temperature, exist in the ground state. That is the lowest electronic and vibrational state (E_0V_0). Absorption of photons, containing the required energy, will excite the molecule to a higher energy state. That is, moving an electron (or electrons) into higher energy states, and also increasing the vibrational energy of the molecule (E_1V_1, E_1V_2 etc.). The rotational energy of the molecule may, also, be increased.

The quantum theory tells us that the energy level of the molecule can exist only in definite energy states and so only light of frequencies corresponding to the energy required to elevate the molecule from the ground state to an excited state will be absorbed. As a result of the number of possible excited states that can be produced, light of a broad range of frequencies will be absorbed. This gives the broad absorption bands seen with most compounds. The absorption process is very fast and takes about 10^{-15} second.

The excited state of the molecule exists for a much longer period of time than that required to excite it. The added energy of the molecule can be lost in a number of ways; including resonant transfer to other molecules (collisions), non-radiative internal conversion to heat, loss of some energy in conversion to the triplet state (change in electron spin), by chemical reaction or by loss of heat.

If the compound is to fluoresce, it will do so after the excited state has existed for about 10^{-8} second. During this time, some energy will have been lost as heat to the surroundings and the molecule will have reached the energy state E_1V_0. Fluorescence emission will occur from this point. Therefore, the energy released in the form of a fluorescence photon must have less energy than the excitation photon. Therefore, the wavelength of the emitted light will always be longer than that required for excitation (Stokes' Law or Stokes' Shift).

Where light is lost from the molecule as fluorescence, the energy level of the molecule can return to different vibrational and rotational energy levels of the electronic ground state (E_0V_0, E_0V_1, E_0V_2 etc.). This tends to give a broad emission band.

As illustrated below, the E_0V_0 to E_1V_0 transition is common to both absorption and fluorescence. This is known as the 0-0 transition and one would expect to find the absorption and fluorescence spectra overlapping at this point.

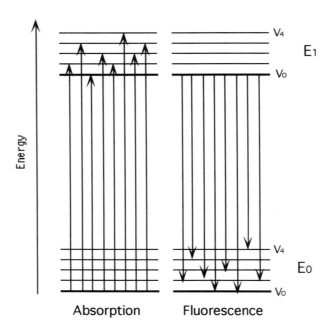

| | | | V4 | |
| | | | V0 | E1 |

Absorption   Fluorescence

However, 0-0 transitions rarely coincide in practice, as small amounts of energy are lost, by the absorbing molecule, to the surrounding solvent molecules.

The ratio of the number of molecules that absorb the light, to the number that fluoresce, is known as the quantum efficiency. The maximum value is 1, but the usual value is less than unity. Any available transition of the excited molecule, that avoids the fluorescence transition, will reduce the quantum efficiency.

Due to the period of time between excitation and emission, the molecule will have changed its orientation in a random fashion. As a result, the emitted light is non-polarised, even if a polarised excitation beam had been used. Light is, therefore, emitted in all directions.

Compounds that contain two or more conjugated double bonds (ie. alternate single and double bonds between carbon atoms) will possess π electrons (ie. electrons not bonded to a single carbon atom, but as a cloud around all the carbon atoms in the conjugated double bond structure). These electrons can be moved into higher energy states by the absorption of light, and the compound may fluoresce when the electron returns to the ground state.

If the freedom of these electrons is increased, say by the substitution of O, N or S into the structure, the chances of fluorescence will also increase. On the other hand, some substituted groups tend to localise the electrons and, therefore, reduce the chance of fluorescence.

Benzene is weakly fluorescent. The attachment of an amino group to the benzene ring, to give aniline, produces a much more fluorescent compound, due to an increase in the freedom of the π electrons. On the other hand, the nitro group of nitrobenzene tends to withdraw the π electrons from the ring. Thus, nitrobenzene is non-fluorescent.

Compounds that are fluorescent are said to possess "native" fluorescence. Many non-fluorescent compounds can be converted, by simple chemical reactions, to products that fluoresce, ie. "chemically induced" fluorescence. An example of this is the conversion of thiamine (Vitamin B_1) to thiochrome by oxidation using an alkaline potassium ferricyanide solution.

The intensity of a fluorescent solution is given by the formula:

$$Q = I_o\,(2.303\ \varepsilon CL)\ \phi f$$

Where,

Q	=	The fluorescence intensity.
I_o	=	The incident light intensity.
ε	=	The molar absorption coefficient.
C	=	The concentration of the fluorescent solute, in mol/1.
L	=	The cuvette path length in centimetres.
ϕf	=	The quantum efficiency of fluorescence.

This formula only applies to very dilute solutions, as discussed under "Factors Affecting Fluorescence Intensity" below.

Note that the fluorescence intensity is directly proportional to the concentration of the solute, as well as to the intensity of the light source and to the quantum efficiency.

FACTORS AFFECTING FLUORESCENCE INTENSITY.

(a)   Concentration.

The intensity of fluorescence is directly proportional to the concentration of the fluorescent compound in very dilute solutions only.

The light emitted by the fluorescent compound has to pass through the solution before reaching the photodetector and, in doing so, some of the light will be absorbed by other molecules of the compound under consideration. The higher the concentration, the greater is the proportion of emitted light absorbed. With very dilute solutions, this re-absorption of light is very small in comparison with the light emitted. The fluorescence intensity of the solution, therefore, increases in direct proportion to the concentration of the fluorescent compound up to a certain point. After this, the increase in fluorescence begins to fall off with increasing concentration. As the concentration increases further, the fluorescence intensity may even start to decrease. Hence, a specific value of fluorescence intensity may, therefore, correspond to two concentrations of the substance; as illustrated below:

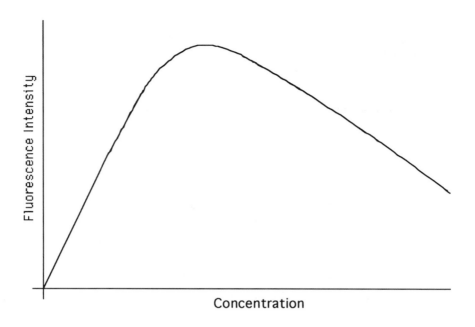

(b)   Solvent.

Changing the solvent can produce a change in fluorescence intensity and a change in the wavelength of the emission maximum. The solvent used must not have a natural fluorescence, nor should it contain fluorescent impurities. Solvents can have an enhancing or quenching effect on the emission intensity. This is due, in part, to the fact that the dielectric constant of the solvent will effect the freedom of the π electrons and thus alter the fluorescence as well as the wavelength of the fluorescence maximum.

Examples of the effect of different solvents on the fluorescence of sulphanilamide is illustrated below:

Solvent	Relative Intensity	Solvent	Relative Intensity
Water	100	Acetone	0
Methanol	72	Chloroform	0
Ethanol	107	Benzene	17
Propanol	89	Toluene	9
n-Pentanol	48	Ether	78

Whilst the emission maximum wavelength remains constant, the maximum excitation wavelength does vary with the solvent used.

When very weak solutions of fluorescent compounds are being measured, Raman spectrum lines from light scattering of the solvent may interfere with the measurement of the emitted light. If this is so, a different solvent should be used or a different excitation wavelength.

Direct light scattering, by the solute or solvent, is known as Rayleigh scattering. Similar scattering, by colloidal particles is known as Tyndall scattering. The scattered light is of the same wavelength as the excitation beam.

2-4

During the Rayleigh scattering process, some energy is absorbed by some of the molecules, to increase their vibrational and/or rotational energies. In this case, the scattered light will, therefore, have less energy. This effect is known as Raman scattering. The same amount of energy is lost, irrespective of the excitation wavelength. Hence, the Raman scattered light will always be the same frequency of light below the excitation frequency. This phenomenon is very weak and, therefore only affects high sensitivity fluorescence measurements. The degree of Raman scatter is different for different solvents.

(c)   Buffers.

Buffers have been shown to have a quenching effect on the fluorescence of some compounds. This effect is usually slight but does increase with the concentration of the buffer.

(d)   pH and Ionisation.

The intensity of fluorescence of an ionisable compound depends on the state of ionisation; which, in turn, is dependent on the pH of the solution containing the sample. As a large number of biological compounds that possess native fluorescence or can have chemically induced fluorescence are ionisable, their ability to fluoresce or the intensity of their fluorescence will depend on their ionised state and hence the pH.

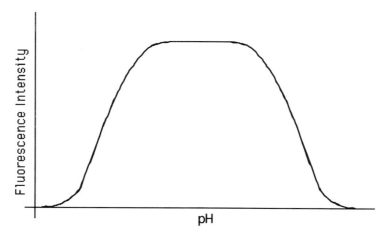

Some compounds, when excited by light, may also ionise. The fluorescence of such compounds is, therefore, that of the ionised form. This means that the fluorescence is being emitted by ions that would not normally be present at the pH under consideration. ie. :

$$MH + h\nu_1 \xrightarrow{\text{Excitation}} M^*H \xrightarrow{\text{Ionisation}} M^{*-} + H^+$$

$$\text{Emission} \downarrow$$

$$MH \xleftarrow{\ + H^+\ } M^- + h\nu_2$$

v_1 = The frequency of the existing light.

v_2 = The frequency of the emitted light.

h = Planck's constant.

MH = A compound capable of ionisation.

* = Excited state.

(e)   Temperature.

The fluorescence intensity of a substance tends to decrease with increasing temperature. This is probably due to the fact that an increase in temperature causes an increase in the velocity of molecules in solution. There is, therefore, an increase in the number of collisions between molecules in any given time. Energy, that could have been lost via fluorescence is, therefore, lost as heat in these collisions.

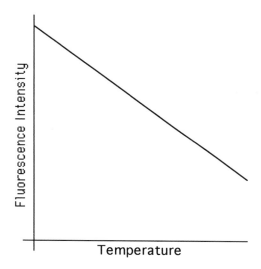

(f)   Photodecomposition.

Some chemicals, even though stable in a strong or weak solution, may decompose in light in the very dilute solutions used in fluorimetry. Needless to say, the rate of photodecomposition will depend on the intensity of the light source. Most photodecompositions result in a conversion of the fluorescent compound to a non-fluorescent product; however, some mildly fluorescent compounds can decompose to products that have a greater fluorescence than the original compound. In the former case the fluorescence will decrease with time; whereas, in the latter case the fluorescence will increase with time. Most fluorimeters are, therefore, fitted with a shutter between the light source and the sample. This is opened just before the fluorescence is measured.

(g)   Quenching.

Quenching takes place when the presence of a compound in the solution diminishes or abolishes the fluorescence of another compound.

The "inner filter" effects, produced by some quenching agents, are the result of the absorption of the exciting light and/or the absorption of the emitted light. Some solvents will exert this effect and must, therefore, be avoided. The problem of self-quenching has been discussed under Section (a).

"True" Quenching occurs when the quenching agent degrades the energy of the fluorescent compound in its excited state. As a result, the energy of the excited state is lost by means other than light emission. This type of quenching is due to collisions between the quencher molecules and the excited fluorescent molecules. This is a diffusion controlled process.

The magnitude of the quenching is dependent upon the fluorescence lifetime of the fluorescent molecule and upon the nature and molar concentration of the quencher.

The energy degradation of the excited state of the fluorescent molecule, by the quencher, can occur in a number of ways. These include the conversion of the excited molecule to its triplet state (reversal of an electron spin), electron transfer and energy transfer.

For energy transfer to take place, the energy level of the excited quencher molecule must be just below that of the excited fluorescent molecule. The quencher molecule is usually converted to its triplet state.

This type of quenching can occur in solids where the fluorescent and quencher molecules are close to each other. For example, anthracene containing 1 part in 1,000 phenazine will lose its blue fluorescence. The energy level of the triplet state of excited phenazine is just below that of the singlet excited state of anthracene.

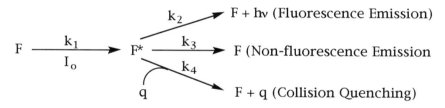

Where:

$$F \quad = \quad \text{The non-excited fluorescence molecule.}$$
$$F^* \quad = \quad \text{The excited fluorescence molecule.}$$
$$h\nu \quad = \quad \text{Light emission.}$$
$$I_0 \quad = \quad \text{The incident light intensity.}$$
$$q \quad = \quad \text{The quencher molecule.}$$
$$k_1, k_2, k_3 \text{ and } k_4 \text{ are rate constants.}$$

Other quenching agents can cause a chemical change in the fluorescent molecule, usually converting it to a non-fluorescent molecule. This type of quenching is usually very specific. An example of this is the effect of pH when H^+ ions are added or removed from the solution (See Section (d)).

Some of the quenching agents can be assayed by their quenching effect on a fluorescent compound in solutions. For example, oxygen can be measured by its quenching of the fluorescence of a borate-benzoin complex.

INSTRUMENTATION.

The basic layout of a fluorimeter is shown below:

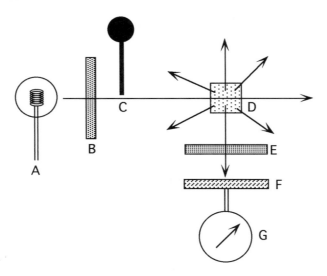

A	:	Light Source.
B	:	Primary Filter or Monochromator.
C	:	Shutter to reduce radiation exposure of the sample (Not present in all instruments. Some have a shutter before the detector.).
D	:	Sample Cuvette.
E	:	Secondary Filter or Monochromator.
F	:	Photodetector.
G	:	Display.

Instruments using filters are known as fluorimeters and those using monochromators are known as spectrofluorimeters or fluorescence spectrometers.

Note that the emitted light is measured at right angles to the excitation beam, so that as little as possible of the excitation light, from reflection or scattering, reaches the photodetector.

Where there is a reasonable difference between the excitation and emission wavelengths, all the excitation light will be absorbed by the secondary filter or monochromator. However, when these wavelengths are close, some Rayleigh scattered light will reach the photodetector.

Light Sources.

As fluorescence intensity is proportional to the intensity of the light striking the sample, more intense light sources are normally used in fluorimeters than those used in spectrophotometers. Most spectrofluorimeters use a xenon arc lamp as their lamp source. This emits a continuous spectrum between 200 and 800 nm, even though the intensity of the emitted light is spectrally uneven. The light intensity drops off fairly rapidly below 300 and above 650 nm.

Most filter fluorimeters use mercury lamps in the UV and tungsten-halogen lamps in the visual range. The mercury lamp has the disadvantage that it does not give a continuous emission; but rather, a series of high intensity emission lines. This makes it suitable for excitation of samples absorbing at, or close to, the mercury emission lines, but not for others. Zinc lamps can also be used but these, again, have narrow emission lines.

Some of the emission lines (in nm), that can be selected by interference filters are:

Mercury	Zinc
253.7	213.9
366.3	330.3
435.8	334.5

Sample Containers.

Most instruments use cuvettes with a square cross-section. The sides of the cuvettes, through which the excitation beam enters and fluorescent emission leaves, are made of optically ground glass or silica (quartz). This usually includes all 4 sides of the cuvettes. Silica cuvettes can be used for all measurements; whereas, glass cannot be used for light below 320 nm.

Some fluorimeters use cylindrical tubes to hold the sample. The excitation beam is focused onto the centre of the sample. The emitted light is collected by means of a lens, before being quantitated.

The fluorimeter cuvettes must be clean to obtain good results and the optical surfaces should not be touched. This is especially important where wavelengths below 300 nm are used.

Detectors.

Photomultiplier tubes are used as detectors in almost all instruments. The output of these can be fed directly into a galvanometer or can be amplified by a DC amplifier before being displayed. Phototubes are often used to monitor the intensity of the excitation beam.

SENSITIVITY.

Fluorimetric techniques are much more sensitive than those using absorption. As the wavelength of the emitted light is longer than the excitation light, a filter or monochromator (the secondary filter) is used to stop the excitation light reaching the photodetector, but allowing the emitted light to pass. This means that estimations of fluorescence emission are carried out against a very low background of light intensity. Modern photomultipliers can detect very low levels of light under these conditions. On the other hand, the lower limit of sensitivity of light absorption techniques depends on the accuracy with which two very similar intensities of light can be compared.

Fluorimetry has the advantages over absorption techniques in that only one measurement has to be made as opposed to two, and that the output of the photodetector is directly proportional to the concentration; whereas, with absorption methods, the output is inversely proportional to the log of the concentration.

Many methods can be designed so that the absorption of the sample is measured at high concentrations and the fluorescence measured at low concentrations. Thus the same technique can be used for a wide range of concentrations.

The lowest concentration of substances that can be determined by absorption methods is of the order of 1 mmol/m^3; whereas, concentration as low as 1 μmol/m^3 can be determined by fluorimetric techniques.

An Example of a Spectrofluorimeter.

The Hitachi F-2000 Fluorescence Spectrophotometer is a typical example of a spectrofluorimeter.

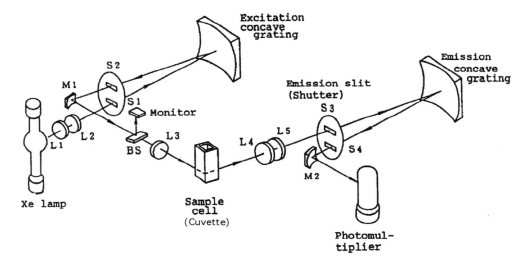

The light source for this instrument in a xenon arc lamp with an ozone self-dissociation function. The monochromators are concave gratings ruled with 900 lines/mm. The excitation monochromator is blazed at 300 nm and the emission monochromator blazed at 400 nm.

A small proportion of the excitation light is monitored using a beam splitter (BS) and a detector to measure its intensity. Hence, correction can be made for variations in the light output of the lamp, and in grating efficiency, with different wavelengths. Hence, a "corrected" excitation spectrum will be similar to an absorption spectrum.

The instrument can measure between 220 and 730 nm to an accuracy of $\pm$ 5 nm. The bandwidth can be set to 10 or 20 nm. Wavelengths can be scanned at 15, 60, 240 or 1,200 nm/min, in either the excitation or emission mode.

Light is detected by a photomultiplier tube and the output fed to a computer for data manipulation. Information is displayed on a CRT and can be printed out on a thermal printer.

EXERCISES

Fluorimeter cuvettes have optical faces on all 4 sides and should only be handled by the top. All 4 faces should be wiped with soft paper tissues before the cuvette is placed in the instrument.

Older spectrofluorimeters are calibrated by setting the instrument to read 100 (Relative Fluorescence Intensity) on the strongest solution (Top Standard) and zero with no light reaching the photomultiplier or on the reagent blank (Zero standard).

Modern spectrofluorimeters are direct reading in relative fluorescence units.

When setting the wavelengths on the older spectrofluorimeters, always come from a lower value.

All glassware must be thoroughly clean.

EXERCISE 2-1

THE EFFECT OF CONCENTRATION ON FLUORESCENCE INTENSITY

In this exercise, phenol will be used to demonstrate that fluorescence intensity is directly proportional to concentration only in low concentration. The "Inner Filter", self-quenching, effects of higher concentrations will be demonstrated.

Phenol has the structure:

REAGENTS AND EQUIPMENT:

1.   400 mg/l Phenol.
            Dissolve 400 mg of phenol in 1 litre of distilled water.

Test Tubes.
Quartz Fluorimeter Cuvettes.
Spectrofluorimeter.
Soft Paper Tissues.
5 ml Pipettes.

PROCEDURE:

1.   Add 5.0 ml of distilled water to 8 clean test tubes, labelled A to H.

2.   Add 5.0 ml of 400 mg/l phenol to tube A, cover with parafilm and mix by inversion.

3.   Transfer 5.0 ml of this solution to tube B and mix well.

4.   Repeat this serial dilution until 5.0 ml of G are added to H and mixed.

5. Read the fluorescence intensity of these solutions, distilled water and the 400 mg/l solution, using quartz fluorimeter cuvettes, with the following wavelength settings:

    Excitation : 270 nm.
    Emission   : 330 nm.

If using an old spectrofluorimeter:

(a) Set the instrument to read 100 on sample D (25 mg/l) and read E, F, G, H and distilled water.

(b) Set the instrument to read 90 on sample C and read the fluorescence of the other samples, distilled water and the 400 mg/l solution.

Plot 2 standard curves. One of 0 to 25 mg/l (a) and one of 0 to 400 mg/l (b).

EXERCISE 2-2

THE EFFECT OF SOLVENT ON FLUORESCENCE INTENSITY

Apart from the presence of fluorescent, or quenching, impurities in a solvent, the solvent itself can affect the fluorescence intensity. This can be due to the dielectric constant of the solvent, hydrogen bonding of solvent and solute, quenching by the solvent and/or due to ionisation of the solute in the solvent.

In this exercise, the effect of solvent on the fluorescence intensity of sulphanilamide will be examined. The maximum emission wavelength does not change with the solvent used. However, there is a shift in the maximum absorption (excitation) wavelength.

Sulphanilamide has the structure:

$$H_2N-\!\!\left\langle\!\!\bigcirc\!\!\right\rangle\!\!-SO_2NH_2$$

REAGENTS AND EQUIPMENT:

1. 2 mg/l Sulphanilamide in Chloroform.
2. Methanol.
3. n-Pentanol.
4. Benzene.

Heating block at 60°C, in a fume cupboard, with Pasteur pipettes connected to a compressed air manifold.
Test Tubes.
5 ml Graduated Pipettes.
Quartz Fluorimeter Cuvettes.
Spectrofluorimeter.
Soft Paper Tissues.

PROCEDURE:

1. Pipette 3.0 ml samples of sulphanilamide (2 mg/l in chloroform) into 4 test tubes, labelled A to D.

2. Evaporate these to dryness in the 60°C heating block, using a gentle stream of compressed air blown onto the surface of the liquid. Make sure that all the chloroform has evaporated.

3. Add 3.0 ml of the solvents as follows:

A	:	Distilled water.
B	:	Methanol.
C	:	n-Pentanol.
D	:	Benzene.

4. Cap the tubes with parafilm and mix on a vortex mixer for 30 seconds to fully dissolve the sulphanilamide.

5. Using quartz fluorimeter cuvettes, read the fluorescence intensity of the solutions including the 2 mg/l in chloroform, using the following wavelengths:

Excitation	:	305 nm.
Emission	:	355 nm.

   If using an old spectrofluorimeter, set the instrument to read 100 on sample A.

Tabulate the results.

EXERCISE 2-3

THE EFFECT OF pH ON FLUORESCENCE INTENSITY

In this exercise the ionisation of aniline will be used to demonstrate the effect of pH on fluorescence intensity.

Aniline
Cation
(Non-fluorescent)

Aniline
(Fluorescent)

Aniline
Anion
(Non-fluorescent)

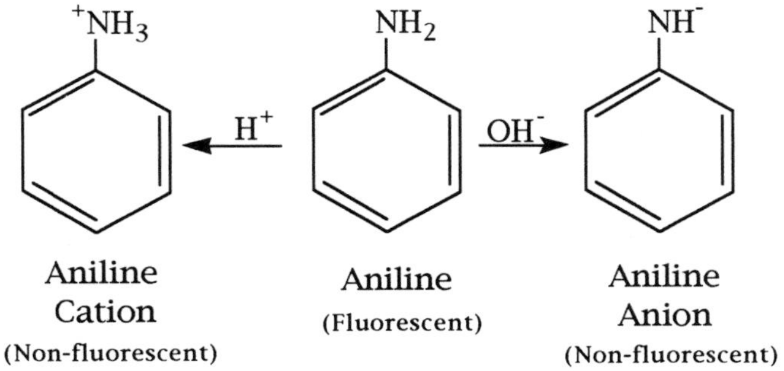

The cation will be produced by adding one drop of concentrated HCl to an aqueous solution of aniline. The anion will be produced by adding 40% NaOH to an aqueous solution of aniline. However, conversion to the anion is not complete with the amounts of NaOH added.

REAGENTS AND EQUIPMENT:

1.   5 mg/l Aniline in Water.
          Add 5 µl of AR Aniline to 1 l of distilled water.
2.   Conc HCl.
3.   40% NaOH.

Quartz Fluorimeter Cuvettes.
Spectrofluorimeter.
Test Tubes.
Soft Paper Tissues.
5 ml Pipettes.
Pasteur Pipettes.

PROCEDURE:

1.   Add 5.0 ml of the aniline solution (5 mg/l in water) to 4 tubes labelled A, B, C and D.

2.   Add the following to their respective tube and mix:

          A   :   Nothing.
          B   :   1 drop of Conc HCl.
          C   :   10 drops of 40% NaOH.
          D   :   20 drops of 40% NaOH.

3.   Read the fluorescence intensity of these solutions, in quartz cuvettes, using the following wavelengths:

          Excitation   :   300 nm.
          Emission   :   355 nm.

   If using an old spectrofluorimeter, set the instrument to 100 on A.

Tabulate the results.

EXERCISE 2-4

THE EFFECT OF TEMPERATURE ON FLUORESCENCE INTENSITY

For most compounds, there is a decrease of about 1% in fluorescence intensity, per degree rise in temperature. However, some compounds show a greater sensitivity to temperature changes. p-Anisidine is one such compound and will be used in this exercise. It has the following structure:

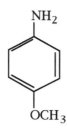

REAGENTS AND EQUIPMENT:

1.   2 mg/l p-Anisidine in water.
           Dissolve 2 mg of AR p-Anisidine in 1 l of distilled water.

Test Tubes.
Glass or Quartz Fluorimeter Cuvettes.
Soft Paper Tissues.
Spectrofluorimeter.
Water Baths at 40°C, 50°C and 60°C.
5 ml Pipettes.

PROCEDURE:

1.   Add 5 ml of the 2 mg/l p-Anisidine solution to 3 test tubes and place one in each of
       40°C, 50°C and 60°C waterbaths.

2.   Allow the solutions to come to temperature (about 5 minutes).

3.   Rapidly measure the fluorescence intensity of each sample, and a room temperature
       sample, in a spectrofluorimeter with the following wavelength settings:

           Excitation   :   320 nm.
           Emission   :   375 nm.

       If using an old spectrofluorimeter, set the instrument to 100 on the room temperature
       sample.

       A rise in fluorescence intensity will be noted as the temperature of the sample falls.

Tabulate the results.

EXERCISE 2-5

THE EFFECT OF QUENCHING ON FLUORESCENCE INTENSITY

(i)   "Inner Filter" Effect.

This is where the excitation and/or emission light is absorbed by the quenching agent. Hence, chloroform can not be used as a solvent below 245 nm, benzene below 280 nm and acetone below 330 nm.

Coloured compounds tend to absorb both the excitation and emission light. This will be demonstrated in this exercise where dimethyl yellow quenches the fluorescence of POPOP. POPOP (1,4-di-(2-(5-phenyloxazolyl))-benzene) was used as a secondary scintillant in liquid scintillation counting of β emitting isotopes. It absorbed light emitted by the primary scintillant and re-emitted the light at a longer wavelength. This longer wavelength was the optimum for the photomultipliers used in old counters.

POPOP has the structure:

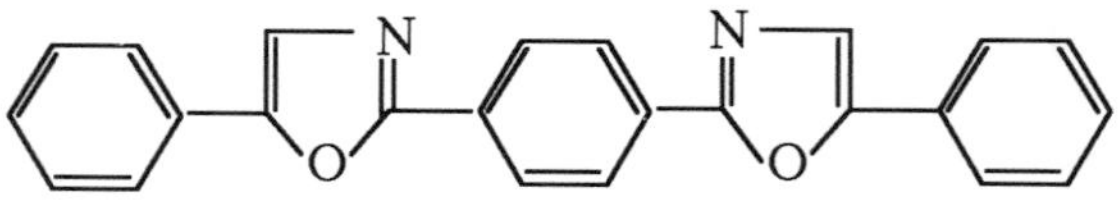

POPOP has an absorption maximum at 365 nm and an emission maximum at 415 nm. It has a fluorescence quantum efficiency, in toluene at 20°C, of 0.85.

REAGENTS AND EQUIPMENT:

1. POPOP Solution (10 mg/l in toluene).
2. Dimethyl Yellow (200 mg/l in toluene).

Test Tubes.
50, 100 and 200 μl Pipettors and Tips.
5 ml Pipettes.
Glass Fluorimeter Cuvettes.
Soft Paper Tissues.
Fluorimeter.

PROCEDURE:

1. Set up 4 test tubes labelled 0, 50, 100 and 200.

2. Pipette 5.0 ml the POPOP solution into each test tube.

3. Add 50, 100 and 200 μl of the dimethyl yellow solution to its appropriate tube and mix well.

4. Measure the fluorescence intensity of the 4 solutions with the following wavelength settings:

   Excitation :   365 nm.
   Emission   :   415 nm.

   If using an old instrument, set it to 100 on the unquenched tube (0) and read the relative fluorescence intensity of the others.

Tabulate your results.

(ii) True Quenching.

This is due to the loss of energy of the excited fluorescent molecule associated with collisions between the fluorescent molecule and the quenching molecule.

The quenching effect of some of the solvents used in Exercise 2-2, especially chloroform, will have been noted.

In this exercise the quenching effect of a halide ion (Cl$^-$) on quinine fluorescence will be demonstrated.

Quinine, in H_2SO_4, has excitation maxima at 250 and 350 nm and an emission maximum at 450 nm.

REAGENTS AND EQUIPMENT:

1.   0.05 M H_2SO_4.
            For reagent 2.
2.   Quinine Solution (1 mg/l in 0.05 M H_2SO_4).
3.   Conc HCl.

Test Tubes.
5 ml Pipettes.
Pasteur Pipettes.
UV (Hg) Lamp.

PROCEDURE:

1.   Pipette 5.0 ml of the quinine solution into 2 test tubes.

2.   Add 1 drop of conc HCl to one of the tubes and mix.

3.   Examine the tubes under a UV (Hg) light in a darkened room.

Comment on the results.

EXERCISE 2-6

THE EMISSION SPECTRUM OF NADH.

As will have been seen in Exercise 1-8, NADH has an absorption peak at 340 nm. This is due to the absorption of light energy by the reduced nicotinamide part of the molecule. Some of the light energy absorbed at 340 nm is emitted as fluorescence with a maximum intensity at 460 nm.

NADH has the structure:

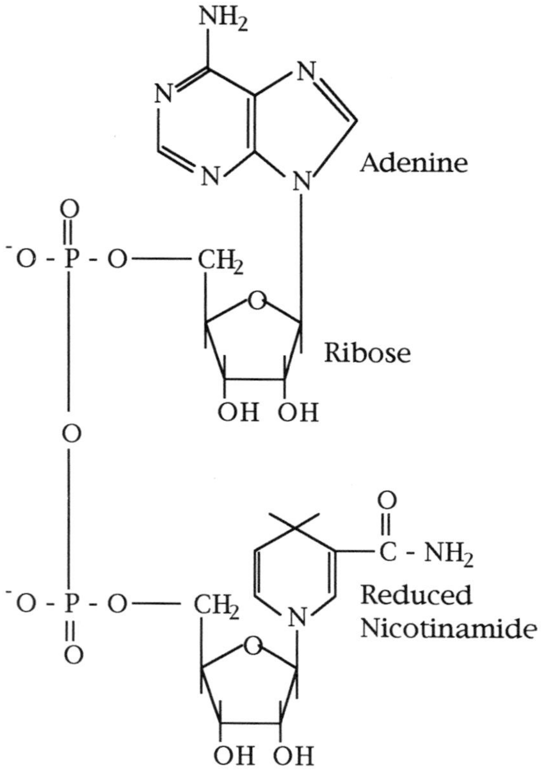

The 260 nm absorption band of NADH can also be used to excite the molecule. The adenine part of the molecule absorbs at this wavelength. Some of this energy is then transferred to the reduced nicotinamide part of the molecule. Energy is then liberated, as fluorescence, when the electronic configuration, around the reduced nicotinamide, returns to the ground state.

As mentioned in Exercise 1-8, the change in absorbance at 340 nm can be used to monitor oxidation-reduction reactions. However, the sensitivity can be increased by monitoring the change in fluorescence intensity at 460 nm.

In this exercise, the emission spectrum of NADH will be examined, using the excitation wavelength of 340 nm.

REAGENTS AND EQUIPMENT:

1.   50 mM NADH.
                Dissolve 7 mg of NADH in 200 ml of distilled water.
                Make up fresh and keep on ice.

Glass or Quartz Fluorimeter Cuvettes.
Soft Paper Tissues.
Spectrofluorimeter capable of recording spectra, or connected to a recorder (Preferably a X/Y recorder, if the fluorimeter has a wavelength output).

PROCEDURE:

1. Set the instrument for an emission scan, with the following wavelength settings:

    Excitation   :   340 nm.
    Emission   :   320-600 nm.

    With an old instrument there is usually a large power surge when the xenon arc is struck. Therefore, do not connect the recorder to the instrument or to the mains until the xenon lamp is on.

    Connect the fluorescence output to an ordinary recorder or to the Y input of a X/Y recorder and the wavelength output to the X input. Adjust the X input so that the chart scale matches the wavelength of the fluorimeter.

2. Place a fluorimeter cuvette containing 50 mM NADH in the instrument.

3. Run a pre-scan on a modern instrument. This sets the intensity scale. Then run the emission scan,

    With the older instruments the intensity scale will have to be set with the emission wavelength on 460 nm and adjusting the instrument output and/or the recorder input.

    Run the emission scan. With a normal recorder, the chart should be started when the instrument scan is started and stopped when the scan is stopped or the chart market at 320 and 600 nm.

Note the small scatter peak at 340 nm. Record the actual emission maximum wavelength.

EXERCISE 2-7

THE DETERMINATION OF THIAMINE IN URINE

Thiamine (Vitamin B_1) was the first of the B group of water-soluble vitamins to be recognised.

The coenzyme of Vitamin B_1 is thiamine pyrophosphate (TPP or cocarboxylase). It is required for the decarboxylation, or oxidative decarboxylation, of a number of α keto acids (eg. pyruvic acid). TPP is also the coenzyme for the two transketolase reactions of the hexose monophosphate shunt.

Lack of the vitamin leads to the disease "Berri-berri". This is associated with loss of weight, fatigue and gastro-intestinal upsets. Cardiac impairment is also seen at a later stage.

Thiamine deficiency can be detected by an elevated blood level of pyruvate, a decreased level of transketolase activity in lysed red cells and/or by a reduced excretion of thiamine in the urine. The last method will be used in this exercise.

Thiamine, itself, does not possess appreciable fluorescence; but it is readily oxidised to thiochrome, which is fluorescent. Thiochrome has a maximum excitation wavelength of 365 nm and a maximum emission at 435 nm.

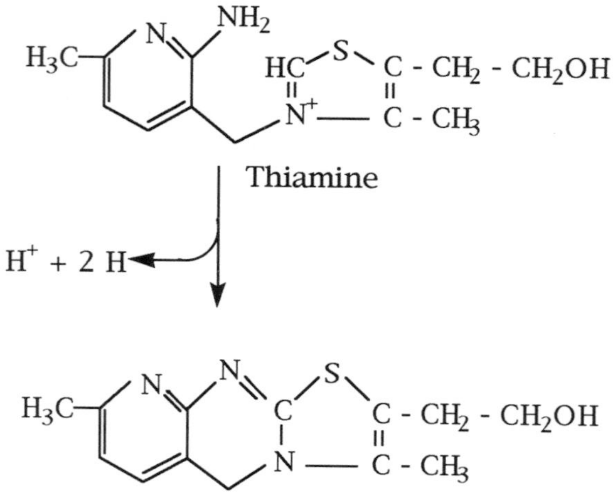

$$H_3C \cdots \begin{array}{c} N \\ NH_2 \end{array} \cdots HC \overset{S}{\underset{N^+}{\parallel}} \begin{array}{c} C - CH_2 - CH_2OH \\ \parallel \\ C - CH_3 \end{array}$$

Thiamine

$$H^+ + 2H \longleftarrow$$

$$H_3C \cdots \begin{array}{c} N \\ \end{array} \cdots \begin{array}{c} N \\ C \\ N \end{array} \overset{S}{\underset{C - CH_3}{\parallel}} \begin{array}{c} C - CH_2 - CH_2OH \\ \parallel \\ C - CH_3 \end{array}$$

Thiochrome

A timed collection of urine is made and the volume measured. If the pH of the urine is not on the acid side of pH 6.0, it is adjusted to pH 3.0 - 6.0 with glacial acetic acid. (This retains the positive charge on thiamine). The thiamine is then adsorbed onto Amberlite IR 120 (strong cation exchange resin). The column is then washed with 0.1 M acetic acid to remove interfering anions and uncharged material. The thiamine is eluted from the column with M KCl and then oxidised to thiochrome with potassium ferricyanide under alkaline conditions.

Even though the oxidation is more complete in alkaline solution, the resultant thiochrome is less stable than in neutral or acid solution. The thiochrome is, therefore, extracted into isobutanol as quickly as possible. The extract is then dried with anhydrous sodium sulphate and its fluorescence determined. Thiamine standards are oxidised and extracted, as above, and their fluorescence determined.

Case History.

The patient is a Vietnamese refugee, of about 50 years of age, who had just arrived in Australia by boat. He was showing signs of mental confusion, anorexia, muscle weakness and ataxia. A nutritional deficiency was suspected.

As part of his assessment, a 4 hour urine sample of 200 ml was collected for thiamine analysis.

The 200 ml urine sample has been passed through an Amberlite IR 120 column. After washing the column with 0.1 M acetic acid, the thiamine was eluted with M KCl in a total volume of 200 ml.

REAGENTS AND EQUIPMENT:

1.   0.1 M Acetic Acid.
            Add 5.5 ml of glacial acetic acid to 1 l of distilled water and mix.
            For reagent 2.

2.   Thiamine Standard (3.0 μM in 0.1 M Acetic Acid).
           Stock :   Dissolve 10 mg of thiamine hydrochloride in 100 ml of reagent 1.
                        Store at 4°C.
           Working Standard : Dilute the stock 1 in 100 with reagent 1.
                        Prepare fresh and keep on ice.
3.   Patient's Urine Eluate.
           Mix 10 ml of working thiamine standard with 190 ml of M KCl.
           Prepare fresh and keep on ice.
4.   M KCl.
           Dissolve 75 g of KCl in 1 l of distilled water.
5.   Alkaline Ferricyanide Reagent.
           Dissolve 150 g of NaOH in distilled water and make to 1 l. Add 300 mg of $K_3Fe(CN)_6$ and
           stir to dissolve.
           Make up fresh and put in dispenser bottle set to 6.0 ml.
6.   Isobutanol.
           In a dispenser set to 5.0 ml.
7.   Anhydrous Sodium Sulphate.

Watch glass containing 2-3 g of sodium sulphate.
           As an indication of weight.
Spatulas.
Test Tubes.
25 ml Stoppered Tubes.
Centrifuge Tubes.
Bench Centrifuge.
Glass or Quartz Fluorimeter Cuvettes.
Soft Paper Tissues.
Spectrofluorimeter.
5 and 10 ml Graduated Pipettes.
Pasteur Pipettes.
Gladwrap.

PROCEDURE:

1.   Pipette duplicate 10 ml aliquots of the patient's urine eluate into test tubes.

2.   Set up the following standards, in test tubes, using the standard thiamine solution (3
       μmol/l in 0.1 M acetic acid).

	A	B	C	D	E
ml of thiamine standard	-	1.0	2.0	3.0	4.0
ml of M KCl	10.0	9.0	8.0	7.0	6.0

3.   To each of the 7 tubes, in turn, carry out the following procedures:

       a.   Decant the entire contents into the 25 ml stoppered tube provided.

       b.   Add 6.0 ml of alkaline ferricyanide reagent (0.03% potassium ferricyanide
              in 15% NaOH), followed by 5.0 ml of isobutanol.

       c.   Shake vigorously for 90 seconds.

d. Pour the contents back into the respective test tubes.

4. As soon as the two layers separate, transfer the organic (top) layer into labelled centrifuge tubes with a Pasteur pipette.

5. Add 2-3 grams of anhydrous sodium sulphate to each tube, cap with "Gladwrap", place your thumb over the top of the tube and shake for about 20 seconds.

6. Centrifuge for about 2-3 minutes and then decant into clean, labelled, test tubes. The supernatant should be free of any suspension.

7. Determine the fluorescence intensity of the isobutanol extracts in the spectrofluorimeter with the following wavelength settings:

Excitation : 365 nm
Emission : 435 nm.

If using an old fluorimeter, set the instrument to read 100 on the strongest standard (E).

Plot a standard curve of the amount of thiamine per tube against fluorescence intensity and read off the amount in the 10 ml of eluate.

Calculate the 24 hour excretion of thiamine.

Reference Range:

0.25 - 1.5 µmol of thiamine are excreted in 24 hours by a person on an adequate diet.

EXERCISE 2-8

THE DETERMINATION OF SERUM MAGNESIUM

Divalent ions from complexes with a number of organic compounds known as chelating agents. One of these, 8-hydroxyquinoline-5-sulphonic acid, has a greater binding capacity for magnesium than for other divalent ions found in serum.

The chelate formed between magnesium and this sulphonic acid derivative exhibits intense fluorescence. Even though the chelates of other divalent ions show fluorescence, they do so with much less intensity than the magnesium chelate, when compared on a molar basis. As a result, the interference caused by calcium and other divalent ions in serum has little effect on this method for magnesium determination.

However, for very low magnesium levels, correction of the fluorescence of other divalent cations can be made by repeating the assay at pH 3.5. At this pH, the chelate of the other divalent cations still fluoresce as their fluorescence is not pH dependent; however, the magnesium chelate does not fluoresce. Hence, fluorescence at pH 3.5 is due to the non-magnesium divalent cations. This value can then be subtracted from the pH 7.0 fluorescence to give the fluorescence due to magnesium only.

The reaction between magnesium and the chelating agent is shown below:

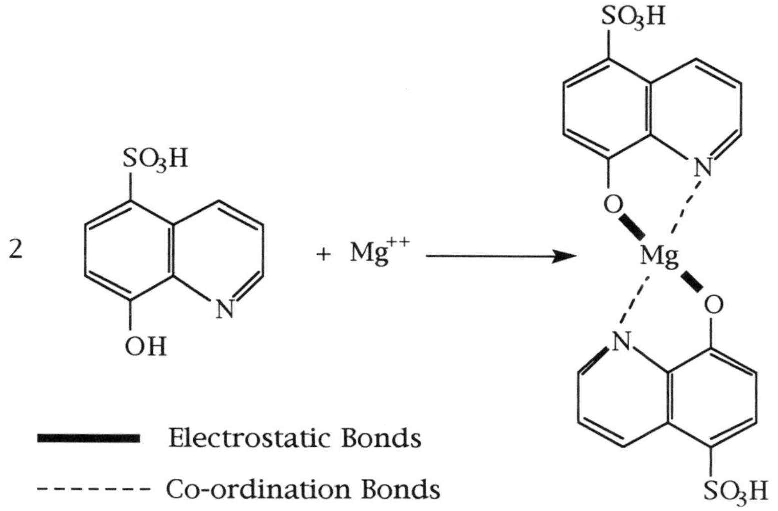

$$2 \quad \text{[8-hydroxyquinoline-5-sulphonic acid]} \quad + \ Mg^{++} \longrightarrow \text{[Mg chelate complex]}$$

────── Electrostatic Bonds

------- Co-ordination Bonds

As mentioned above, the fluorescence of the complex is pH dependent. A pH of 7.0 is optimum, with the fluorescence being abolished below pH 5.0. The fluorescent chelate has a maximum excitation wavelength of 375 nm and a maximum emission wavelength of 505 nm.

Magnesium standards and the unknown are, therefore, diluted with buffered 8-hydroxyquinoline-5-sulphonic acid. The fluorescence of these solutions is read in a spectrofluorimeter. A serum blank is also included so that a correction can be made for any residual fluorescence in the serum.

Case Study.

A 26 year-old man, known to suffer from Crohn's disease (ulceration of the terminal ileum) presented with severe diarrhoea and abdominal pain. He showed signs of increased neuromuscular excitability (abnormal reflexes). A serum calcium determination had yielded a normal result. Hence, the request for a serum magnesium determination.

REAGENTS AND EQUIPMENT:

1. Tris Buffer (0.1 M, pH 7.0)
    Dissolve 12.1 g of Tris (hydroxymethyl) methylamine in about 800 ml of distilled water. Adjust the pH to 7.0 with HCl and make to 1 l.
2. Chelating Reagent.
    Dissolve 304 mg of 8-hydroxyquinoline-5-sulphonic acid monohydrate in 500 ml of reagent 1.
3. Magnesium Standard (2.0 mM)
    Dissolve 240 mg of dry $MgSO_4$ (or 492 mg of $MgSO_4 \text{:} 7H_2O$) in about 800 ml of distilled water and make to 1 l.
4. 0.9% NaCl.
    For reagent 5.
5. Patient's Serum.
    Mix equal volumes of serum, or heparinised plasma, with 0.9% NaCl. Keep on ice.

Wassermann Tubes.
Test Tubes.
Graduated 1 and 5 ml Pipettes.
100 µl Pipettors.
Glass or Quartz Fluorimeter Cuvettes.
Soft Paper Tissues.
Spectrofluorimeter.

PROCEDURE:

1. Make up the following stock standards, in **Wassermann tubes**, using the 2 mmol/l magnesium standard:

	A	B	C	D	E
ml of Mg standard (2 mmol/l)		0.2	0.4	0.6	0.8
ml of H$_2$O	≅ 1	0.8	0.6	0.4	0.2

2. Pipette 100 µl aliquots of the above magnesium standards into **test tubes**, labelled as above.

3. Pipette 100 µl samples of serum into 4 test tubes.

4. Add 4.0 ml of Tris buffer (0.1 M, pH 7.0) to two serum tubes from step 3 and mix (serum blanks).

5. Add 4.0 ml of chelating reagent (2.5 mM 8-hydroxyquinoline-5-sulphonic acid in 0.1 M Tris buffer, pH 7.0) to all other tubes (7) and mix.

6. Read the fluorescence intensity of all these solutions (9) and Tris buffer, using the following wavelength settings:

Excitation   :   375 nm.
Emission   :   505 nm.

If using an old instrument, set the instrument to read 100 on the top standard (E).

Subtract the Tris buffer value from the average "serum blank" value. Then subtract this value from the average value of the serum chelate.

Plot a standard curve, and read off the serum magnesium concentration.

Reference Range:

Serum magnesium concentration: 0.8 - 1.0 mmol/l.

EXERCISE 2-9

THE DETERMINATION OF PHENYLALANINE IN PLASMA

An elevated plasma phenylalanine level is used to confirm the diagnosis of phenylketonuria (PKU) in a neonate.

PKU is an autosomal recessive disease with a frequency of about 1 in 12,000 births in a population of Western European origin.

97-99% of PKU patients have a deficiency in the enzyme phenylalanine hydroxylase. This converts phenylalanine to tyrosine.

The enzyme requires the coenzyme tetrahydrobiopterin and the enzyme dihydropteridine reductase to keep the coenzyme reduced. 1-3% of PKU patients have a defect in dihydropteridine reductase or another enzyme involved with biopterin synthesis.

Untreated PKU leads to mental retardation, seizures, and hypopigmentation. Brain damage starts in the second or third week of life, hence early diagnosis is essential.

The elevation in blood phenylalanine levels leads to the formation of phenylpyruvic acid, which is excreted in the urine. This reacts with "Phenistix" or $FeCl_3$ to give a greenish-blue colour.

The standard screening test for PKU is the Guthrie test. When the baby is 1 to 2 weeks old, a blood sample is collected onto filter paper, dried and a standard sized disc is cut from the paper.

The disc is placed on an agar plate containing *Bacillus subtilis* and β-2-thienylalanine and the plate is incubated. High levels of phenylalanine in the blood disc diminish or abolish the action of the growth inhibitor (β-2-thienylalanine) on the bacterial growth. Hence, in cases of PKU, there will be bacterial growth around the filter paper disc.

If this test is positive, a plasma estimation of the level of phenylalanine must be carried out to confirm the diagnosis of PKU.

Treatment is to reduce the phenylalanine in the diet. This must be continued until the brain is fully developed and probably should be continued for life. This is essential for pregnant women to avoid damage to the foetal brain.

In this method, the sample is deproteinised with TCA and centrifuged. The phenylalanine in the supernatant then reacts with ninhydrin and copper to form a fluorescent complex. The reaction is enhanced by the inclusion of L-leucyl-L-alanine in the reaction mixture.

Standards are made up in an albumin solution, as this reduces the interference caused by the TCA. Fluorescence emission is measured at 490 nm with an excitation wavelength of 365 nm. A plasma blank is carried out to correct for any fluorescent substances in the plasma.

Case History.

The plasma is from a 2 week old infant, who's blood had given a positive result for phenylalanine in the Guthrie test.

REAGENTS AND EQUIPMENT:

1. 5 M NaOH.
    Dissolve 20 g of NaOH in about 70 ml of distilled water and make to 100 ml.
    For reagent 2.
2. Succinate Buffer (pH 5.88, 0.6 M).
    Dissolve 7.1 g of succinic acid in about 60 ml of distilled water (Heat to dissolve). Adjust the pH to 5.88 with 5 M NaOH. Make to 100 ml with distilled water.
    For reagents 5 and 6.
    Store at 4°C.
3. Ninhydrin Solution (0.03 M).
    Dissolve 134 mg of ninhydrin in 25 ml of distilled water.
    For reagents 5 and 6.
    Store at 4°C.
4. L-Leucyl-L-Alanine (5 mM).
    Dissolve 25.3 mg of Leu-Ala in 250 ml of distilled water.
    For reagent 5.
    Store at 4°C.
5. Buffered Ninhydrin-Peptide Solution.
    Mix 25 ml of reagent 2,
         10 ml of reagent 3
    and   5 ml of reagent 4.
    Make up fresh and keep on ice.
6. Buffered Ninhydrin-H_2O Solution.
    Mix 25 ml reagent 2,
         10 ml reagent 3
    and   5 ml of distilled water.
    Make up fresh and keep on ice.
7. Sodium Carbonate - NaK Tartrate Solution.
    Dissolve 1.33 g of Na_2CO_3 and 57 mg of NaK tartrate : 4 H_2O in 500 ml of distilled water.
    For reagent 9.
8. Copper Sulphate Solution (0.8 mM).
    Dissolve 200 mg of $CuSO_4:5H_2O$ in 1 l of distilled water.
    For reagent 9.
9. Copper Carbonate - Tartrate Solution.
    Mix 300 ml of reagent 7
    with 200 ml of reagent 8.
    Prepare fresh.
10. Trichloroacetic Acid (0.6 M).
    Dissolve 9.8 g of AR grade TCA in about 80 ml of distilled water and make to 100 ml.

11. 7.5% Albumin Solution.
        Dissolve 7.5 g of bovine serum albumin in about 80 ml of distilled water and make to
        100 ml.
12. Phenylalanine Standard (1.0 mM).
        Dissolve 8.3 mg of phenylalanine in 50 ml of reagent 11.
        Make up fresh and keep in ice.
13. Patient's Plasma.
        Mix 10 ml of reagent 11 with 10 ml of plasma.
        Make up fresh and keep in ice.

Eppendorf Centrifuge.
Eppendorf Tubes.
Wassermann (13 x 110 mm) Tubes.
Parafilm.
20, 100, 200 and 300 μl Pipettors and tips.
5 ml Pipettes.
60°C Water bath.
Glass or Quartz Cuvettes.
Spectrofluorimeter.
Soft Paper Tissues.

PROCEDURE:

1. Set up the following in Eppendorf centrifuge tubes:

	0	0.5	1.0	U_1	U_2
μl of 7.5% albumin	200	100		-	-
μl of 1.0 mmol/l phenylalanine in albumin standard	-	100	200	-	-
μl of patient's plasma	-	-	-	200	200

2. Add 200 μl of 0.6 M TCA to each tube, mix and stand for 10 minutes.

3. Centrifuge at 13,000 r.p.m. for 2 minutes.

4. Set up the following in Wassermann tubes:

	0	0.5	1.0	U_1	U_2	U_A	U_B
μl of ninhydrin-peptide solution	300	300	300	300	300	-	-
μl of ninhydrin - H_2O solution	-	-	-	-	-	300	300
μl of supernatant from Step 3	20	20	20	20	20	20 (U_1 or U_2)	-

5.   Cap the tubes with parafilm, mix well and incubate for 2 hours in a 60°C water-bath.

6.   Cool, add 5.0 ml of copper carbonate-tartrate solution to each tube and mix well.

7.   Read the fluorescence intensity of each sample, using the following wavelength settings:

Excitation : 365 nm
Emission   : 490 nm

If using an old instrument, set the instrument to read 100 on the top standard (1.0).

Subtract U_B from U_A and then subtract this value from the average U_1 and U_2 values.

Plot a standard curve and read off the patient's values.

Reference Values:

Most healthy infants have between 50 and 150 $\mu mol/l$. A phenylalanine level above 250 $\mu mol/l$ indicates PKU.

The treatment of PKU patients requires regular monitoring of their blood phenylalanine levels. These should be maintained between 100 and 200 $\mu mol/l$.

EXERCISE 2-10

THE DETERMINATION OF BLOOD PORPHYRINS

The dermatological problems associated with all types of porphyria, except acute intermittent porphyria, are due to an increased production of porphyrins.

When the skin is exposed to sunlight, the porphyrins catalyse an increased production of H_2O_2. This damages the skin, producing blistering, scaring and hirsutism.

Patients presenting with increased skin sensitivity to sunlight should undergo quantitative blood and urine porphyrin determinations to exclude porphyria.

High levels of total blood porphyrins are seen in the erythropoietic porphyrias (erythropoietic protoporphyria and congenital erythropoietic porphyria).

High total blood porphyrin levels are also seen in chronic lead poisoning and in iron deficiency anaemia. However, in these conditions, the porphyrins are bound to zinc. In this form, the porphyrins do not diffuse out of the erythrocytes and so these patients do not suffer from skin problems when exposed to sunlight.

Case History.

A 24 year-old woman presented with a generalised body rash that did not respond to non-steroidal anti-inflammatory creams. In order to rule out erythropoietic porphyria, a blood porphyrin determination has been requested.

In this method, the blood is diluted with isotonic saline and the porphyrins extracted into a glacial acetic acid - ether mixture. The porphyrins are then re-extracted into dilute HCl and their fluorescence compared with that of a uroporphyrin standard. The method estimates both free and zinc bound porphyrins. The presence of zinc-protoporphyrin can then be detected, in elevated samples, by an emission peak at 587 nm.

REAGENTS AND EQUIPMENT:

1. 0.9% NaCl.
2. Diethyl Ether : Glacial Acetic Acid (4:1) Mixture.
        80 ml of diethyl ether plus 20 ml of glacial acetic acid.
3. Ethanol : Water (19:1) Mixture.
        95 ml of ethanol plus 5 ml of distilled water.
4. 2.74 M HCl.
        Add 47 ml of conc HCl to about 100 ml of distilled water and make to 200 ml.
5. Uroporphyrin Standard (20 nmol/l).
        Dilute commercially available standardised uroporphyrin (600 nmol/l), 1 in 30 with reagent 4.
        Prepare fresh and keep on ice.
6. Patients Blood.
        Freshly collected blood in a heparinised tube.
        Keep on ice.

Haematocrit Centrifuge and Tubes.
Vortex Mixers.
Centrifuge Tubes.
Bench Centrifuge.
50 and 200 µl Pipettors and Tips.
5 ml Graduated Pipettes.
Test Tubes.
Wassermann Tubes.
Glass or Quartz Fluorimeter Cuvettes.
Soft Paper Tissues.
Spectrofluorimeter Capable of Excitation and Emission Scans.

PROCEDURE:

1. Determine the haematocrit of the blood sample.

2. Pipette 0.45 ml of 0.9% NaCl and 50 µl of blood into a centrifuge tube and mix well.

3. Add 5 ml of diethyl ether : glacial acetic acid (4:1) mixture in 1 ml portions and "vortex" for about 10 seconds after each addition. "Vortex" for a further 15 seconds when all the mixture is added.

4. Centrifuge for about 3 minutes and decant the ether supernatant into a test tube.

5. Add 3.0 ml of 2.74 M HCl to the tube and "vortex" for 1 minute.

6. Stand for 5 minutes and then transfer about 2.5 ml of the aqueous (bottom) layer to a fluorimeter cuvette.

7. Set the emission wavelength to 602 nm on the spectrofluorimeter.

8. Carry out an **excitation** scan on the 20 nmol/l (in 2.74 M HCl) uroporphyrin standard (that has reached room temperature), between 380 and 440 nm. (Carry out a pre-scan, prior to the excitation scan, to set the intensity scale).

9. Using the same settings, carry out a similar **excitation** scan on the extract.

From the excitation scans draw base lines between 380 and 440 nm.

Measure the highest point of the excitation peaks from the baseline, in mm.

Calculate the porphyrin concentration, in μmol/l packed red cells, using the following formula:

$$3.45 \times \frac{\text{Sample peak height}}{\text{Standard peak height}} \times \frac{1}{\text{Haematocrit}}$$

Reference Values:

Healthy adult : 0.4 - 1.4 μmol/l packed red cells.

Protoporphyria : > 5 μmol/l packed red cells.
Mild lead exposure : 2-5 μmol/l packed cells.
Severe lead exposure : > 10 μmol/l packed cells.
Iron deficiency anaemia : 1.4-10 μmol/l packed cells.

10. Add 50 μl of the blood sample to 200 μl of de-ionised H_2O, in a Wassermann tube, and "vortex".

11. Add 50 μl of the resultant haemolysate to 3 ml of ethanol : H_2O (19:1, v/v) in a centrifuge tube and "vortex".

12. Centrifuge and carry out an **emission** scan on the supernatant between 550 and 670 nm using an excitation wavelength of 415 nm (Carry out a pre-scan, prior to the emission scan).

The presence of a significant peak of 587 nm indicates the presence of zinc-protoporphyrin (Compare the intensity scales of the 2 scans when considering the peaks significance). The presence of such a peak excludes protoporphyria. Some zinc-protoporphyrin is seen in congenital erythropoietic porphyria, along with uroporphyrin, coproporphyrin and straight protoporphyrin.

GENERAL REFERENCES:

Udenfriend, S., (1962 and 1969), Molecular Biology Monographs, Volumes 1 and 2, "Fluorescence Assay in Biology and Medicine", Academic Press, New York.

Williams, R. T. and Bridges, J. W., (1964), "Fluorescence of Solutions: A Review", J. Clin. Path., 17, 371-394.

Rubin, M., (1970), "Fluorimetry and Phosphorimetry in Clinical Chemistry", Advances in Clin. Chem., 13, 161.

Penzer, G. R., (1980), In "An Introduction to Spectroscopy for Biochemists", Ed. Brown, S. B., Chapter 3, p 70-114, Academic Press.

Tiffany, T. O., (1994), "Fluorimetry, Nephelometry and Turbidimetry", In "Tietz Textbook of Clinical Chemistry", 2nd Edition, Eds. Burtis, C. A. and Ashwood, E. R., p 132-158, W. B. Saunders, Philadelphia.

Rhys Williams, A. T., (1973), "Clinical Chemistry using Fluorescence Spectroscopy. A Review of Current Techniques". A booklet published by Perkin-Elmer.

"Introduction to Fluorescence Spectroscopy", (Undated). A booklet published by Perkin-Elmer.

Determination of thiamine in urine.

Freed, M., (1966), Chairman, "Methods of Vitamin Assay", 3rd Ed., p. 123-145, Interscience Publishers.

Determination of serum magnesium.

Thiers, R. E., (1965), Standard Methods in Clinical Chemistry, 5, 131, Academic Press.

Determination of plasma phenylalanine.

Hsia, D. Y. Y., Berman, J. L. and Slatis, H. M., (1964), J. Am. Med. Assoc., 188, 203.

Determination of blood porphyrins.

Poulos, V. and Lockwood, W. H., (1980), Int. J. Biochem., 12, 1049-1050.

Elder, G. H., Smith, S. G. and Smyth, S. J., (1990), Ann. Clin. Biochem., 27, 395-412.

CHAPTER 3

ENZYMOLOGY

ENZYMES.

Enzymes are biological catalysts. They speed the attainment of equilibrium in a chemical system.

The name "Enzyme" was first used by Kuhne, in 1878, for the unknown substance, in yeast, that was responsible for fermentation. The word is Greek and means "In yeast".

The first enzyme isolated in the pure state was urease. This was crystallised by Sumner, in 1926, and shown to be a pure protein.

Enzymes can be pure proteins, or a protein plus a non-protein component. In the latter case, the whole enzyme is the "Holoenzyme", with the protein part called the "Apoenzyme" and the non-protein part called the "Prosthetic group".

The prosthetic group is an enzyme "Cofactor". Where the cofactor is not an integral part of the enzyme, it acts as a substrate and is known as a "Coenzyme". However, the term "Coenzyme" is often used to include the prosthetic group.

Haem, biotin, pyridoxal phosphate and flavin, for example, function as prosthetic groups. NAD, Coenzyme A and THFA, on the other hand, function as coenzymes. Many of these cofactors are derived from the B Vitamins.

Most metal ion cofactors (Eg. Fe^{++}, Cu^{++}, Mo^{+++} and Zn^{++}) serve as prosthetic groups. Mg^{++}, however, acts as a coenzyme in many reactions in which it is involved. Free metal ion cofactors are also known as "Activators".

Catalysis.

Chemicals in solution frequently exist in a "Metastable" state, which is not the equilibrium state. For equilibrium to be reached, additional "Activation" energy has to be added to the system in order to activate the reaction. Heating the reaction mixture is an obvious example.

A catalyst lowers the amount of activation energy that has to be added to the system to get the reaction to take place.

The free energy of the initial system, and of the equilibrium system, are the same whether a catalyst is used or not.

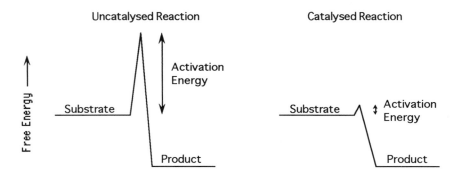

If one looks at the energy distribution of the molecules, we have:

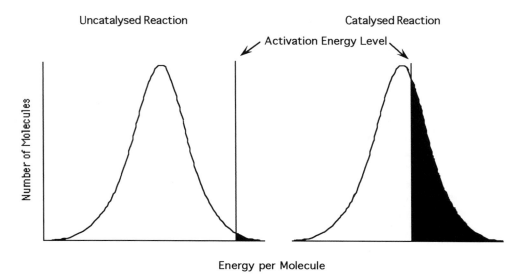

As can be seen, only a few molecules have the necessary energy to react in the uncatalysed system. However, in the catalysed system, a very much larger number of molecules have the required energy.

Classification and Nomenclature.

Prior to 1964, enzyme nomenclature was very confusing, with the same enzyme known by different names and the same name used for different enzymes. The International Union of Pure and Applied Chemistry and the International Union of Biochemistry, in 1964, produced some recommendations for classification and nomenclature. These recommendations were slightly modified in 1972 and 1978. In 1964 there were 875 enzymes listed. This had risen to 1,770 in 1972 and 2,122 in 1978. The recommendations are:

1. The ending "-ase" be applied to single enzymes only. The word "System" to be added where more than one enzyme is involved.

2. Enzymes to be classified according to the reaction catalysed.

3. Enzymes to be divided into groups based on the type of reaction catalysed, together with the name of the substrate(s).

The direction to be the same for all enzymes in the class, even though only the reverse reaction has been demonstrated.

4. The enzyme be given a systematic name, formulated according to definite rules. The enzyme also be given a working (recommended) name, usually the current name.

5. The source of the enzyme be given.

6. The enzyme given a code number, consisting of 4 components:

> One of 6 main classes.
> A sub-class.
> A sub-sub-class.
> The serial number in the sub-sub-class.
>
> The main classes are: 1. Oxidoreductases, 2. Transferases, 3. Hydrolases, 4. Lyases, 5. Isomerases and 6. Ligases.

For example, lactate dehydrogenase is the recommended name for the enzyme that has the systematic name of L-lactate:NAD^+ oxidoreductase and the EC number of 1. 1. 1. 27. The sub-class of 1 indicates that the enzyme acts on CH.OH as the hydrogen donor. The sub-sub-class of 1 indicates that NAD^+ or $NADP^+$ is the hydrogen acceptor.

Enzyme Units.

Prior to 1961, a whole variety of units were used for enzyme activity. These were often named after the originators of the method. For example, a King and Armstrong unit of alkaline phosphatase was 1 μg of phenol released from phenyl phosphate in 15 minutes by 100 ml of serum at 37^oC.

The Enzyme Commission, in 1961, recommended that the unit be defined as "The amount of enzyme which will catalyse the transformation of one micromole of the substrate per minute under standard conditions". This became known as "The International Unit".

A number of problems exist with this unit. The precise meaning of "amount of enzyme" was not defined and there was a move to SI units. Therefore, the International Unit was replaced by "The Katal", in 1972. The katal is the amount of activity that converts one mole of substrate per second. These units can be inter-converted as follows:

$$1 \text{ kat} = 6 \times 10^7 \text{ IU.}$$
$$1 \text{ IU} = 16.67 \text{ nkat.}$$

The "Specific Activity" of an enzyme is expressed as katals per kilogram and the "Molar Activity" as katals per mole. For enzymes in solution, activity can be expressed as katals per litre.

These units do not include conditions of the assay, such as temperature, pH and substrate concentration. These should be kept in mind when comparing enzyme activities.

Mode of Action.

The mode of enzyme action was first put forward by Michaelis and Menten in 1913. This is summarised below:

Enzyme + Substrate $\rightleftharpoons$ Enzyme-Substrate Complex $\rightleftharpoons$ Enzyme + Products

$$E + S \rightleftharpoons ES \rightleftharpoons E + P$$

The Active Site.

This is the site on the enzyme to which the substrate is bound. It is a specific group of amino acids. These amino acids need not be adjacent, as the secondary and tertiary structure of the enzyme can bring amino acids from different chains, or parts of the same chain, close together. The binding of substrate can also alter the configuration of the enzyme to bring another group of amino acids close to their binding site to take part in the catalytic reaction. Prosthetic groups are also involved in the active site.

Most active sites have amino acids containing reactive groups in their side chains. Eg.

Serine	:	-OH
Cysteine	:	-SH
Histidine	:	Imidazole -N=

Substrate Specificity.

Unlike inorganic catalysts, enzymes are much more specific for their substrate.

Carboxyl esterases have low specificity, as they act on a variety of esters of carboxylic acid:

$$R\text{-}COOR' + H_2O \rightleftharpoons R\text{-}COOH + HO\text{-}R'$$

On the other hand, L-arginase acts only on L-arginine and not on the D form:

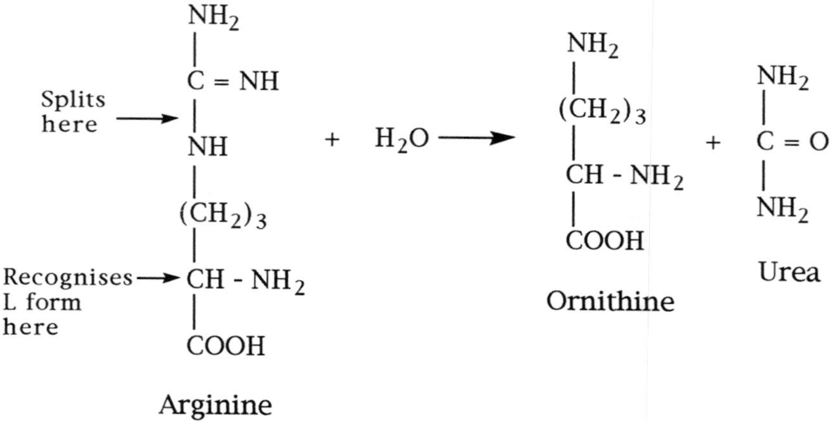

pH.

All enzymes have an optimum pH. However, the actual value can alter with temperature, concentration of substrate and time.

For maximum activity, the enzyme requires a certain degree of protonisation. Proton binding to the active site and/or to other sites effects the ability of the enzyme to bind the substrate and catalyse the reaction. Eg.

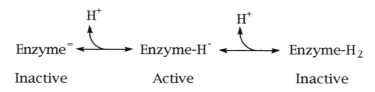

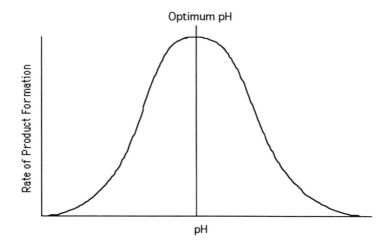

Protons can also bind to some substrates. They are also substrates themselves in some reactions and, therefore, their concentration effects the rate of reaction.

Temperature.

All enzymes have an optimum temperature. The rate of reaction increases with increasing temperature due to a general kinetic effect. However, increasing temperature starts to denature the enzyme. The rate of denaturation increases as the temperature increases. These two effects work against each other and most enzymes are inactivated by 70-80°C.

The effect of temperature on enzyme activity is illustrated below:

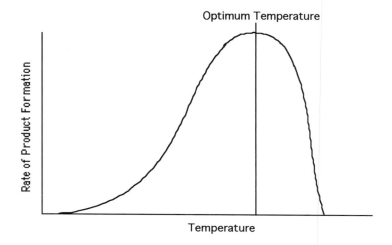

Enzyme Kinetics.

Illustrated below is a plot of the concentration of a product, in an enzyme catalysed reaction, against time:

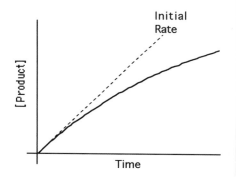

The fall in the rate of product formation is due to:

1. The fall in substrate concentration.
2. Reversal of the reaction due to product formation (Approach to equilibrium).

Additional factors can be:

3. Poisoning or denaturation of the enzyme.
4. Change in pH due to product formation.

Due to these factors, the "Initial Rate" or "Initial Velocity" is considered when enzyme kinetics is discussed.

It is also preferable to use a kinetic method, as illustrated above, when determining enzyme activity, rather than measuring the level of product after a fixed time.

The effect of substrate concentration.

The following curve is produced when the initial velocity is plotted against substrate concentration, for a fixed amount of enzyme:

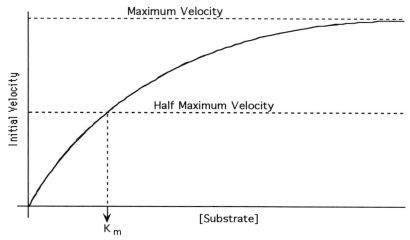

The substrate concentration that produces half maximum velocity is known as the Michaelis constant (K_m). It is derived from the rate constants from the initial velocity equation:

$$E + S \xrightleftharpoons[k_2]{k_1} ES \xrightarrow{k_3} P + E$$

$$K_m = \frac{k_2 + k_3}{k_1}$$

The relationship between substrate concentration and initial velocity is tabulated below:

Substrate concentration in multiples of K_m	1	2	10	20	50	100
Initial velocity as a percentage of V_{max}	50	67	91	95	98	99

For enzyme assays, it is desirable to have zero order reactions as regards substrate concentration. That is, the initial reaction rate is independent of the substrate concentration. As can be seen above, this is only achieved at very high substrate concentrations. However, most enzyme assays settle for substrate concentrations of 10 to 20 times K_m.

For example:

An enzyme assay is set up using a substrate concentration of 10 x K_m and the product is determined after 10% of the substrate is used up.

3-7

During this time the reaction rate will have fallen from 91% of V_{max} to 90%.

This fall is only due to the change in substrate concentration. However, product will have been formed and the reverse reaction will be having an effect on the reaction rate. The drop in reaction rate may, therefore, be 5% or more.

Hence, it is desirable to measure the initial rate rather than the level of product after a fixed time.

One should be aware that some enzymes are inhibited by high substrate concentrations. Hence, there will be an optimum substrate concentration and that should be used in the assay.

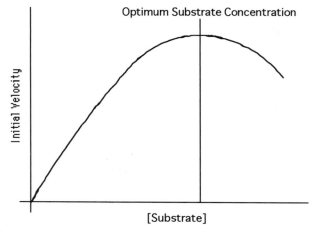

Determination of K_m.

The easiest way of determining the K_m of an enzyme is to carry out a series of assays using a fixed amount of enzyme and different substrate concentrations. The reciprocals of initial velocity (V) and substrate concentration ([S]) are then plotted (A Lineweaver-Burk plot):

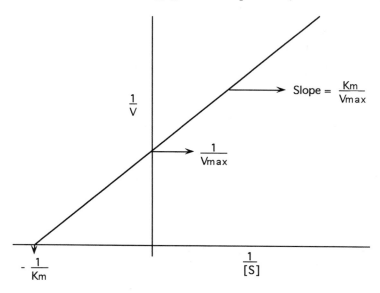

Enzyme Inhibition.

The rate of enzyme catalysed reactions may be altered, in a specific manner, by compounds other than the substrate. Activators increase the rate; whereas, inhibitors and inactivators decrease it.

Inhibition and activation of enzymes, by key metabolites, provides the normal means of fine control over enzyme activity within the cell.

External interference with metabolism by drugs, pesticides or toxic agents often depends on the inhibition of enzymes.

Enzyme activity may be reduced by a variety of non-specific inactivators such as acids, alkalis, heavy metals and proteases. These act by disrupting the protein structure.

Inhibitors interact with the enzyme at a small number of sites and do not markedly change the three-dimensional structure of the enzyme. In most cases the action is specific (ie. the inhibitor will react with a specific enzyme) and is reversible.

There are three types of inhibitor:

1.   Competitive.

In this type of inhibition, the inhibitor competes with the substrate for the active site. Hence, we have:

$$E + I \rightleftharpoons EI$$

as well as:

$$E + S \rightleftharpoons ES \longrightarrow E + P$$

For example, malonic acid acts as an inhibitor of succinate dehydrogenase. The enzyme will form a complex with either succinic acid or malonic acid. However, the dehydrogenation reaction only takes place with succinic acid ($COOH.CH_2.CH_2.COOH$) not with malonic acid ($COOH.CH_2.COOH$).

This type of inhibition can also include the case where the inhibitor binds to a site on the enzyme, other than the active site. In doing so, it alters the shape of the enzyme so that it can no longer bind the substrate. This type of inhibitor will usually have a dissimilar structure to the substrate; whereas, the usual competitive inhibitor has a similar structure.

Increasing the substrate concentration, with a fixed amount of inhibitor, will eventually swamp the inhibitor and the original V_{max} will be reached:

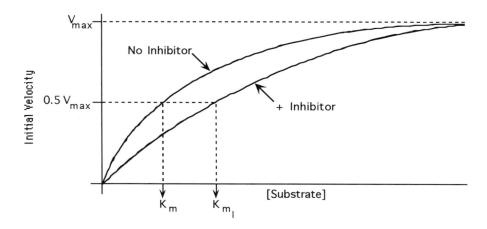

As can be seen, V_{max} remains the same, but the apparent K_m is increased.

2.   Uncompetitive.

In this type of inhibition, the inhibitor will bind to the enzyme-substrate complex but not to the free enzyme:

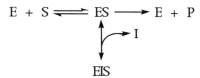

As ES now has a third direction in which to go, V_{max} will be decreased.

K_m is essentially a measure of how hard you have to push the substrate to get the enzyme from the substrate-free state to the substrate-bound state. As the inhibitor combines with ES, it pulls the enzyme over to the substrate-bound state. K_m is, therefore, lowered:

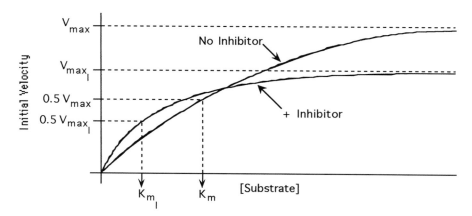

Uncompetitive inhibition is extremely rare in single substrate reactions. However, L-phenylalanine is an uncompetitive inhibitor of rat intestinal alkaline phosphatase.

3.   Non-competitive.

In this type of inhibition, the inhibitor can combine with the enzyme or the enzyme-substrate complex:

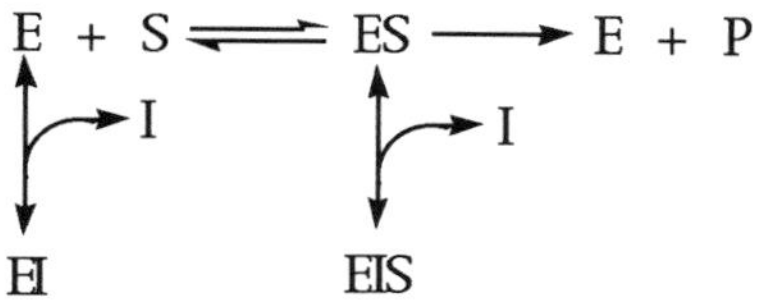

$$E + S \rightleftharpoons ES \longrightarrow E + P$$

$$\downarrow I \qquad \downarrow I$$

$$EI \qquad EIS$$

In simple non-competitive inhibition, the inhibitor has equal affinity for E and ES.

As with uncompetitive inhibition, V_{max} will be reduced.

In the case of K_m, I is pulling on E as well as ES. This pull is equal and so K_m is not altered.

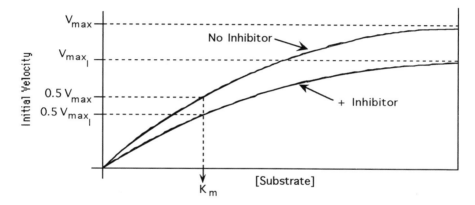

An example of a non-competitive inhibitor is 5,5'-dithiobis-(2-nitrobenzoic acid). It acts by combining reversibly with available -SH groups of cysteine side chains of enzymes:

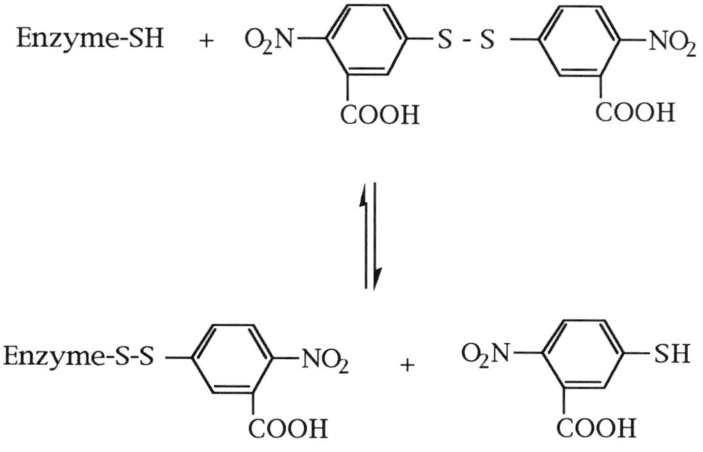

Enzyme-SH + O_2N—⟨benzene⟩—S - S—⟨benzene⟩—NO_2
                                    | |
                                  COOH   COOH

⇅

Enzyme-S-S—⟨benzene⟩—NO_2 + O_2N—⟨benzene⟩—SH
                          | |
                        COOH   COOH

3-11

The Lineweaver-Burk plots, for all three types of inhibitor, are shown below:

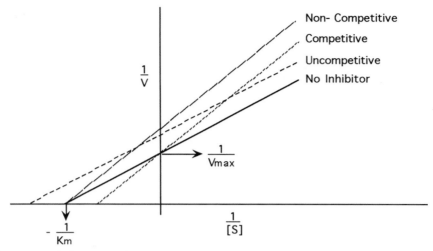

The characteristics of these inhibitors can be summed up in the following table:

Type of inhibition	Inhibitor combines with	Effect on V_{max}	Effect on K_m	Effect on Lineweaver-Burk plot
Competitive	E	Unchanged	Increased	Convergence on the $\frac{1}{v}$ axis
Uncompetitive	ES	Decreased	Decreased	Parallel lines
Simple non-competitive	E and ES	Decreased	Unchanged	Convergence on the $\frac{1}{[S]}$ axis

Allosteric Enzymes.

For classical enzymes, a plot of the initial velocity against substrate gives a hyperbolic curve. However, for allosteric enzymes, the plot is sigmoidal:

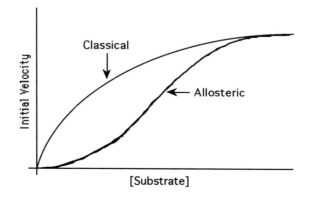

3-12

Most allosteric enzymes have a quaternary structure; ie. they are made up of subunits. The binding of one molecule of substrate to the enzyme makes it easier for the second molecule of substrate to bind. This is similar to the binding of oxygen to haemoglobin, and hence the sigmoidal oxygen dissociation curve of haemoglobin.

In addition to the effect of substrate concentration, some allosteric enzymes respond to "Effector" molecules. These are small molecules that bind to sites on the enzyme other than the active site. These molecules can act as activators and shift the curve to the left (more hyperbolic), or as inhibitors and shift the curve to the right (more sigmoidal).

Isoenzymes.

Isoenzymes are enzymatically active proteins, catalysing the same reaction, and occurring in the same species, but differing in certain physico-chemical properties. They usually have different K_m values.

The term "Heteroenzyme" has been used for enzymes from different species that catalyse the same reactions.

Isoenzymes are coded for by different genes. These may be on the same chromosome or on different chromosomes. For example, human pancreatic and salivary amylase genes are on chromosome 1; whereas, cytoplasmic and mitochondrial genes for malate dehydrogenase are on chromosomes 2 and 7, respectively.

Isoenzymes can be monomeric or oligomeric (ie. made up of subunits). Amylase isoenzymes are monomeric; whereas, lactate dehydrogenase (LD) isoenzymes consist of four subunits. LD is a cytoplasmic enzyme and its tetramer structure is required for enzyme activity. Each subunit can be either H (Heart) or M (Muscle). The subunits have a molecular weight of about 33,000 each. Each type of subunit is encoded by a separate gene:

$$H \quad : \quad \text{Human Chromosome 12.}$$
$$M \quad : \quad \text{Human Chromosome 11.}$$

Hence, there are 5 types of isoenzyme:

○ Heart Sub-unit

● Muscle Sub-unit

The H sub-unit has more acidic, and less basic, amino acids than the M sub-unit. Hence, the H sub-unit carries more negative charge and the H_4 isoenzyme will migrate fastest towards the anode on electrophoresis.

The isoenzymes are classified in order of their electrophoretic mobility:

$$LD_1 \quad : \quad H_4$$
$$LD_2 \quad : \quad H_3M$$

$$LD_3 \quad : \quad H_2M_2$$
$$LD_4 \quad : \quad HM_3$$
$$LD_5 \quad : \quad M_4$$

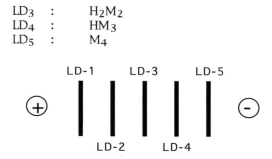

The approximate distribution in some human tissues is tabulated below:

Tissue	LD_1	LD_2	LD_3	LD_4	LD_5
Heart	50	30	15	5	
Skeletal Muscle			10	20	70
Liver		5	15	20	60
Kidney	35	30	25	10	
Erythrocytes	40	40	15	5	
Plasma	25	35	25	10	5

In addition to their different electrophoretic mobility, the LD isoenzymes respond differently to inhibitors, heat and alternative substrates. For example, LD_1 is heat (65°C) stable, not inhibited by 0.8 M sodium perchlorate and will catalyse the inter-conversion of α-hydroxybutyrate and α-ketobutyrate. On the other hand, LD_5 is destroyed at 65°C, inhibited by the sodium perchlorate and will not act on α-hydroxybutyrate.

Clinical Enzymology.

The intracellular concentration of most enzymes is 1,000 to 10,000 times that in the extracellular fluid. Hence, when the tissue is damaged, enzymes leak out and the plasma concentration rises. This rise can be used in the diagnosis of the tissue damaged and the extent of the damage.

Enzymes within the cell are unevenly distributed between the organelles and the cytoplasm. Hence, the degree of cellular damage can produce different patterns of enzyme release:

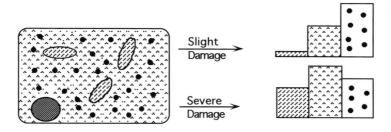

● is only present in the cytoplasm.

⫶⫶ is only present in mitochondria.

ʌ⌃ʌ is present in both mitochondria and the cytoplasm.

The intracellular distribution of some clinically useful enzymes is tabulated below:

Plasma Membrane	Alkaline Phosphatase γ-Glutamyl Transferase
Endoplasmic Reticulum	γ-Glutamyl Transferase
Cytoplasm	Aspartate Aminotransferase Alanine Aminotransferase Lactate Dehydrogenase Creatine Kinase
Mitochondria	Aspartate Aminotransferase Creatine Kinase
Lysosomes	Acid Phosphatase

Following damage to the tissue, there is a delay before the plasma level of the enzyme starts to rise. This delay depends on how close the damaged cells are to capillaries and on the thickness of their basal membranes. For example, there is a delay of a few minutes when liver cells are damaged. This delay can be several hours following myocardial infarction and may be several days with skeletal muscular damage.

Under normal conditions, there is a constant leakage of enzymes into the blood. This is balanced by the removal of these enzymes. This is mainly by proteolytic activity in the blood and in the reticuloendothelial system. Most of the enzymes in plasma are too large to pass through the glomeruli and be removed in the urine. As a result, a steady state level of enzymes is found in the plasma.

Plasma enzymes can be divided into 3 groups:

1.   Enzymes for use in the blood.
        Eg.   Coagulation enzymes.
               Cholinesterase.

2.   Leakage of cellular enzymes.
         Eg.   Lactate dehydrogenase.
                Aspartate aminotransferase.

3.   Leakage of digestive enzymes.
         Eg.   Amylase.
                Lipase.

Damage to the organ producing these enzymes will produce a fall in the level of group 1 and a rise in the level of groups 2 and 3.

As an example, the events following a myocardial infarction are illustrated below:

Time	Event	Result
Zero	Dysfunction of membrane-bound ion pumps.	Loss of intracellular ions: K^+, Zn^{++}, $HPO_4^=$, Mg^{++} etc.
	Dysregulation of intracellular metabolism.	Loss of intracellular metabolites: Lactate, Adenosine etc.
60 Min	Modification of cell membranes.	Loss of intracellular macromolecules: Troponin T, Troponin I, Myoglobin, Enzymes, etc.
	Cell necrosis.	Loss of all cell contents.

The changes in the plasma level of some diagnostic enzymes, after a myocardial infarction, is diagrammatically illustrated below:

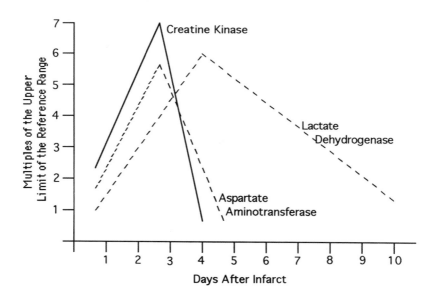

The Diagnostic use of Isoenzymes.

High levels of lactate dehydrogenase (LD) are found in the plasma with damage to skeletal muscle, heart, liver, kidneys and erythrocytes. Hence a rise in total LD is non -specific. However, the plasma LD isoenzyme pattern is much more specific. For example, isoenzymes can be used to separate heart from liver:

$$\text{Heart}: \quad LD_1 \text{ and } LD_2 \text{ elevated.}$$
$$\text{Liver}: \quad LD_5 \text{ elevated.}$$

These can easily be differentiated using electrophoresis:

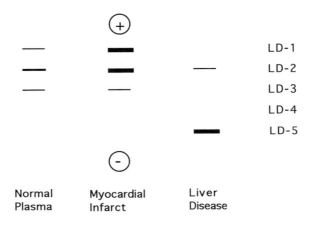

Extra electrophoretic bands can be seen in the plasma of some patients with cancer. These are due to a modified M subunit produced by the tumour. These modified isoenzymes carry slightly more negative charge and, therefore, run ahead of the normal bands.

Creatine kinase (CK) is mainly found in muscle (skeletal and cardiac) and brain. Hence, the plasma level is raised in muscular dystrophy, cerebrovascular disease and following a myocardial infarct.

CK is a dimeric enzyme. ie. it is made up of 2 different subunits with molecular weights of about 40,000 each:

$$M: \quad \text{Muscle} \quad \text{(Gene on chromosome 19).}$$
$$B: \quad \text{Brain} \quad \text{(Gene on chromosome 14).}$$

Therefore, there are 3 isoenzyme types:

CK_1 BB   :   The only component in brain.
CK_2 MB   :   Mainly found in heart muscle.
CK_3 MM   :   The major component of skeletal muscle.

MB is often monitored following myocardial infarction. As with LD, the CK isoenzymes can be assayed using electrophoresis. CK_2 (CK-MB) is the isoenzyme of interest:

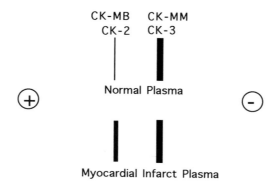

Normal Plasma

Myocardial Infarct Plasma

Creatine Kinase Isoforms.

Isoforms are enzymes that have been modified after protein synthesis. ie. they derive from the same enzyme gene(s). The C terminal lysine of the M subunit of creatine kinase is slowly hydrolysed from the enzyme, in plasma, to M^*. This product has less positive charge and will migrate faster towards the anode. Hence, CK_2 has 2 isoforms (MB and M^*B) and CK_3 has 3 isoforms (MM, M^*M and M^*M^*).

These isoforms can be resolved on high-grade electrophoresis to give a better and earlier diagnosis of myocardial infarction, than standard electrophoresis.

Normally, most CK_2 in the plasma is M^*B (MB_1). Following myocardial infarction, MB (MB_2) is released from the heart. Hence, the ratio MB_2/MB_1 is raised. This is used in diagnosis:

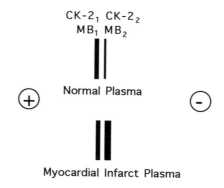

Normal Plasma

Myocardial Infarct Plasma

Enzymes as Reagents.

Due to the specificity of enzymes, they can be used to assay components in complex mixtures, without prior purification steps. Plasma proteins tend to interfere in most colorimetric methods and so a deproteinisation step is required in the assay. Hence, a number of methods using enzymes as reagents can be carried out directly on plasma or serum.

The first method, using enzymes as reagents, was one for plasma (or whole blood) glucose. This method dates from 1956 and uses glucose oxidase to generate hydrogen peroxide, which oxidises a chromogen. However, this method still requires deproteinisation. The old methods for glucose determination used the reducing power of glucose, in alkaline solutions, to reduce cupric copper or ferricyanide. There are a number of components in whole blood and plasma that act as reducing agents and so the old methods were not specific. Hence, the new method greatly improved the specificity.

One should be aware that enzyme specificity can occasionally be a drawback. For example, yeast alcohol dehydrogenase can not be used to assay methanol, as the oxidation of methanol is not catalysed by this enzyme.

Most methods allow the reaction to go to completion (equilibrium). Where the equilibrium does not favour the removal of the analyte, the reaction can be driven in the desired direction by removing the product. For example, pyruvate formed in the assay of plasma lactate, using lactate dehydrogenase, can be removed by the inclusion of hydrazine in the reaction mixture.

A number of kinetic methods are used in which the rate of product formation is proportional to the analyte concentration. For these one needs enzymes with high K_m values, so that the analyte concentration is less than $0.2 \times K_m$. This means that small variations in the enzyme concentration does not affect the rate.

Enzymes are commonly used in carrier bound systems where the enzymes are attached to a solid matrix. This attachment has the added advantage in that it tends to stabilise the enzyme. An example is seen in the Ames (Bayer Diagnostics) "Glucostix":

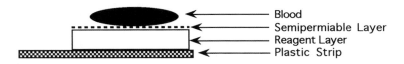

Blood
Semipermiable Layer
Reagent Layer
Plastic Strip

The blood is filtered by the semipermeable layer and the plasma enters the reagent layer where the following reactions take place:

$$\text{Glucose} + O_2 \xrightarrow{\textit{Matrix Bound Glucose Oxidase}} \text{Gluconic Acid} + H_2O_2$$

$$H_2O_2 + \text{o-tolidine} \xrightarrow{\textit{Matrix Bound Peroxidase}} \text{o-tolidine blue} + H_2O$$

After 30 seconds the excess blood is blotted off and the green colour (blue on a yellow background) read in a reflectance meter, 90 seconds later.

Immobilised enzymes are also used in biosensors. For example, some glucose biosensors use glucose oxidase and a miniaturised oxygen electrode:

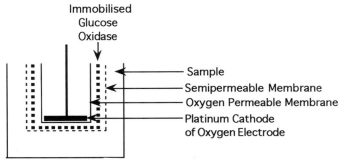

Immobilised
Glucose
Oxidase

Sample
Semipermeable Membrane
Oxygen Permeable Membrane
Platinum Cathode
of Oxygen Electrode

The glucose concentration in the sample is inversely proportional to the partial pressure of oxygen at the platinum cathode of the oxygen electrode.

EXERCISES

EXERCISE 3-1

THE POLARIMETRIC DETERMINATION OF INVERTASE

Invertase has the recommended name of β-D-fructofuranosidase and the systematic name of β-D-fructofuranoside fructohydrolase. It has the number: 3.2.1.26. The class 3 is a hydrolase, the sub-class 2 is a glycosidase and the sub-sub-class 1 relates to the hydrolysis of o-glycosyl compounds. The enzyme hydrolyses sucrose to glucose and fructose:

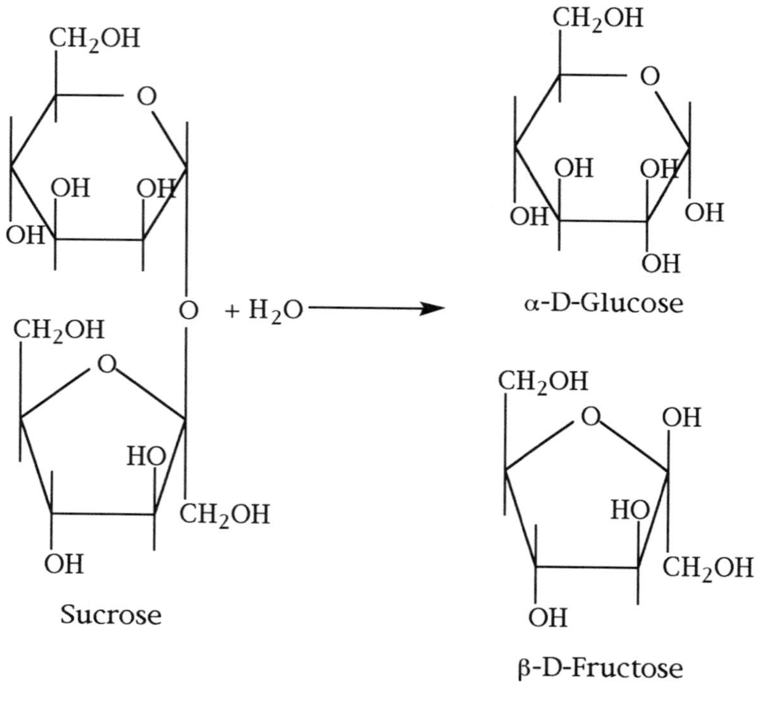

The enzyme used in this exercise is of yeast origin and is a glycoprotein containing about 50% carbohydrate.

Sucrose, glucose and fructose, like other sugars, will rotate polarised light. The rotation of sodium D light (589.0 and 589.6 nm), at 20°C, of these sugars in a 100 mm light path, is tabulated below:

Sugar	Molar Rotation (Degrees)
Sucrose	+ 22.7
Glucose	+ 9.5
Fructose	- 16.7

Sucrose rotates the light clockwise (dextrorotatory (d) or +) and the products (glucose + fructose) rotate the light anticlockwise (laevorotatory (l) or -). Hence, when a molar solution of sucrose is fully converted to glucose and fructose, the optical rotation will go from + 22.7 to - 7.2º (A change of 29.9º). Hence, "Invertase" is the commonly used name for the enzyme. This change can be monitored using a polarimeter.

Polarimetry is one of the oldest instrumental techniques, going back to the beginning of last century. It involves the measurement of the change in direction of polarised light after its interaction with optically active substances.

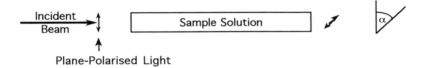

In the diagram, above, the beam has been rotated clockwise through an angle, α. The compound in solution is, therefore, dextrorotatory (d) and α has a positive sign.

Ordinary light behaves as though it consists of a large number of electromagnetic waves, vibrating in all possible orientations around the direction of propagation.

Plane-polarised light is obtained by removing all the rays except those vibrating in a particular plane. This is achieved, in most polarimeters, by using a quartz or calcite crystal cut diagonally and cemented together, as shown below:

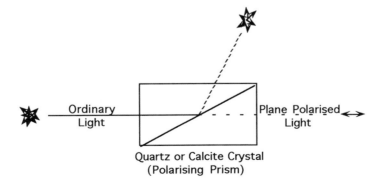

The refractive index of quartz and calcite depends on the direction of vibration of the electromagnetic wave. The horizontal component, shown above, has the smallest refractive index and the vertical component the largest. By adjusting the angle of the cut and using a cement that has a refractive index as close as possible to the refractive index of the horizontal component of light in quartz or calcite, all the electromagnetic waves, except the horizontal ones, are reflected.

The light entering the polarising prism must be a parallel beam and so a collimating lens is included in the instrument between the light source and the prism.

Plane polarised light can be considered as the vector sum of two circularly polarised rays, one moving clockwise and the other anticlockwise; each having the same magnitude of vibration.

Optically active substances have different refractive indices to these two components. This means that one of these circularly polarised rays is slowed down more than the other. As a result, the plane of polarisation of the ray is rotated. In the example below, the solution has a higher refractive index for the anticlockwise (L) component. Hence, the emergent beam is rotated clockwise:

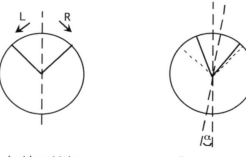

Incident Light   Emergent Light

The amount of rotation is directly proportional to the concentration of the optically active component and to the path length of the light through the solution.

The optical rotation of a solution changes considerably with the wavelength of light used. Therefore, most polarimeters use light sources producing light of specific wavelengths.

Mercury or sodium vapour lamps are the most common light sources. Mercury has a number of emission lines and the desired one can be selected by an appropriate filter. A sodium lamp has a double emission line of 589.0 and 589.6 nm (the D lines).

Optically active compounds are asymmetrical, ie. they do not possess a plane, or centre, of symmetry. Examples of such compounds are amino acids, polypeptides, sugars, steroids and antibiotics. Polarimetry is mainly used to determine the concentration of optically active compounds, but can also be used as one of the techniques for determining the structure of unknown compounds.

The temperature of the solution effects the optical rotation and a standard temperature of $20^{\circ}C$ is adopted for most measurements. The notation of the specific rotation usually includes the temperature as a superscript and the wavelength as a subscript. Where a sodium lamp is used, the subscript D is often used to denote the D lines of sodium. Eg:

$$[\alpha]_D^{20}$$

[α] is the specific rotation (degrees rotation of a 1 kg/l solution with a path length of 100 mm).

The rotation of the polarised light is determined by using a second polarising prism (the analyser). The amount of light passing through the analyser prism depends on the rotation of the prism in relation to the plane of the polarised light.

The analyser is attached to a graduated circle so that the angle of rotation can be measured. The graduated circle is fitted with a vernier so that the rotation can be determined to the nearest 0.05°.

In the cheaper instruments, the analyser prism is rotated until minimum light intensity is reached and the scale is then read. Two readings are usually taken, one with the polarimeter tube filled with solvent and then the second with the tube filled with the solution to be analysed. The difference in the two readings is the optical rotation.

Judging the point of minimum light intensity is difficult and so most instruments are fitted with a half-shade device.

The half-shade device rotates part of the field a small amount in relation to the rest of the field. The balance point is then achieved when the analyser prism is rotated so that the whole field is equally illuminated. The eye finds it much easier to balance light intensities than to judge minimum light intensity.

A common half-shade device is an additional small polarising prism covering part of the field and rotated slightly in relation to the main polarising prism. Another common half-shade device is a quartz half-wave plate. This is cut to thickness, so that the slow ray lags one half wavelength behind the fast ray. This results in a small amount of rotation of the part of the field, covered by the quartz, in relation to the rest of the field.

A typical instrument and its basic layout of the instrument are shown below:

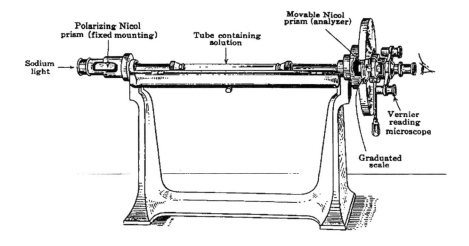

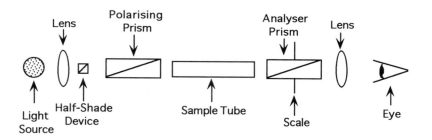

The light intensity, with and without the half-shade device, and the view through the eye-piece, are illustrated below:

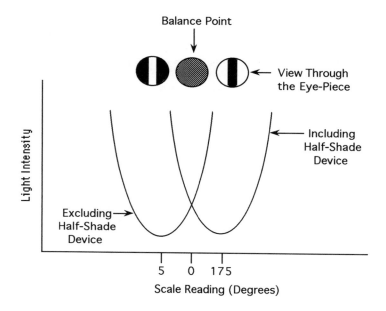

In this exercise, 5.0 mg of invertase will be added to 100 ml of a 1.0 molar solution of sucrose in McIlvaine's citrate-phosphate buffer, pH 4.6. The optical rotation will be recorded at approximately 10 minute intervals, using a 200 mm sample tube.

REAGENTS AND EQUIPMENT:

1.   Citrate-Phosphate Buffer, pH 4.6.
      (a)   Dissolve 1.92 g of citric acid (or 2.10 g of citric acid monohydrate) in distilled water and make to 100 ml (0.1 M).
      (b)   Dissolve 1.78 g of $Na_2HPO_4{:}2H_2O$ (or 3.58 g of $Na_2HPO_4{:}12H_2O$) in distilled water and make to 50 ml (0.2 M).
      Mix 53 ml of (a) with 47 ml of (b) and check the pH.
2.   1.0 Molar Glucose in Citrate-Phosphate Buffer, pH 4.6.
      Dissolve 34.2 g of AR sucrose in about 80 ml of reagent 1 and make to 100 ml with reagent 1.
3.   Invertase (β-D-Fructofuranosidase).
      Sigma grade VII (From baker's yeast).

Polarimeter with Na Lamp.
200 mm Polarimeter Tube.
Balance.
250 ml Flask.
Stop Clock.

PROCEDURE:

1. Accurately weigh out about 5 mg of invertase. Record the exact weight to the nearest 0.1 mg.

2. Add the invertase to 100 ml of 1.0 M sucrose in citrate-phosphate buffer (pH 4.6) in a 250 ml flask, mix well to dissolve the invertase and start the clock.

3. Flush out the weighing vessel with the mixture and return it to the 250 ml flask, so that all the invertase is used.

4. Fill the 200 mm polarimeter tube to the brim and slide the glass window across the top, excluding any air bubbles. Place the rubber washer and screw fitting and tighten moderately. Wash and dry the end windows of the tube and place it in the polarimeter.

5. Approximately every 10 minutes (for 2 hours), make 3 readings of the optical rotation to the nearest 0.1° and record the exact time of the 2nd reading.

    The initial set of readings should be close to $+ 22.5^{\circ}$.

6. Plot the average of the 3 readings, for each set of readings, against time.

CALCULATIONS:

1. Read off the rate of change in optical rotation in degrees per 100 minutes. The response should be a straight line. If not, draw a straight line at the initial (maximum) rate.

2. Divide by 0.0598 to obtain the number of millimoles per litre of sucrose hydrolysed per 100 minutes.

3. Divide by 6 to obtain the number of micromoles per litre hydrolysed in one second.

4. Divide by 10 x the weight of invertase used, to obtain the number of micromoles hydrolysed per second per milligram of enzyme.

    This is the specific activity of the enzyme (kat/kg). The temperature and pH (4.6) should also be stated when quoting enzyme activity.

EXERCISE 3-2

THE EFFECT OF TEMPERATURE ON INVERTASE ACTIVITY

As with exercise 1, the enzyme invertase will be used in this exercise. However, the release of reducing sugars will be used to monitor enzyme activity. Sucrose is not a reducing sugar as the reducing groups of glucose and fructose are involved in the linkage between the 2 sugars. The C_1 carbon of glucose is linked to the C_2 carbon of fructose. Hydrolysis of sucrose liberates the two reducing sugars.

The action of invertase can, therefore, be followed by observing the rate of reducing sugar formation. The amount of reducing sugar formed will be determined by the Somogyi-Nelson method. In this method, glucose and fructose (in an alkaline solution), reduce cupric copper to cuprous copper. The cuprous copper then reduces arsenomolybdate to molybdenum blue, the absorbance of which is measured at 680 nm.

The reducing sugars form enediol salts, in alkaline solution. These are strong reducing agents. On oxidation, the double bond is ruptured and shorter chain acids are formed. Eg:

$$
\begin{array}{c}
H - C - OH \\
\parallel \\
C - OH \\
\mid \\
R \\
\text{Enediol}
\end{array}
\quad \xrightarrow{NaOH} \quad
\begin{array}{c}
H - C - O^- \ Na^+ \\
\mid \\
C - OH \\
\mid \\
R \\
\text{Enediol} \\
\text{Salt}
\end{array}
\quad \xrightarrow{Oxid} \quad
\begin{array}{c}
H - COO^- \ Na^+ \\
+ \\
R - COO^- \ Na^+
\end{array}
$$

The effect of 6 temperatures, between 0 and 70°C, on invertase activity will be examined in this exercise.

REAGENTS AND EQUIPMENT:

1. Citrate- Phosphate Buffer, pH 4.6.
    Prepare as for reagent 1, exercise 3-1.
2. 25 mM Sucrose in Citrate-Phosphate Buffer, pH 4.6.
    Dissolve 856 mg in 100 ml of reagent 1.
    Prepare fresh.
3. Invertase Stock Solution (100 mg/l).
    Dissolve 10 mg of invertase (Sigma grade VII) in 100 ml of distilled water.
    Refrigerate or keep on ice.
4. Working Invertase Solution (0.5 mg/l).
    Dilute 500 µl of reagent 3 to 100 ml with distilled water.
    Make up fresh and keep on ice.
5. Somogyi's Alkaline Copper Reagent.
    (1) (a) Dissolve 48 g of anhydrous sodium carbonate and 24 g of Na K tartrate in 500 ml of distilled water.
         (b) Dissolve 8 g of $CuSO_4:5H_2O$ in 100 ml of distilled water.
         Add (b) to (a) with constant stirring.
    (2) Dissolve 360 g of anhydrous sodium sulphate in 1 l of distilled water. Boil to expel the air and then allow to cool.
    Mix (1) and (2) and dilute to 2 l with distilled water.
    Keep for at least a month at room temperature and filter, if necessary. (Like a good red wine, this reagent needs time to mature!)

6.   This reagent can be put out in a dispenser set to 2.0 ml.

Nelson's Arsenomolybdate Reagent.

Dissolve 100 g of ammonium molybdate in 1,800 ml of distilled water. Add 84 ml of concentrated sulphuric acid and stir to dissolve the molybdate. Add 12 g of sodium arsenate ($Na_2HAsO_4:7H_2O$) and keep stirring until dissolved. Make to 2 l with distilled water. Incubate at $37^{\circ}C$ for 24 to 48 hours.

Store in a dark bottle.

This reagent can be put out in a dispenser set to 1.0 ml.

7.   1.0 mM Glucose Standard.

Dissolve 90 mg of AR glucose in 500 ml of distilled water.

Make up fresh or use saturated benzoic acid (2 g/1 H_2O), instead of distilled water in the preparation.

8.   Ice.

Water Baths at 40°, 50°, 60°, 70° and $100^{\circ}C$.

Thermometers.

Spectrophotometer(s), Cuvettes and Paper Tissues.

Test Tubes.

Beakers.

Stop Clocks.

1 and 2 ml Graduated Pipettes.

$200\,\mu l$ Pipettors with Tips.

Aluminium Foil Squares ($\approx 50 \times 50$ mm).

Parafilm Squares ($\approx 50 \times 50$ mm).

PROCEDURE:

1.   Set up an ice bath by mixing approximately equal amounts of water and ice, in a beaker.

2.   Pipette 1.0 ml samples of 25 mM sucrose in citrate-phosphate buffer (pH 4.6) into 6 test tubes labelled 0, RT, 40, 50, 60 and 70.

3.   Place the tubes in their appropriate water bath, leaving the RT tube at room temperature. Check that the water baths are close to the stated temperatures and record the actual temperatures and the room temperature.

4.   After a 5 minute equilibration period, add 200 µl of invertase solution (0.5 mg/1) to each tube, at 30 second intervals, mix and note the time.

5.   After exactly 15 minutes incubation, pipette 2.0 ml of Somogyi's alkaline copper reagent into each tube and mix well (This stops the enzyme reaction). Remove the tubes from the water baths.

6.   Set up the following standards, in test tubes, and mix well:

	A	B	C	D
ml of glucose standard (1 mmol/l)	0	0.3	0.6	0.9
ml of distilled water	1.2	0.9	0.6	0.3
ml of copper reagent	2.0	2.0	2.0	2.0

7. Cap all tubes with aluminium foil and incubate in a boiling water bath for 10 minutes.

8. Cool quickly in a cold water bath for 1 minute. Add 1.0 ml of Nelson's arsenomolybdate reagent and mix well.

9. Add 20 ml of distilled water to each tube and mix by inversion, placing a piece of parafilm between thumb and tube.

10. Read the absorbance of all tubes, against tube A at 680 nm, in a spectrophotometer. If any tube has an absorbance greater than the top standard, dilute 5.0 ml with 5.0 ml of distilled water and read again.

Plot a standard curve of absorbance against the amount (μmol) of glucose per tube and read off the unknowns.

Plot a graph of μmol of reducing sugar formed against temperature.

Calculate the specific activity of the enzyme at room temperature (pH 4.6) in kat/kg. Remember that one catalytic event yields 2 molecules of reducing sugar.

EXERCISE 3-3

THE EFFECT OF pH ON INVERTASE ACTIVITY

In this exercise, a similar enzyme system will be used to that used in 3-2, except that the pH will be altered and the temperature kept constant.

Sucrose tends to hydrolyse in acid. The rate of hydrolysis is proportional to the hydrogen ion concentration. However, this effect is negligible at room temperature with the acid pHs used in this exercise. pHs between 3.0 and 7.8 will be examined.

REAGENTS AND EQUIPMENT:

The same reagents and equipment are required, as for exercise 3-2, except:

Delete reagents 1, 2, 4 and 8 and replace with:

1. Invertase Solution (2.0 mg/l).
       Dilute 2.0 ml of the invertase stock (100 mg/l) to 100 ml with distilled water.
       Prepare fresh.
2. Citrate-Phosphate Buffers.
       (a)   0.2 M Citric Acid.
               Dissolve 38.4 g of citric acid (or 42.0 g of the monohydrate) in distilled water and make to 1 l.
       (b)   0.4 M Phosphate.
               Dissolve 71.2 g of $Na_2HPO_4:2H_2O$ (or 143.2 g of $Na_2HPO_4:12H_2O$) in distilled water and make to 1 l.
       Mix (a) and (b) in the following proportions and check the pH:

pH	ml of 0.2 M Citric Acid	ml of 0.4 M Phosphate
3.0	159	41
3.8	129	71
4.8	101	99
5.8	79	121
6.8	45	155
7.8	9	191

3.   50 mM Sucrose.
          Dissolve 17.1 g of AR sucrose in distilled water and make to 1 l.
          Prepare fresh.

Water Bath at 100°C only.
200 and 500 μl Pipettors and Tips.

PROCEDURE:

1.   Set up a series of 6 test tubes, labelled 1 to 6.

2.   Add 500 μl of 50 mM sucrose to each tube.

3.   Add 500 μl of the appropriate citrate-phosphate buffer to its tube, as indicated below:

Tube	Buffer pH
1	3.0
2	3.8
3	4.8
4	5.8
5	6.8
6	7.8

4.   Add 200 μl of invertase solution (2.0 mg/l) to each tube, in the order 1 to 6, and mix well.

5.   After exactly 15 minutes, at room temperature, add 2.0 ml of Somogyi's alkaline copper reagent to each tube, in the order 1 to 6, and mix well.

6.   Set up the following standards, in test tubes, and mix well:

	A	B	C	D
ml of glucose standard (1 mmol/l)	0	0.3	0.6	0.9
ml of distilled water	1.2	0.9	0.6	0.3
ml of copper reagent	2.0	2.0	2.0	2.0

7.   Cap all tubes with aluminium foil and incubate in a boiling water bath for 10 minutes.

8.   Cool quickly in a cold water bath for 1 minute. Add 1.0 ml of Nelson's arsenomolybdate reagent and mix well.

9.   Add 20 ml of distilled water to each tube and mix by inversion, placing a piece of parafilm between thumb and tube.

10. Read the absorbance of all tubes, against tube A at 680 nm, in a spectrophotometer. If any tube has an absorbance greater than the top standard, dilute 5.0 ml with 5.0 ml of distilled water and read again.

Plot a standard curve of absorbance against the amount (μmol) of glucose per tube and read off the unknowns.

Plot a graph of μmol of reducing sugar formed against pH.

EXERCISE 3-4

THE EFFECT OF SUBSTRATE ON YEAST ALCOHOL DEHYDROGENASE ACTIVITY

Compared with inorganic catalysts, enzymes are very specific as to the reactions that they catalyse. However, different groups of enzymes vary considerably in their degree of specificity. Dehydrogenases, kinases and synthetases show a high degree of specificity; whereas, esterases, phosphatases and peptidases are fairly non-specific. Enzymes that act at the asymmetric centre of an optically active compound show absolute specificity for one of the optical isomers. The other isomer is not acted upon by the enzyme.

Alcohol dehydrogenase has the systematic name of Alcohol:NAD^+ oxidoreductase. It catalyses the reaction between two substrates, if one considers NAD to be a substrate. The enzyme has the Enzyme Commission number of 1.1.1.1. The class 1 is an oxidoreductase, sub-class 1 is the CH-OH group acting as hydrogen donor and sub-sub-class 1 is NAD or NADP acting as hydrogen acceptor. This enzyme has an absolute specificity for NAD, but is less specific for the alcohol. EC No 1.1.1.2 has absolute specificity for NADP.

EC No 1.1.1.1 has been isolated from a number of different organisms and from different mammalian tissues. These enzymes have different specificities for different alcohols. Some are specific for low molecular weight alcohols and others specific for alcoholic groups on larger molecules such as steroids and vitamins etc.

Alcohol dehydrogenases have a quaternary structure and, hence, have a number of subunits. The yeast enzyme has four subunits, each of which has a molecular weight of about 37,000. Each subunit contains two atoms of zinc. One of these stabilises the tertiary structure and the other plays a role in the reaction taking place at the active site. The cycle of events involving alcohol dehydrogenase is shown below:

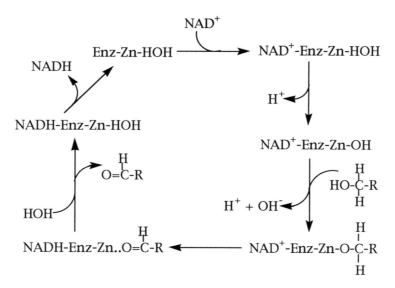

As will be seen in this exercise, yeast alcohol dehydrogenase is fairly specific for ethanol. This is the normal product of fermentation. Ethanol is produced from acetaldehyde, which is the product of the decarboxylation of pyruvic acid.

The following alcohols will be used in this exercise:

$CH_3.OH$ $CH_3.CH_2.OH$ $CH_3.CH_2.CH_2.OH$

Methyl Alcohol   Ethyl Alcohol   n-Propyl Alcohol

$\begin{matrix} CH_3 \\ CH_3 \end{matrix}\!\!>\!CH.OH$ $CH_3.CH_2.CH_2.CH_2.OH$

                                      n-Butyl Alcohol

Iso-Propyl Alcohol

The enzyme activity will be monitored by observing the reaction:

Alcohol + NAD⁺ ⟶ Aldehyde (or Ketone) + NADH + H⁺

This will be measured by observing the increase in absorbance at 340 nm as the NAD is converted to NADH. Similar molar amounts of alcohol, together with a fixed amount of enzyme and an excess of NAD will be used in the assay, from which the initial rate will be determined. The assay will be carried out using a pyrophosphate buffer of pH 8.5.

REAGENTS AND EQUIPMENT:

1.   50% H_3PO_4.
            Slowly add 25 ml of concentrated phosphoric acid to 25 ml of distilled water, with
            constant mixing.
            For reagent 2.

2.   60 mM Pyrophosphate Buffer, pH 8.5.

        Dissolve 26.8 g of $Na_2P_2O_7:10H_2O$ in 900 ml of distilled water. Adjust the pH to 8.5 with reagent 1 and make to 1 l with distilled water.

        Aliquots can be frozen. Use fresh, or only refrigerate for a few days.

3.   15 mM NAD.

        Dissolve 1.0 g of NAD in 100 ml of distilled water.

        Aliquots can be frozen. Use fresh or thawed aliquot.

        Keep in ice.

4.   300 mM Ethanol.

        Dilute 1.76 ml of AR ethanol to 100 ml with distilled water.

5.   300 mM Methanol.

        Dilute 1.21 ml of AR methanol to 100 ml with distilled water.

6.   300 mM Iso-Propanol.

        Dilute 2.29 ml of iso-propanol to 100 ml with distilled water.

7.   300 mM n-Propanol.

        Dilute 2.24 ml of n-propanol to 100 ml with distilled water.

8.   300 mM n-Butanol.

        Dilute 2.74 ml of n-butanol to 100 ml with distilled water.

9.   1% Albumin.

        Dissolve 1 g of BSA in 100 ml of distilled water.

        Required for reagent 10.

10.   Yeast Alcohol Dehydrogenase.

        Dissolve enough alcohol dehydrogenase in reagent 9 to give an activity of about 50 μkat/l ($\approx$ 10 mg/l).

        Make up fresh and keep on ice.

Double Beam spectrophotometer with a Kinetic Function or a spectrophotometer (single or double beam) connected to a recorder. If the latter, adjust the recorder so that full scale equals 1 absorbance unit and that the recorder units correspond to the spectrophotometer readings.

Glass cuvettes, plastic cuvette stirrers and paper tissues.

100, 500 μl and 1 ml Pipettors with tips.

2 ml graduated Pipettes.

PROCEDURE:

1.   Set the spectrophotometer to 340 nm and to read zero with nothing in the light path and to the kinetic mode, if using an instrument with this function.

2.   To a glass cuvette, add the following:

Distilled water	:	1.3 ml
60 mM pyrophosphate buffer, pH 8.5	:	500 μl
*Yeast alcohol dehydrogenase (Approx 50 μkat/l stabilised in 1% albumin)	:	100 μl
*15 mM NAD	:	100 μl

    * These reagents should be kept in ice.

3.   Place the cuvette in the light path, and start the recorder with a chart speed of 1 cm/min, if using a recorder.

4.   Add 1.0 ml of 300 mM n-propanol and rapidly mix with a plastic cuvette stirrer. Close the cell housing and start the run (if using a spectrophotometer in the kinetic mode) as soon as possible.

5.   Record for about 5 minutes, or until a good trace is obtained of the initial velocity (whichever is shorter in time).

6. Repeat the assay with the 4 other alcohols (each is 300 mM). Use a chart speed of 5 cm/min for the ethanol assay.

CALCULATIONS:

1. For each assay, obtain the initial (maximum) rate of change in absorbance units per minute. If using a chart recorder draw a straight line at the maximum rate and take 2 well separated readings so that an accurate value can be calculated.

2. NADH has a molar absorption coefficient of 6,220 at 340 nm. That is, a molar solution of NADH would have an absorbance of 6,220.

    Hence, 1 mM NADH would have an absorbance of 6.22. Therefore, 1 μmol/ml would have an absorbance of 6.22.

    Calculate the enzyme activity for each alcohol, in μkat/l of enzyme solution, as follows:

    (i)   Divide the values obtained in step 1 by 60 to obtain the absorbance change per second.

    (ii)   Divide by 6.22 to give μmol/ml NADH formed per second.

    (iii) Multiply by 3.0 to obtain the number of μmol of NADH formed per second in the cuvette.

    (iv)   Multiply by 10,000 to give μmol of NADH formed per second by one litre of enzyme solution (ie. μkat/l).

    ie. Multiply the values obtained in Step 1 by 80.4.

3. Tabulate these results.

EXERCISE 3-5

THE EFFECT OF SUBSTRATE CONCENTRATION ON YEAST ALCOHOL DEHYDROGENASE ACTIVITY

In this exercise, similar assays will be carried out to those used in exercise 3-4. However, a range of ethanol concentrations will be used with 2 different enzyme concentrations. From these results, the K_m value for yeast alcohol dehydrogenase will be read from a Lineweaver-Burk plot.

REAGENTS AND EQUIPMENT:

The reagents and equipment are the same as for exercise 3-4, except that reagents 4, 5, 6, 7 and 8 should be replaced with:

1. 6.0 M Ethanol.
        Dilute 35.1 ml of ethanol to 100 ml with distilled water.

Wassermann Tubes.
1.0 ml Pipettors and Tips.
Beakers of ice.
Thermometers.

PROCEDURE:

1. Add 1.0 ml of distilled water to 6 Wassermann tubes labelled B to G.

2. Add 1.0 ml of 6.0 M ethanol to tube B and mix. Also add about 1 ml of 6.0 M ethanol to a tube labelled A.

3. Pipette 1.0 ml of the contents of tube B into tube C and mix.

4. Repeat this serial dilution with tubes C, D, E and F. You should end up with 2.0 ml in tube G and 1.0 in the rest.

5. Set the spectrophotometer to 340 nm and to read zero with nothing in the light path and to the kinetic mode, if using an instrument with this function.

6. To a glass cuvette, add the following:

Distilled water	:	2.2 ml
60 mM pyrophosphate buffer, pH 8.5	:	500 µl
*Yeast alcohol dehydrogenase (Approx 50 µkat/l stabilised in 1% albumin)	:	100 µl
*15 mM NAD	:	100 µl

    * These reagents should be kept in ice.

7. Place the cuvette in the light path, and start the recorder with a chart speed of 5 cm/min, if using a recorder.

8. Add 100 µl of 6.0 M ethanol and rapidly mix with a plastic cuvette stirrer. Close the cell housing and start the run (if using a spectrophotometer in the kinetic mode) as soon as possible.

9. Record for about 5 minutes, or until a good trace is obtained of the initial velocity (whichever is shorter in time).

10. Repeat the assay with each of the diluted ethanols. Use a slower chart speed for the more dilute solutions.

11. Dilute 500 µl of the yeast alcohol dehydrogenase with 500 µl of distilled water. Keep the tube in a beaker of ice.

12. Repeat the 7 assays, using 100 µl of this diluted enzyme, instead of the original enzyme solution.

13. Record the temperature of the cell housing of the spectrophotometer.

CALCULATIONS:

1. Calculate the initial enzyme velocity (V) in μkat/l, as in exercise 3-4, for all 14 assays.

2. Calculate the ethanol concentration in the cuvette for each of the 7 concentrations used ([S]), in mol/l.

3. Plot V against [S], for each alcohol dehydrogenase solution, on the same graph.

4. Calculate the reciprocals of V and [S].

5. Plot $\frac{1}{V}$ against $\frac{1}{[S]}$ for each alcohol dehydrogenase solution, on the same graph. You should have a similar graph to the one below:

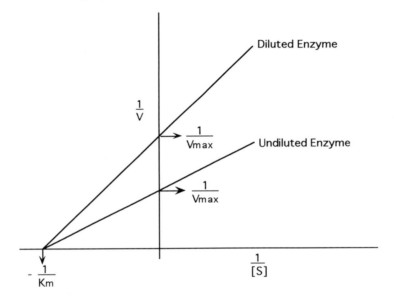

Read off $\frac{1}{K_m}$ and calculate K_m.

Read off $\frac{1}{V_{max}}$ for each enzyme solution and calculate V_{max} for each.

Indicate K_m and the two V_{max} values on the graph of V against [S].

Reference Value:

The K_m, for ethanol, of yeast alcohol dehydrogenase = 24 mM at pH 7.4 and 26°C.

EXERCISE 3-6

THE EFFECT OF AN INHIBITOR ON YEAST ALCOHOL DEHYDROGENASE ACTIVITY

Hydroxylamine ($NH_2.OH$) is an inhibitor of alcohol dehydrogenase. In this exercise, this inhibition will be examined by studying the effect of 3 different inhibitor concentrations (100, 10 and 1 mmol/l) on the initial velocity of the enzyme catalysed reaction, using a series of substrate concentrations.

From a direct plot of initial velocity against substrate concentration (for each inhibitor concentration) and from the Lineweaver-Burk plot, the type of inhibition will be determined.

The assay system will be the same as that used in exercises 3-4 and 3-5 to which hydroxylamine will be added. Hydroxylamine is unstable and is, therefore, prepared by neutralising hydroxylamine hydrochloride with sodium hydroxide, in the cuvette, just before use.

REAGENTS AND EQUIPMENT:

Reagents and equipment as for exercise 3-5, plus:

1. Hydroxylamine Hydrochloride (300 mM).
   Dissolve 2.08 g of hydroxylamine hydrochloride in distilled water and make to 100 ml.
   Prepare fresh.
2. NaOH (3.0 M).
   Dissolve 12.0 g of NaOH in distilled water and make to 100 ml.

10 ml Graduated Pipettes.
Test Tubes.

PROCEDURE:

Preparation of Ethanol Solutions:

1. Add 1.0 ml of distilled water to 5 Wassermann tubes labelled B to F.

2. Add 1.0 ml of 6.0 M ethanol to tube B and mix. Also add about 1 ml of 6.0 M ethanol to a tube labelled A.

3. Pipette 1.0 ml of the contents of tube B into tube C and mix.

4. Repeat this serial dilution with tubes C, D, and E. You should end up with 2.0 ml in tube F and 1.0 in the other tubes.

Preparation of Hydroxylamine Hydrochloride Solutions:

5. Add 1.0 ml of the 300 mM hydroxylamine hydrochloride to 9.0 ml of distilled water, in a test tube, and mix well to give a 30 mM solution.

6. Add 1.0 ml of the above 30 mM solution to 9.0 ml of distilled water and mix well to give a 3 mM solution.

Preparation of the Neutralising Solutions:

7. Add 1.0 ml of 3.0 M NaOH to 9.0 ml of distilled water and mix well to give a 300 mM solution.

8. Add 1.0 ml of the above 300 mM solution to 9.0 ml of distilled water and mix well to give a 30 mM solution.

Assay:

9. Set the spectrophotometer to 340 nm and to read zero with nothing in the light path and to the kinetic mode, if using an instrument with this function.

10. To a glass cuvette, add the following:

Distilled water	:	1.1 ml
300 mM Hydroxylamine HCl	:	1.0 ml
3.0 M NaOH	:	100 µl
60 mM pyrophosphate buffer, pH 8.5	:	500 µl
*Yeast alcohol dehydrogenase (Approx 50 µkat/l stabilised in 1% albumin)	:	100 µl
*15 mM NAD	:	100 µl

* These reagents should be kept in ice.

11. Place the cuvette in the light path, and start the recorder with a chart speed of 1 cm/min, if using a recorder.

12. Add 100 µl of 6.0 M ethanol (Tube A) and rapidly mix with a plastic cuvette stirrer. Close the cell housing and start the run (if using a spectrophotometer in the kinetic mode) as soon as possible.

13. Record for about 5 minutes.

14. Repeat the assay with each of the diluted ethanols.

15. Repeat the 6 assays, but use the 30 mM hydroxylamine hydrochloride and 300 mM NaOH, instead of the 300 mM and 3.0 M solutions.

16. Repeat the 6 assays, but use the 3.0 mM hydroxylamine hydrochloride and 30 mM NaOH. A faster chart speed may be needed for the higher ethanol concentrations.

CALCULATIONS:

1. Calculate the initial enzyme velocity (V) in µkat/l, as in exercise 3-4, for all 18 assays.

2. Calculate the ethanol concentration in the cuvette for each of the 6 concentrations used ([S]), in mol/l.

3. Plot V against [S], for each inhibitor concentration, on the same graph.

4. Calculate the reciprocals of V and [S].

5. Plot $\frac{1}{V}$ against $\frac{1}{[S]}$ for each inhibitor concentration, on the same graph.

6. From these plots, and the information on pages 3-9 to 3-12, decide which type of inhibitor is present.

EXERCISE 3-7

THE DETERMINATION OF PROSTATIC ACID PHOSPHATASE IN PLASMA

Determinations of plasma levels of specific enzymes plays a very important role in disease diagnosis and in monitoring the progress of that disease. Where possible, it is better to use a kinetic method for enzyme determination, than measuring the amount of product formed after a fixed time. However, in this exercise, the amount of product will be determined after a 30 minute incubation period.

Acid phosphatases have the systematic name of Orthophosphoric-monoester phosphohydrolase (acid optimum) and the EC number of 3.1.3.2. The class 3 is for a hydrolase, the sub-class of 1 is for hydrolysis of an ester and the sub-sub-class of 3 is for a phosphoric acid monoester. As the name implies, the enzyme hydrolyses phosphate from its monoesters, in acid conditions. The optimum pH for plasma acid phosphatases is about 5.0, with a broad plateau from pH 4 to 6.

The acid phosphatases in normal plasma originate from platelets and leucocytes and from erythrocytes, should there be any haemolysis (red cells have a concentration of acid phosphatase about 100 times greater than that in plasma). The acid phosphatase level in plasma is considerably increased when bone metastases develop from prostatic carcinoma.

Therefore, the determination of acid phosphatase, in plasma or serum, is almost always for the detection, or monitoring, of carcinoma of the prostate. Hence, it is the prostatic isoenzyme that is determined. The use of inhibitors, such as formaldehyde and L-tartrate, can be employed for this purpose. Formaldehyde inhibits acid phosphatases from blood cells and platelets; whereas, L-tartrate inhibits prostatic acid phosphatase. However, the use of substrates that are only acted upon by the prostatic enzyme offers greater simplicity in the method.

In this exercise, thymolphthalein monophosphate is used as the substrate. Even though the non-prostatic enzymes act on this substrate, they do so at a very much slower rate than the prostatic enzyme.

The sample is incubated with the substrate, in an acetate buffer of pH 5.4, for 30 minutes at 37°C. The liberated thymolphthalein is then determined by the addition of alkali. This produces a red colour, which is read at 595 nm. The alkali also stops the enzymic reaction.

Prostatic acid phosphatase is very unstable in neutral or alkaline solutions. Hence, sodium citrate or acetic acid is added to the sample to bring the pH to about 5.4. Under these conditions, the activity is maintained at room temperature for several hours or for up to a week in a refrigerator.

Elevations in prostatic acid phosphatase are seen in most cases where there are metastases from a prostatic tumour. Highest levels are seen when the metastases are in bone. Where the carcinoma is localised in the prostate, enzyme levels are usually normal. Where treatment is successful, enzyme levels slowly return to the normal.

Case Study:

The plasma is from a 69 year-old man who was experiencing difficulty urinating. A rectal examination demonstrated an enlarged prostate. As part of the follow up, a blood sample was collected for plasma prostate specific antigen and prostatic acid phosphatase determinations.

REAGENTS AND EQUIPMENT:

1. Buffered Substrate.
   (i)   Dilute 3.24 ml of Brij-35 (300 g/l) to 300 ml.
   (ii) Dissolve 83 mg of disodium thymolphthalein monophosphate:11H_2O in 100 ml of the diluted Brij.
   (iii) Dissolve 1.92 g of sodium acetate trihydrate in this solution.
   (iv) Adjust the pH to 5.4 with 0.1 M HCl.
   Refrigerate.
2. Alkaline Reagent.
   Dissolve 5.3 g of anhydrous sodium carbonate and 2.0 g of sodium hydroxide in distilled water and make to 500 ml.
3. Patient's Plasma.
   Mix 50 mg of disodium citrate monohydrate with 5 ml of plasma or serum from freshly collected blood.
   Refrigerate.

Test Tubes.
Spectrophotometers, Glass Cuvettes and Paper Tissues.
100 µl Pipettors and Tips.
2 ml Graduated Pipettes.
37°C Water Bath.
Stop Clocks.

PROCEDURE:

1. Pipette 1.1 ml of buffered substrate into a test tube and allow to come to temperature in a 37°C water bath.

2. Add 100 µl of the patient's plasma, mix and return the tube to the water bath. (Note the time).

3. Prepare a blank by adding 2.0 ml of alkaline reagent, 1.1 ml of buffered substrate and then 100 µl of plasma to a test tube and mix well.

4. Exactly 30 minutes after step 2, add 2.0 ml of the alkaline reagent to the sample tube and mix well.

5. Read the absorbance of the sample tube against the blank tube, at 595 nm in a spectrophotometer.

From the molar absorption coefficient of thymolphthalein (in alkali) at 595 nm (39,200), calculate the enzyme activity in nkat/l at 37°C.

Reference Range:

The range for healthy men, using this method, is:

8 - 32 nkat/l at 37°C.

EXERCISE 3-8

THE DETERMINATION OF PLASMA LACTATE DEHYDROGENASE AND ITS ISOENZYMES

As mentioned previously, it is better to use a kinetic method for enzyme determination, than measuring the amount of product formed after a fixed time. The oxidoreductases can be determined by monitoring the rate of change in absorbance at 340 nm. Other enzyme classes can often be linked to an oxidoreductase system and thus measured by observing absorbance changes at 340 nm. For example, alanine aminotransferase (AlAT) can be measured with the following coupled reactions:

$$\text{Alanine} + \alpha\text{-Ketoglutaric Acid} \xrightarrow{AlAT} \text{Pyruvic Acid} + \text{Glutamic Acid}$$

$$\text{Pyruvic Acid} + \text{NADH} \xrightarrow{LD} \text{Lactic Acid} + \text{NAD}$$

Lactate dehydrogenase (LD) is added in excess so that pyruvic acid is converted to lactic acid as soon as it is formed. Hence, the rate of the coupled reactions is dependent upon the level of AlAT.

In this exercise, plasma lactate dehydrogenase, and some of its isoenzymes, will be measured by a kinetic method. LD catalyses the reaction:

$$
\begin{array}{l}
\text{CH}_3 \\
|\\
\text{CH - OH} + \text{NAD}^+ \rightleftharpoons \\
|\\
\text{COOH} \\
\text{Lactic Acid}
\end{array}
\qquad
\begin{array}{l}
\text{CH}_3 \\
|\\
\text{C}= \text{O} + \text{NADH} + \text{H}^+ \\
|\\
\text{COOH} \\
\text{Pyruvic Acid}
\end{array}
$$

It has the systematic name of L-lactate:NAD$^+$ oxidoreductase and the EC number of 1.1.1.27.

The enzyme is non-specific and will act on other α-keto on α-diketo acids. The equilibrium of the above reaction (at neutral pH) lies well to the left and so pyruvate and NADH are used as substrates in most of the spectrophotometric methods. Optimal activity of the enzyme system occurs at pyruvate concentrations between 0.5 and 1.0 mM and between pH 7.3 and 7.7.

Endogenous α-keto and α-diketo acids, in plasma, are removed by incubation with NADH, at room temperature. The fall in absorbance at 340 nm is then measured after the addition of pyruvate.

Red cells contain about 100 times the enzyme concentration of plasma and so haemolysed samples should not be used. The enzyme is quite stable and keeps at room temperature for at least 8 days. Oxalate inhibits the enzyme and should, therefore, not be used as an anticoagulant.

As mentioned on pages 3-13 and 3-14, lactate dehydrogenase isoenzymes can be separated by electrophoresis, the use of inhibitors, heat or by using a different substrate. In this exercise, heat and an inhibitor will be used.

All the isoenzymes are stable at room temperature. However, LD_5 is destroyed by incubation for 30 minutes at 57°C. All the isoenzymes, except LD_1, are destroyed by incubation for 30 minutes at 65°C.

0.8 M sodium perchlorate inhibits the isoenzymes that contain a M subunit. Therefore, only LD_1 has enzyme activity.

Elevated plasma levels of total LD are seen following myocardial infarction, renal disease, liver disease, diseases of skeletal muscle and with some anaemias and leukaemias. These diseases can be differentiated, to some extent, by looking at the LD isoenzymes.

Case History.

The sample is heparinised plasma from a 52 year-old man who informed his GP that he had suffered severe chest pains, one week ago, that had lasted several hours. LD isoenzymes have been requested to investigate the possibility of a myocardial infarction. One week after a myocardial infarction, most of the cardiac enzymes will have returned to their basal levels. However, LD (mainly LD_1 and LD_2) remains elevated for 10 days or so.

REAGENTS AND EQUIPMENT:

1.   0.1 M Potassium Phosphate Buffer, pH 7.4.
             Dissolve 7.0 g of anhydrous K_2PO_4 and 1.35 g of anhydrous KH_2PO_4 in distilled water
             and make to 500 ml.
2.   0.25% NADH in Phosphate Buffer.
             Dissolve 125 mg of NADH in 50 ml of reagent 1.
             Make fresh and keep on ice.
3.   0.25% Sodium Pyruvate in Phosphate Buffer.
             Dissolve 125 mg of sodium pyruvate in 50 ml of reagent 1.
             Keep on ice.
4.   1 M Perchlorate in Phosphate Buffer.
             Dissolve 7.0 g of sodium perchlorate monohydrate in 50 ml of reagent 1.
5.   Patient's Plasma.
             Heparinised (or EDTA) plasma or serum.
             Keep on ice.

Water Baths at 57 and 65°C.
$200\,\mu l$ and 1 ml Pipettors and Tips.
5 ml Graduated Pipettes.
Test Tubes.
Double Beam spectrophotometer with a Kinetic Function or a spectrophotometer (single or double beam) connected to a recorder. If the latter, adjust the recorder so that full scale equals 1 absorbance unit and that the recorder units correspond to the spectrophotometer readings. Set the recorder to 1 cm/min.
Glass cuvettes, plastic cuvette stirrers and paper tissues.

PROCEDURE:

1. Dilute 1.0 ml of the patient's plasma with 1.0 ml of 0.1 M potassium phosphate buffer (pH 7.4) and mix well.

2. Divide this solution into 3 tubes. Leave one tube at room temperature, place one in the 57°C water bath and the other in the 65°C bath.

   (Make sure that the water baths are at these temperatures.)

3. Incubate for 30 minutes and remove the tubes.

4. Assay the LD activity as follows:

   Add the following to a glass cuvette:

0.1 M phosphate buffer, pH 7.4	:	2.5 ml.
0.25% NADH in phosphate buffer	:	200 μl.
Diluted sample	:	200 μl.

   Stand the cuvettes at room temperature for 20 minutes to remove any keto acids.

   Place the cuvette in a spectrophotometer set to 340 nm (Set the spectrophotometer to 6.0 Absorbance units, if using a single beam spectrophotometer) and note the absorbance for 1 minute (There should be no change).

   Add 100 μl of the sodium pyruvate solution (0.25% in phosphate buffer), mix with a plastic cuvette stirrer and record the absorbance for about 5 minutes. (Allow a 30 second lag phase, if using a spectrophotometer with kinetic mode.)

   Note the temperature of the cell housing.

5. Repeat the assay with the following:

1 M perchlorate - phosphate buffer, pH 7.4 :		2.5 ml
0.25% NADH	:	200 μl.
Diluted RT plasma sample	:	200 μl.

Calculate the rate of absorbance change and convert to 25°C by multiplying by the factor below:

20°	:	1.47
21°	:	1.36
22°	:	1.26
23°	:	1.17
24°	:	1.08
26°	:	0.93
27°	:	0.86

Calculate the enzyme activity of the total LD, LD_1 (temperature and inhibitor methods), LD_5 and LD_{2-4} in μkat/l at 25°C (The molar absorption coefficient of NADH, at 340 nm, is 6,220).

Reference Range:

Total LD : 1.7 - 4.2 μkat/l at 25°C.

Average Isoenzyme Distribution in Plasma:

1 : 27%
2 : 35%
3 : 20%
4 : 7%
5 : 11%

Plasma lactate dehydrogenase levels start to rise about 12 hours after a myocardial infarction. They peak around 30 hours and then decline to the basal level over the next 2 weeks. LD_1 becomes the predominant isoenzyme and the normal plasma ratio of LD_1/LD_2 is reversed.

EXERCISE 3-9

THE FLUORIMETRIC DETERMINATION OF ALKALINE PHOSPHATASE IN PLASMA.

Plasma alkaline phosphatase levels are useful in the assessment of liver and bone diseases and to assess placental function in the third trimester of pregnancy.

High levels of the enzyme are seen with obstruction to bile flow. Hepatic cell damage also gives rise to an elevation, but to a lesser extent than that seen with biliary obstruction.

Increased activity of osteoblasts leads to a rise in plasma alkaline phosphatase. Paget's disease, hyperparathyroidism, hypervitaminosis D and bone tumours are all associated with increased levels.

The placenta produces large amounts of alkaline phosphatase and some of this enters the maternal circulation. Therefore, there is a rise in maternal plasma alkaline phosphatase in the last trimester of pregnancy. With decreased placental function this rise is reduced or absent.

Alkaline phosphatases catalyse the hydrolysis of monoesters of orthophosphoric acid. They have the systematic name of orthophosphoric-monoester phosphohydrolase (alkaline optimum) and the Enzyme Commission number of 3.1.3.1. As the name suggests, they have optimum activity under alkaline conditions. The actual optimum pH depends on the substrate used, the substrate concentration, the temperature and the source of enzyme. The optimum pH usually lies in the range 8.6-10.3.

In this exercise, the substrate used is 4-methyl umbelliferyl phosphate.

This has the structure:

It also has the name of 4-methyl-7-phospho-coumarin.

This substrate is hydrolysed, under the influence of alkaline phosphatase, to 4-methyl umbelliferone, which has a phenolic -OH group on position 7. At pH 10, 4 methyl-umbelliferone ionises to give the phenolic anion. This phenolic anion has maximum absorption at 360 nm and a fluorescence emission maximum of 450 nm.

As the substrate concentration used in this assay is less than the K_m value, this is not a suitable method for clinical work. It is included to illustrate a fluorimetric enzyme assay. Umbelliferone derivatives are used by a number of commercial companies in their immunoassay systems.

REAGENTS AND EQUIPMENT:

1.   Carbonate-Bicarbonate Buffer Containing Mg (pH 10.0).
        Dissolve 6.36 g of anhydrous Na_2CO_3, 3.36 g of anhydrous $NaHCO_3$ and 240 mg of anhydrous $MgSO_4$ in distilled water and make to 1 l.
        For reagents 2 and 3 and to be put out.
2.   50 nM 4-Methyl Umbelliferone.
        Stock (1 mM).   Dissolve 19.8 mg of the Na salt in 100 ml of methanol.
                           Store in a dark bottle at 4°C.
        Working Standard.   Add 5 µl of the stock to 100 ml of reagent 1 and mix well.
                           Make fresh and keep in ice.
3.   Substrate Solution.
        Dissolve 6.4 mg of 4-methyl umbelliferyl phosphate in 500 ml of reagent 1.
        Make fresh and keep on ice.
4.   Plasma.
        Keep on ice.

5 ml Graduate Pipettes.
50 µl Pipettors and Tips.
Glass or Quartz Fluorimeter Cuvettes.
Soft Paper Tissues.
Plastic Cuvette Stirrers.
Spectrofluorimeter capable of time scans. This exercise can be done with an old spectrofluorimeter connected to a recorder. However, fluorimeter output and recorder input will have to be adjusted to give optimum results.

PROCEDURE:

1.   Select "Time scan" and 200 seconds on the instrument and set for an excitation wavelength of 360 nm and an emission wavelength of 450 nm.

2.   Pipette 3.0 ml of the substrate solution (50 µM 4-methyl umbelliferyl phosphate in 0.1 M carbonate-bicarbonate buffer, pH 10.0, containing 1 mM $MgSO_4$) into a fluorimeter cuvette.

4.   Add 50 µl of plasma, mix with a plastic cuvette stirrer, place in the instrument and start the run as soon as possible. Start the chart on the recorder if using an old fluorimeter.

5.   After the run, re-scale the trace so that it starts in the bottom left hand corner of the screen and finishes in the top right. Obtain a print of this trace.

6.   Then obtain calibration values by measuring the fluorescence intensity of the buffer-Mg solution and 50 nM 4-methyl umbelliferone in the buffer-Mg solution. These can be marked on the chart, if an old fluorimeter is used with a recorder.

Obtain the initial, or maximal, velocity of the enzyme reaction from the trace.

Convert this into pmoles of 4-methyl umbelliferone produced per second and then calculate the enzyme activity of the plasma in nkat/l.

Reference Range:

About 15-50 nkat/l at room temperature.

EXERCISE 3-10

THE DETERMINATION OF PLASMA GLUCOSE BY AN ENZYMATIC METHOD

This, and the next 2 exercises, are examples of enzymes used as reagents in analytical methods.

In this exercise, a method for glucose determination will be used. The glucose is phosphorylated, using hexokinase, to glucose-6-phosphate. After phosphorylation, it is oxidised to 6-phosphogluconolactone using a bacterial glucose-6-phosphate dehydrogenase and NAD^+. The gluconolactone rapidly reacts with water to produce 6-phosphogluconic acid. The reaction sequence is outlined below:

CH$_2$OH — O — OH OH OH OH α-D-Glucose + ATP $\xrightarrow[\text{Mg}^{++}]{\text{Hexokinase}}$ CH$_2$O-PO$_3^=$ — O — OH OH OH OH α-D-Glucose-6-Phosphate + ADP

G-6-PD / Mg^{++} / NAD$^+$ → NADH + H$^+$

CH$_2$O-PO$_3^=$ — O — =O OH OH OH OH 6-Phosphogluconolactone + H$_2$O ⟶ CH$_2$O-PO$_3^=$ — OH OH — =O OH OH OH 6-Phosphogluconic Acid

As NADH absorbs light at 340 nm, the absorbance at this wavelength, after the reaction has gone to completion, can be used to determine the amount of glucose present in the original sample.

The hexokinase is extracted from baker's yeast. Even though the reaction catalysed by this enzyme is essentially irreversible, it is not specific for glucose. The enzyme also catalyses the phosphorylation of other sugars such as fructose and mannose.

The glucose-6-phosphate dehydrogenase is extracted from *Leuconostoc Mesenteroides*. This enzyme is specific for glucose-6-phosphate, as other phosphorylated sugars do not react. Hence, this enzyme gives the method its specificity. This enzyme, unlike the one isolated from yeast or present in human erythrocytes, can use either NAD$^+$ or NADP$^+$ as coenzyme. The enzyme reaction is about twice as fast if NAD$^+$ is used, compared with NADP$^+$. Hence, NAD$^+$ is used in this method.

Should either enzyme be contaminated with phosphoglucose isomerase, any fructose or fructose-6-phosphate present in the sample, can be converted through to glucose-6-phosphate and give elevated results. Significant amounts of fructose are not found in normal plasma. However, high levels can be found in essential fructosuria and in hereditary fructose intolerance.

Even though the reactions can be carried out on serum or plasma, more reliable results are obtained if the sample is deproteinised before the assay. Hence, a barium hydroxide-zinc sulphate deproteinisation step is included and the assay carried out on the supernatant after centrifugation.

Case History.

The sample is plasma from a 78 year-old woman. She has been admitted to hospital with a respiratory infection and left ventricular failure. She was found to have glycosuria and so a plasma glucose has been requested. Moderate hyperglycaemia is often seen in serious illness.

REAGENTS AND EQUIPMENT:

1. Zinc Sulphate Solution.
    Dissolve 22 g of $ZnSO_4:7H_2O$ in distilled water and make to 1 l.
2. Saturated Barium Hydroxide Solution.
    Dissolve 80 g of $Ba(OH)_2:8H_2O$ in about 950 ml of hot distilled water. Cool to room temperature and make to 1 l. Leave overnight at room temperature.
    For reagent 3.
3. Approx 55 mM Barium Hydroxide.
    Without disturbing the precipitate, transfer 245 ml of reagent 2 and make to 1 l with distilled water.
    Equal volumes of this reagent and reagent 1 should produce a neutral pH solution. If not, add either reagent 2 (or distilled water) to reagent 3, until it does.
4. 0.1 M Tris Buffer, pH 7.5, containing 4 mM Magnesium Acetate.
    (i)   Dissolve 15.8 g of Tris-HCl in distilled water and make to 1 l.
    (ii) Dissolve 6.1 g of Tris-base in distilled water and make to 500 ml.
    Mix 800 ml of (i) with 200 ml of (ii) and dissolve 1.1 g of magnesium acetate in this mixture. Check the pH and add (i) or (ii), if necessary.
    For reagent 5.
5. Enzyme Reagent.
    Add 400 mg of NAD, 330 mg of ATP, 4 μkat of hexokinase and 4 μkat of *Leuconostoc Mesenteroides* glucose-6-phosphate dehydrogenase in 500 ml of reagent 4.
    Prepare fresh or freeze and thaw prior to use.
    Add to dispenser set to 2.0 ml.
6. 20 mM Glucose Standard.
    Dissolve 1.80 g of AR glucose in distilled water and make to 500 ml.
    Prepare fresh, or use saturated (2 g/l) benzoic acid to prepare the reagent.
7. Patient's Plasma.
    Dissolve 70 mg of AR glucose in 50 ml of plasma or serum.

Centrifuge Tubes.
Test Tubes.
Vortex Mixers.
50, 100, 200, 500 μl and 2 ml Pipettors and Tips.
Spectrophotometers, Cuvettes and Paper Tissues.

PROCEDURE:

1. Set up the following, in centrifuge tubes, using the 20 mmol/l glucose standard:

	0	5	10	15	20	S	S
ml of H$_2$O	≈4	4.15	4.1	4.05	4.0	-	-
ml of ~ 55 mM Ba(OH)$_2$	-	-	-	-	-	2.0	2.0
μl of glucose standard (20 mmol/l)	-	50	100	150	200	-	-
μl of Patient's plasma	-	-	-	-	-	200	200

2. Mix on a vortex mixer for 5 seconds and then add 2.0 ml of 2.2% $ZnSO_4(7H_2O)$ to the S tubes only.

3. Mix the S tubes on a vortex mixer and then centrifuge for 5 minutes.

4. Set up a series of test tubes, labelled as in step 1, and add 2.5 ml of enzyme reagent to each tube. (The enzyme reagent contains NAD^+, ATP, hexokinase and glucose-6-phosphate dehydrogenase in a 0.1 M Tris-magnesium buffer, pH 7.5).

5. Add 500 µl of supernatant (or diluted standard) to its appropriate tube and mix on a vortex mixer for 5 seconds.

6. Stand the tubes at room temperature for at least 15 minutes.

7. Read the absorbance of each tube, at 340 nm, against tube 0.

Fasting Reference Range:

Serum or plasma glucose concentration : 3.6-5.8 mmol/l.

EXERCISE 3-11

THE DETERMINATION OF PLASMA CHOLESTEROL

This exercise is an example of the use of a commercial enzyme reagent mixture on untreated plasma or serum.

About 75% of cholesterol in plasma exists as esters between fatty acids and the hydroxyl group on position 3 of the cholesterol steroid ring. Cholesterol, with the other plasma lipids, exist in plasma as lipoprotein complexes.

The old methods for cholesterol determination involved the extraction of cholesterol, from its lipoprotein complex, into FeCl - glacial acetic acid. Concentrated sulphuric acid was then added to give a purple product. The use of enzyme reagents avoids the use of concentrated acids and also produces a more specific method.

In this method, cholesterol esters are hydrolysed by cholesterol esterase. The cholesterol is then oxidised to cholest-4-en-3-one by cholesterol oxidase. This last reaction also produces hydrogen peroxide. This is then reacts with phenol and 4-aminophenazone (4-aminoantipyrine) to give a red quinone-imine dye. This latter reaction is catalysed by peroxidase.

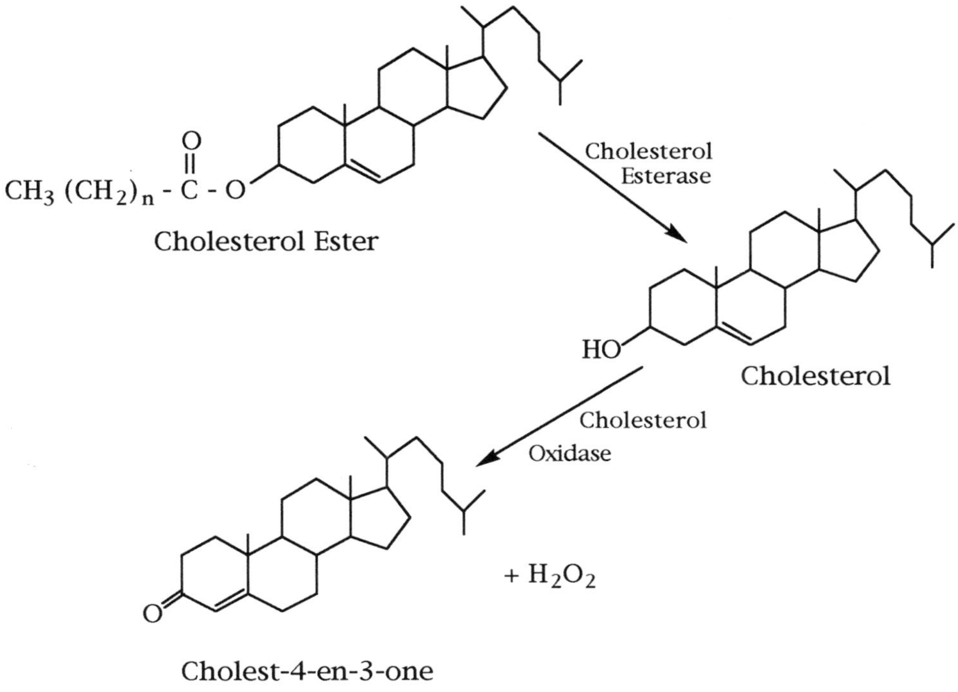

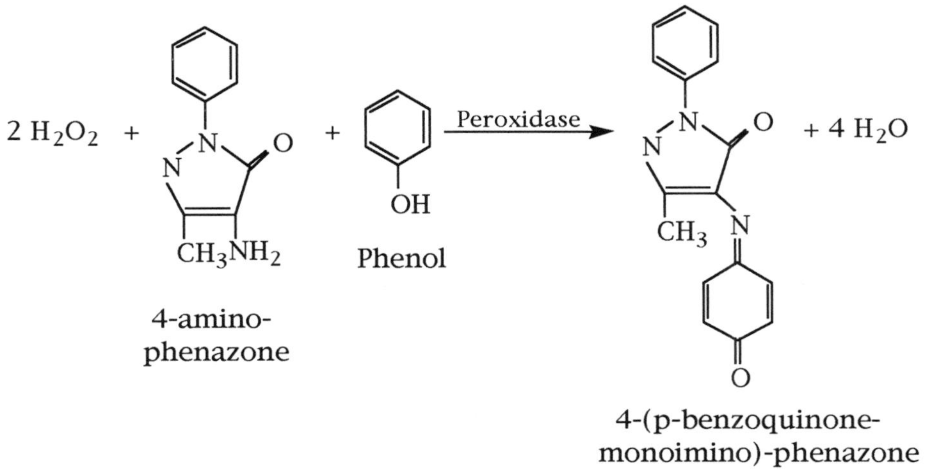

Cholesterol Ester

Cholesterol Esterase

Cholesterol

Cholesterol Oxidase

$+ H_2O_2$

Cholest-4-en-3-one

$2 H_2O_2 +$ 4-amino-phenazone $+$ Phenol $\xrightarrow{\text{Peroxidase}}$ 4-(p-benzoquinone-monoimino)-phenazone $+ 4 H_2O$

Plasma cholesterol levels are one of the risk indicators of coronary heart disease. The National Heart Foundation has recommended that total cholesterol levels should be below 5.5 mmol/l. Cholesterol levels above 6.5 mmol/l are associated with an exponential increase in the mortality risk of heart disease.

An indication of the mortality risk for middle-aged men is illustrated below:

Plasma Total Cholesterol (mmol/l)	Relative Risk
4.0	1.00
4.5	1.05
5.0	1.3
5.5	1,8
6.0	2.3
6.5	2.8
7.5	4.1

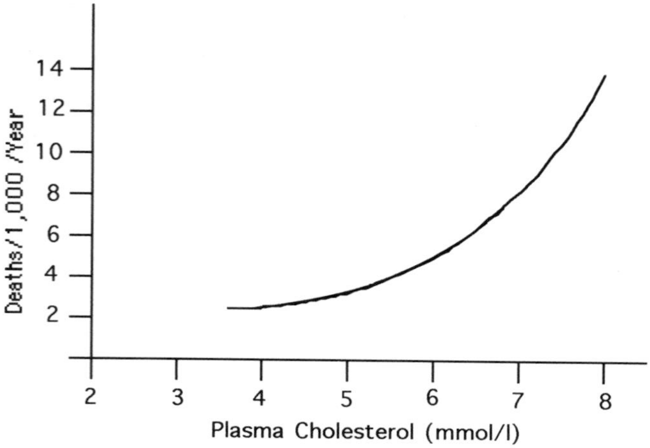

Case History.

The sample is plasma from a 55 year-old man who has given a blood sample as part of a regular medical check.

REAGENTS AND EQUIPMENT:

1.   Enzyme Reagent.
            Dissolve the contents of Boehringer-Mannheim Monotest Cholesterol bottle (Cat No 290319) in 32 ml of distilled water.
2.   Patient's Plasma.
            Heparinised or EDTA plasma or serum.

Spectrophotometers, Cuvettes, Plastic Cuvette Stirrers and Paper Tissues.
            Check that the light beam is fully covered by the 2 ml used in the assay. If not, use narrow microcuvettes.
$20\,\mu$l and 2 ml Pipettors.

PROCEDURE:

1.   Add 2.0 ml of the enzyme reagent to a cuvette and read its absorbance at 500 nm, against a distilled water blank (Abs 1).

2.   Add 20 µl of the patient's plasma and mix with a plastic cuvette stirrer.

3.   Stand at room temperature for 10 minutes.

4.   Read the absorbance of the sample against the water blank, at 500 nm, within the next 60 minutes (Abs 2).

CALCULATIONS:

1.   Subtract Abs 1 from Abs 2.

2.   Multiply the result by 14.9 to obtain the patient's cholesterol level in mmol/l.

Bilirubin above 70 μmol/l interferes to give low results.

Reference Range:

Total cholesterol in serum or plasma from fasting adults:

3.5 - 6.5 mmol/l.

The level increases with age and when the subject is on a high-fat diet. There is only about a 3% increase after a meal.

EXERCISE 3-12

THE DETERMINATION OF BLOOD GLUCOSE USING "GLUCOSTIX"

This exercise is an example of the use of carrier bound enzymes. Ames "Glucostix" will be used to determine your capillary blood glucose level.

"Glucostix" are plastic strips with 2 reagent areas attached to one end of the strip. The reagent areas contain glucose oxidase, peroxidase and chromogenic oxygen acceptors. For the lower concentration range the acceptor is o-tolidine and for the higher concentration area the acceptors are 4-aminoantipyrine (4-aminophenazone) and 3,5-dichloro-2-hydroxybenzene-sulphonate. The reagent areas are covered by a semipermeable membrane to prevent blood cells from entering. Both reagent blocks are yellow and so the blue product in the low range appears green and the red products in the high range appear orange.

Blood is applied to the reagent areas and allowed to stand for 30 seconds. The blood is then blotted off and the colour of reagent area is compared with the charts on the bottle 90 seconds later.

Quantitative results can be obtained when "Glucostix" are used in conjunction with a reflectance meter.

REAGENTS AND EQUIPMENT:

1.   70% Ethanol.

Ames (Bayer Diagnostics) "Glucostix".
Blood Lancets or "Glucolets".
Stop Clocks or Watches.
Paper Tissues.

PROCEDURE:

1.   Thoroughly wash your hands.

2.   Swab a finger with 70% ethanol and allow to dry.

3.   Jab the finger with a sterile blood lancet or a finger puncture device such as "Glucolet".

4.   Allow a large drop of blood to form.

5.   Hold a "Glucostix" horizontally and apply the drop of blood to cover both yellow reagent areas. Start the stopwatch.

6.   Place the strip on a sheet of paper tissue and after exactly 30 seconds, blot the strip with a paper tissue to remove the blood.

7.   After another 90 seconds, compare the colour of the reagent areas with the charts on the bottle. If the green area is darker than the 6 mmol/l block, read the value from the orange area. If the green area indicates a value of 6 mmol/l or less, ignore any orange change in the higher concentration reagent area. Interpolate the results if the colour falls between the two colour blocks on the chart.

Fasting Reference Range:

Fasting capillary blood glucose   :   3.6 - 5.3 mmol/l.

GENERAL REFERENCES.

Enzyme Nomenclature. Recommendations (1972) of the International Union of Pure and Applied Chemistry and the International Union of Biochemistry. Elsevier Scientific Publishing Company, (1973).

Enzyme Nomenclature. Recommendations (1978) of the Nomenclature Committee of the International Union of Biochemistry. Academic Press, (1979).

Engel, P.C., (1981), Enzyme Kinetics - The Steady-State Approach. 2nd Edition. Outline Studies in Biology. Chapman and Hall.

Cornish-Bowden, A. and Warton, C.W., (1988), Enzyme kinetics. In Focus. IRL Press.

Schmidt, E. and Schmidt, F.W., (1976), Brief Guide to Practical Enzyme Diagnosis. 2nd Revised Edition. Boehringer Mannheim GmbH.

Clinical Enzymology, (1986). A booklet published by BioMerieux.

Moss, D.W. and Henderson, A.R., (1994), Enzymes. Chapter 20 of Tietz Textbook of Clinical Chemistry, 2nd Edition, Eds. Burtis, C.A. and Ashwood, E.R., W.B. Saunders Co.

Determination of plasma prostatic acid phosphatase.

Ewen, L.M. and Spitzer, R.W., (1976), Clin. Chem. <u>22</u>, 627.

Determination of plasma lactate dehydrogenase.

Henry, R.J., Chiamori, N., Golub, O.J. and Berkman, S., (1960), Am. J. Clin. Path. **34**, 381.

Fluorimetric Determination of plasma alkaline phosphatase.

Cornish, C.J., Neale, F.C. and Posen, S., (1970), Am. J. Clin. Path. **53**, 68-76.

Determination of plasma glucose.

Schmidt, A.M., (1974), Federal Register (U.S.A.), Vol. 39, No. 126, p 24,146 - 24,147.

Determination of plasma cholesterol.

Siedel, J., Hagele, E.O., Ziegenhorn, J. and Wahlefeld, A.W., (1983), Reagent for the Enzymatic Determination of Serum Total Cholesterol with Improved Lipolytic Efficiency, Clin. Chem. **29**, 1075 - 1080.

Determination of blood glucose.

"Glucostix" information sheets, published by Bayer Diagnostics.

CHAPTER 4

ATOMIC EMISSION AND ABSORPTION

If heat energy is applied to a salt, the salt will dissociate into neutral atoms. A small proportion of these atoms will absorb some of the energy and, in doing so, move into higher energy states ($M \rightarrow M^*$). In falling back to the "ground state" some of the energy will be emitted as light ($M^* \rightarrow M + h\nu$, where h is Planck's constant and ν is the frequency of the emitted light). These changes are associated with an electron moving between different orbitals.

The quantum theory tells us that each atom has definite energy states in which its electrons can exist and so, in falling back from one energy level to the next, a fixed quantity of energy is released corresponding to light of a specific wavelength. Atoms, therefore, have specific line emission spectra. If the exciting energy is increased, the number of energy states may be increased with the net result of an increase in the number of emission lines. However, if enough energy is applied to the atom, an electron may be ejected and the atom will become ionised ($M \rightarrow M^+ + e^-$). The emission spectrum of the ion will be different to the original atom, but similar to the preceding element in the periodic table.

The events following the aspiration of a calcium chloride solution into a flame are illustrated below:

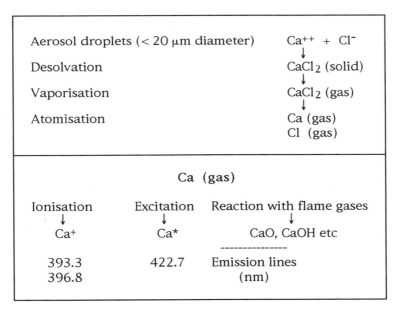

The intensity of the 422.7 nm emission line can be used to measure the amount of calcium in a solution.

The energy levels (and emission lines, in nm) associated with a sodium atom are illustrated below:

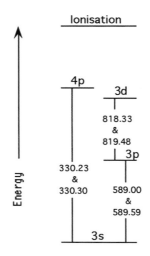

The double emission lines are due to the two different angular momentums of the electron, produced in the excited state of the atom.

As discussed above, a salt solution, aspirated into a flame, will cause the flame to emit light at wavelengths characteristic of the salt being aspirated. Only a very small proportion of the element under consideration reaches the excited atomic state, whereas a larger proportion of the element exists in the flame as ground state atoms. These ground state atoms can absorb light of exactly the same wavelengths as those emitted by the excited atoms that return directly to the ground state. Hence, ground state sodium atoms will absorb light of 330.3 and 589.0 nm but not 818.3 nm. The lines associated with transfer from, or to, the ground state are known as resonance lines. It is these resonance lines that are used in atomic absorption spectrophotometry.

FLAME EMISSION TECHNIQUES.

FLAME PHOTOMETRY.

In the flame photometer, the sample solution is nebulised and fed into a flame. In the flame the liquid vaporises and the salt dissociates into atoms, a portion of which reach a high energy level and, on falling back to the "ground state", emit light of a characteristic wavelength. The emitted light is isolated by a monochromator (usually a filter) and focused onto a photodetector, the output of which is then fed into a galvanometer or digital read-out. The layout of a basic instrument is illustrated below:

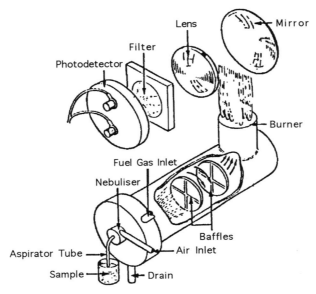

EEL Flame Photometer

The sample to be analysed is drawn up into the nebuliser by the venturi effect of a high-speed air jet, at a rate of about 5 ml/min. On mixing with air, the sample stream is broken up into a fine mist and directed at a series of baffles, which trap large droplets. These then drain away. The fine spray is mixed with the fuel gas and fed into a burner. The emitted light is collected by a mirror and lens, and then passed via a filter to a photodetector.

Flame photometry is subject to a number of potential sources of error. These are:

A.   Instrumental Sources.

Stable operation of a flame photometer requires a steady flame of constant thermal output. For this reason the air, or O_2, and fuel gas pressure must be kept constant and so must the fuel composition. The next requirement is a constant rate and physical state of the sample supplied to the flame. Droplet size, surface tension and viscosity of the samples must be constant and solutes must not be allowed to deposit in the nebuliser nozzle, as this decreases the bore size.

B.   Interference Present in the Sample.

(i)   Bandwidth Interference.

This occurs when the emission from an interfering element is read by the instrument in conjunction with the desired element. This is reduced by increasing the resolution of the spectral isolation unit, using a different wavelength, at which there is no interference, or by measuring the interfering metal separately and making a correction.

(ii)   Background Interference.

This occurs when a substance is present in the unknown sample that produces a continuous emission over extended portions of the spectrum. Burning plasma proteins, for example.

(iii) Radiation Interference.

This occurs when an element or ion causes the desired metal to emit more or less light. This can cause enhancement or depression of the emission and is known as positive or negative interference. Cations tend to produce negative interference on Na, K or Ca emission in low temperature flames, but positive interference in high temperature flames. Anion interference is almost always negative.

Partly to reduce these interferences, biological fluids are usually diluted before being aspirated into a flame photometer. The only elements, present in diluted biological fluids, that produce an intense light output with the standard air/propane flame are sodium and potassium. Hence, flame photometry is mainly used to assay these two elements. Calcium determination usually requires its precipitation as calcium oxalate. This is then redissolved in perchloric acid before aspiration into the flame.

Sodium has a more intense light output than an equimolar amount of potassium and the plasma level of sodium is about 35 times that of potassium. Hence, there is considerable sodium interference in the measurement of potassium in plasma or serum. Interference includes bandwidth interference, where the sodium light passes through the potassium filter, and radiation interference where the sodium decreases the potassium emission. Therefore, the potassium standards have to be made up in a sodium chloride solution of equivalent strength to normal plasma (140 mmol/l).

This problem is overcome when an internal standard is used. Lithium is the most common internal standard material. The instrument, therefore, has 3 detectors (Na, K and Li).

This system has two main advantages. Firstly, it negates the radiation interference of sodium in potassium assays. The samples are diluted with a high concentration of lithium. The lithium has a fixed interference on the potassium emission, which is thus independent of the sodium concentration; ie. the fixed lithium concentration causes a constant amount of radiation interference which, because of the high lithium concentration, overrides the sodium interference.

The second advantage is that it eliminates the instrumental sources of error. These include variations in the thermal output of the flame and variations in the sample aspiration rate.

The emission intensities of sodium and potassium are ratioed electronically against that from the lithium. This is done by balancing the output from the sodium and potassium detectors against the output from the lithium detector:

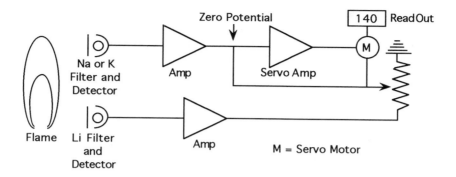

The instrument is set to read zero with just the lithium diluent being aspirated. When sodium (and/or potassium) enters the flame, current flows in the circuit. This causes the servo motor to move the contact on the variable resistor until no current flows (Zero potential at the point illustrated). The servo motor also moves the read out display. The instrument is calibrated with solutions of known sodium and potassium concentration.

As a result of using an internal standard, any change in flame temperature, or nebulisation rate etc., will effect the three emissions to a proportional extent and so the two ratios will remain constant. This provides much greater stability in the instrument with improved accuracy and precision.

The use of interference filters (Na : 589 nm, Li : 671 nm and K : 766 nm) greatly reduces bandwidth interference.

INDUCTIVELY COUPLED PLASMA (ICP) EMISSION.

A plasma is generated by using radio frequency energy to excite and ionise argon gas. This, together with the heat of recombination (Ar^+ + an electron) produce very high temperatures. 10,000 K can be generated at the core of the plasma. This is considerably higher than the standard air-propane flame of only 2,200 K.

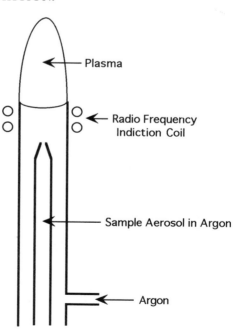

Sample compounds, aspirated into the plasma, dissociate into atoms that become excited and ionise. Emission lines can be seen from both excited atoms and ions. Sample emission is usually observed near the top of the plasma cone. At this point, ionisation of sample analyte atoms is suppressed, due to the large number of electrons from the ionised argon. Hence, most sample emission is from excited atoms.

Due to the relatively low temperatures of flame photometry, only the alkali and alkaline earth metals (Eg. Na, K and Ca) are excited. ICP provides enough energy to excite all metals.

ICP provides lower inter-element interference, simultaneous multi-element determination, greater compound dissociation and a greater concentration range than seen with flame emission.

Individual elements can be determined by using a high speed, high resolution scanning monochromator or by arranging detectors at the specific wavelengths, corresponding to the elements of interest, as the emitted light is reflected from a diffraction grating:

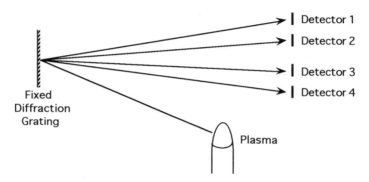

Inductively Coupled Plasma - Mass Spectrometry (ICP-MS).

The energy imparted by the plasma of an ICP is adequate to produce ions with a single positive charge. These are ideal samples to be fed into a mass spectrometer.

Elemental ions are collected from the plasma by a water-cooled metal cone (sample cone) with an entry diameter of 0.5 to 1.0 mm. These are then fed through a second cone (skimmer cone) and ion extraction lenses. The system is evacuated by vacuum pumps, illustrated below, to remove most of the argon. Finally, elemental ions are fed into a quadrupole mass spectrometer. Specific ion masses are then determined. These include all of the natural isotopes of the element of interest.

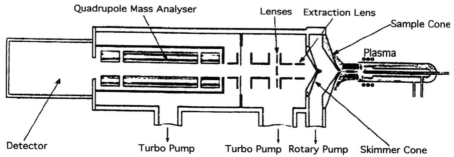

Detection limits are in the order of sub-parts per billion (< 1 µg/l) with a range of 5-6 orders of magnitude.

Rare isotopes can be used as internal standards. For example, Isotope dilution ICP-MS is the Centres for Disease Control reference method for blood lead. In this method, a known amount of a lead isotope (not present in the sample) is added to the standards and samples. The ratio of two isotopes (from the ICP-MS results on the standards) is then plotted against the standard concentrations and the sample values read from the plot using the sample isotope ratios. Alternatively, a different element can be used as an internal standard.

Advantages	Disadvantages
Low detection limits.	High instrument cost.
Wide linear range.	Matrix interference.
Simultaneous multi-element analysis.	Some element carry-over - Eg. Hg.
Simple sample preparation.	
High sample through-put.	
The use of internal standards.	

Mass spectrometry (as GC-MS) is discused in more detail in chapter 9 (pages 9-13 to 9-15).

ATOMIC ABSORPTION TECHNIQUES.

Absorption of light, by ground state atoms, is the basis of atomic absorption techniques.

High energy is not needed to generate ground state atoms. Hence, all metals generate ground state atoms in conventional flames. Therefore, atomic absorption techniques can be used for most elements; whereas, flame emission is limited to a few elements only.

Emission lines have bandwidths of about 0.001 nm. Atomic absorption lines are considerably broader. This is due to the collision of the absorbing atoms with other atoms or molecules and to the rapid motion of the absorbing atoms.

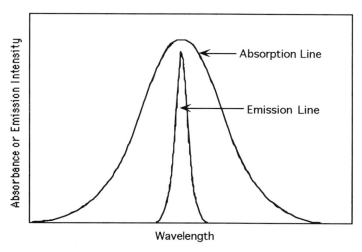

FLAME ATOMIC ABSORPTION.

As discussed above, a salt solution, aspirated into a flame, will cause the flame to emit light at wavelengths characteristic of the salt being aspirated. As discussed, this is due to the returning to the ground state of the excited atoms produced in the flame. Only a very small proportion of the element under consideration reaches the excited atomic state, whereas a larger proportion of the element exists in the flame as ground state atoms. It is absorption of light by these ground state atoms that is used in atomic absorption spectrophotometry.

When an excited atom returns to the ground state, a fixed quantum of energy is released and this is seen as the emission of a specific wavelength of light. It, therefore, follows that if the same quantum of energy is applied to the ground state atom it will be elevated to its excited state. This quantum of energy can, therefore, be applied to ground state atoms by exposing them to light of the same wavelength as that emitted when the excited atom returns to its ground state. A proportion of this light will, therefore, be absorbed by ground state atoms. The amount of light absorbed follows Beer - Lambert's Law.

The purpose of the flame is, therefore, to provide a supply of ground state atoms. Other means can be used to introduce ground state atoms into the light path.

The sample to be analysed is drawn up into the nebuliser by the venturi effect of a high speed air jet, at a rate of about 5 ml/min. On mixing with air, the sample stream is broken up into a fine mist and directed at a glass bead, or baffles, which traps large droplets. These run, via a drain tube, to a liquid trap and then drain away. The fine spray enters the spray chamber where it is mixed with the fuel gas. This mixture is then fed into a long narrow burner where it is mixed with more air and burnt. The shape of the burner is such, so that a long path length of sample is provided.

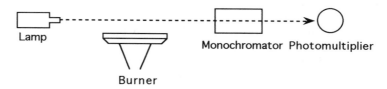

As only light of specific wavelengths will be absorbed by the sample, it is essential to use a light source that produces only these wavelengths. As a result, the emission lines of the metal under consideration are used. To achieve this, hollow cathode lamps are usually used.

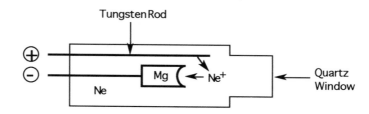

The cathode (negative electrode) consists of a hollow cup made of the element to be determined (Eg. Mg) and the anode (positive electrode) usually consists of a tungsten wire or disc. The two electrodes are sealed in a glass envelope, filled with an inert gas (Eg. Neon) at low pressure. The window of the lamp is made of quartz to allow UV light transmission.

When a large voltage is applied across the electrodes, inert gas molecules become charged at the anode and move at high velocity to the cathode. On striking the cathode, the inert gas molecules knock metal atoms off the surface of the cathode and into the space within the cathode cup. Collisions between some of these atoms and other high speed inert gas ions cause the metal atoms to be elevated to higher electronic energy levels (the excited state). These, in falling back to their ground state, emit light of the desired wavelengths.

The shape of the cathode is such that a fairly directional beam of light is produced. This beam is focused by a lens so that a parallel beam passes through the flame. This beam is then focused onto the entry slit of the monochromator. The size of the slit is adjustable. A wider slit lets through more light but decreases the resolution.

Light of the desired wavelength, leaving the monochromator, is then directed onto a photomultiplier. This produces an electrical current. Light is produced, not only by the hollow cathode lamp, but also by the flame itself and by emission of excited atoms in the flame.

To eliminate the latter two sources of error, the hollow cathode lamp is powered by a square wave AC current. The output of the photomultiplier is fed into an AC amplifier and so only light from the hollow cathode lamp is amplified. The output of the AC amplifier is demodulated and fed, as a DC current, to the galvanometer or digital display, which then reads in transmission. The DC output can also be fed to a logarithmic amplifier and then into the galvanometer or digital display. The instrument then reads in absorbance units.

Interferences:

As with flame emission, a number of factors can affect atomic absorption measurements. These can be subdivided into:

(a)   Matrix effects that influence the amount of sample reaching the flame,

(b)   Chemical interferences affecting the number of atoms formed in the flame and

(c)   Non-atomic absorption by undissociated molecules in the flame.

Matrix Effects.

a.   Precipitation of the element being analysed must be avoided.

b.   Viscosity, surface tension, density and solvent vapour pressure of standards and unknowns must be the same, unless internal standardisation is used.

Chemical Interferences

a.   Incomplete dissociation of compounds. (Most common).

This is due to the formation of compounds, such as calcium phosphate, which do not dissociate in the flame, hence reducing the number of ground state atoms in the flame. This can be overcome by the following methods:

i.   The fuel composition of the flame can be adjusted so that the element with, and without, the interfering material gives approximately the same absorption.

ii.   Higher flame temperatures will increase the amount of dissociation of the compound (but see b, below).

Approximate temperatures of commonly used flames:

Propane - Air	:	1,900°C
Propane - Oxygen	:	2,800°C
Acetylene - Air	:	2,400°C
Acetylene - Nitrous oxide	:	2,900°C

iii. The element can be extracted into an organic solvent using a chelating agent. Ketone and ester solvents are the best.

iv.   The interferent can sometimes be extracted into an organic solvent.

Both these extraction methods separate the element from the interferent.

v.   The last method that can be used is to add a releasing agent. Lanthanum or strontium nitrate can be added to a calcium solution containing phosphate. These combine with phosphate more strongly than does calcium, thus allowing more ground state calcium atoms to be formed in the flame.

b.   Ionisation.

High temperature flames can impart enough energy to the atoms to eject electrons, thus producing ions. These ions do not absorb light of the same wavelengths as the free atoms.

For example, calcium is 3% ionised in an air-acetylene flame and 43% ionised in a N_2O-acetylene flame. Magnesium is not ionised in an air-acetylene flame.

This interference can be overcome by adding a cation with a lower ionisation potential than the element being analysed (Eg. Cs (3.9 eV) or K (4.3 eV) can be added for Ca (6.1 eV) determination. These produce an excess of electrons in the flame, which suppress the ionisation of calcium).

Non-Atomic Absorption.

This is due to light absorption by undissociated molecules in the flame and light scattering by particles in the flame. Both these effects occur over broad wave bands and have more effect in the far UV than in the visual and near UV regions. These can often be overcome by using a higher wavelength resonance line or a higher temperature flame to increase the amount of molecular dissociation.

Correction for non-atomic absorption can be made by using the absorption of light of the same wavelength from a continuum lamp (Eg. a deuterium lamp). Light absorbed in this case is due mainly to non-atomic absorption with little direct atomic absorption, because the monochromator selects a broad band when compared with the narrow atomic absorption line:

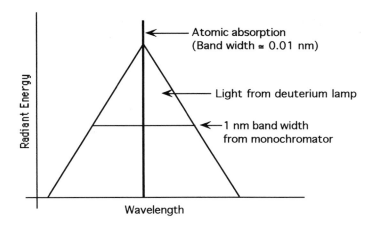

This deuterium lamp value (non-atomic absorption) is then subtracted from the hollow cathode lamp value (atomic + non-atomic absorption).

The optics are usually aligned so that beams of light from both light sources pass through the sample:

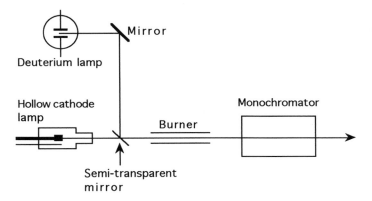

Zeeman Correction.

An alternative method for non-atomic absorption correction involves the Zeeman effect. When ground state atoms are exposed to a strong magnetic field, the absorption of light (polarised perpendicular to the magnetic flux [⊥]) is altered. 2 absorption lines are seen, one on each side of the original line. Light polarised parallel to the magnetic flux (//) is not affected. A magnetic strength of about 1.0 Tesla (10,000 Gauss) is required to separate the components:

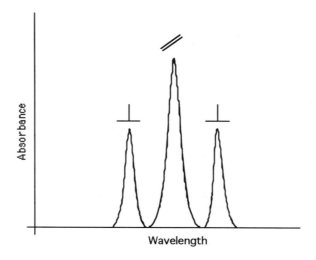

Hence, atoms in the light path will absorb the parallel polarised light but not the perpendicular polarised light. Non-atomic absorption is broad spectrum and so both types of polarised light are absorbed. Therefore, the perpendicular polarised light absorption (non-atomic) is subtracted from the parallel polarised light absorption (atomic + non-atomic) to give the corrected atomic absorption. With most instruments, the light is polarised after it leaves the monochromator:

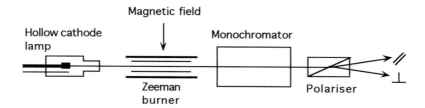

Zeeman correction has the advantage over the use of a deuterium lamp, as it can be used for all wavelengths. The emission from a deuterium lamp is only continuous up to about 360 nm. Hence, it can not be used for non-atomic absorption above this wavelength.

The optical layout, as shown above, also gives the instrument a double beam effect, as the ⊥ component is continuously subtracted from the // component. This corrects for variations in the lamp output.

OTHER METHODS OF GENERATING GROUND STATE ATOMS

ELECTROTHERMAL.

In this method, the sample is heated in a graphite tube or cup. The diagram below is of the Varian carbon rod. The sample (1-5 μl) is added to the small graphite tube, held in place by 2 graphite rods. These graphite rods also act as electrical conductors.

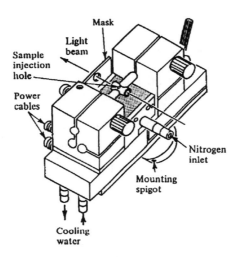

The heating involves 3 stages. Firstly, the solvent is evaporated. Secondly, organic matter is decomposed and volatilised. Finally the inorganic material is volatilised, generating ground state atoms. These three stages are obtained by increasing the current flow through the graphite tube, and hence its temperature.

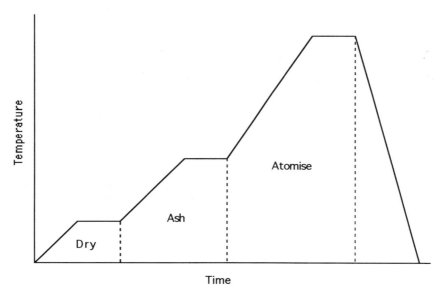

During the atomise phase, ground state atoms enter the light path as a short pulse and are then swept away by the nitrogen stream that bathes the tube. This produces a brief signal that can be recorded on a chart recorder or as an integrated digital reading.

The advantages over flame techniques are:

1. Only small volumes of sample are required.
2. Pre-digestion of complex samples is not necessary.
3. Increased sensitivity.

VAPOUR GENERATION.

This method is used for mercury determination.

The sample is cold digested to convert all mercury compounds to Hg^{++}. The digest is then treated with stannous chloride to produce metallic mercury:

$$Hg^{++} + Sn^{++} \rightarrow Hg^{0} + Sn^{++++}$$

Some of this mercury vaporises and is swept through a quartz tube, in the light path, by a stream of nitrogen. This produces a brief signal that can be recorded on a chart recorder or as an integrated digital reading.

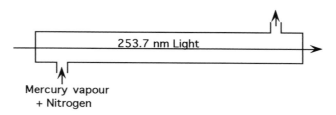

253.7 nm Light

Mercury vapour
+ Nitrogen

HYDRIDE GENERATION.

This method is used for the determination of arsenic and for low levels of selenium.

For arsenic determination, the sample is cold digested to yield As^{+++} and As^{+++++}. The As^{+++++} is reduced to As^{+++} with potassium iodide. Sodium borohydride is then added to form arsine gas:

$$As + NaBH_4 \rightarrow AsH_3$$

The arsine gas is then swept, by a stream of nitrogen, into an open-ended quartz tube that is heated by a flame. The arsine decomposes in the heated tube to yield arsenic atoms and hydrogen. The absorption of 193.7 nm light, by the arsenic atoms, produces a brief signal that can be recorded on a chart recorder or as an integrated digital reading.

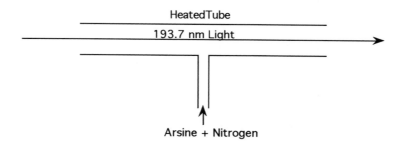

HeatedTube

193.7 nm Light

Arsine + Nitrogen

EXERCISES

Most of these exercises are written for specific instruments. Therefore, some modifications may have to be made for different instruments.

EXERCISE 4-1

THE DETERMINATION OF SODIUM AND POTASSIUM IN SERUM OR PLASMA BY FLAME PHOTOMETRY (1)

This exercise involves the use of a basic flame photometer with glass filters for the determination of both Na and K. As glass filters have fairly wide spectral bandwidths, bandwidth interference can be a problem.

However, the plasma level of potassium has very little effect on the sodium emission, and so straight sodium standards can be used for the determination of sodium. The plasma level of sodium is about 35 times that of potassium. Sodium also has a much higher intensity emission than potassium. Hence, there is considerable sodium interference in the measurement of potassium in plasma or serum. Therefore, the potassium standards have to be made up in a sodium chloride solution of equivalent strength to normal plasma (140 mmol/l).

Serum or plasma is diluted 1/500 for the determination of sodium and 1/50 for the determination of potassium.

These solutions are then read in the flame photometer using the appropriate filter (590 nm [yellow] for sodium and 765 nm [red] for potassium). The instrument is set to read zero with distilled water being aspirated into the flame and to read 100 with the top standard.

Case Study.

The sample is plasma from a 60 year-old woman with Addison's disease. She had been unwell for about 1 month. She was experiencing extreme muscle weakness and confusion. She was admitted to hospital after collapsing on the street.

Addison's disease is associated with hypofunction of the adrenal cortex. Low levels of the cortical hormone aldosterone lead to a loss of sodium and retention of potassium by the kidneys.

REAGENTS AND EQUIPMENT:

1. 40 mmol/l Na Standard.
        Dissolve 4.68g of dry Analar NaCl in distilled water and make to 2 litres.
2. 5 mmol/l K Standard.
        Dissolve 373 mg of dry Analar KCl in distilled water and make to 1 litre.
3. Patient's Plasma or Serum.
        120 ml of normal plasma or serum + 20 ml of 0.1% KCl.
        Keep on ice.
4. Distilled Water.

EEL Flame Photometer Mark II or similar basic flame photometer. Some modifications may be needed, if a different flame photometer is used.
100 ml Volumetric Flasks.
Test Tubes.

PROCEDURE:

1. Pipette 2.0 ml of the patient's sample into a 100 ml volumetric flask and make to the mark with distilled H_2O (1/50 dilution for the K determination). (Save a test tube full for the assay).

2. Dilute 10.0 ml of the above solution to 100 ml with distilled H_2O, using the volumetric flask. (1/500 dilution for the Na determination). (Save a test tube full for the assay).

3. You are provided with a 40 mmol/l Na solution. Set up the following standards using the 100 ml volumetric flask. (Save a test tube full of each standard for the assay).

	A	B	C	D	E	F
µl of 40 mmol/l Na	500	600	700	800	900	1 000
H$_2$O			To the 100 ml mark			
Equivalent plasma [Na] (mmol/l)	100	120	140	160	180	200

4. You are also provided with a 5 mmol/l K solution. Set up the following standards using the 100 ml volumetric flask. (Save a test tube full of each standard for the assay).

	G	H	I	J	K
ml of 5 mmol/l K	0.8	1.6	2.4	3.6	4.0
ml of 40 mmol/l Na	7.0	7.0	7.0	7.0	7.0
H$_2$O			To the 100 ml mark		
Equivalent plasma [K] (mmol/l)	2	4	6	8	10

5. Switch on the EEL flame photometer, open the chimney cover and check that the correct filter is in position.

6. Turn on and ignite the gas.

7. Turn on the compressed air and adjust the air pressure to its maximum ($\cong$ 18 p.s.i.).

8. With the flame as one big cone, turn down the gas supply until separate steady cones are just formed. Shut down the chimney cover.

9. Unclamp the galvanometer and aspirate the strongest standard. Adjust for full scale deflection with the "Sensitivity Control" (front of the instrument).

10. Aspirate distilled (or deionised) water and adjust to zero with the "zero control" (right hand side of the instrument).

11. Again, aspirate the strongest standard and readjust to 100.

12. Repeat steps 10 and 11, until the instrument is stable.

13. Aspirate the remaining standards and record the readings.

14. Check zero with distilled H_2O.

15. Aspirate the unknowns and record the readings.

16. Aspirate distilled H2O to remove any remaining sample from the nebuliser and burner.

17. When finished: clamp the galvanometer, turn off the air, then the gas and switch off the instrument.

Plot the percentage emission against the standard concentration, for both the sodium and potassium assays, and read off the values of the unknown.

Reference Ranges:

Sodium   :   135 - 150 mmol/l
Potassium :   3.6 - 5.0 mmol/l

EXERCISE 4-2

THE DETERMINATION OF SODIUM AND POTASSIUM IN SERUM OR PLASMA BY FLAME PHOTOMETRY (2)

In this exercise, sodium and potassium levels will be determined by a method using lithium as an internal standard. This involves a dedicated instrument, such as the Instrumentation Laboratory IL 143.

The instrument is set to zero with 15 mM lithium nitrate and calibrated with a 140 mM Na and 5.0 mM K solution, diluted 1 in 200 with the lithium nitrate. The sample is diluted 1 in 200 with the lithium nitrate and the Na and K values read from the scale.

Case Study:

The sample is plasma from a 60 year-old woman with Addison's disease. She had been unwell for about 1 month. She was experiencing extreme muscle weakness and confusion. She was admitted to hospital after collapsing on the street.

Addison's disease is associated with hypofunction of the adrenal cortex. Low levels of the cortical hormone aldosterone lead to a loss of sodium and retention of potassium by the kidneys.

REAGENTS AND EQUIPMENT:

1. 140 mmol/l Na, 5 mmol/l K Standard.
    Dissolve 373 mg of dry Analar KCl and 8.19 g of dry Analar NaCl in distilled water and make to 1 litre.
2. 15 mmol/l Lithium Nitrate.
    Dissolve 1.034 g of $LiNO_3$ in about 800 ml of distilled water. Add 2 ml of 30% Brij 35 and make to 1 litre with distilled water.
3. Patient' Plasma or Serum.
    120 ml of normal plasma or serum + 20 ml of 0.1% KCl.
    Keep on ice.

IL 143, or similar flame photometer that uses lithium internal standardisation, together with a propane cylinder. This exercise is written for the IL 143 and some modifications may be required for other instruments.
Sample Beakers.
100 ml Pipettors.
10 ml Graduated Pipettes.
20 ml Volumetric Flasks.

PROCEDURE:

1. Dilute the standard (140 mmol/l Na, 5 mmol/l K) and the patient's sample 1/200 with the 15 mmol/l lithium nitrate. (100 µl to 20 ml, using a pipettor and a 20 ml volumetric flask).

2. Open the propane valve, one quarter turn anti-clockwise, on the propane cylinder beside the instrument.

3. Turn the "On/Off" switch to "On". The flame should ignite within a few seconds and the "Flame On" indicator should come on. If either the propane or air pressure (or both) is low, the gas valves will shut off and the appropriate light will come on, as well as the "Flame On" light going out.

4. Aspirate 15 mmol/l lithium nitrate for a few minutes to fully wet the nebuliser.

5. Adjust the "Set Li" so that the needle is in the centre of the level indicator (ie. at the apex of the triangle).

6. Set the Na and K digital displays to 0000 with their respective "Zero" controls.

7. Set the "K Range" to 20.

8. Aspirate the diluted standard and set the "Na" digital display to 1400 with the Na "Balance" and set the "K" digital display to 050 with the K "Balance".

9. Recheck the zero with the lithium nitrate and then recheck the standard.

10. Aspirate the diluted patient's sample. Read and record its Na and K concentration.

11. Aspirate the 15 mmol/l $LiNO_3$ to wash out the nebuliser.

12. After the last assay, aspirate distilled water then close the propane valve. Wait until the "Flame On" light goes out.

13. Wait one further minute and switch the "On/Off" switch to "Off".

Reference Ranges:

Sodium   :   135 - 150 mmol/l
Potassium :   3.6 - 5.0 mmol/l

EXERCISE 4-3

THE DETERMINATION OF SERUM CALCIUM AND MAGNESIUM BY FLAME ATOMIC ABSORPTION

The only interferences associated with atomic absorption of magnesium in an air/acetylene flame, using diluted serum, are due to the presence of protein and to a small extent, the presence of phosphate. Phosphate, however, has a much greater interference in calcium determinations, due to the stability of calcium phosphate. Protein also interferes with calcium determinations. Both these interferences are overcome by the use of a 20 mmol/l strontium chloride releasing agent containing 0.6% perchloric acid (Pybus' reagent).

Serum is diluted 1/50 with Pybus' reagent and standards are made up in a similar fashion. The atomic absorption of both standards and unknowns are determined for both magnesium and calcium.

Case Study.

A 26 year-old man, known to suffer from Crohn's disease (ulceration of the terminal ileum) presented with severe diarrhoea and abdominal pain. He showed signs of increased neuromuscular excitability (abnormal reflexes). The physician has requested serum calcium and magnesium determinations.

REAGENTS AND EQUIPMENT:

1. 2.0 mmol/l Magnesium Standard.
   Dissolve 493 mg of dry Analar $MgSO_4:7H_2O$ (or 241 mg of $MgSO_4$) in distilled water and make to 1 litre.
2. 4.0 mmol/l Calcium Standard.
   Dissolve 400 mg of dry Analar $CaCO_3$ in about 500 ml of distilled water, plus 5 ml of M HCl, and then make to 1 litre with distilled water.

3.   Pybus' Reagent.

        Dissolve 3.8 g of $SrCl_2:6H_2O$ to about 500 ml of distilled water. Add 9 ml of 70% (SG 1.67)
        $HClO_4$ and make to 1 litre with distilled water.

        Add to dispenser set to 4.9 ml.

4.   2.5 mmol/l Calcium Gluconate.

        Dissolve 110 mg of calcium gluconate in 100 ml of distilled water.

        For reagent 5.

5.   Patient's Serum.

        Dilute 20 ml of normal serum, or heparinised plasma, with 20 ml of reagent 4.

        Keep on ice.

6.   Control Serum.

        Reconstitute a control serum with known Ca and Mg values.

        Keep on ice.

Atomic Absorption Spectrophotometer with burner plus air and acetylene cylinders. This exercise
is written for a Varian Model 1100 or 1200. Some modifications may be needed for other
instruments.

100 μl Pipettors and Tips.

50 ml Measuring Cylinders.

100 ml Beakers.

PROCEDURE:

1.   Locate all the controls on the atomic absorption instrument and make sure that you
        understand the function of each.

2.   Dilute duplicate samples of 100 μl of the unknown and the control serums to 5.0 ml
        with 4.9 ml of Pybus' reagent.

3.   Make up the following standards using the standard magnesium solution (2.0 mmol/l):

	A	B	C	D	E
ml of Mg standard (2.0 mmol/l)	0.2	0.4	0.6	0.8	1.0
ml of H_2O	0.8	0.6	0.4	0.2	0

4.   Make up the following standards using the standard calcium solution (4.0 mmol/l):

	F	G	H	I	J
ml of Ca standard (4.0 mmol/l)	0.2	0.4	0.6	0.8	1.0
ml of H_2O	0.8	0.6	0.4	0.2	0

5.   Dilute 100 ml of each of the above standards to 5.0 ml with 4.9 ml of Pybus' reagent.

6.   Dilute 1.0 ml of the magnesium standard (2.0 mmol/l) to 50 ml with distilled water to
        give the 40 μmol/l Mg setting solution.

4-20

7.   Dilute 2.5 ml of the calcium standard (4.0 mmol/l) to 50 ml with distilled water to give the 200 μmol/l Ca setting solution.

8.   Set up the atomic absorption spectrophotometer as described below:

i.   Make sure all gas valves are closed, that both lamp controls are off and that the spray chamber liquid trap is full.

ii.   Make sure that the lamps to be used are in the appropriate quadrants and the lamp required is in the operating position.

iii. Press the "Power" button.

iv.   Switch on the lamps and set the recommended current (Ca: 3 mA, Mg: 3 mA, combined Ca/Mg: 5 mA).

v.   Set the recommended slit width (Ca: 0.5 nm, Mg: 0.5 nm).

vi.   Set the monochromator to the required wavelength (Ca: 422.7 nm, Mg: 285.2 nm).

vii. Set the burner vertical control to read 4. (This ensures that the burner is clear of the optical path).

viii. Set the wavelength exactly, as follows: Press "Trans" and "Damp A". Adjust "λ Scan" to produce maximum signal. If the meter goes off scale, bring back onto the scale with the "Gain" control.

ix.   Adjust the position of the lamp, with the adjusting screw on the quadrant, to give maximum signal.

x.   Turn the "Support Gas" selector to "Off".

xi.   Set the gas line pressures:

        Acetylene   :   70 kPa   (10 p.s.i.)
        Air   :   400 kPa   (60 p.s.i.)

xii. Turn the "Support Gas" selector to "Air".

xiii. Adjust the "Support" control to read "6" on the flow meter.

xiv. Adjust the "Fuel" control to read between "3" and "4" on the flow meter.

xv.   Press the "Ignite" button and adjust the "Fuel" control to produce the desired flame. (An oxidising flame, fuel-lean, is best for these two assays).

xvi. Press "Abs".

xvii. Aspirate distilled H_2O and press "Zero" until the light goes out. The meter or digital display should read 0.000 ± 0.001. Zero absorbance adjustment can be made with the "Gain" control.

xviii. Aspirate the 200 µmol/l Ca setting solution for the Ca assay or 40 µmol/l Mg setting solution for the Mg assay.

xix. Make sure the burner is parallel to the optical path. Adjust "Horiz" for maximum absorbance and then "Vert" for maximum absorbance. (On this last adjustment, make sure the burner does not enter the optical path, ie. do not have it too high).

xx. Aspirate distilled H_2O and press "Zero".

xxi. Aspirate the solution used in Step xviii. Adjust the "Fuel" flow to give maximum absorbance without interference.

xxii. The position of the glass bead in the nebuliser can be adjusted with a screwdriver, if necessary.

xxiii. Aspirate distilled H_2O.

9. Read the atomic absorption of Pybus' reagent (zero standard), the magnesium standards, controls and unknowns, as follows:

i. Aspirate the blank solution (Pybus' reagent).

ii. Press "Zero" and hold until the light goes out, or zero the instrument with the "Gain" control.

iii. Aspirate the standards and unknowns and record each reading on the meter or digital display.

iv. Aspirate distilled H_2O to clean out the system.

10. Set the instrument for "Expanded Scale Absorbance Measurements" and use the "Int 3" mode for the calcium determinations. Read the atomic absorption of Pybus' reagent, the calcium standards, controls and unknowns, as follows:

i. Press "Conc" and "Damp A".

ii. Turn both "Low" and "High" controls fully anti-clockwise.

iii. Aspirate the blank solution (Pybus' reagent).

iv. Press "Zero" and hold until the light goes out, or use the "Gain" control to zero the instrument.

v. Aspirate the highest standard.

vi. Adjust the "Low" control so that the meter to read 0.9 or about 9000 on the digital read-out.

vii.   Recheck the "Zero" and "Low" settings.

viii.   Select "Int 3".

ix.   Aspirate all the standards (including the blank - Pybus' reagent) and the unknowns, in turn.

x.   Press "Read", once the sample is aspirating. The reading is made while the "Read" button is illuminated.

xi.   Remove the sample after the "Read" light goes out, and record the value.

Plot standard curves for both calcium and magnesium and read off the unknowns and controls.

11.   After the last sample, flush the burner, by aspirating distilled water, for 5 minutes. Then turn of the instrument as follows:

i.   Turn off the acetylene at the cylinder.

ii.   Wait until the flame goes out, then turn off the air at the cylinder.

iii.   Turn off the power at the instrument at the mains.

Reference Ranges:

Magnesium   :   0.8 - 1.0 mmol/l.
Calcium   :   2.1 - 2.6 mmol/l.

Notes:

1.   Erythrocytes contain about three times the concentration of magnesium, when compared with plasma, but only about 1/50 the concentration of calcium, and so haemolysed blood or blood stored before the separation of the serum should not be used.

2.   Do not use blood treated with oxalate or EDTA for the determination of plasma calcium and/or magnesium as these chemicals strongly chelate both calcium and magnesium.

EXERCISE 4-4

THE DETERMINATION OF BLOOD LEAD BY FLAME AND ELECTROTHERMAL ATOMIC ABSORPTION

Lead in whole blood is predominantly in the erythrocyte, with less than 5% in the plasma. Hence, whole blood is used for the determination of lead in cases of lead exposure.

In this method, lead is chelated with ammonium pyrrolidone dithiocarbamate (APDC) and extracted into n-butyl acetate. After centrifugation, the organic layer is aspirated into the flame of an atomic absorption spectrophotometer and samples added to the graphite tube of a second instrument for electrothermal determination of lead.

As lead exposure produces a microcytic anaemia, results expressed per litre of blood can be misleading. Hence, results are expressed per standard haematocrit for men, women or children.

Case Study:

Blood was collected from a lead worker who was feeling tired and irritable. He complained of constipation and a metallic taste in his mouth. There was a blue line on his gums and when he raised his arm, the hand hung limply down.

REAGENTS AND EQUIPMENT:

1. 2% APDC.
    Dissolve 2.0g of ammonium pyrrolidine dithiocarbamate in 100 ml of distilled water, with warming. Filter, if necessary.
2. n-Butyl Acetate.
    Shake 1 l with 100 ml of distilled water and discard the water layer.
3. 1.0 mmol/l Lead Stock Standard.
    Dissolve 331 mg of dry (2 hours at 80°C) anhydrous lead nitrate ($Pb(NO_3)_2$) in about 500 ml of distilled water, plus 10 ml of concentrated nitric acid. Make to 1 litre with distilled water.
    For reagent 4.
4. 10 μmol/l Lead Standard.
    Dilute 1.0 ml of reagent 3 to 100 ml with distilled water and mix.
    Prepare fresh.
5. Lead Worker's Blood.
    Dissolve 1.0 g of lead nitrate in distilled water and make to 1 litre.
    Add 200 μl of this solution to 100 ml of normal blood and mix well.
    Keep on ice.

Screw-capped Polypropylene Centrifuge Tubes.
Bench Centrifuges.
Wassermann Tubes.
5 ml Graduated Pipettes.
2 μl Capillary Pipettors.
Atomic Absorption Spectrophotometers, one fitted with a burner and the other with a graphite furnace. This session is written for the Varian Model 1100 or 1200 and the Model CRA-63 carbon-rod with power pack. A chart recorder is fitted to the instrument with the carbon-rod. Some modifications will probably be necessary if other instruments are used. The top standard should be run with the electrothermal method to adjust the instrument scale expansion (Low) and/or the recorder input voltage to give good peak heights.
Haematocrit Centrifuge and Tubes.

PROCEDURE:

1. Set up the following in screw-capped polypropylene centrifuge tubes:

	A	B	C	D	E	F
ml of patient's blood.	4.0	-	-	-	-	-
ml of 10 μmol/l Pb.	-	0	1.0	2.0	3.0	4.0
ml of H$_2$O.	-	4.0	3.0	2.0	1.0	0
Equivalent blood lead concentration in μmol/l.	-	0	2.5	5.0	7.5	10.0

2. Add 0.5 ml of 2% APDC to each tube, cap and mix by inversion 15 times.

3. Allow the tubes to stand for 5 minutes.

4. Add 5.0 ml of n-butyl acetate to each tube, cap and shake for 3 minutes.

5. Centrifuge the tubes for 5 minutes.

6. Transfer the organic layers to labelled Wassermann tubes.

7. Check the following on the instrument (Varian Model 1100 or 1200).

      Flame   :   Air/acetylene - very lean.
      Lamp   :   Pb.
      Lamp current   :   8 mA.
      Wavelength   :   283.3 nm - check the peak.
      Mode   :   Conc.
      Low   :   ≈ 9
      Slit width   :   0.5 nm.

8. Zero the instrument on "Damp A" while aspirating n-butyl acetate.

9. Select the "Int 10" function.

10. Aspirate the samples, in ascending order, and press the "Read" button.

11. Remove the sample as soon as the "Read" light goes out and record the digital reading.

12. Check the following on the Varian Model 1100 or 1200, fitted with the Model CRA-63 carbon-rod workhead and power pack, and connected to a chart recorder:

    AA:
      Wavelength   :   283.3 nm - check the peak.
      Slit width   :   0.5 nm.
      Lamp   :   Pb.
      Lamp current   :   7 mA.

| Mode | : | Conc. |
| Low | : | ≈ 5 |

Nitrogen flow:
   About 4 l/min (6.5 units on the flow meter).

Cooling water flow:
   About 500 ml/min.

Power Pack:

Dry	:	15 seconds at 2.5 V (100°C).
Ash	:	6 seconds at 7.25 V (800°C).
Step Atomise	:	1.5 seconds at 9.2 V (2,800°C).
Step/Ramp	:	Step.

Chart recorder:

| Speed | : | 5 mm/minute. |
| Input | : | 10mV. |

13.   Lift the chart pen off the paper and apply 2 μl of the n-butyl acetate extracts to the carbon rod via its central hole.

14.   Place the chart pen on the paper and press "Start" on the carbon-rod power pack.

15.   Carry out duplicate assays on each n-butyl acetate extract.

16.   Measure the peak heights on the chart paper.

17.   Determine the haematocrit of the sample.

Plot graphs of expanded scale absorbance values (Flame) and peak heights (Electrothermal) against the equivalent blood lead concentration. Read off the lead concentration of the blood sample. Correct this blood level to a standard haematocrit, assuming the sample is from a man. (ie. divide by the % haematocrit and multiply by 45).

Reference Values:

Below 1.2 μmol/l	:	Normal.
1.2 - 3.6 μmol/l	:	Acceptable.
Above 3.6 μmol/l	:	Lead exposure - remove to a lead free area.
Above 5.2 μmol/l	:	Remove from work until the lead is below 3.6 μmol/l.

These figures are lead levels corrected to a standard haematocrit. ie. the determined valve is divided by the subject's % haematocrit and multiplied by 45 for men, 42 for women and 38 for children.

Notes:

1.   Lead also has an absorbance line at 217.0 nm. This is more sensitive than the 283.3 nm line, but is more prone to give a "noisy" background.

2.   When using electrothermal atomisation, an alternative approach is to apply blood directly to the carbon-rod. However, residues tend to build up on the rod and background correction may become necessary.

3.   The method of standard additions is often used for lead determination. In this method, known amounts of lead are added to the sample. A standard curve of response (Absorbance or Peak height) against added lead is plotted and the amount of lead in the sample is determined by extrapolation back to the zero response (Absorbance or Peak height). Eg:

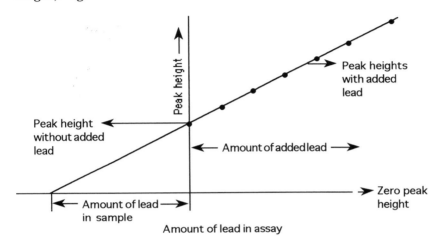

EXERCISE 4-5

THE DETERMINATION OF LIVER COPPER BY ELECTROTHERMAL ATOMIC ABSORPTION

Liver copper measurements aid in the diagnosis of Wilson's disease (elevated) and in Menkes' disease (low). Liver biopsy samples are usually less than 5 mg wet weight. Hence, electrothermal methods of atomic absorption are essential.

Wilson's disease is an autosomal recessive disease affecting about 1 in 100,000 live births. It is due to a defect in the removal of copper from the body, by the liver. Copper is deposited in the liver, brain and other tissues, leading to severe neurological symptoms and liver failure. Symptoms usually appear between the ages of 6 and 40.

Menkes' disease is X-linked and affects about 1 in 100,000 live births. The absorption of copper, via the intestine, is defective, leading to low tissue copper levels. The disease is associated with kinky, twisted hair, mental retardation, abnormal bone formation, temperature instability and susceptibility to infection. Symptoms usually appear by 3 months.

In this exercise, a graphite furnace will be used with an atomic absorption spectrophotometer.

A sample of liver will be digested with concentrated nitric acid. 2 µl of the digest will be applied to the graphite furnace. The cycle of dry, ash and atomise will then be initiated. Standards will be treated in a similar manner.

A portion of the liver will be weighed and dried in a vacuum oven, so that the wet/dry weight can be calculated.

REAGENTS AND EQUIPMENT:

1.   Analar Concentrated Nitric Acid.
2.   Copper Stock Standard.
        Accurately weigh about 1.0 g of copper. Dissolve in about 10 ml of AR concentrated nitric acid and make to 100 ml with distilled water. Calculate the exact concentration in mmol/l (6.355 g/100 ml = 1 mol/l).
        For reagent 3.
3.   10 µmol/l Copper Standard.
        Accurately dilute reagent 2, with distilled water, to give a 10 µmol/l solution.
        Dilute fresh and store in a plastic bottle. Copper tends to bind to glass storage vessels.
4.   Liver.
        Ox or sheep.
        Keep cool.

Stainless Steel Dissecting Instruments.
Filter Paper.
Analytical Balance.
60°C Heating Block.
10 ml Beakers.
Vacuum Oven.
500 µl Pipettors.
2 µl Capillary Pipettors.
1 and 2 ml Graduated Pipettes.
Acid-washed Glass Wassermann Tubes.
Plastic Wassermann Tubes.
Parafilm.
Atomic Absorption Spectrophotometer, fitted with graphite furnace and recorder. This exercise is written for a Varian Model 1100 or 1200, fitted with a CRA-63 carbon rod. The scale expansion (Low) and/or the recorder input voltage may need to be altered to give good peak heights. Modifications will probably have to be made if a different instrument is used.

PROCEDURE:

1.   Blot a piece of fresh liver with filter paper to remove excess blood.

2.   Using stainless steel instruments, remove about 5 mg of liver.

3.   Accurately weigh the liver (to the nearest 0.1 mg) by using a pre-weighed acid-washed glass Wassermann tube or by using the torsion balance and then adding the liver to an acid-washed glass Wassermann tube.

4.   Add 500 µl of conc HNO_3 and cap the tube with parafilm.

5.   Incubate the tube at 60°C in the heating block and agitate occasionally until the liver has dissolved.

6.   Using plastic Wassermann tubes, dilute the 10 µmol/l Cu standard with deionised water to give 4.0 ml of 2.5, 5.0 and 7.5 µmol/l standards.

7. Check the following:

AA:

Lamp	:	Cu.
Lamp current	:	6 mA.
Wavelength	:	324.7 nm (Peak the monochromator).
Slit width	:	0.5 nm.
Mode	:	Conc.
Low	:	~ 5.

Nitrogen flow:
    4 1/min (6.5 units on the flow meter).

Cooling water flow rate:
    About 500 ml/min.

Power pack:

Dry	:	5 V (120°C) for 20 sec.
Ash	:	7 V (750°C) (for 15 sec.
Atomise	:	6 V (1,800°C) for 5 sec.
Step/Ramp	:	Step.

Chart recorder:

Speed	:	10 mm/min.
Input	:	10 mV.

8. Apply 2 μl of the following (in duplicate):
    Deionised water
    2.5 μmol/l Cu
    5.0 μmol/l Cu
    7.5 μmol/l Cu
    10.0 μmol/l Cu
    Digest
    Conc HNO_3 to the rod and press the start button.

Lift the chart pen when you add the sample to the rod and lower the pen at the end of the "Dry" stage.

9. Add about 1/2 gram of liver to a weighed 10 ml beaker and re-weigh.

10. Dry the beaker, overnight, in the vacuum oven and re-weigh.

Calculate the wet/dry ratio.

Measure the peak heights and plot a standard curve. Read off the unknown and calculate the copper content in mmol/kg dry weight.

Human Reference Ranges:

Neonates (0-7 weeks)	:	2.8 - 3.9 mmol/kg dry weight
Adults	:	0.1 - 0.6 mmol/kg dry weight

EXERCISE 4-6

THE DETERMINATION OF URINARY MERCURY BY VAPOUR GENERATION ATOMIC ABSORPTION

The organic material present in the sample is digested at room temperature, by the oxidising action of potassium permanganate and sulphuric acid. This action converts all mercury compounds to mercuric ions.

The diluted digest is then transferred to a reaction tube and stannous chloride is added to reduce the mercuric ions to free mercury.

$$Hg^{++} + Sn^{++} \longrightarrow Hg + Sn^{++++}$$

After a period of stirring, to allow the mercury to vaporise into the space above the liquid in the reaction tube, the vapour is then swept through an absorption cell in the light path of an atomic absorption spectrophotometer. This is achieved by applying a stream of nitrogen, as illustrated below:

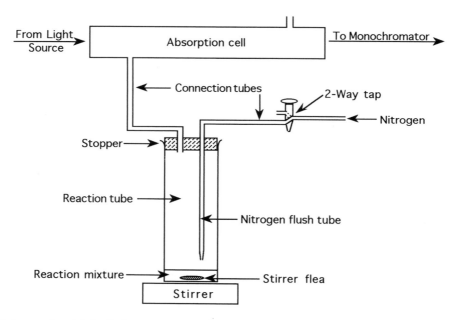

The output from the AA is fed to a computing integrator and the area of the peak generated by the burst of mercury vapour is calculated.

A series of standards are treated in the same manner, with the exception of the digestion, and a standard curve plotted.

Case Study:

The urine sample is part of a 24 hour collection of 1.24 l from a 45 year old woman who had eaten bread made from cereal grain seed that had been treated with alkyl-mercury fungicides. She had shown signs of emotional instability, loss of hearing, visual impairment and tremor of the hands.

REAGENTS AND EQUIPMENT:

1.   AR Concentrated Sulphuric Acid.
2.   6% Potassium Permanganate.
         Dissolve 6 g of $KMnO_4$ in 100 ml of distilled water.
3.   1.0 mmol/l Stock Mercury Standard.
         Dissolve 272 mg of dry Analar $HgCl_2$ in about 500 ml of distilled water plus 20 ml of AR concentrated nitric acid. Make to 1 litre with distilled water.
         For reagents 4 and 7.
4.   6.0 µmol/l Mercury Standard in 50% H_2SO_4.
         Add 1.2 ml of reagent 3 to about 40 ml of distilled water and mix. Slowly add 100 ml of AR concentrated sulphuric acid, with cooling. Make up to 200 ml with distilled water.
         Dilute fresh and store in a plastic bottle.
5.   50% Sulphuric Acid.
6.   10% Stannous Chloride.
         Dissolve 20 g of $SnCl_2:2H_2O$ in 40 ml of concentrated HCl. Add 1 granule of AR tin and dilute to 200 ml.
         Make up fresh.
7.   Patient's Urine.
         Add 40 ml of reagent 3 to 200 ml of normal urine.
         Keep on ice.

Ice Buckets.
250 ml conical flasks.
5 and 10 ml Graduated Pipettes.
Parafilm.
50 ml Volumetric Flasks.
Test Tubes.
Atomic Absorption Spectrophotometer with integrator attached to the recorder output. The procedure is written for a Varian Model 1100 or 1200 with a Hewlett Packard 3390A integrator. Modifications will probably have to be made with other equipment.
Varian Mercury Analyser Kit Model 64 or home made apparatus, as in the diagram above; with nitrogen cylinder, flow meter and 2-way tap.

PROCEDURE:

1.   Pipette 5.0 ml of the urine into a 250 ml conical flask, in a beaker of ice.

2.   Add 5.0 ml of AR concentrated H_2SO_4, slowly with swirling, so that the reaction mixture remains cool.

3.   Slowly add 10.0 ml of 6% $KMnO_4$ with swirling.

4.   Cap the flask with parafilm and leave in the fume cupboard over night or until the next class. (Make sure that your name is on the flask).

****

5. After the digestion is complete ($\cong$12 hours), dilute the sample to 50 ml with distilled water, in a volumetric flask.

6. Set up the following standards using the mercury standard (6 µmol/l in 50% H_2SO_4):

	A	B	C	D	E
ml of Hg standard (6 µmol/l).	-	1.0	2.0	2.5	3.0
ml of 50% H_2SO_4	3.0	2.0	1.0	0.5	-

7. Check the following parameters on the AA:

Lamp	:	Hg.
Lamp current	:	3 mA.
Wavelength	:	253.7 nm (Check the peak).
Band width	:	0.5 nm.
Mode	:	Conc.
Low	:	$\cong$ 1.0

Nitrogen cylinder pressure:		100 kPa.
Nitrogen flow rate	:	2 l/min (3.5 on ball flow meter).

8. Check the following parameters on the computing integrator:

Chart speed	:	10 mm/min (1 cm/min).
Attenuation	:	4.
Threshold	:	4.
Peak Width	:	0.16
Zero	:	10.

9. Pipette 1.0 ml of each standard, in turn, into the reaction tube.

10. Add 7.0 ml of deionised water.

11. Make sure that the nitrogen is not flowing via the nitrogen flush tube.

12. Add 2.0 ml of 10% $SnCl_2$ (in 20% HCl) to the reaction tube and rapidly insert the stopper assembly.

13. Switch on the stirrer (fairly rapid stirring) and start the stop clock.

14. Zero the instrument and start the integrator.

15. Switch off the stirrer at exactly 90 seconds.

16. When the stirrer flea has stopped, switch the nitrogen flow to the reaction tube.

17. When the peak trace has returned to the baseline, stop the integrator. Remove and clean the reaction tube and stirrer flea. Turn the nitrogen flow to the atmosphere.

18. After the last standard, add 5.0 ml of the diluted digest and 3.0 ml of deionised water to the reaction tube and repeat steps 11 to 17.

19. Repeat the assay on 2.0 ml of diluted digest and 6.0 ml of deionised water.

Plot a standard curve of peak area against the amount of mercury in the assay, read off the unknown values and calculate the mercury excretion in µmol/24 hours.

Reference Values:

The majority of people, not exposed to mercury, have urine values below 150 nmol/24 hours, with very few above 250 nmol/24 hours.

Industrial workers with levels between 250 and 750 nmol/24 hours have been reported to be symptom-free.

Values up to 13 µmol/24 hours have been reported in workers with clinical symptoms of mercury poisoning.

GENERAL REFERENCES:

Evenson, M. A., (1994), Flame Photometry and Atomic Absorption Spectrophotometry. In Tietz Textbook of Clinical Chemistry, 2nd. Ed., Eds: Burtis, C. A. and Ashwood, E. R., 125 - 131, W. B. Saunders Co.

Willard, H. H., Merritt, L. L., Dean, J. A. and Settle, F. A., (1981), Flame Emission and Atomic Absorption Spectrometry, and Atomic Emission Spectroscopy. In Instrumental Methods of Analysis, 6th Ed., 127 - 176, Wadsworth Publishing Co.

Dawson, J. B., (1978), Analytical Atomic Spectroscopy. In Scientific Foundations of Clinical Biochemistry, Volume 1, Eds; Williams, D. L., Nunn, R. F. and Marks, V., 95 - 120, William Heinemann Medical Books Ltd.

Duncan, L., (1976), Clinical Analysis by Atomic Absorption Spectroscopy, Varian Techtron.

Anon., (1979), Analytical Methods for Flame Spectroscopy, Varian Techtron.

Jarvis, K. E., Gray, A. L. and Houak, R. S., (1992), Handbook of Inductively Coupled Plasma Spectrometry, Blackie & Sons Ltd.

Determination of serum sodium.

Varley, M., (1967), Practical Clinical Biochemistry, 4th edition, 490-495.

Operator's manual for the IL 143 flame photometer, (1974), Instrumentation Laboratory Inc., Mass., USA.

Determination of serum calcium and magnesium.

Pybus, J., (1969), Clin. Chim. Acta., <u>23</u>, 309-317.

Determination of blood lead.

Standards Association of Australia, (1980), The Determination of Lead in Venous Blood, Australian Standard AS 2411.

Standards Association of Australia, (1985), Whole Blood Determination of Lead - Electrothermal Atomic Absorption Spectrometric Method, Australian Standard AS 2787.

Determination of liver copper.

Stevens, B. J., (1972), Clin. Chem., 18, 1379.

Determination of urine mercury.

Least, C. J., Rejent, T. A. and Lees, H., (1974), Atomic Absorption Newsletter <u>13</u>, 4.

CHAPTER 5

ION SELECTIVE ELECTRODES

Ion selective electrodes are based on the generation of an electric potential across a membrane that interacts with the ion of interest. The potential generated is dependent upon the "Activity" of the ion on each side of the membrane. Ion activity is the effective free ion concentration.

In a pure salt solution of 10^{-4} M, or less, ion activity is very close to ion concentration. As the concentration increases, interaction between ions (like charged ions repelling and opposite charged ions attracting) reduce the ability of the ions to take part in equilibrium reactions, etc.

Ion activity (a) is related to concentration by the formula:

$$a = \gamma c$$

Where:

γ = The activity coefficient.
c = The concentration.

However, the activity coefficient depends upon the ionic strength (I) of the sample.

$$I = \tfrac{1}{2} \sum c\, z^2$$

Where:

$\sum$ = The sum of cz^2 for all ions in solution.
z = The ionic charge on the ion.

This effect is illustrated below:

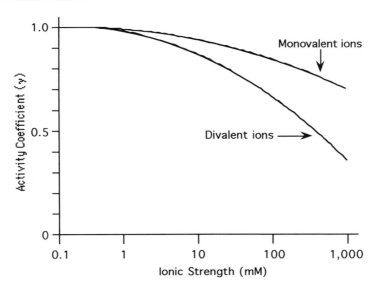

The activity coefficient of monovalent ions, in plasma, is about 0.75. However, it is only about 0.36 for divalent ions.

The potential (E) generated across an ion selective membrane is given by the Nernst equation:

$$E = \frac{RT}{F} \, Ln \, \frac{a_i}{a_{ii}}$$

Where:

R	=	The gas constant (8.315 Joules/$^{\circ}$).
T	=	The absolute temperature (Kelvin).
F	=	Faraday's constant (96,500 Coulombs).
Ln	=	The natural logarithm.
a_i	=	The ion activity on one side of the membrane.
a_{ii}	=	The ion activity on the other side of the membrane.

Ion selective membranes, used in clinical biochemistry, are usually glass of a specific composition (Eg. H^+ and Na^+) or PVC into which an ionophore has been incorporated (Eg. K^+ and Ca^{++}).

Ion selective membranes are incorporated into electrode assemblies that also contain 2 reference electrodes, as illustrated below:

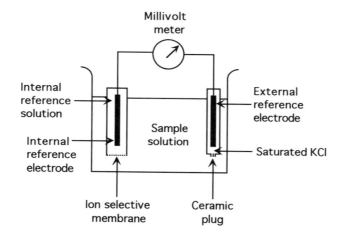

The internal reference electrode is usually a silver/silver chloride electrode. The external one is usually silver/silver chloride or calomel (Hg/Hg_2Cl_2).

Electrical potentials are generated by the 2 reference electrodes, across the ceramic plug and across the ion selective membrane. Within certain limits, the first 3 are constant and the voltage generated across the ion selective membrane increases with the activity of the ion being measured in the sample solution. This is recorded by the millivolt meter.

The 2 electrodes are usually combined, as illustrated below for a pH electrode:

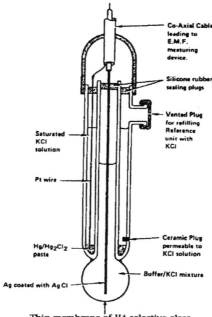

Co-Axial Cable
leading to
E.M.F.
measuring
device.

Silicone rubber
sealing plugs

Vented Plug
for refilling
Reference
unit with
KCl

Saturated
KCl
solution

Pt wire

Ceramic Plug
permeable to
KCl solution

Hg/Hg₂Cl₂
paste

Buffer/KCl mixture

Ag coated with AgCl

Thin membrane of H⁺ selective glass

Reference Electrodes.

Electrical potentials are generated due to the "Solution Tensions" of silver and mercury in aqueous solution:

$$Ag \rightleftharpoons e^- + Ag^+$$

$$Hg \rightleftharpoons e^- + Hg^+$$

The electrons remain on the surface of the metal to give it a negative charge and the metal ions accumulate in a liquid film at the surface of the metal thus producing an electric double layer. Eg:

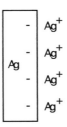

Chloride ions in the solution drive the reaction to the right, as undissociated AgCl or Hg_2Cl_2 is formed. This gives the metal a greater negative charge.

Glass Ion Selective Membranes.

Glass membranes have been developed which respond to a number of monovalent cations. The responsiveness of the electrode to a particular cation can be changed by altering the composition of the glass. Glass electrodes are commonly used for the measurement of the ionic activity of H^+, Na^+, Ag^+ and Li^+.

The glass electrode consists of a cation-responsive glass membrane sealed to the stem of non-responsive, high resistance glass. Both surfaces of the glass membrane are cation responsive.

The surface of the glass, in contact with water, becomes hydrated and swells. This leads to the slow dissolution of the glass. The glass membrane can be represented as:

Internal solution	Hydrated gel layer	Dry glass layer	Hydrated gel layer	External solution
	5-100 nm	≈ 50 μm	5-100 nm	

The thickness of the hydrated gel layer and the rate of dissolution depend on the composition of the glass. The composition of the external solution will also determine the rate of glass dissolution and hence the practical lifetime of the electrode.

The mechanism by which cations affect the potential of the glass membrane is not fully understood. The outer part of the hydrated layer acts as a semipermeable membrane and allows the cation to enter the hydrated layer where an ion exchange takes place. This produces a phase-boundary potential. Cations do not pass through the dry glass layer.

The pH Electrode.

The H^+ sensitive glass membranes are made mainly of lithium silicate. Lanthanum and barium are added as lattice tighteners. The tight lattice allows only the free movement of H^+ into and out of the gel layers. It also reduces the rate of hydrolysis and dissolution of the glass.

The Li^+ in the gel layers tend to be replaced by H^+ from the aqueous solution on both sides of the membrane.

With any particular electrode system, the potentials generated by the reference electrodes will be constant. Between pHs 2 and 12, and with ionic strengths below 0.1, the PD across the ceramic plug will also be constant.

Different glass membranes may show different potentials from that expressed in the Nernst equation. However, for any single electrode system this difference can be corrected by adding a constant specific to that system.

This constant together with those of the electric potentials, described above, can be collected into one general constant (E'). The electrode equation then becomes:

$$E = E' + \frac{RT}{F} \, Ln \, \frac{[H^+]i}{[H^+]e}$$

Where:

  E = The voltage generated by the electrode system.
  $[H^+]i$ = The internal H^+ activity.
  $[H^+]e$ = The external (Sample) H^+ activity.

If M HCl is used in the internal solution, the formula becomes:

$$E = E' + \frac{RT}{F} \, Ln \, \frac{1}{[H^+]e}$$

$$\therefore \ E = E' + \frac{RT}{F} \, Ln \, 1 \ - \ \frac{RT}{F} \, Ln \, [H^+]e$$

$$\therefore \ E = E' + 0 \ - \ \frac{RT}{F} \, Ln \, [H^+]e$$

$Ln = 2.303 \times Log_{10}$

$$\therefore \ E = E' - \frac{RT}{F} \times 2.303 \, Log_{10} \, [H^+]e$$

$$\therefore \ E = E' - \frac{8.313 \times T \times 2.303}{96500} \, Log_{10} \, [H^+]e$$

$$\therefore \ E = E' - 0.0001984 \, T \, Log_{10} \, [H^+]e$$

$pH = -Log_{10} [H^+]$

$$\therefore \ E = E' + 0.0001984 \, T \, pH$$

At 25°C we have:

$$pH = \frac{E - E'}{0.0001984 \times 298}$$

$$\therefore \ pH = \frac{E - E'}{0.0592}$$

E' can be obtained by measuring E with a solution of known pH. In actual practice a pH meter can be calibrated against 2 buffers of known pH, if it is to be used for pH measurement only. Other instrument are calibrated in millivolts as well as pH, these instruments only require one buffer of known pH in the calibration, to correct for the E' of the electrodes used.

For high accuracy measurements, temperature is very important. This not only effects the Nernst equation, but also the dissociation equilibria and junction potentials.

Some pH meter manufacturers use a temperature-compensating electrode. This electrode is a resistance thermometer, which is inserted into the solution under test together with the glass and reference electrodes. The electronics of the amplifier is so designed that, when the pH meter is used in conjunction with the resistance thermometer, the pH is independent of temperature, so long as the temperature range is within about ± 5°C of the temperature of the buffers used to calibrate the instrument.

Other instruments use electronic compensation for temperature changes. In these instruments the temperature of the calibration buffers or sample is set with a control knob on the instrument.

As less than one pA flows through the electrode system, a galvanometer can not be used to measure the PD. Hence, a DC amplifying potentiometer is used. Changes in the electric potential of the outer membrane surface of the H^+ selective glass membrane are measured by comparison with the reference electrode system.

The glass electrode can be used with strong acids and with strong oxidising and reducing agents. However, it is attacked by strong alkaline solutions. Glass electrodes should, therefore, never be left in alkaline solutions for longer than is necessary to measure the pH. The electrodes should be left in distilled water or dilute HCl.

The glass electrode responds rapidly to large pH changes in buffered solutions. However, the response is slower in poorly, or unbuffered, solutions. Equilibrium is reached slowly and may require several seconds. Poorly buffered solutions should be vigorously stirred during measurement to prevent stagnation at the electrode. The electrodes should be thoroughly washed with distilled water after each measurement and rinsed with several portions of the next test solution before making a final reading. Suspensions and colloidal material should be wiped from the glass surface with a soft tissue.

Measurements can be made in partly aqueous solutions, but the degree of hydration of the outer surface of the membrane will alter the potential across the membrane. Hence, values obtained in non-aqueous, or high ionic, solutions will be incorrect.

At very high pHs, the electrode starts to respond to other cations (Eg. Na^+) and the observed pH values will be too low. At very low pHs the rate of dissolution of the glass is altered and the activity of the water in the gel is changed to give a negative error.

The general-purpose electrode can be used between pH 2 and 12 and between -5°C and 60°C to obtain accurate results.

For correct pH measurement it is essential that KCl-calomel half-cell reference electrodes contain KCl at the correct strength. This is saturated KCl in most instruments. In this case it is desirable that the electrode contains about 2 mm of crystalline KCl and that the electrode be gently shaken before each series of measurements.

The Sodium Electrode.

The sodium electrode is a glass electrode. This is similar in design to the H^+ electrode. However, the glass composition is different.

The substitution of Al (III) for Si (IV) in alkali metal-silicate glass greatly increases the selectivity of the glass for cations other than H^+.

The inclusion of the Al leaves the glass with an excess of negativity charged sites that attract cations with the appropriate charge-to-size ratio.

A glass mixture of 11% Na_2O, 18% Al_2O_3 and 71% SiO_2 is very selective for sodium ions. Even so, the glass also responds to [H^+] changes and so the pH of samples should be kept the same.

The function of the Na^+ sensitive glass is similar to that of H^+ sensitive glass, ie. there is a Na^+ exchange with the hydrated layers of the glass. In higher [Na^+] solutions, more Na^+ moves into this layer to produce a higher potential.

PVC Ion Selective Membranes.

The Potassium Electrode.

The K^+ selective membrane consists of the antibiotic valinomycin incorporated into a PVC support. This membrane acts in the same way as H^+ or Na^+ sensitive glass.

Valinomycin has the structure shown below:

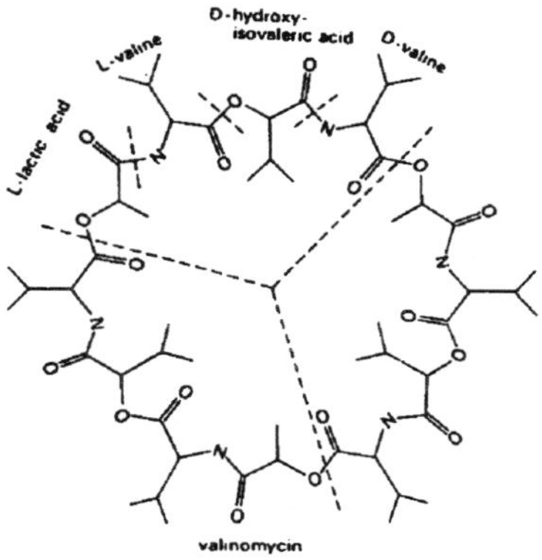

valinomycin

The oxygen in the ring structure forms ion-dipole interactions with K^+ to replace its hydration shell. Sodium has a larger hydration shell than potassium and can not easily enter the valinomycin ring. This electrode is 3,800 times more selective for K^+ than Na^+ and 18,000 times more selective for K^+ than H^+. Hence, pH variations are not a problem with this electrode.

The valinomycin acts as a K^+ exchanger, in much the same way as the glass acts as H^+ or Na^+ exchangers.

The potential generated across the membrane is proportional to the ratio of the internal [K^+] to the sample [K^+].

The Calcium Electrode.

The synthetic compound ETH 1001 is commonly used as the ionophore in calcium selective membranes. This has the structure:

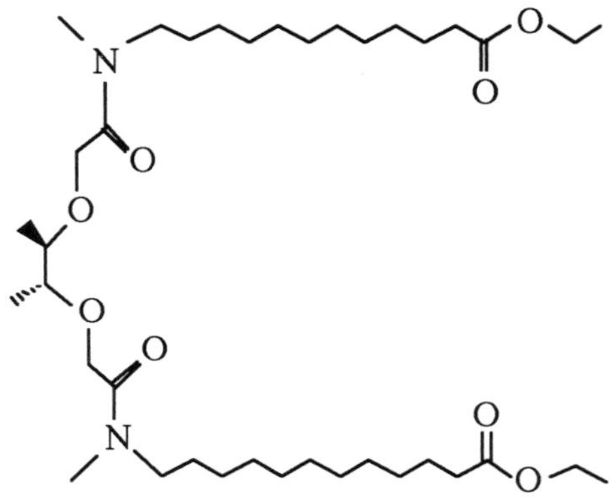

These membranes respond to calcium in solution in much the same way as valinomycin membranes respond to potassium.

About 50% of calcium in plasma is in the ionised form (Ca^{++}). It is this form that will interact with a calcium selective membrane. The remainder is complexed with plasma proteins (mainly albumin) and plasma acids. Some anticoagulants, such as citric acid, will complex with the ionised calcium in plasma and produce low Ca^{++} results.

As the physiological activity of calcium is dependent upon the activity of Ca^{++}, calcium ISEs provide more useful information than total calcium determinations.

CLINICAL SODIUM AND POTASSIUM ION SELECTIVE ELECTRODE ANALYSERS.

Clinical Na and K ISEs are designed to be used with undiluted blood, plasma or serum. This, together with their ease of use, has made them more convenient than flame photometers for the determination of sodium and potassium.

Values obtained with direct reading ISEs are higher than those obtained using dilution flame photometry. This is due to the fact that plasma sodium tends to be more diluted than an aqueous standard, when prepared for flame photometry. This is because plasma contains about 7% protein, leaving only 93% to dissolve the sodium etc. Hence, 1.0 ml of plasma only contains 930 μl of electrolyte solution. Therefore, 1.0 ml diluted to 100 ml is actually 930 μl diluted to 100 ml. This is a 1 in 107.5 dilution, compared with a 1 in 100 dilution of the standard.

This effect is reduced due to Na and K binding to protein and to bicarbonate, etc. This reduces the Na and K activity in undiluted samples.

Due to the above, the calibrators (used to standardise the ISE instruments) are made up in solutions with an ionic strength of 160 mmol/kg (plasma strength) and containing an inert polymer. This produces calibrators with a similar activity coefficient and matrix to plasma.

The layout of the electrodes in most clinical Na/K ISEs is shown below:

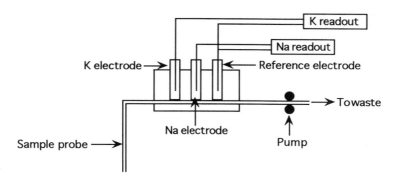

The reference electrode acts as a half-cell and is usually a Ag/AgCl electrode in a strong KCl solution. It contacts the sample via a porous plug or dialysis membrane. It generates a constant potential, when in contact with the sample, which is compared with the output of the ISEs. This difference is amplified, by factors set during calibration, and fed to the appropriate display in concentration units.

EXERCISES

EXERCISE 5-1

DETERMINATION OF THE pH OF URINE

Accurate pH measurements depend upon the accurate calibration of the meter and upon correction for any variation in temperature. It must be remembered that the actual pH of the buffers, used to calibrate the instrument, depends upon the temperature. Hence, the pH at the temperature of the buffer must be used in the calibration.

In this exercise, a pH meter will be calibrated. It will then be used to determine the pH of 2 urine samples. The average pH of urine from an individual on a normal diet is about 6.0. However, the pH may be above 7.0 in urine from vegetarians. The kidneys are able to produce urine with a pH range of 4.4 to 8.2. The low value being found in severe metabolic acidosis and the high value after the administration of bicarbonate etc.

Fresh urine should be used for pH measurement as bacteria rapidly convert urea to ammonia and, hence, increases the pH. Most urine preservatives change the pH and, therefore, can not be used.

Case Study.

The urine samples are from a 22 year-old woman. She was found unconcious by her husband, on returning home from work. 2 empty aspirin bottles were found by her bed. Her urine gave a strongly positive result for salicylates.

She was given forced alkaline diuresis treatment to speed the elimination of the salicylate. This involved the intravenous infusion of 1.5 l of glucose and sodium bicarbonate. By making the kidney tubular fluid alkaline, salicylic acid remains in the ionised form. In the normal acid environment of the distil tubule, salicylate would associate with a proton and be reabsorbed back into the body. With alkaline diuresis, salicylates (and other acidic drugs) are rapidly removed by the kidneys.

Part of the treatment involves regular monitoring of the urine pH. The 2 samples are: 1. On admission and 2. Two hours after the commencement of treatment.

REAGENTS AND EQUIPMENT:

1. pH Meter Calibration Buffers.
   These can be obtained commercially (pH 7.00 and 10.00 at 20°C), or made up as follows:
   1. 3.39 g of dry (110°C for 2 hours) KH_2PO_4 and 3.53 g of dry Na_2HPO_4 dissolved in distilled water and made to 1 litre.
   2. 3.80 g of $Na_2B_4O_7:10H_2O$ dissolved in distilled water and made to 1 litre.
   These buffers have the following pHs:

Temp (°C)	1	2
15	6.900	9.276
20	6.881	9.225
25	6.865	9.180

2. Urine 1.
   Fresh normal urine.
   Check that the pH is around 6.0.
3. Urine 2.
   50 ml of urine 1 + 200 ml of distilled water.
   Add 5% $NaHCO_3$, with stirring, to bring the pH to about 8.0.

pH Meter with Electrodes.
Paper Tissues.
Wash Bottle with Distilled Water.
100 ml Beakers.
Thermometer.

PROCEDURE:

1. Calibrate the pH meter as follows:

   (a) Switch on the instrument and allow to stabilise for about 10 minutes.

   (b) Set the instrument to the temperature of the buffers and samples (Room temperature).

   (c) Wash and blot dry the electrodes, and place them in the 7.00 buffer or Buffer 1.

   (d) Switch the instrument to "pH" and/or "Read", and adjust the instrument to read the correct pH at the current temperature, with the "Calibrate", "Asymm" or "Buffer" control.

   (e) Wash and blot dry the electrodes, and place them in the 10.00 buffer or Buffer 2.

   (f) Set the instrument to read the correct pH for the current temperature, with the "Slope" or "Sens" control.

2.   Read the pH of the 2 urines as follows:

     (a)   Wash and blot dry the electrodes and place them in the sample.

     (b)   Record the pH value.

     (c)   Switch to "Standby". Wash and blot dry the electrodes.

     (d)   Place the electrodes in distilled water or 0.1 M HCl.

EXERCISE 5-2

THE DETERMINATION OF SODIUM AND POTASSIUM IN SERUM OR PLASMA

This exercise is written for the Instrumentation Laboratory IL 501 Na^+/K^+ Analyser. Modifications may be needed for the use of other instruments.

This IL 501 is illustrated below:

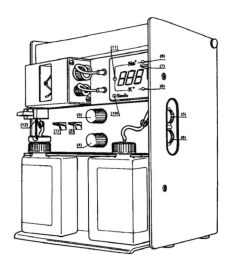

(1) : Aspiration lever.
(2) : Flush lever.
(3) : Na balance control.
(4) : K balance control.
(5) : Na slope control.
(6) : K slope control.
(7) : Display.
(8) : Na LED.
(9) : K LED.
(10) : Standby LED.
(11) : Analysis LED
(12) : Probe.

Sample is aspirated via the probe needle on the left, to a point just above the measuring tank (see diagram below):

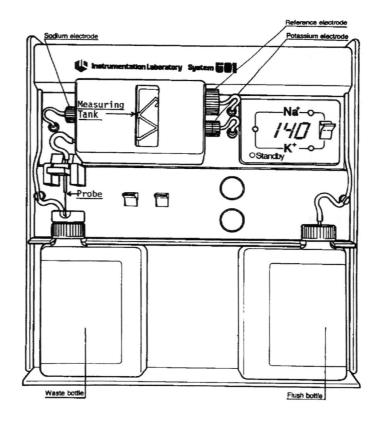

After the measurements have been made, the sample is flushed to waste, via the probe needle, by the flush solution on the right.

Aspiration takes 12 seconds. Stable readings are obtained in 30-40 seconds. If no further readings are made, the measuring tank is automatically flushed after 3 minutes. The flush takes 12 seconds.

After a 30 minute warm up, the instrument is calibrated using 2 calibration solutions (Cal 1, Na: 140 mmol/l, K: 5.0 mmol/l and Cal 2, Na: 80 mmol/l, K: 8.0 mmol/l).

Case Study:

The sample is plasma from a 60 year-old woman with Addison's disease. She had been unwell for about 1 month. She was experiencing extreme muscle weakness and confusion. She was admitted to hospital after collapsing on the street.

Addison's disease is associated with hypofunction of the adrenal cortex. Low levels of the cortical hormone aldosterone lead to a loss of sodium and retention of potassium by the kidneys.

REAGENTS AND EQUIPMENT:

1.   Patient' Plasma or Serum.
         120 ml of normal plasma or serum + 20 ml of 0.1% KCl in 7% albumin.
         Keep on ice.

II. 501 Na^+/K^+ Analyser with flush solution and calibrators Cal 1 and Cal 2.

PROCEDURE:

1.   Turn on the instrument with the switch at the rear and wait 30 minutes to "Warm up".

2.   Carry our the following full calibration, before the first set of readings:

     (1)   Tilt forward the aspirator needle and insert in Cal 1.

     (2)   Press "asp".

     (3)   After the 12 second aspiration (the analysis LED stops flashing), remove Cal 1 and return the aspiration needle to the vertical position.

     (4)   After about 40 seconds, select "K^+" on the display and adjust the "K balance" to display 5.0.

     (5)   Select "Na^+" on the display and adjust the "Na balance" to display 140.

     (6)   Press the "flush" lever.

     (7)   Aspirate Cal 2, as above, (steps 1 - 3).

     (8)   After about 40 seconds, select "K^+" on the display and adjust the "K slope" (side of instrument) to display 8.0.

     (9)   Select "Na^+" on the display and adjust "Na slope" to display 080.

     (10)   Press the "flush" lever.

     (11)   Repeat steps 1 - 10 until no slope adjustment is needed (Steps 8 and 9).

     Or carry out the balance calibration (Before a batch of samples):

     Carry out steps 1 to 6 of the full calibration.

2.   Measure the sample values as follows:

     (1)   Tilt forward the aspiration needle, wipe the needle, and insert in the sample.

     (2)   Press "asp".

     (3)   When the analysis LED stops flashing and remains lit, remove the sample.

     (4)   Return the aspiration needle to the vertical position.

(5)   When the values stabilise, (30 - 40 seconds), record the Na and K values.

(6)   Press the "flush" lever.

(7)   Leave the instrument "On" ("Standby").

Reference Ranges:

Sodium   :   135 - 150 mmol/l
Potassium :   3.6 - 5.0 mmol/l

GENERAL REFERENCES:

Willard, H. H., Merritt, L. L., Dean, J. A. and Settle, F. A., (1981), pH and Ion Selective Potentiometry. In Instrumental Methods of Analysis, (6th Ed), 640 - 663, Wadsworth Publishing Co.

Myerhoff, M. E. and Opdyche, W. N., (1986), Ion-selective Electrodes, Advances in Clin, Chem., 25, 1 - 47, Academic Press.

Russell, L. J. and Buckley, B. M., (1988), Ion-selective Electrodes. In Principles of Clinical Biochemistry, Scientific Foundations, (2nd Ed), Eds: Williams, D. L. and Marks, V., 201 - 214, Heinemann Medical Books.

Determination of serum or plasma sodium and potassium.

Instrumentation Laboratory IL 501 Instrument Manual, (1981).

CHAPTER 6

OXYGEN AND CARBON DIOXIDE ELECTRODES

The Oxygen Electrode.

Oxygen electrodes (probes) are based on the original design of Clarke et al (1953):

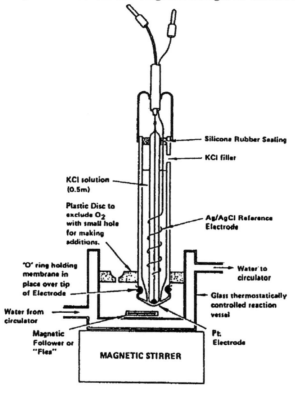

The oxygen probe contains 2 electrodes. These are a silver wire (anode) which is coated with AgCl and wound round the insulation of the other electrode (cathode), which is a platinum or gold disc. Both electrodes are immersed in a 0.1 M KCl solution. When the silver electrode is made positive with respect to the platinum or gold electrode, the following reactions take place in the presence of oxygen:

At the platinum or gold cathode:

$$O_2 + 2\ e^- + 2\ H^+ \rightarrow H_2O_2$$
$$H_2O_2 + 2\ e^- + 2\ H^+ \rightarrow 2\ H_2O$$

The 4 electrons are provided by the cathode of the battery.

At the silver anode:

$$4 \text{ Ag} \rightarrow 4 \text{ Ag}^+ + 4e^-$$

The 4 electrons flow to the anode of the battery and Cl⁻ reacts with the Ag⁺ to deposit AgCl on the silver wire.

Therefore, for each oxygen molecule consumed, 4 electrons flow through the circuit.

The actual current flowing in the circuit is dependent upon the rate at which oxygen reaches the cathode, and the polarising voltage.

As the polarising voltage is increased, so does the current flowing for a given oxygen pressure. However, this response levels out to a plateau between the voltages of 0.5 and 0.8, so that there is very little change in current flow between these voltages:

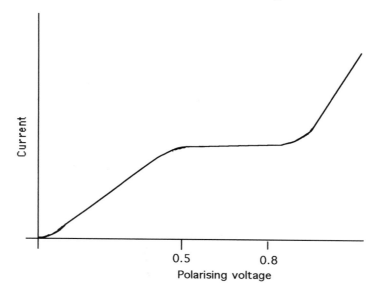

Therefore, the probe is polarised with the voltage of about 0.8 volts, so that small reductions in voltage will not alter the current flow in the circuit. This voltage is set by the "Set potential" trim tab within the polariser unit illustrated below.

The current flowing, therefore, becomes proportional to the oxygen tension of a solution flowing past the cathode, even though the actual flow of current will produce a small drop in potential between the electrodes. If the solution is allowed to stagnate, the oxygen in the solution next to the cathode will be rapidly used up and hence the current will fall. The solution must, therefore, be stirred rapidly to avoid this stagnation. This is achieved with a magnetic stirrer or by the use of vibrating electrodes.

Most electrodes include a thermistor in the oxygen probe to compensate for temperature changes. The circuit layout of the oxygen probe and polariser unit is shown below:

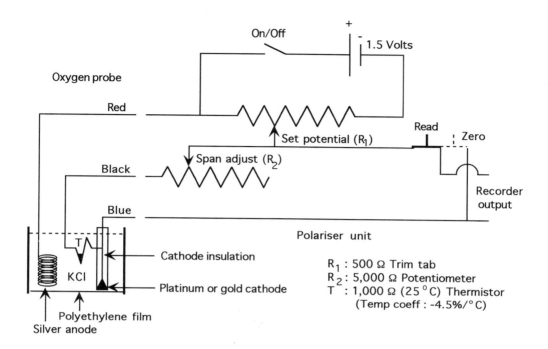

Rather than measuring the current flow in the circuit, the voltage drop across the thermistor is used to detect the current flow. The larger the current flow, the greater the voltage. An additional variable resistor (Span adjust) is included in this part of the circuit to control the voltage output to the recorder. The polariser unit also has a shorting switch, which will short the input to recorder, so that the recorder can be zeroed. An on-off switch from the battery is also included.

Due to the fact that protein and other chemicals will coat the platinum or gold cathode and reduce its effective surface area (the current flow, for a fixed oxygen tension, is proportional to the surface area of the cathode), the cathode is separated from the test solution by a 25 mm thick, intermediate density, polyethylene membrane. This membrane is freely permeable to oxygen. It is close to the cathode so that the oxygen tension in the KCl solution, between the cathode and the membrane, rapidly reaches equilibrium with the test solution. No current flows in the medium being analysed and so this type of electrode can be used to measure the oxygen tension in non-conducting liquids or gases.

The Carbon Dioxide Electrode.

The carbon dioxide electrode consists of a H^+ electrode in a bicarbonate solution (Usually 5 mM) that is separated from the sample by a carbon dioxide permeable membrane. Silicone rubber is commonly used as the membrane material. Only a thin layer of bicarbonate separates the H^+ sensitive glass from the CO_2 membrane. Hence, a rapid equilibrium exists between the test solution and the bicarbonate.

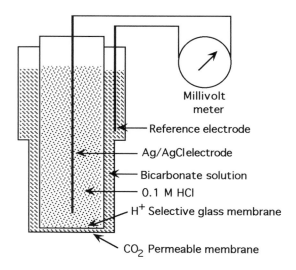

Millivolt meter

— Reference electrode

— Ag/AgCl electrode

— Bicarbonate solution

— 0.1 M HCl

— H^+ Selective glass membrane

— CO_2 Permeable membrane

When CO_2 passes through the membrane the following reactions takes place:

$$CO_2 + H_2O \rightarrow H_2CO_3 \rightarrow H^+ + HCO_3^-$$

As a result, the H^+ activity rises. This is measured, using a millivolt meter, as with a pH meter.

CLINICAL BLOOD GAS ANALYSERS.

Basic clinical blood gas instruments contain a pH electrode as well as CO_2 and O_2 electrodes. The inclusion of the pH electrode means that the acid-base status of the patient can be determined as well as the blood gases. The instrument provides the pCO_2 and pH values. The bicarbonate concentration can be calculated from the Henderson-Hasselbalch equation:

$$pH = pK' + Log_{10} \frac{[HCO_3^-]}{[CO_2]}$$

pK' varies with temperature, pH and ionic strength, but can be taken as 6.10 for plasma at 37°C.

$[CO_2] = \alpha pCO_2$, where α (0.030) is derived from the solubility coefficient of CO_2 at 37°C and the fact that plasma has a water content of 918 g/l.

Some instruments also contain a photometric system for measuring the haemoglobin content of the blood sample and the percentage oxygen saturation. The total haemoglobin content can be calculated from the absorbance at 805 nm. This is an isobestic point for both oxy- and deoxy-haemoglobin. The oxyhaemoglobin content can then be calculated from the absorbance at 650 nm. There is a large difference between the absorbance of oxy- and deoxy-haemoglobin at this wavelength.

Arterial blood pO_2 measurements are an indication of the uptake of oxygen by the blood. The value is reduced with impaired lung function. Results can be used to decide if oxygen is required by the patient and to optimise mechanical ventilation.

Oxygen delivery and release to the tissues are also important components in assessing tissue hypoxia in critically ill patients.

Oxygen delivery is the product of cardiac output and the oxygen content of the blood. The oxygen content of blood is the oxyhaemoglobin oxygen plus the dissolved oxygen. This, in ml of oxygen/l of blood, is given by the formula:

$$(0.0134 \times [Hb] \times sO_2) + (0.03 \times pO_2)$$

Where   : $[Hb]$ =   The haemoglobin concentration in g/l.
                            sO_2 =   The oxygen saturation (%Sat ÷ 100).
                            and pO_2 is in mm Hg.

Oxygen release depends on the shape and position of the oxygen dissociation curve. An indication of this is the p50 value. This is the pO_2 required to give 50% saturation. Modern blood gas instruments will provide this value.

The pCO_2 is proportional to the ratio of the rate of CO_2 production to the ventilation rate. At rest, the rate of CO_2 production is constant and so the pCO_2 is inversely proportional to the ventilation rate. Values below 36 mm Hg are associated with hyperventilation. Values above 44 mm Hg are associated with hypoventilation and above 50 mm Hg are associated with ventilatory failure.

Acidosis is due to an increased production of H^+. If the pH falls below 7.35 a state of acidaemia exists. Alkalosis is associated with a decreased production, or increased removal, of H^+. If the pH rises above 7.45 a state of alkalaemia exists.

Samples for blood gas analysis are usually arterial. These can be collected fairly easily from the radial artery, using microsamplers. These samples provide the only true picture of the blood gas status of the patient. However, arterialised capillary blood can be used. In this case, the collection site is warmed to 42°C and a capillary sample collected. These samples are similar in composition to arterial blood samples, except that the pO_2 is a bit lower.

EXERCISES

EXERCISE 6-1

THE RESPONSE TIME AND LEAKAGE CURRENT OF AN OXYGEN ELECTRODE

The response time is the time required for the electrode to respond to an instantaneous change in oxygen tension. This is usually expressed as 90% response time and 100% response time. The factors that effect the response time are the thickness and composition of the membrane, the surface area of the cathode and the rate of stirring.

Response time can be measured by turning off the stirrer and allowing the oxygen in the vicinity of the electrode to be used up. The stirrer is then switched on again and the time taken to return to 90% and 100% of the original value is determined.

In theory, the current flowing in the circuit (and hence the voltage across the thermistor) will be zero when all the oxygen is removed. However, small faults in the sealing of the platinum or gold cathode to its insulator can occur and will result in current flow when no oxygen is present.

All the oxygen can be removed from a solution by adding a few crystals of sodium dithionite (hydrosulphite). Any current still flowing is due to leakage currents.

REAGENTS AND EQUIPMENT:

1.   Aerated Distilled Water.
              Bubble compressed air through distilled water, at room temperature.
Oxygen Probe.
              Titron type 500MB or similar.

Power Supply.
              Commercial unit or "home made" as illustrated on page 6-3.
Recorder.
              Connected to power supply, as in diagram on page 6-3, and set for a 10 mV input.
Reaction Vessel.
              Commercial unit or "home made" from Perspex sheet and tubes as illustrated below:

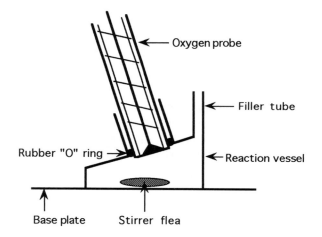

Magnetic Stirrer Motor.
Aspiration Tube.
              Connected to water vacuum pump, to aspirate the reaction vessel.
5 ml Hypodermic Syringes with Needles.
Sodium Dithionite (Sodium Hydrosulphite).
Small Spatulas.
Distilled Water.
              In wash bottles.

PROCEDURE:

1.   Fill the reaction vessel with aerated distilled water, using a hypodermic syringe and needle, until the level reaches half way up the filler tube. Tilt the front of the vessel to make sure that no air bubbles are trapped below the oxygen probe or in the reaction vessel.

2. With the stirrer flea rotating fairly rapidly, adjust the position of the reaction vessel over the stirrer motor so that the stirrer flea is rotating evenly.

3. Switch on the polarising voltage.

4. With the polariser unit switched to "Zero", adjust the recorder to read zero.

5. Switch the polariser to "Read" and adjust the recorder to read about 90 with the "Span adjust" on the polariser.

6. Allow the electrode to stabilise for about 5 minutes.

7. Set the recorder to read 100 (top of the chart), with the "Span adjust", and run the chart at 5 cm/min.

8. Turn of the stirrer and allow the pen to fall to 70.

9. Switch on the stirrer again and record until the pen reaches 100.

10. Switch the recorder to 1 cm/min.

11. Add a few crystals of sodium dithionite, via the filling tube.

12. Record until the voltage reaches a minimum. Any leakage current will be associated with a value greater than zero.

13. Suck out the solution with an aspirator tube connected to a water vacuum pump.

14. Wash the reaction vessel with distilled water, from a wash bottle, 4 times and suck out between each wash.

From the response time (5 cm/min) trace, record the time taken for the pen to reach 97 (90% response time) and 100 (100% response time) after the stirrer was switched on.

Most electrodes have a 90% response time of about 10 seconds and a 100% response time of 1 or 2 minutes.

EXERCISE 6-2

THE DETERMINATION OF SERUM GLUCOSE BY A GLUCOSE OXIDASE - OXYGEN ELECTRODE METHOD

Glucose oxidase catalyses the conversion of β-D-glucose to gluconic acid. D-glucose in solution exists as 36% in the β form, 64% in the α form and a trace in chain form. These three forms exist in equilibrium with each other and, as only the β form can be oxidised by the enzyme, the α form must be converted to the β type during the incubation period. Interconversion is fairly rapid but can be speeded up by adding the enzyme mutarotase with the oxidase.

The oxidase catalyses the oxidation of the β form to gluconolactone and hydrogen peroxide. The gluconolactone then reacts with water to produce gluconic acid.

The hydrogen peroxide, produced during the oxidation of glucose, can then be used to oxidise a chromogenic oxygen acceptor in the presence of the enzyme peroxidase.

$$\text{Peroxidase} + H_2O_2 \longrightarrow \text{Peroxidase-}H_2O_2$$

$$\text{Peroxidase-}H_2O_2 + \text{Chromogenic } O_2 \text{ acceptor} \longrightarrow \text{Chromogen} + H_2O$$

The inclusion of the chromogen removes the hydrogen peroxide and provides a useful monitor of the reaction.

The final oxygen tension, after the reaction has gone to completion, is used in a number of methods. However, these methods take longer than methods that measure the rate of oxygen consumption. Exclusion of oxygen, in methods that measure the final oxygen tension, must also be better than that in the equipment used in this exercise.

When examining the removal of oxygen from a reaction mixture, it is essential that the solution should not be in direct contact with the atmosphere, as oxygen will readily dissolve in the solution and nullify the results. The reaction vessel used in this exercise holds about 5 ml of liquid, which is added via a narrow tube, which limits the surface contact to a very small area. The reaction vessel is filled so that the liquid comes about half way up this tube.

The oxygen tension of an aqueous solution, through which air has been bubbled, is dependent upon the temperature and the atmospheric pressure. The oxygen tension rises with increasing pressure and falls with increasing temperature. No appreciable atmospheric pressure change is likely to take place during the exercise. However, the temperature of the reaction mixture may slightly increase during measurement, due to heat generated by the stirrer motor. The thermistor in the oxygen probe compensates for small changes in temperature during measurement.

As the final glucose concentration is well below the K_m value for the glucose oxidase (4.2 mM for glucose in air saturated aqueous media) and the initial oxygen concentration is the same for all assays, the initial rate of oxygen utilisation is directly proportional to the glucose concentration.

This method is one of the most specific for glucose. Other sugars only give negligible results. However, enzyme inhibition will produce low results, as the reaction rate will be slower in the unknowns than in the standards.

Similar methods, using the rate of oxygen consumption, are used in a number of commercial glucose analysers.

Case History.

The sample is plasma from a 78 year-old woman. She has been admitted to hospital with a respiratory infection and left ventricular failure. She was found to have glycosuria and so a plasma glucose has been requested. Moderate hyperglycaemia is often seen in serious illness.

REAGENTS AND EQUIPMENT:

1.   Aerated Glucose Oxidase Reagent.
            (a)   0.15 M Acetate Buffer, pH 5.0.
                        (i)   0.15 M Acetic Acid.
                                Add 2.56 ml of glacial acetic acid to 300 ml of distilled water and mix.
                        (ii)   0.15 M Sodium Acetate.
                                Dissolve 14.3 g of $CH_3COONa:3H_2O$ in 700 ml of distilled water.
                        Add (i) to (ii) and mix well.
            (b)   1% o-Tolidine in Ethanol.
                    Dissolve 100 mg of o-tolidine (3,3'-dimethylbenzidine) in 10 ml of ethanol.
            Prepare up fresh:
                        400 ml of acetate buffer.
                        5 ml of 1% o-tolidine.
                        50 mg of glucose oxidase ($\approx$ 200 U/mg).
                        4 mg of peroxidase ($\approx$ 200 U/mg).
                        Make up to 500 ml with acetate buffer.
            Bubble compressed air through the above reagent, at room temperature.
2.   Patient's Serum.
            Dissolve 70 mg of AR glucose in 50 ml of serum or plasma.
3.   20 mmol/l Glucose Standard.
            Dissolve 1.80 g of AR glucose in distilled water and make to 500 ml.
            Make up fresh, or make up in saturated ($\approx$ 0.25%) benzoic acid instead of water.

Oxygen Probe.
        Titron type 500MB or similar.
Power Supply.
        Commercial unit or "home made" as illustrated on page 6-3.
Recorder.
        Connected to power supply, as in diagram on page 6-3, and set for a 10 mV input.
Reaction Vessel.
        Commercial unit or "home made" from Perspex sheet and tubes as illustrated on page 6-6.
Magnetic Stirrer Motor.
Aspiration Tube.
        Connected to water vacuum pump, to aspirate the reaction vessel.
1 and 5 ml Hypodermic Syringes with Needles.

Distilled Water.
    In wash bottle.
Wassermann Tubes.

PROCEDURE:

1. Set up the following standards, using the glucose standard (20 mmol/l), in Wassermann tubes:

	A	B	C	D	E
μl of glucose standard (20 mmol/l)	0	250	500	750	≅ 1,000
μl of H$_2$O	≅ 1,000	750	500	250	0

2. Using a hypodermic syringe and needle, fill the reaction vessel with aerated glucose oxidase solution until the filler tube is half full. Tilt the front of the reaction vessel upward to dislodge any trapped air bubbles.

3. Adjust the stirrer and position of the reaction vessel to give fast, even stirring.

4. Switch the polariser unit to "Zero" and adjust the zero on the recorder. Then switch the unit to "Read".

5. Adjust the "Span adj" to give a reading of about 90 on the recorder. Allow the system to stabilise for about 5 minutes and readjust the "Span adjust" to a reading of 90. Leave this "Span adjust" setting for the rest of the exercise.

6. Using a 1 ml hypodermic syringe, inject exactly 0.2 ml of standard E with the tip of the needle near the bottom of the reaction vessel. Aspirate the excess enzyme solution from the top of the filler tube.

7. Record for 6 or 7 minutes at 1 cm/min. There may be a delay before the electrode responds to the drop in oxygen tension.

8. Suck out the solution, wash twice with distilled water (sucking out after each wash) and refill with glucose oxidase solution.

9. When the voltage has stabilised, start the recorder and repeat steps 6, 7 and 8 for glucose standards D, C and B in turn and then with 0.2 ml of the patient's serum sample.

10. Wash out the reaction vessel 4 times with distilled water, aspirating after each wash.

Draw a straight line of best fit along the recorder trace, for each sample, to cover 5 minutes on the chart.

Note the recorder values at the start and end of the 5 minute period.

Subtract the lower values from the higher values.

Plot a standard curve of these values against the glucose concentration of the standards used and read off the patient's serum glucose concentration.

Fasting Reference Range:

Serum or plasma glucose concentration : 3.6-5.8 mmol/l.

EXERCISE 6-3

BLOOD GAS DETERMINATIONS

In this exercise, the partial pressures of oxygen and carbon dioxide will be determined on your capillary blood samples and on blood samples equilibrated with known gas mixtures. The pH will also be determined on these samples.

The general layout of a modern blood gas instrument is shown below:

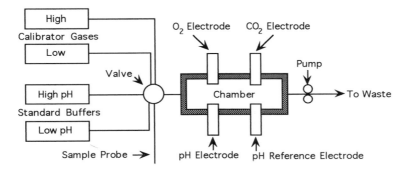

Calibration buffers and gases, as well as the samples, enter a thermostatically heated (37°C) chamber via a selector valve. After a short equilibration period, the pH, pO_2 and pCO_2 can be measured. The pH electrode is calibrated using standard buffers and the O_2 and CO_2 electrodes are calibrated with the calibrator gases.

The buffers are usually 7.386 and 6.841 (at 37°C). The calibrator gas mixtures contain oxygen and carbon dioxide at physiological partial pressures. The residual gas is nitrogen. An example of the gas mixtures is:

Gas	O_2	CO_2	N_2
Low	0%	5%	95%
High	20%	10%	70%

The pO_2 and pCO_2 values are then calculated from the atmospheric pressure. Some instruments contain a barometer and do the calculations automatically. Otherwise, the atmospheric pressure must be measured and the value entered into the instrument. Most modern instruments are self-calibrating.

Samples are aspirated into the chamber. After an equilibrium period, during which the sample reaches 37°C, the pO_2, pCO_2 and pH are recorded. The sample is then pumped to waste and the chamber washed.

As the calibrating and operating instructions for blood gas analysers vary between instruments, no operating instructions will be included in this book.

REAGENTS AND EQUIPMENT:

1. Skin Prep.
    0.5% chlorhexidine in 70% ethanol.
2. Blood.
    Expired transfusion blood can be used.

Blood Gas Analyser.
    With calibrator buffers and gases.
Barometer.
    If not a component of the blood gas analyser.
Heparinised Capillaries.
Paper or Plastic Bags.
Bucket of Warm (≈ 45°C) Water.
Paper Towels.
Blood Lancets.
Cotton Wool.
100 ml Separating Funnels.

PROCEDURE:

1. Set up the blood gas analyser for the analysis of capillary samples.

2. Analyse a resting blood sample by warming the hand in warm water and drying it. Wipe the collection finger with the skin prep and blot dry. Draw a large drop of blood using the blood lancet. Fill a heparinised capillary and analyse this blood.

3. Repeat step 2 after running up several flights of stairs.

4. Repeat step 2 after rebreathing a lung full of air into a paper or plastic bag, a few times.

5. Add about 2 ml of blood to a 100 ml separating funnel.

6. Displace the air in the funnel with one of the gases used to calibrate the analyser and replace the stopper.

7. Rotate the funnel so that the blood is evenly spread, as a thin film, over the surface of the funnel.

8. Analyse the sample.

9. Repeat steps 5 to 8 with the other calibrating gas.

Record the results and calculate the bicarbonate concentration using the Henderson-Hasselbalch equation, if this is not done by the analyser.

Reference Ranges:

Arterial Blood	Oxygen	:	85 - 108 mm Hg.
	Carbon dioxide	:	32 - 48 mm Hg.
	pH	:	7.35 - 7.45.
	Bicarbonate	:	18 -23 mmol/l.

The oxygen partial pressure tends to fall with age. It also decreases with altitude.

Carbon dioxide partial pressure is slightly lower in women than in men.

GENERAL REFERENCES:

Adams, A. P. and Hahn, C. E., (1979), Principles and Practice of Blood-Gas Analysis, Franklin Scientific Projects Ltd.

Anonymous, (1991), The Deep Picture, Radiometer A/S.

Mollard, J.-F., (1994), Pre-analytical Considerations in pH/Blood Gas Analysis, AVL Medical Instruments AG.

Walmsley, R. N., Koay, E. S. C. and Watkinson, L. R., (1989), Acid-Base Disorders, Singapore University Press.

The Oxygen Electrode:

Clarke, L. C., Wolf, R., Granger, D. and Taylor, Z., (1953), Continuous Reading of Blood Oxygen Tensions by Polarography, J. App. Physiol., **6**, 189.

O_2 and CO_2 Electrodes:

Severinghaus, J. W. and Bradley, A. F., (1958), Electrodes for blood pO_2 and pCO_2 Determinations, J. App. Physiol., **13**, 515.

Clinical Blood Gas Instruments:

Pruden, E. L., Siggaard-Anderson, O. and Tietz, N. W., (1994), part of Blood Gases and pH, Chapter 30 in Tietz Textbook of Clinical Chemistry, 2nd. Ed., Eds. Burtis, C. A. and Ashwood, E. R., p 1393-1410, W. B. Saunders Co.

CHAPTER 7

CHROMATOGRAPHY

Chromatography is the most commonly used method for separating endogenous compounds and therapeutic drugs. Chemical analysis often requires that the compound of interest be separated from a complex mixture, such as plasma, before analysis. This can be achieved by one, or more of the chromatographic techniques.

Compounds can be separated on the basis of their surface adsorptive groups, partition coefficients, size and/or surface charge.

In general terms, chromatography involves the differential distribution of components in a mixture between a stationary phase and a mobile phase. The stationary phase is either a solid or a liquid held to a support. The mobile phase is either a liquid or a gas. Due to the different distribution, some components of the mixture are retarded by the stationary phase more than others. Hence, the further the mixture migrates through the stationary phase, the greater the degree of separation.

Adsorption Chromatography.

Adsorption chromatography is the oldest form of chromatography. It was first used by Michel Tswett, a Russian botanist, to separate plant pigments at the end of the 19th century.

The adsorption process involves the formation of hydrogen bonds and Van der Waals interactions between the components in the mixture and the adsorbent. These interactions are displaced by the action of solvent in liquid chromatography (Chapter 8) or by increased temperature in gas-solid chromatography (Chapter 9).

The most common adsorbents are silica gel, alumina, magnesium oxide and charcoal. The strength of adsorbents can be increased by "Activation". This usually involves heating the adsorbent to remove water from the adsorption binding sites.

Non-polar solvents have little eluting power, when compared with polar solvents. Polar groups on the solvent compete with polar groups on the solute for the binding sites. Hence, solutes can be adsorbed onto a column of adsorbent from a non-polar solvent, such as hexane, and eluted with a polar solvent such as ethanol. As the solute elutes from the column it continually redissolves in the solvent and is adsorbed further down the column. Different solutes have different degrees of adsorption and, hence, are eluted at different times.

Affinity Chromatography.

Affinity chromatography is a specific type of adsorption chromatography. However, ionic bonds and hydrophobic interactions are also formed as well as hydrogen bonds and Van der Walls interactions. Affinity chromatography involves the specific interaction between biological molecules (Eg. Antibody and antigen, hormone and receptor, enzyme and substrate analogue etc.). One component of the pair is attached to a solid support. The other, in a complex mixture, will then bind. The other components of the mixture can then be removed by washing and the compound of interest eluted from the support.

Due to the strong interaction between the two molecules, elution usually involves a gross change in the eluting fluid such as a change in pH or ionic strength. Strong urea solutions have also been used to elute strongly adsorbed proteins.

Partition Chromatography.

This technique involves the use of the different solubilities of components of a mixture in the stationary and mobile phases. Both phases are "Liquid" and molecules (Solutes) partition themselves between the two phases depending on their relative solubility in each phase. Hence, molecules that have a high solubility for the stationary phase (relative to the mobile phase) will be retarded as the mobile phase moves over the stationary phase. Molecules with a high solubility in the mobile phase will elute rapidly.

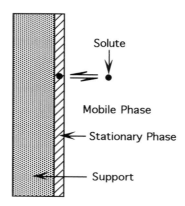

Normal Phase Partition Chromatography.

In this case, the stationary phase is aqueous and the mobile phase is an organic solvent that is non-miscible with water. Paper chromatography is an example, where water is held to the paper fibres and the solvent moves up or down the paper. Polar compounds are retarded in this type of chromatography.

Reverse Phase Partition Chromatography.

As the name implies, this is a reversal of normal phase. The organic phase is held stationary and an aqueous (or polar solvent) mobile phase is used. This technique is commonly used in high performance liquid chromatography where C_{18} chains are attached to silica particles and a buffer is used as the mobile phase (Chapter 10). Non-polar compounds are retained by the stationary phase.

Steric Exclusion (Gel Permeation, Gel Filtration) Chromatography.

In this type of chromatography, a gel with a fine pore structure is used as the stationary phase. Small molecules can penetrate into these pores and, therefore, will be retarded. Large molecules cannot enter the pores and will elute rapidly.

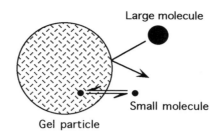

Large molecule

Small molecule

Gel particle

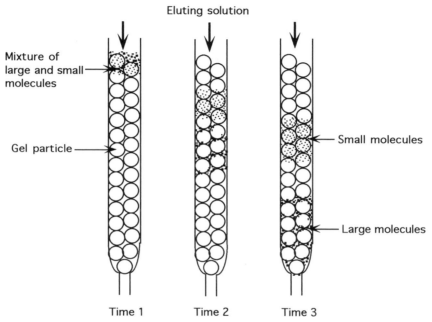

Eluting solution

Mixture of large and small molecules

Gel particle

Small molecules

Large molecules

Time 1   Time 2   Time 3

Within certain limits, molecules with a similar shape can be separated on the basis of their molecular weight:

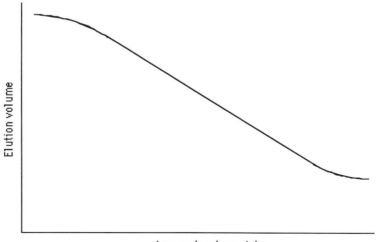

Elution volume

Log molecular weight

7-3

Ion Exchange Chromatography.

The stationary phase is a support to which ionic groups are attached. These groups will attract compounds containing groups with the opposite charge. These compounds can then be displaced by mobile phases containing ions with the same charge as that on the compound of interest.

Strong cation exchange groups are usually sulphonic acid ($-SO_3^-$); whereas, weak cation exchange ones are carboxylic acid groups ($-COO^-$). Strong anion exchange groups are usually quaternary amines and weak ones are tertiary amines. As the name implies, binding to weak exchange groups is weaker than to strong exchange groups. Organic compounds bind more weakly to the groups than inorganic ions. Hence, they can be easily displaced by inorganic ions.

A variety of supports can be used for ion exchange chromatography. These include polystyrene resins, Sephadex, Sepharose, silica particles and cellulose.

QUALITATIVE AND QUANTITATIVE ANALYSIS.

In addition to the purification of compounds from a complex mixture, chromatographic techniques can be used for qualitative and/or quantitative analysis.

Qualitative analysis relies on the extent of migration of the compounds of interest in the chromatographic system. This can be the distance migrated in paper or thin layer chromatography or the retention time in the system in gas and high performance liquid chromatography (HPLC).

Compounds have characteristic R_f values in planar chromatography (Eg. paper and thin layer chromatography). R_f is defined as the ratio of the distance migrated by the compound to the distance migrated by the solvent front. R_f values in a particular system do vary slightly with:

> Dimensions of the tank.
> Grade of the paper or thin layer plate.
> Ascending or descending flow with paper chromatography.
> Length of flow.
> Volume of solvent used.
> Equilibrium time (if any).
> Temperature.
> The nature of the sample and any pre-treatment.
> Any other volatile material added to the tank.

Biological compounds and drugs are usually colourless and, therefore, have to be located. This is achieved by using their physical and/or chemical properties. Eg. fluorescence, absorption of UV light, reaction to give a coloured product.

The size of the spot, in comparison with a series of standards can give a semi-quantitative indication of the amount present.

With column chromatography (Eg. gas chromatography and HPLC), retention time in the system can be used for qualitative identification of the components in a mixture. Quantitative information can be obtained by the extent of response of the detector (See Chapters 9 and 10).

EXERCISES

EXERCISE 7-1

DEMONSTRATION OF REVERSE PHASE PARTITION CHROMATOGRAPHY USING A C_{18} COLUMN

The Waters Sep-Pak C_{18} cartridge is a short column containing C_{18} chains linked to silica support beads. It is normally used for the concentration and/or clean up of a sample, prior to separation on an analytical chromatography column. In this exercise, the cartridge will be used to concentrate a mixture of two dyes. The dyes will then be eluted, in turn, from the column by decreasing the polarity of the eluting solvent. The two dyes are methyl green and methyl orange.

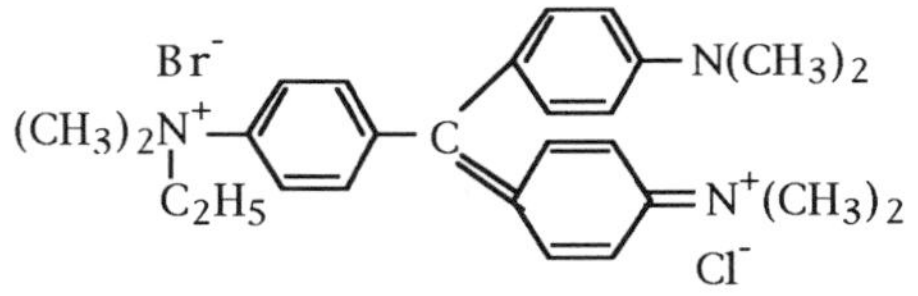

Methyl Green 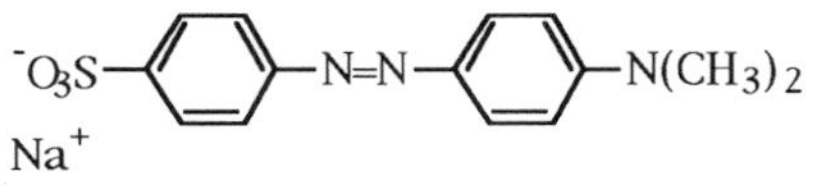   Methyl Orange

The column has to be conditioned with a solvent lower in the elutropic series than water. This is because the C_{18} chains tend to lie flat on the silica when the column is dry. The solvent allows them to "fluff-up" into their useful position. Water is too polar to allow them to do this.

Dry C_{18} Wet C_{18}

REAGENTS AND EQUIPMENT:

1. Distilled Water.
2. 100% Ethanol.
3. 20% Ethanol.
4. 50% Ethanol.
        Add 1 drop of 6 M HCl to 500 ml.
5. Dye Solution.
        5 mg of methyl orange.
        5 mg of methyl green.
            Dissolve in 1 litre of distilled water.

Waters Sep-Pak C_{18} Columns.
    Waters Cat. No. WAT051910.
    These columns can be reused.
10 ml Syringes.
Test Tubes.

PROCEDURE:

NB.   With all injections through the Sep-Pak C_{18}, make sure the syringe is connected to the long end of the Sep-Pak.

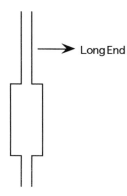

Long End

1.   Aspirate about 5 ml of ethanol into a 10 ml syringe and remove any air bubbles.

2.   Connect the syringe to the Sep-Pak and slowly expel the ethanol through the Sep-Pak, to displace the air.

3.   Wash out the syringe with distilled water and then use it to slowly flush three lots of 10 ml of distilled water through the Sep-Pak.

4.   Load 10 ml of dye solution (5 mg/l of methyl green and 5 mg/l of methyl orange in distilled water) into the syringe and slowly pass through the Sep-Pak. Note the retention of the dye.

5.   Wash the Sep-Pak with 10 ml of distilled water, using the syringe. Collect the eluate in a test tube and note the colour.

6.   Wash the Sep-Pak with 10 ml of 20% ethanol, using the syringe. Collect the eluate in a tube and note the colour.

7.   Wash the Sep-Pak with 10 ml of 50% ethanol, using the syringe. Collect the eluate in a tube and note the colour.

8.   Wash the Sep-Pak with 10 ml of 100% ethanol and return it to the container used for storing the Sep-Paks.

From your results, comment on which dye is the most polar.

EXERCISE 7-2

2-WAY PAPER CHROMATOGRAPHY OF URINARY AMINO ACIDS

Chromatography paper contains about 15% water, held to the paper fibres. This water acts as the stationary phase in paper chromatography. Paper chromatography is, therefore, normal phase partition chromatography. Amino acids are separated according to their solubility in the water and in an organic solvent (the mobile phase) moving up the paper. The most non-polar amino acids migrate the furthest, due to their greater solubility in the organic solvent. The amino acids are identified by their migration (R_f) in 2 solvents run at right angles to each other. R_fs are calculated from the following formula:

$$R_f = \frac{\text{The distance travelled by the component from the origin}}{\text{The distance travelled by the solvent front from the origin}}$$

Paper also acts as an adsorbent, having an affinity for polar groups. The structure also contains about one carboxyl group per 120 glucose residues in the cellulose structure. Therefore, it also shows weak cation exchange properties.

One way chromatography will not always separate all the components of a complex mixture. Therefore, 2-way chromatography can be used where the second solvent has different properties to the first. This can often resolve a complex mixture.

In this exercise, the first solvent is butanol/acetic acid and is run overnight. The solvent front will reach the end of the paper and, to a small extent, evaporate from there. This gives slightly increased R_f values. After drying, the second solvent, phenol/ammonia, is run overnight at right angles to the first. Movement of this solvent stops when the solvent front reaches the end of the paper.

The phenol/ammonia has to be used second, as it is hard to remove all traces of phenol from the paper during the drying process.

R_f values tend to be increased by increasing the water content of the solvent. Ammonia, added to phenol, increases the R_f values of the basic amino acids, due to the deprotonisation of amino groups.

Inorganic salts, present in the sample, tend to give poor chromatograms. This is probably due to their hydrophilic effect of extracting water from the solvent to form "pools". A number of methods of desalting have been used. These include electrolytic desalting, electrodialysis, ion exchange, adsorption, steric exclusion and solvent extraction techniques. The urine, used in this exercise, has already been desalted by an ion exchange procedure.

After the chromatograms have been run and dried, the positions of the amino acids must be visualised. This is done by treating the paper with a location reagent. The location reagent used for amino acids is ninhydrin. This reacts with amino acids as described below.

The first step in the reaction is oxidative deamination of the amino acid and reduction of ninhydrin to hydrindantin:

$$\text{Ninhydrin} + \text{R.CH.COO} \longrightarrow \text{Hydrindantin} + CO_2 + NH_3 + \text{R.CH=0}$$
$$\underset{NH_2}{|}$$

Ninhydrin   Hydrindantin

The hydrindantin then condenses with the ammonia and another molecule of ninhydrin:

$$\text{Hydrindantin} + 2\ NH_3 + \text{Ninhydrin} \longrightarrow \text{Ruhemann's purple}^- + NH_4^+ + 3\ H_2O$$

Ruhemann's Purple

Heat is required to drive the reactions and they are probably coupled in some way, as ammonia will only react slowly with ninhydrin and hydrindantin. Most amino acids produce Ruhemann's purple. However, proline and hydroxy proline give a different reaction product (yellow) with ninhydrin. This product has an absorption maximum at 440 nm. Ruhemann's purple has a maximum absorbance at 570 nm.

Ruhemann's purple tends to fade rapidly in the atmosphere. Therefore, pyridine is added to the reagent to stabilise the colour of the spots.

The colour of the spots is then further stabilised by copper nitrate, which changes the colour from purple to red-orange. These chromotograms can then be stored.

Case History.

The sample is urine from a 2 month old infant suffering from Maple Syrup Urine Disease. In this disease the amino acids valine, leucine and isoleucine cannot be metabolised beyond the aminotransferase stage. As a result, large amounts of these amino acids, and their α-keto acids, appear in the urine.

Maple Syrup Urine Disease is an autosomal recessive disease affecting about 1 in 200,000. The urine may smell of maple syrup from about the 5th day after birth. The infant feeds poorly and may vomit as well as showing muscular hypertonicity and convulsions. If not treated, brain damage and death will result. Treatment is by the reduction of the branched-chain amino acids in the diet to a level needed to just meet the body's requirements.

Normal urine contains amino acids in small amounts. These will also be seen in the patient's urine. In the 20 μl used in this exercise, glycine will give a medium spot and alanine, serine, glutamic acid and histidine will give small spots. The Rf values of all these amino acids are shown below:

Amino Acid	Butanol/Acetic	Phenol/Ammonia
Histidine	0.11	0.72
Glutamic acid	0.28	0.26
Serine	0.22	0.35
Glycine	0.23	0.42
Alanine	0.30	0.58
Valine	0.51	0.78
Iso-leucine	0.67	0.85
Leucine	0.70	0.85

As can be seen from the Rf values, leucine and isoleucine separation is not complete.

REAGENTS AND EQUIPMENT:

1. Desalted Patient's Urine.
   Dissolve:
   100 mg of valine,   100 mg of iso-leucine,
   100 mg of leucine,   50 mg of glycine,
   20 mg of alanine,   20 mg of serine,
   20 mg of glutamic acid   and   20 mg of histidine
   in 100 ml of distilled water containing 1 drop of 6 M HCl.
2. Butanol : Glacial Acetic Acid : Water (12:3:5).
   Mix together:
   600 ml of butanol,
   150 ml of glacial acetic acid
   and 250 ml of distilled water.
3. Phenol : Water : Ammonia (160:40:1).
   Mix together:
   800 g of phenol,
   200 ml of distilled water
   and 5 ml of 0.880 (SG) ammonia.
4. 0.2% Ninhydrin in Acetone.
5. 2% Pyridine in Distilled Water.
6. Copper Nitrate Solution.

   Add 200 μl of 10% nitric acid to 100 ml of ethanol.
   To this solution, add 500 μl of a saturated copper nitrate solution and mix well.

Cover the bench in clean brown paper.
5 µl Capillary Samplers + Capillaries.
Whatman No. 1 Paper, 250 x 250 mm (10" x 10").
        Pattern A - Cat No. 3001-960.
Chromatogram Frames + Spacers.
        Shandon Scientific Co. or equivalent.
Chromatography Tanks and Lids.
        Shandon Scientific Co. or equivalent.
Petroleum Gel.
Thin Perspex plates (250 x 250 mm), each with a 20 mm square hole cut 20 mm from one corner:

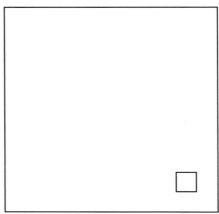

Wide Rubber Stoppers.
Hair Dryers.
Chromatogram Dip Trays.
        Shandon Scientific Co. or equivalent.
Oven at 105°C with Clean Paper on the Shelves.
String with Clothes Pegs under Fume Hood.
Forceps.
Disposable gloves.

PROCEDURE:

1.   Thoroughly wash and dry your hands before handling the chromatography paper or
        wear disposable gloves.

2.   Place a sheet of Whatman No. 1, 250 x 250 mm, chromatography paper on the paper
        covered bench.

3.   Write your name, in pencil, on the top centre of the paper.

4.   Rule, in pencil, a line 30 mm from the bottom and write under it
        "← Phenol/ammonia".

5.   Rotate the paper 90° clockwise and make another line 30 mm from the bottom and
        write under it "Butanol/acetic acid →".

6.   Rotate the paper back to the original position and place between two Perspex sheets so
        that the intersection of the two lines appears in the window of the two sheets. Place this
        assembly on 4 rubber stoppers.

7.	Apply 20 µl of desalted patient's urine using a 5 µl sampler, to the intersection. This must be done in small aliquots so that the spot size does not exceed 5 mm in diameter. After each addition, dry the spot with warm air from a hair drier before applying the next.

8.	Place the chromatogram on the frame with your name away from you. Make sure that spacers separate your chromatogram from the others.

9.	When the frame is full, attach the top plate of the frame.

10.	At about 6.00 pm., make sure that the chromatography tank is on a level bench and add 300 ml of the butanol : glacial acetic acid : water mixture.

11.	Place a layer of petroleum gel on the top rim of the tank.

12.	Add the chromatography frame so that the student names are on the top (Butanol/acetic acid arrow pointing upwards).

13.	Place the lid on the tank and rotate slightly so that the petroleum gel forms a good seal.

14.	Run overnight.

15.	Remove the chromatography frame, keep uptight and place on a wad of paper towels in a fume cupboard to dry.

16.	Clean and dry the tank.

17.	At about 6.00 pm., add 300 ml of the phenol : water : ammonia to the tank.

18.	Place a layer of petroleum gel on the top rim of the tank.

19.	Add the chromatography frame so that the Phenol/ammonia arrow points upwards.

20.	Place the lid on the tank and rotate slightly so that the petroleum gel forms a seal.

21.	Run overnight.

22.	Remove the chromatography frame, keep upright and place on a wad of paper towels in a fume cupboard to dry.

LOCATION PROCEDURE:

1.	Thoroughly wash and dry your hands before handling the chromatograms or wear disposable gloves.

2.	Run the chromatogram through the dip tray containing the 0.2% ninhydrin in acetone to which 2 or 3 ml of 2% pyridine has been added just before dipping. Use forceps or wear disposable gloves to dip the chromatograms.

3.	Hang the chromatogram up to dry under the fume hood for 2 or 3 minutes.

4.   Place them in the 105°C oven, on clean paper, for 2-3 minutes.

5.   Dip the chromatogram in the copper nitrate solution and hang up to dry under the fume hood.

Identify the amino acids present from their R_f values (they may not be exactly the same as the quoted values) and from the diagram below:

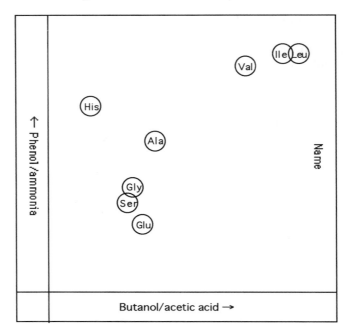

Label the amino acids on the chromatogram, with a pencil, and place a dot at the centre of each spot. Use these dots to calculate the R_f values for each amino acid in the 2 solvent systems. Compare these values with those on page 7-9.

EXERCISE 7-3

THE DETERMINATION OF PLASMA HAPTOGLOBIN BY STERIC EXCLUSION CHROMATOGRAPHY

The stationary phase used in this exercise is Sephadex, a product of Pharmacia Fine Chemicals. It is obtained by cross-linking linear dextrans (Glucose units joined by α 1-6 linkages) to give a three dimensional network of polysaccharide chains. This network carries no charge, but is strongly hydrophilic due to the large number of hydroxyl groups that it contains.

When Sephadex is added to water, it swells considerably to form a gel with the water molecules entering the internal network of the Sephadex particles. The gel, therefore, consists of a porous three-dimensional network with the size of the pores determined by the degree of cross-linking.

In a column filled with the swollen gel, small molecules are able to move in the water within the gel as well as in the water outside the gel. However, molecules that are larger than the largest pores can not enter the gel and remain in the water outside the gel. Smaller molecules can penetrate the gel to an extent depending on their size and the pore sizes in the gel. Hence, the large molecules elute from the column first. Molecules are then eluted in order of decreasing size.

Haptoglobins are glycoproteins in the plasma that combine with free haemoglobin. These proteins occupy about 25% of the α_2 globulin fraction of plasma. Haemoglobin, released into the plasma by haemolysis, is prevented from loss via the kidneys by forming a complex of high molecular weight with haptoglobins. Haptoglobins are made up of 2 subunits, α and β. These can be associated with each other in a number of combinations. The smallest consists of 2 molecules of α and one of β. This has a molecular weight of 83,000.

Sephadex G-100, used in this exercise, has an exclusion limit of 100,000, which means that molecules with a molecular weight above about 100,000 will not enter the gel particles. The haptoglobin-haemoglobin complex has a molecular weight in excess of 100,000 and will, therefore, pass through the column in the space outside the Sephadex particles. Free haemoglobin, on the other hand, will enter the intra-molecular space of the Sephadex granules, as it has a molecular weight of 64,500. The passage of haemoglobin through the column is, therefore, slowed down. Hence, free haemoglobin can be separated from the haptoglobin complex.

In this exercise, an excess of haemoglobin is added to plasma and the haptoglobin complex separated on the Sephadex column. The bound haemoglobin is then determined by its absorption at 415 nm (the λ_{max} of the Soret band of oxyhaemoglobin).

Haemoglobin, released into the plasma by intravascular or extravascular haemolysis, is very rapidly complexed with plasma haptoglobins. The complex is destroyed by the reticuloendothelial system at a faster rate than new haptoglobins can be synthesised by the liver. Therefore, as a net result of haemolysis, the plasma level of haptoglobins is reduced. A fall in the haemoglobin binding power of plasma is, therefore, a valuable aid in the diagnosis of haemolysis. Ahaptoglobinaemia is seen in conditions associated with a shortened life span of erythrocytes and in pernicious anaemia. Low levels of haptoglobins are also found in acute hepatitis, as the liver is the site of haptoglobin synthesis.

High plasma levels of haptoglobins are seen in conditions that produce an increase in glycoproteins. Therefore, high levels are found in malignancy, during chronic infection and in pregnancy.

REAGENTS AND EQUIPMENT:

1. 2% NaCl.
2. 0.9% NaCl.
    Dilute 900 ml of Reagent 1 to 2.0 l with distilled water.
    For Reagent 3.
3. 1.5 mmol/l Haemoglobin Solution.
    Centrifuge 20 ml of expired transfusion service blood to sediment the cells.
    Wash the cells (3 times) with 50 ml of 0.9% NaCl and centrifuge, each time.
    Lyse the cells by adding 10 ml of distilled water and 5 ml of chloroform and gently mixing by inversion, for 2 minutes.
    Centrifuge for 5 minutes and keep the top layer (Hb solution).
    Standardisation:

Reagent:

    100 mg of sodium bicarbonate,

    5 mg of potassium cyanide,

    20 mg of potassium ferricyanide.

    Dissolve in distilled water and make to 100 ml.

        Make fresh and keep in a dark bottle at $4^{\circ}C$.

Mix 100 µl of the Hb solution with 24.9 ml of the reagent and stand for 30 minutes.

Read the absorbance, at 540 nm, against a reagent blank.

Multiply this value by 5.63 to give the concentration of the Hb solution in mmol/l.

Dilute the Hb solution to give a final strength of 1.5 mmol/l.

Make fresh, each week, and store at $4^{\circ}C$.

4.   1.5 µmol/l Haemoglobin Standard.

    Dilute 100 µl of Reagent 3 to 100 ml with Reagent 1.

    Make fresh and keep on ice.

5.   Plasma or Serum.

6.   Sephadex Suspension.

    Sephadex G-100 - Pharmacia Biotech Cat. No. 17-0060-02.

    Suspend the Sephadex in an excess of 2% NaCl, at room temperature, for 3 days. 1 g swells to occupy about 20 ml. Store at $4^{\circ}C$.

    Remove the fines by swirling the suspension and letting the mixture stand for a few minutes and then pouring off the top liquid plus some of the top suspension. Repeat this 2 more time.

        The Sephadex can be reused after washing with 2% NaCl. It can also be stored for the following year after washing and drying. Washing is easily carried out in a large conical flask. The washes should be with 3 lots of distilled water, one of 50% ethanol and one of 100% ethanol. The Sephadex can then be spread out on clean paper to dry.

Glass Columns, 370 (to tap) x 25 mm (Internal diameter), with Taps.

Stands, Bosses and Clamps.

Teflon Wool.

Glass Rods or Tubes ($\approx$ 450 mm long).

Filter Paper Discs (24 mm diameter).

Pasteur Pipettes with Rubber Teats.

25 and 10 ml Measuring Cylinders.

Wassermann Tubes.

20 and 500 µl Pipettors with Tips.

Test Tubes.

10 ml Graduated Pipettes.

Spectrophotometers and Cuvettes.

Paper Tissues.

PROCEDURE:

You will be provided with a suspension of Sephadex G-100. This has been allowed to swell, for 3 days, in an excess of 2% NaCl. The smallest particles of Sephadex, which tend to lower the flow rate, will have been removed by decantation.

Preparation of the column.

1.   Fill the column to a depth of about 50 mm with 2% NaCl.

2.   Add a very small Teflon wool plug to the bottom of the column. (Push it down with a glass rod or tube).

3.   Pour off most of the liquid from above the Sephadex.

4.   Stir the Sephadex suspension and fill the column right to the top.

5.   Allow the Sephadex to pack under its own weight for about 15 minutes.

6.   Open the tap and the Sephadex will pack down even more.

7.   When the liquid layer above the Sephadex starts to decrease, the column is fully packed. Turn off the tap at this point. You should now be left with about 100 mm of headspace. (If you have more than 100 mm of head space, drain the liquid down to just above the top of the Sephadex, then add more Sephadex, as described above.)

8.   Place a filter paper disc on top of the Sephadex gel and allow the liquid level to just reach the filter paper.

NB: NEVER LET THE FLUID LEVEL FALL BELOW THE TOP OF THE GEL.

Separation of the haemoglobin-haptoglobin complex.

1.   Place a 25 ml measuring cylinder under the column.

2.   Mix 20 μl of the strong haemoglobin solution (1.5 mmol/l) with 500 μl of plasma, in a Wassermann tube.

3.   Quantitatively apply this mixture to the column, opening the tap to allow the sample to enter the column. Add one drop at a time, from a Pasteur pipette, to keep the filter paper disc just moist. Turn off the tap when all the sample has been added.

4.   Add a few drops of 2% NaCl to the Wassermann tube and then add these drops to the column, opening the tap just before the addition. Repeat this until all the sample is on the column. (Turn off the tap when no sample is being added).

5.   Add a few drops of 2% NaCl dropwise to the column until the red band has penetrated about 20 mm into the column. Then carefully add the 2% NaCl to fill the headspace.

6.   When the first orange-yellow drop reaches the bottom of the column, turn off the tap and record the volume of fluid in the measuring cylinder. This is the **Void Volume** (V_O); the volume of fluid outside the Sephadex gel particles. Discard this sample and replace the measuring cylinder with a 10 ml cylinder.

7.   Collect the whole of the first orange-yellow band to come off the column, make to 10.0 ml with 2% NaCl and mix well. This is the haptoglobin complex.

8.   Continue eluting the column until all the free haemoglobin has been removed, discard this eluate and then return the Sephadex to the original flask or beaker This can be achieved by applying distilled or deionised water, under pressure, to the bottom of the column, with the tap open. Remove the Teflon wool and filter paper disc.

9.   You will be provided with a standard solution of haemoglobin (1.5 μmol/l) in 2% saline; set up the following dilutions:

	A	B	C	D	E
ml of Hb standard (1.5 μmol/l)	2.0	4.0	6.0	8.0	10
ml of 2% NaCl	8.0	6.0	4.0	2.0	0
Haemoglobin conc in nmol/ml	0.3	0.6	0.9	1.2	1.5

10. Read the absorbance of the above standard solutions, and of the collected sample of complex, at 415 nm in a spectrophotometer, using 2% NaCl as the blank.

Plot a standard curve and read off the haemoglobin concentration of the eluate. Calculate the total amount of haemoglobin bound to the haptoglobin of the 0.5 ml of plasma sample.

Calculate the haptoglobin concentration as μmol haemoglobin bound per litre of plasma.

Reference Range:

Plasma from a healthy adult will bind 6-25 μmol haemoglobin/l.

EXERCISE 7-4

THE DETERMINATION OF TOTAL AMINO ACIDS IN URINE USING ION EXCHANGE CHROMATOGRAPHY

Urine contains urea, peptides and proteins as well as ammonia and other metabolic products, which produce interference in amino acid estimations. These substances can be removed by using ion exchange columns. As amino acids are ampholytes, either cation or anion exchange can be used. An anion exchange resin has the advantage that urea is not bound by the resin, as is the case with cation exchange resins.

Amberlite IRA 400 is a strong anion exchange resin and will be used in this exercise. It is a polystyrene resin with all the positively charged quaternary amine groups balanced by Cl^- ions. The Cl^- ions will be replaced with OH^- ions by passing NaOH down the column. After washing out the excess NaOH, the urine sample is applied to the column. The anions in the urine displace the OH^- ions and remain bound to the column. Cations and uncharged material will pass through the column.

Amino acids, because they contain both positively and negatively charged groups, can be displaced from the column with greater ease than inorganic anions. Dilute acetic acid is used to elute the amino acids from the column, leaving behind the inorganic anions.

The ion displacement on the resin can be illustrated as follows:

(1) Initial resin form.

$+$ • Cl^- — OH^-

Cl^-

(2) Cl^- displaced by OH^-.

$+$ • OH^-

$$^-OOC\overset{\overset{\displaystyle H}{|}}{\underset{\underset{\displaystyle NH_3^+}{|}}{C}}-R$$

OH^-

(3) The amino acid displaces the OH^- group, which removes the proton from the $-NH_3^+$ on the amino acid.

$+$ •

$$^-OOC\overset{\overset{\displaystyle H}{|}}{\underset{\underset{\displaystyle NH_2}{|}}{C}}-R$$

$$^-OOC\text{-}CH_3$$
$$H^+$$

$$^-OOC\overset{\overset{\displaystyle H}{|}}{\underset{\underset{\displaystyle NH_3^+}{|}}{C}}-R$$

(4) The acetate ion replaces the amino acid on the resin. This is assisted by the proton of acetic acid binding to the $-NH_2$ group of the amino acid to give the molecule a + charge. This decreases the amino acid affinity for the resin.

$+$ • $^-OOC\text{-}CH_3$

The • represents a loose electrostatic bond between the resin and the anion.

After dissociation from the resin, the $-COO^-$ group on a proportion of amino acids will associate with an additional proton to give the $-COOH$ group. The amino acid now has a net positive charge and is repelled by the resin.

$$^-OOC\overset{\overset{\displaystyle H}{|}}{\underset{\underset{\displaystyle NH_3^+}{|}}{C}}-R \quad \xrightarrow{\quad H^+ \quad} \quad HOOC\overset{\overset{\displaystyle H}{|}}{\underset{\underset{\displaystyle NH_3^+}{|}}{C}}-R$$

The amino acid content of the eluate is determined using the ninhydrin reaction described in Exercise 7-1. The reaction mixture is buffered at pH 5 and includes cyanide to prevent re-oxidation of the hydrindantin by atmospheric oxygen.

Under these conditions most amino acids, with the exception of proline and hydroxyproline, give a 100% yield of Ruhemann's purple. The absorbance of the Ruhemann's purple is measured at 570 nm.

Abnormally high levels of urinary amino acids are found in a number of conditions. These can be classified into two categories, "overflow" and "renal".

The "overflow" category is associated with high plasma levels of amino acids. Severe liver damage is associated with a reduced capacity by the liver to metabolise amino acids and so the plasma level of amino acids builds up.

Inborn errors of metabolism associated with amino acids are seen as a plasma increase of a single amino acid or group of amino acids. For example, phenylalanine in phenylketonuria and leucine, isoleucine and valine in maple syrup urine disease. The plasma level of the amino acids exceeds the renal threshold and the proximal tubules of the kidney are unable to reabsorb all the amino acids filtered by the glomeruli.

In the "renal" category, the plasma amino acid concentrations are normal, but the proximal tubules are failing to reabsorb all the filtered amino acid. This can be due to heavy metal poisoning (eg. lead or mercury poisoning), in which case, an increase in all amino acids is found in the urine.

Genetic diseases, such as Fanconi syndrome, are also associated with increased urinary amino acids. In Fanconi syndrome, part of the proximal tubule does not fully develop and remains in a degenerate form. All amino acids, along with glucose, phosphate and other products are lost in the urine.

In cystinuria, the reabsorption mechanism for cystine and the basic amino acids is not operative and so these amino acids appear in the urine.

In Hartnup disease, all amino acids except proline, hydroxyproline and glycine are not reabsorbed and are, therefore, found in large amounts in the urine.

REAGENTS AND EQUIPMENT:

1. 2% NaCl.
    Dissolve 40 g of NaCl in 2 l of distilled water.
2. 2 M NaOH.
    Dissolve 80 g of NaOH in distilled water and make to 1 l.
3. 1 M Acetic Acid.
    Add 57 ml of glacial acetic acid to distilled water and make to 1 l.
4. 0.25 mmol/l Amino Acid Standard.
    Stock (10 mmol/l):
    Dissolve 75 mg of glycine and 147 mg of glutamic acid in 200 ml of distilled water.
    Store at 4°C or frozen.
    Working standard:
    Dilute 2.5 ml to 100 ml with distilled water.
5. 0.2 mmol/l NaCN in Acetate Buffer (pH 5.3).
    (a) 0.2 M sodium acetate - acetic acid buffer.
        (i) Dissolve 8.2 g of sodium acetate (or 13.6 g of the trihydrate) in distilled water and make to 500 ml.
        (ii) Dilute 20 ml of reagent 2 to 100 ml with distilled water.
        Mix 413 ml of (i) with 87 ml of (ii). Check the pH and adjust with (i) or (ii), if necessary.
    (b) Dissolve 60 mg of NaCN in 100 ml of distilled water.
        (Note that, due to its instability, more NaCN is needed than the theoretical amount (49 mg)).
    Mix 490 ml of (a) with 10 ml of (b).
    Store at 4°C.

6.   Place in dispenser set to 0.5 ml.
       3% Ninhydrin in Methyl Cellosolve.
       Dissolve 9 g of ninhydrin in 300 ml of methyl cellosolve (2-methoxyethanol).
       Store at 4°C.
       Place in a dispenser set to 0.5 ml.
7.   Isopropanol/Water.
       Mix 1 l of isopropanol and 1 l of distilled water.
       Place in a dispenser set to 5.0 ml.
8.   Fresh Urine.
       Keep on ice.
9.   Amberlite IRA 400.
       Aldrich Cat No. 24,766-9 or equivalent.
       Suspend in distilled water and allow to settle. Decant and repeat the washing, twice.
       Leave the resin in an excess of water.
       Store at 4°C.
       The resin can be washed and reused.

200 mm x 10 mm (Internal diameter) Glass Columns with Taps.
       These can be made from glass tubing with one end constricted by heating in a flame and
       drawing out the tube. This is then cut and the sharp edge blunted in a flame. To this is
       attached a piece of plastic tubing with a screw clamp. A cut down Pasteur pipette is then
       inserted into the plastic tube.
Stands, Bosses and Clamps.
Teflon Wool.
Glass Rods or Tubes ($\simeq$ 250 mm long).
25 ml Measuring Cylinders.
Broad Range pH Paper.
5, 10 and 20 ml Pipettes.
100, 200, 500 and 1,000 μl Pipettors and Tips.
100 ml Volumetric Flasks.
Test Tubes.
Vortex Mixers.
Aluminium Foil Square (50 x 50 mm).
Boiling Water Bath.
Spectrophotometers and Cuvettes.
Paper Tissues.

PROCEDURE:

You will be provided with Amberlite IRA 400 in water.

Preparation of the column.

1.   Fill the column to a depth of about 25 mm with distilled or deionised H_2O.

2.   Insert a small Teflon wool plug at the bottom of the column.

3.   Pour off most of the water above the Amberlite.

4.   Stir the Amberlite into a suspension and quickly pour into the column to give a column
       height of about 75-100 mm.

5.   Insert another small Teflon wool plug above the resin.

       NB: NEVER LET THE FLUID LEVEL DROP BELOW THIS UPPER PLUG.

6.   Open the tap and allow the water level to drop to the upper plug.

7. Over a period of about 10 minutes, allow about 25 ml of 2 M NaOH to run through the column. (A slight colour change will occur in the resin.)

8. Wash with water until the pH is neutral to pH paper.

Isolation of urinary amino acids.

1. Slowly apply 10.0 ml of normal urine to the column, followed by 20 ml of H_2O after all the urine has entered the resin.

2. When the last of the water enters the upper plug, slowly apply 25 ml of M acetic acid to the column.

3. Discard the first 4 ml to come off the column. (A slight colour change will occur in the resin as the acid passes down the column.)

4. Collect exactly 20 ml (this will contain the amino acids) and mix.

5. Wash the column with 10 ml of H_2O, then 20 ml of 2% NaCl and finally with 10 ml of H_2O and return the resin to the original container. This can be achieved by applying distilled or deionised water, under pressure, to the bottom of the column, with the tap open. Remove the Teflon wool.

Determination of the total amino acids.

1. Transfer 5.0 ml of the 20 ml eluate to a 100 ml volumetric flask and make to the mark with H_2O.

2. Set up the following tubes, using the diluted eluate and the amino acid standard (0.25 mmol/l), and mix well.

	A	B	C	D	E	F	G
ml of diluted eluate	1.0	1.0	-	-	-	-	-
µl of a-a standard (0.25 mmol/l)	0	0	0	200	400	600	800
µl of H_2O	0	0	1,000	800	600	400	200
ml of 0.2 mM NaCN in acetate buffer, pH 5.3	0.5	0.5	0.5	0.5	0.5	0.5	0.5
ml of 3% ninhydrin in methyl cellosolve	0.5	0.5	0.5	0.5	0.5	0.5	0.5
nmol of amino acid per tube	-	-	0	50	100	150	200

3. Cap the tubes with aluminium foil and heat in a boiling water bath for exactly 15 minutes.

4. Immediately on removing the tubes from the bath, rapidly add 5.0 ml of isopropanol : water (1:1) mixture and mix on the vortex mixer.

5. Allow to cool to room temperature.

6. Read the absorbance of all tubes against tube C at 570 nm.

Plot a standard curve and read off the value of the diluted eluate.

Assume the urine to be part of a 24 hour sample of 1,450 ml. Calculate the amino acid output/24 hours.

The assay tubes contain the equivalent of 25 μl of urine.

Reference Range:

Normal urinary output of amino acids is 5-15 mmol/24 hours. The actual amount varies directly with the protein input.

EXERCISE 7-5

THE DETERMINATION OF BLOOD HAEMOGLOBIN A_{1C} USING ION EXCHANGE CHROMATOGRAPHY

97% of adult haemoglobin is HbA. Part of this HbA consists of glycated haemoglobin, HbA_1. About 80% of the HbA_1 is HbA_{1C}.

HbA_{1C} is formed by the condensation of glucose with the N-terminal amino groups of the β globin chains. This initially produces a Schiff base, which then undergoes an Amadori rearrangement to form a stable ketoamine. The reaction of glucose with proteins is illustrated below:

Glucose

$$H-\overset{\overset{\displaystyle H}{|}}{\underset{\underset{\displaystyle R}{|}}{C}}\overset{\nearrow O}{\underset{}{C}}-OH$$

+

Protein (Hb)

$$\overset{|}{\underset{H\quad H}{N}}$$

Rapid ⇌

Schiff Base

$$H-\overset{\overset{\displaystyle N}{\|}}{\underset{\underset{\displaystyle R}{|}}{C}}\qquad H-\overset{}{\underset{}{C}}-OH$$

Moderate

Very slow

Amadori Product
Ketoamine

$$HbA_{1C}$$

$$H-\overset{\overset{\displaystyle N-H}{|}}{\underset{\underset{\underset{\displaystyle R}{|}}{C=O}}{C}}-H$$

"AGEs"

$$R'-C \cdots N^{+} \quad C \atop C-R''$$

Advanced Glycated End Products

Significant amounts of AGEs do not form from HbA_{1C} due to the very slow reaction rate and the relatively short life span of erythrocytes. The formation of HbA_{1C} is non-enzymatic and is taking place over the whole life span of the erythrocyte. The rate of formation is proportional to the blood glucose concentration.

Therefore, the amount of HbA_{1C} in the blood reflects the average blood glucose concentration over the previous 2-3 months. It is therefore, a good means of long term monitoring of the control of blood glucose concentration in diabetics.

Good glucose control reduces the risk of retinopathy, nephropathy, neuropathy and general micro and macrovascular complications. These complications are initially due to the formation of AGEs.

Interpretation of results is based on the normal life span of erythrocytes. Hence, patients with haemolytic disorders will have low HbA_{1C} values.

A number of companies make HbA_{1C} kits. The "BioSystems" kit is described for use in this exercise. This consists of disposable columns containing a weak cation exchange resin, buffer and reagents.

The erythrocytes are first haemolysed in an acidic buffer reagent that breaks down the Schiff base. (The actual concentration of Schiff base is dependent upon the current blood glucose concentration, rather than the long term level).

An aliquot of the haemolysate is then applied to a cation exchange resin column, which binds the haemoglobin.

The HbA_{1a} and HbA_{1b} are then eluted from the column with a low ionic strength borate/phosphate buffer. These haemoglobins are the product of the reaction between haemoglobin and phosphorylated sugars. Therefore, they contain an extra negative charge. Hence, they are less strongly bound to the negatively charged resin and are, therefore, more easily eluted.

The HbA_{1C} is then eluted with a higher strength buffer. The other haemoglobins (mainly HbA_0) remain on the column due to the fact that they still have all their terminal $-NH_3^+$s.

The HbA_{1C} is collected and its absorbance measured at 415 nm (the λ_{max} of the Soret band of oxyhaemoglobin). The absorbance of an aliquot of diluted haemolysate is also measured at 415 nm and the HbA_{1C} percentage of the total Hb is calculated.

The results are very dependent upon the temperature of the column during sample loading and elution. As the assay is designed to be run 23°C, the sample and the eluting reagents are maintained at this temperature prior to loading and use.

REAGENTS AND EQUIPMENT:

1.   BioSystems HbA_{1C} Kit.
            Cat. No. 11023.
            Kit reagents 2 and 5 should be kept in the 23°C water bath, prior to their use.
2.   Fresh Blood.

Stands, Bosses and Clamps.
25, 50, 100 and 200 μl Pipettors and Tips.
2 and 5 ml Graduated Pipettes.
Wassermann Tubes.
Parafilm.
Test Tubes.
Spectrophotometers and Cuvettes.
Paper Tissues.
Water Baths at 23 and 37°C.

PROCEDURE:

1.   Invert the columns a few times to suspend the resin and then place them in the clamps. Allow to stand for 10 minutes.

2.   Pipette 100 μl of Reagent 4 (Schiff base eliminator) and 25 μl of blood into a Wassermann tube. Cap the tube with parafilm.

3.   Mix and incubate at 37°C for 20 minutes. Transfer the tube to the 23°C water bath and allow it to cool to this temperature.

4.   Remove the upper cap and bottom closure from a cation exchange column.

5.   Gently push the upper filter down to the resin surface with the rounded end of a pipette.

6.   Allow the column to drain completely to waste.

7. Carefully add 50 µl of the haemolysate to the upper filter and let the column drain completely to waste.

8. Add 200 µl of Reagent 5 (HbA$_{1a+b}$ eluting solution) and allow the column to fully drain to waste.

9. Add 2.0 ml of Reagent 5 and allow the column to fully drain to waste.

10. Place a Wassermann tube under the column, add 4.0 ml of Reagent 2 (Buffer), collect all the eluate and mix well.

11. Mix 50 µl of haemolysate, from step 3, 4.0 ml of Reagent 3 (NaCl solution) and 8.0 ml of distilled or deionised water in a test-tube.

12. Read the absorbance of the solutions produced in steps 10 and 11, against distilled or deionised water, at 415 nm.

Calculate the percentage the HbA$_{1C}$ is of the total Hb.

HbA$_{1C}$:   10 µl of blood in 4.0 ml.
Total Hb   :   10 µl of blood in 12.05 ml.

Reference Values:

4.2 - 5.9%

% HbA$_{1C}$	Degree of glucose control
>10%	Poor
9-10%	Fair
8-9%	Good
7-8%	Excellent
6-7%	Non-normal glycaemia
< 6%	Non-diabetic level

EXERCISE 7-6

THE REMOVAL OF ALBUMIN FROM PLASMA BY AFFINITY CHROMATOGRAPHY

Even though affinity chromatography can be traced back to 1910, it is only since 1968 that it has been regularly used for protein purification etc.

The technique uses the specific interaction between biological molecules. A ligand (molecule to be bound) is attached to a support (matrix) and, usually, added to a column. The mixture containing the molecule of interest is then passed through the column, where the molecule of interest is bound. The column is then washed to remove the other components of the mixture. Finally, the compound of interest is eluted from the column.

The affinity constant of the binder for its ligand should be in the order of 10^4 to 10^8. With lower affinity constants, the binding is too weak for successful separation. If the affinity constant is too high, it is difficult to separate the complex and release the component of interest from the column.

Where the attached ligand is small, binding may not be able to take place due to the steric hindrance of the support. In this case a spacer arm has to be included between the support and the ligand.

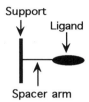

The support material used in this exercise is Sepharose CL, a cross-linked agarose. Agarose is a linear polymer of alternating D-galactose and 3,6-anhydro-L-galactose. The ligand (Cibacron Blue F3G-A) is attached to the Sepharose by a C_3H_6 spacer arm using ether links. This product is known as Blue Sepharose CL-6B.

Albumin will bind a whole variety of chemicals, such as fatty acids, bilirubin, calcium, citrate and a number of dyes. One of these dyes is Cibacron Blue F3G-A. Hence, when plasma or serum is passed down a Blue Sepharose CL-6B column, albumin will be retained by the column and the eluate will be albumin free.

The dye is not specific for albumin. Some lipoproteins, coagulation factors and NAD and NADP requiring enzymes will also be bound. Most of the proteins can be eluted with strong KCl solutions and the lipoproteins removed with 6 M urea.

The removal of albumin from plasma or serum is often the first step in the purification of one of the globulin proteins.

In this exercise, Pasteur pipettes will be used as columns to which about 1 ml of swollen Blue Sepharose CL-6B will be added. 1 ml of 1/10 plasma will be passed through the column and the eluate collected. After washing, the albumin will be desorbed with 1.5 M KCl and collected. Finally the lipoproteins will be desorbed with 6 M urea and the column re-equilibrated with the original buffer. The protein content of the eluates will be determined by the method of Lowry et al. (1951).

This method involves two reactions. The first is the biuret reaction, whereby cupric copper, in alkaline solution, bonds with peptide nitrogen. The second is a reduction of phosphotungstic and phosphomolybdic acids to molybdenum blue and tungsten blue. This second reaction is carried out mainly by the copper protein complex, but the mildly reducing amino acids in the protein: tyrosine, tryptophan and cysteine also reduce the two acids.

Folin and Ciocalteu's reagent contains the two acids and lithium sulphate. The coloured products that form on reduction of the two acids are of unknown composition and the lithium produces a more soluble product than that produced in the presence of sodium (sodium salts are used in the preparation of the two acids). Turbidity is, therefore, less likely to develop in Folin-Ciocalteu's reagent than in the reagent without the lithium salt.

The absorbance of the resultant blue-grey colour is measured at 750 nm against a reagent blank.

REAGENTS AND EQUIPMENT:

1.   50 mmol/l Tris-HCl Buffer, pH 8.0.
        (i)   Dissolve 12.1 g of Tris (Tris(hydroxymethyl)amino-methane) in distilled water and make to 1 l.
        (ii)   0.1 M HCl.
        Mix 500 ml of (i) with 292 ml of (ii) and mix well.
        For reagents 2 and 3.
2.   Buffered NaCl Solution.
        Dissolve 1.17 g of Analar NaCl in 400 ml of reagent 1.
3.   Buffered KCl Solution.
        Dissolve 11.18 g of Analar KCl in 100 ml of reagent 1.
4.   6 M Urea.
        Dissolve 36 g of urea in distilled water and make to 100 ml.
5.   2% Sodium Carbonate in 0.1 M NaOH.
        Dissolve 4.0 g of NaOH in 1 l of distilled water, add 20 g of sodium carbonate and stir until dissolved.
6.   2% Sodium (or Potassium) Tartrate.
        Dissolve 2 g in 100 ml of distilled water.
7.   1% Copper Sulphate.
        Dissolve 1.0 g of $CuSO_4:5H_2O$ in 100 ml of distilled water.
8.   Diluted Folin-Ciocalteu's Reagent.
        Dilute the reagent (Sigma Cat. No. F 9252) with an equal volume of distilled water.
9.   Plasma or Serum.
        Keep in ice.
10.   250 mg/l Protein Standard.
        Dissolve 25 mg of human (or bovine) albumin in distilled water and make to 100 ml.
        Store in a plastic bottle and keep in ice.

Columns.
    Cut off the bottom of Pasteur pipettes and fit about 50 mm of 1-2 mm (internal diameter) plastic tubing. Place a screw clamp over the tubing.
Stands, Bosses and Clamps.
Pasteur Pipettes.
Teflon Wool.
Blue Sepharose CL-6B.
    Pharmacia Biotech Cat. No. 17-0830-01.
    Allow the gel to swell in an excess of distilled water for about 15 minutes. Decant the water and wash the gel with more distilled water. Then decant most of the water.
    1 g swells to about 3.5 ml.
    It can also be stored for the following year after washing and drying. Washing is easily carried out in a conical flask. The washes should be with 3 lots of distilled water, one of 50% ethanol and one of 100% ethanol. The Sepharose can then be spread out on clean paper to dry.
Rubber Tube Connected to Compressed Air Tap.
100, 200, 500 and 1,000 μl Pipettors and Tips.
2 and 5 ml Graduated Pipettes.
Wassermann Tubes.
Test Tubes.
50 ml Measuring Cylinders.
50 ml Volumetric Flasks.
100 ml Beakers.
500 ml Beaker for Used Sepharose.
Spectrophotometers and Cuvettes.
Paper Tissues.

PROCEDURE:

Preparation of the Column.

You will be provided with Pasteur pipettes fitted with capillary tubing and screw clamps.

Compressed air can be used carefully to speed up the column flow rate, but do not let the Sepharose run dry.

1.   With the tap closed, add distilled water to the column to the depth of about 10 mm.

2.   Add a very small piece of Teflon wool to the base of the column. Push it down with another Pasteur pipette.

3.   Swirl the Blue Sepharose into a suspension and pour into the column until a gel depth of about 40 mm is reached.

4.   Place a very small piece of Teflon wool on top of the Sepharose.

     NB: NEVER LET THE FLUID LEVEL DROP BELOW THIS UPPER PLUG.

5.   Run the liquid down to the top of the Teflon wool plug.

6.   Slowly run 5 ml of 50 mmol/1 NaCl in 50 mmol/1 Tris-HCl buffer, pH 8.0, through the column.

Preparation of Albumin Depleted Plasma.

7.   Dilute 500 µl of plasma with 4.5 ml of buffered NaCl.

8.   Add 1.0 ml of this solution to the column.

9.   Place a Wassermann tube under the column. Open the tap gently and run the liquid down to the upper plug.

10. Add 3 lots of 1.0 ml of buffered NaCl and collect each eluate in the Wassermann tube and mix well.

Desorption of the Albumin.

11. Wash the column with a further 5 ml of buffered NaCl and discard the eluate.

12. Place another Wassermann tube under the column and add 2.0 ml of 1.5 M KCl in 50 mmol/1 Tris-HCl buffer, pH 8.0, to the column.

13. Slowly allow the liquid to flow through the column.

14. Repeat the elution with another 2.0 ml of buffered KCl, collecting the eluate in the same tube, and mix well.

Regeneration of the Blue Sepharose.

15. Slowly run 5 ml of 6 M urea through the column and discard the eluate.

16. When the last of the urea reaches the top plug, slowly run two lots of 5 ml of buffered NaCl through the column and discard the eluate,

17. Use distilled or deionised water (under pressure) to expel the gel into a collection beaker. Remove the pieces of Teflon wool.

Protein Determination.

18. Make up the alkaline copper reagent as follows:

    a. To 50 ml of 2% Na_2CO_3 in 0.1 M NaOH, add 500 ml of 2% Na (or K) tartrate and mix well.

    b. To the above solution, add 500 ml 1% $CuSO_4 : 5 H_2O$ and mix well.

    This solution should be used within 2 hours of preparation.

19. Dilute the 1/10 plasma (from Step 7) to 1/500 (2.0 ml to 50 ml with distilled H_2O).

20. Dilute the eluates (from Steps 10 and 14), 1/5 with distilled water.

21. Pipette duplicate 1.0 ml samples of these solutions (from Steps 19 and 20) into clean test tubes.

22. Set up the following test tubes, using the 250 mg/l protein standard:

	A	B	C	D	E	F
μl of standard protein (250 mg/l)	0	200	400	600	800	1,000
μl of distilled H_2O	1,000	800	600	400	200	0
μg of protein	0	50	100	150	200	250

23. Add 5.0 ml of the alkaline copper reagent to each tube and mix well.

24. Stand for 10 minutes.

25. Add 500 ml of diluted Folin-Ciocalteu's reagent to each tube and mix well. (The F-C reagent has been diluted 1:1 with H_2O).

26. Read the absorbance of the tubes against tube A at 750 nm, after exactly 30 minutes. (Time is important as the colour increases with time).

Plot the standard curve (there is a slight deviation from Beer's law) and read off the amount of protein in the sample tubes. Average the duplicate results and calculate total plasma protein in g/l and the eluted proteins as g/l of original plasma.

Reference Ranges:

Total serum protein	:	59-82 g/l
Total plasma protein	:	62-85 g/l
Albumin	:	35-52 g/l

GENERAL REFERENCES:

Tabor, M. W., (1989), Chromatography: Theory and Practice. Chapter 4 in Clinical Chemistry - Theory, Analysis and Correlation (2nd. Ed.), Eds. Kaplan, L. A. and Pesce, A. J., 73-93, The C. V. Mosby Company.

Pharmacia Biotech publish the following useful booklets:

Gel Filtration: Principles and Methods (6th Ed.).
   Cat. No. 18-1022-18.
Ion Exchange Chromatography: Principles and Methods.
   Cat. No. 18-1114-21.
Affinity Chromatography: Principles and Methods.
   Cat. No. 18-1022-29.

2-Way Urine Amino Acid Paper Chromatography.

Smith, I., (1969), "Aminoacids, Amines and Related Compounds" in Chromatographic and Electrophoretic Techniques, Volume 1 (Chromatography), 3rd Edition, Ed. Smith, I., 104-169, Heineman.

Plasma Haptoglobins, using Steric Exclusion Chromatography.

Ratcliff, A. P. and Hardwick, J., (1964), J. Clin. Path. 17, 676.

Total Amino Acids in Urine, using Ion Exchange Chromatography.

Morris, C. J. O. R. and Morris, P., (1963), Separation Methods in Biochemistry, Pitman, London, p 229.

Rosen, H., (1957), Arch. Biochem. Biophys., 67, 10.

Blood HbA$_{1C}$, using Ion Exchange Chromatography.

BioSystems Glycosylated Haemoglobins HbA$_1$ - HbA$_{1C}$ Information Booklet.

Goldstein, D. E., Little, R. R., Wiedmeyer, H.-M., England., J. D. and McKenzie, E. M., (1986), Glycated Hemoglobin: Methodologies and Clinical Applications, Clin. Chem., 32, B66 - B70.

Plasma Albumin, using Affinity Chromatography.

Travis, J., Bowen, J., Tewksbury, D., Johnson, D. and Pannell, R., (1976), Isolation of Albumin from Whole Human Plasma and the Fractionation of Albumin-Depleted Plasma, Biochem, J., 157, 301-306.

Lowry, O. H., Rosebrough, N. J., Farr, A. L. and Randall, R. J., (1951), Protein Measurement with Folin Phenol Reagent, J. Biol. Chem., 193, 265.

CHAPTER 8

THIN LAYER CHROMATOGRAPHY
AND
EXTRACTION TECHNIQUES

THIN LAYER CHROMATOGRAPHY

Thin layer chromatography (TLC) is best suited to separating non-ionic compounds that are soluble in organic solvents. Most TLC involves the use of adsorbents; and is, therefore, a form of adsorption chromatography.

The adsorbents are usually silica or alumina, bonded to the surface of glass, plastic or aluminium sheets by a bonding agent (usually gypsum [$CaSO_4$] or an organic binder in the case of commercial ready-to-use plates). This layer of adsorbent is usually about 0.2 mm thick. Adsorption is a surface phenomenon and so it is important that the particle size is small, so as to provide a large surface area. On the other hand, the rate of migration of the mobile phase through the adsorbent decreases as the particle size decreases. The optimum particle size is 3-18 μm. Adsorbers are very porous. These pores greatly increase the surface area. Silica gel has pores with diameters of 3 to 13 nm which gives a surface area of about 400 m^2/g. Alumina has a surface area of about 80 m^2/g.

The hydroxyl groups of silica particles tend to be slightly acid and so the particles have a greater affinity for basic compounds. Alumina, on the other hand, has oxide groups which tend to be basic. Alumina, therefore, has a greater affinity for acid compounds. It is mainly the polar functional groups of the solute that interact with the stationary phase. The surface adsorbing groups of a silica gel particle are shown below:

The hydrogen bonded -OH groups act as the strongest adsorbers. These are present in "Activated" adsorbers, where the adsorber has been heated to dive off water. This highly active surface is required for separating non-polar solutes with a non-polar solvent.

Hence, the water content of the stationary phase is very important as this determines the strength of the binding sites. Separating polar solutes with a non-polar solvent requires a deactivated surface with a fair amount of water in the particle.

Sample component migration is dependent upon the mobile phase. This competes for the binding sites on the stationary phase. More polar solvents have a greater affinity for the binding site than the less polar ones. In general, the more polar the solvent is, the greater its eluting power. Hence, the greater the degree of migration in the plate. Some common solvents are listed below in order of polarity:

n-Pentane	Least Polarity
n-Hexane	
Iso-octane	
Cyclohexane	
Diethyl ether	
Chloroform	
Dichloromethane	
Acetone	
Dioxane	
Tetrahydrofuran	
Ethyl acetate	
Acetonitrile	
Iso-propanol	
Ethanol	
Methanol	
Ethylene glycol	
Water	Greatest Polarity

Mixtures of solvents can be used. The elutropic strength of a mixture of two solvents is not necessarily a linear relation with the percentage of one in the other, due to solvent-solvent and solvent-surface interaction. The elutropic strength can be the same using different solvent mixtures, but resolution can be improved with certain solvents. This means that it is often better to use a solvent mixture rather than a pure solvent. Water is not miscible with dichloromethane and those above it in the list.

A fluorescent indicator can be added to the adsorbent to assist in the localisation of the sample components after the chromatogram has been run. This usually absorbs light at 254 nm (one of the mercury emission lines) and emits light in the visible region. The components appear as dark spots on a fluorescent background when the plate is exposed to a mercury lamp.

Samples are applied in the form of spots at points (origin) near the bottom of the plate. The plates are then placed in a tank containing the mobile phase at a depth of about 3-5 mm. The mobile phase moves up the plate by capillary action. This motion is countered by gravity and, hence, the flow rate decreases as the liquid moves up the plate.

Due to a lack of equilibrium between the stationary and mobile phases, variable flow rates, and variable evaporation rates of the mobile phase, R_f values (the ratio of the distance moved by the solute to the distance moved by the solvent front) will vary with the equipment used. Hence, it is essential to run standards with the unknowns.

After the mobile phase has migrated most of the way up the plate, the solvent front is marked and the plate is dried. Components are then localised by non-destructive methods (UV absorption or fluorescence, or adsorption of iodine vapour). Specific (destructive) reagents can then be used to localise components of interest.

Components in unknown mixtures are identified by comparing their migration, and response to localising agents, with known standards.

Tables of R_f values in the TLC system used and the response to localising agents can also help in identification, if the unknown contains components not included in the standards run on the TLC.

Many compounds may have the same R_f value on a single system. Hence, all unknown compounds with the same R_f value as a known standard may not be the same compound. However, if it also has the same R_f value as the standard in a completely different TLC system, the chances of the two compounds being the same is much greater. Even so, mass spectrophotometry of the two compounds would be needed for complete confirmation.

EXTRACTION TECHNIQUES

Due to the complex nature of biological fluids, it is often difficult to identify and quantitate endogenous or exogenous compounds that are present in small amounts. Hence, means of isolating, and often concentrating, the compound of interest are necessary before both qualitative and quantitative analysis. This is often true when a mixture of xenobiotics (compounds foreign to the body. Eg. drugs and poisons) are present in the sample.

A systematic series of extractions will separate acidic, neutral and basic compounds. Hence, analysis can be carried out on each of these groups, which are now partially purified.

The basis of extraction is the use of a non-polar organic solvent (eg. diethyl ether or chloroform. Chloroform has the advantage, over ether, as it is non-flammable) and an aqueous phase. Polar compounds will remain in the aqueous layer; whereas, non-polar compounds will be extracted into the organic layer.

The polarity of ionisable organic compounds depends upon their ionisation. An organic acid will be non-ionised (hence, non-polar) at low pH; and will, therefore, move into the organic phase. If the pH is increased, the compound will ionise ($HA \rightarrow H^+ + A^-$) and become polar (ie. water soluble) and will move into the aqueous phase. The reverse situation applies to basic compounds. These are non-ionised (organic soluble) at high pH, and ionised (water soluble) at low pH.

Weak acids and bases require greater pH shifts to ionise them and render them water soluble. Hence, weak bases can be separated from strong bases, and weak acids from strong acids, by pH adjustment.

Extraction systems can be used for urine, gastric aspirate, tissue homogenates, plasma and whole blood. These involve shaking the sample, after pH adjustment, with a water immiscible organic solvent in a separating funnel. However, emulsions are often formed due the presence of proteins. Emulsions are more likely to form when chloroform is used as the extracting solvent, rather than ether. Protein precipitation can be used prior to extraction. In this case, protein bound components will be removed and the assay may result in reporting a false negative or producing low quantitative values.

The problem of emulsions can be avoided by using adsorptive columns such as Merck's "Extrelut". These columns are designed to replace, and to improve upon, the standard extraction techniques.

The basis of the system is that the aqueous phase is held to an inert support and given a large surface area. The organic phase flows over the immobilised aqueous phase and solutes distribute themselves between the two phases, depending on their solubility. The organic phase is then collected and used for analysis.

The inert support of "Extrelut" is a large-pore kieselguhr. This is packed into a column between two filter plates. Flow through the column is controlled by the diameter of a hypodermic needle or cannula fitted to the base of the column.

The amount of kieselguhr in the pre-packed columns is designed to adsorb between 0.1 and 20 ml of aqueous solution, depending on the column. Elution is then carried out using a water immiscible solvent (eg. chloroform, ether, etc.). The aqueous phase remains held on the kieselguhr and non-polar substances are extracted into the organic solvent.

The pH of the aqueous solution can range from 1 to 13 and can be changed on the column, after an initial extraction. After extraction, the aqueous solution may be eluted from the column with saturated NaCl.

Eluting solutions can include a small proportion of a water miscible solvent (eg. 10% methanol in chloroform). However, if too much water miscible solvent is included in the eluting mixture, the volume of the aqueous phase is expanded and some aqueous phase will be lost from the column.

"Extrelut" has a number of advantages over the usual extraction procedures. It avoids the need to shake the solvents in a separating funnel; which, in turn, avoids the formation of emulsions. The eluted organic phase does not require drying with anhydrous sodium sulphate, as is the case with the standard procedure. Extraction using "Extrelut" columns is as complete as a series of shakings with the organic solvent. This saves on time and solvent volumes.

The larger "Extrelut" columns can be repacked with fresh kieselguhr for future use.

EXERCISES

EXERCISE 8-1

THE IDENTIFICATION OF A BASIC DRUG IN A TABLET

In 1982, Stead et al. published the evaluation of 29 TLC systems for the identification of drugs. From these, 4 systems were chosen for basic drugs and four for acidic and neutral drugs. R_f data on 794 drugs (594 basic, 48 neutral and 152 acidic) on all 8 systems are included in the paper. They recommend that the best system for screening basic drugs is to use a KOH treated silica stationary phase and chloroform : methanol (9:1) as the mobile phase. For acidic and neutral drugs, they recommend a silica stationary phase and chloroform : acetone (4:1) as the mobile phase.

In the identification of drugs in a tablet it is often essential to isolate the drugs from filling material (eg. starch) in the tablet. This usually involves extraction into an organic solvent.

Drugs can be classified into acidic (eg. aspirin), neutral (eg. digoxin) or basic (eg. diazepam). The class into which a drug falls depends upon the ionisable groups in its chemical structure.

In an alkaline solution, the acidic drugs will ionise and become water soluble. The neutral and basic drugs will be non-ionised and soluble in a water immiscible organic solvent (eg. ether).

In acid solutions, basic drugs become ionised and will be water soluble. Acidic and neutral drugs will now be soluble in the water immiscible solvent.

Hence, the component drugs can be separated into groups according to their acid/base characteristics before other means of identification are used.

TLC is a rapid and simple first step in identification. This should then be followed by a confirmatory technique such as high performance liquid chromatography (HPLC) and examining the ultra violet (UV) spectrum of the drug.

In this exercise, the extraction of the basic drug in a tablet will be followed by TLC and UV spectra determination.

The tablet will be treated with 0.45 M NaOH to remove the charge on the basic drugs. The mixture will then be extracted with diethyl ether. This extract (containing the basic drug) will then be dried with anhydrous sodium sulphate and evaporated to dryness.

The residue will then be redissolved in methanol and 2 µl applied to a 0.2 mm thick silica (Average pore diameter = 6 nm) gel TLC plate. These plates will have been pre-treated by dipping them into a methanolic solution of KOH (0.1 M) and allowing them to air dry. This neutralises the acidic groups on the silica and, hence, increases the migration of the basic drugs. The remainder of the methanol solution will be used for a UV scan.

A selection of basic drug standards will be applied to the TLC plate, along with the unknown. The plate will then be run in Methanol : Ammonia (100 : 1.5) and dried.

The silica gel contains a compound that fluoresces green when excited by a 254 nm light source (Hg lamp with 254 nm filter). Hence, the plate will be examined under UV light of 254 nm. The drugs should appear as dark spots on a green background.

The plate will also be examined under UV light of 366 nm (Hg lamp with 366 nm filter). Any fluorescent drug will fluoresce against a dark blue background.

The plates will then be sprayed with acidified iodoplatinate. Most basic drugs give brown, grey, blue or purple spots.

The standard drugs are:

Codeine	-	An analgesic
Diazepam (Valium)	-	A tranquilliser
Imipramine	-	An antidepressant
Morphine	-	An analgesic
Oxazepam (Serepax)	-	A tranquilliser
Primaquine	-	An antimalarial
Quinine	-	An antimalarial

These are made up as 5g/l solutions in methanol.

Most drugs absorb in the UV and have characteristic spectra. If the drug can ionise in acid or alkaline media, it will often show different spectra at pHs 2 and 13. These facts can be used in confirming the identification of the drug.

REAGENTS AND EQUIPMENT:

1. 0.45 M NaOH.
    Dissolve 18 g of NaOH in distilled water and make to 1 l.
2. Analar Diethyl Ether.
3. HPLC Grade Methanol
4. Anhydrous Sodium Sulphate.
5. Drug Standards.
    Separately, dissolve 10 mg of the following in 2 ml aliquots of methanol:
    Codeine.
    Diazepam.
    Imipramine.
    Morphine.
    Oxazepam.
    Primaquine.
    Quinine.
6. 0.1 M HCl.
7. Developing Solution.
    Add 15 ml of ammonium hydroxide (SG 0.880) to 1 l of methanol and mix well.
8. Treated TLC Plates.
    Dissolve 2.8 g of KOH in 500 ml of methanol. Dip the TLC plates (Merck Cat No 5554 - Silica Gel G 60 F_{254} on aluminium foil, 200 x 200 mm) in this solution and allow to air dry.
9. Tablets.
    Preferably Codis (8 mg codeine phosphate + 500 mg aspirin - soluble tablet).
    Other tablets can be used if they contain one of the standard drugs.
10. Acidified Iodoplatinate Reagent.
    Dissolve 250 mg of platinic chloride (chloroplatinic acid) and 5 g of KI in distilled water and make to 100 ml. Add 2 ml of concentrated HCl and mix well.

Steam Bath with Adjacent Compressed Air Connected to Rubber Tubes and Pasteur Pipettes.
Pestles and Mortars (If using a non-soluble tablet).
100 ml Beakers.
250 ml Separating Funnels with Stoppers.
Stands, Bosses and Rings.

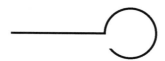

50 ml Measuring Cylinders.
Spatulas.
    Weigh out 3 - 4 g of anhydrous sodium sulphate as an indication of the required amount.
1 ml Pipettes.
Pencils and Rulers.
2 μl Capillary Samplers and Capillaries.
TLC Tanks with Lids.
254 nm UV Lamp in a Darkened Room.
366 nm UV Lamp in a Darkened Room.
Spray Gun Connected to Compressed Air, in a Fume Cupboard.
    Add acidified iodoplatinate reagent.
Test Tubes.
100 μl Pipettors and Tips.

5 ml Graduated Pipettes.
UV Scanning Spectrophotometer with Quartz Cuvettes.
Paper Tissues.

EXTRACTION PROCEDURE:

Note that ether is highly flammable. There must be no flames or sparks in the laboratory.

1. Add 1 tablet to 30 ml of 0.45 M NaOH in a beaker and stir to dissolve as much as possible of the tablet.

    If a non water-soluble tablet is used, it will have to be ground up in a pestle and mortar before treatment with the NaOH.

2. Transfer the contents of the beaker to a separating funnel.

3. Add 50 ml of ether and shake vigorously for 2 minutes to extract the neutral and basic drugs. Make sure that the stopper is held in place and release the pressure in the funnel, after the first few shakes, by inverting the funnel and opening the tap.

4. Stand until the 2 layers separate and then drain off, and discard, the bottom layer.

5. Transfer the top (ether) layer to a beaker and add 3-4 grams of anhydrous sodium sulphate.

6. Stir the liquid in the beaker to adsorb any aqueous liquid onto the sodium sulphate.

7. Transfer the ether extract to another beaker and evaporate to dryness on the steam bath, using a gentle stream of compressed air to speed the evaporation. Remove the beaker as soon as the ether has evaporated.

8. Redissolve the residue in 1 ml of methanol.

TLC IDENTIFICATION:

1. With a pencil, mark out a 200 x 200 mm KOH treated TLC plate as shown below (The spots should be 20 mm apart and 15 mm from the bottom of the plate):

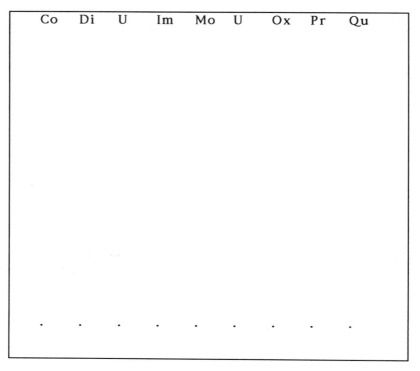

2. Using a 2 µl capillary sampler, apply 2 µl of the standards and unknown (U) to spots as indicated above. Wash the capillary, 3 or 4 times with methanol, between each sample.

3. Add 100 ml of developing solution (Methanol : Ammonia) to a TLC tank.

4. Add the plates (Up to 2 per tank).

5. Allow the solvent to run about 3/4 of the way up the plate.

6. Remove the plate and mark the solvent front with a pencil.

7. Allow the plate to dry in the fume cupboard.

8. Examine the plate under the 254 nm lamp and mark the position of any spots with a pencil mark in the centre of the spot.

9. Examine the plate under the 366 nm lamp and note any fluorescent spots.

10. Carefully spray the plate with acidified iodoplatinate and note the colour of the spots. Compare the colour of the unknown with that of the standards having similar migration. (The colours fade fairly rapidly).

11. Measure the distance from the origin to the centre of each spot and to the solvent front.

12. Calculate the Rf values for all spots and compare the standard drugs with the published values below:

Codeine	:	0.33
Diazepam	:	0.75
Imipramine	:	0.48
Morphine	:	0.37
Oxazepam	:	0.56
Primaquine	:	0.19
Quinine	:	0.50

13.   From the calculated R_f values, the colour of the spots and their fluorescent characteristics, identify the basic (or neutral) drug in the tablet.

UV SPECTRUM:

1.   Dilute 100 µl of the original methanol solution with 2.5 ml of 0.45 M NaOH to give a pH 13 solution.

2.   Dilute 100 µl of the methanol solution with 2.5 ml of 0.1 M HCl to give a pH 2 solution.

3.   Scan both these solutions, against their appropriate blank, between 220 and 350 nm, using overlay scans. Obtain the peak wavelengths (λ_{max}), under data processing, for each scan.

4.   Compare the λ_{max} values (nm) with those listed below:

Drug	pH 2	pH 13
Codeine	285	285
Diazepam	242, 284, 366	-
Imipramine	251	252
Morphine	285	298
Oxazepam	234, 280	233, 344
Primaquine	265, 574, 334	-
Quinine	250, 317, 346	280, 330

5.   Compare the shape of UV spectra with those of the standard drugs:

—————— pH 2 spectra.
.... pH 13 spectra.

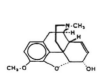

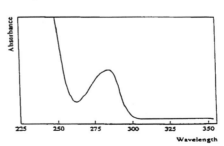

Codeine

8-9

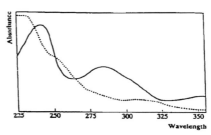

Diazepam

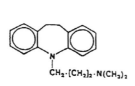
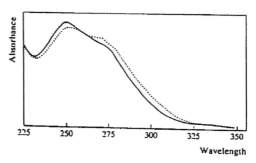

Imipramine

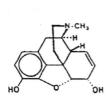
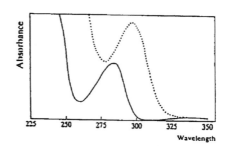

Morphine

The difference in codeine and morphine spectra is due to the phenolic -OH on morphine that ionises at pH 13 to the phenolic anion -O⁻. Codeine contains a methyl ether at this site. This does not ionise. Hence, there is no alkaline shift in the codeine spectrum.

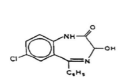
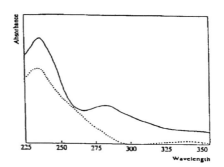

Oxazepam

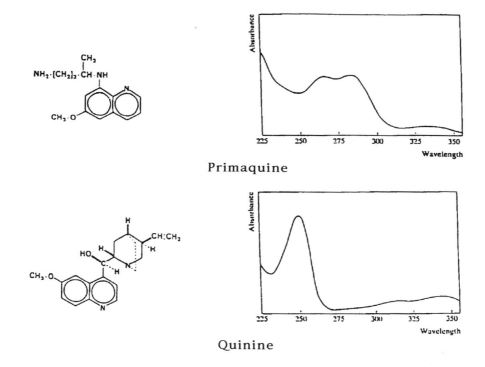

Primaquine

Quinine

EXERCISE 8-2

TOXI-LAB

"Toxi-Lab" is a commercial TLC drug identification system. It is currently produced by ANSYS Diagnostics Incorporated of the USA.

The basis of the system is extraction of the drugs into an organic solvent, concentration and then TLC separation. Separate systems are available for acidic and basic drugs.

The system is mainly designed to screen urine, but gastric aspirant and various body fluids can be used.

"Toxi-lab A" is the system for neutral and basic drugs. Urine is added to "Toxi-Tube A" which contains an organic solvent and buffers to bring the pH to 9. After extraction, the organic phase is concentrated onto a "Toxi-Disc". The "Toxi-Disc" is then inserted into a "Toxi-Gram A". This is a thin glass-fibre sheet impregnated with silica gel, and a vanadium salt. The "Toxi-Gram" also contains a number of standard drugs, in positions 1, 2, 3 and 4.

The chromatogram is then developed with methanol : water : ethyl acetate (2:1:58), plus a specified volume of NH_4OH.

After drying the chromatogram, the drugs are visualised by a staining technique. The chromatogram is initially exposed to formaldehyde vapour and then dipped in concentrated H_2SO_4. This is followed by a water wash and then examination under UV light. The final dip is in a modified Dragendorff's reagent (Iodinated bismuth subnitrate). The unknown is compared with the various standards, at all stages of the visualisation process. The results are recorded on a "Toxi-Lab A Worksheet".

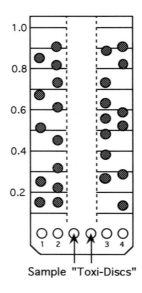

Sample "Toxi-Discs"

Toxi-Lab A Drug Standards and Order of Migration:

Position	Name	Use	Approx R_f
A-1	Propoxyphene	Analgesic	0.86
	Methadone	"	0.67
	Meperidine	"	0.51
	Codeine	"	0.26
	Morphine	"	0.15
A-2	Diazepam	Tranquilliser	0.92
	Cocaine	Stimulant	0.83
	Paracetamol (Acetaminophen)	Analgesic	0.73
	Caffeine	Stimulant	0.60
	Nicotine	"	0.45
	Amphetamine	"	0.31
	Methamphetamine	"	0.22
	Pseudoephedrine	Decongestant	0.16
A-3	Phencyclidine	Hallucinogen	0.89
	Trimeprazine	Antihistamine	0.73
	Triflupromazine	Tranquilliser	0.64
	Chrorpromazine	"	0.57
	Imipramine	Antidepressant	0.48
	Trifluoperazine	Tranquilliser	0.38
	Quinine	Antimalarial	0.27
A-4	Methaqualone	Hypnotic	0.92
	Meprobamate	Tranquilliser	0.83
	Amitriptyline	Antidepressant	0.59
	Doxepin	"	0.53
	Nortriptyline	"	0.28
	Strychnine	Stimulant	0.14

"Toxi-Lab B" is the system for neutral and acidic drugs. Urine is added to the "Toxi-Tube B", which contains an organic solvent and buffers to bring the pH to 4.5. After extraction, the organic phase is concentrated onto a "Toxi-Disc", which is then inserted into a "Toxi-Gram B". "Toxi-Gram B" is a glass fibre sheet impregnated with silica gel, but without vanadium.

The chromatogram is developed with dichloromethane : ethyl acetate (60:40), plus a specified volume of NH_4OH. The amount of ammonium hydroxide used in the developing solutions is adjusted to allow the migration of the drugs to be the same as that on the worksheets. The NH_4OH increases the migration of the basic drugs (Toxi-Gram A) and reduces the migration of the acidic drugs (Toxi-Gram B). However, it is important to compare the unknown with the reference drugs, rather than with the R_f values indicated on the worksheets.

After development and drying, the chromatograms are dipped into diphenylcarbazone (in dichloromethane) solution and allowed to dry. They are then dipped into silver nitrate and into mercuric sulphate solutions. The chromatograms are finally viewed under UV light. The unknown is compared with the various standards, at the colour developing stages. The results are recorded on a "Toxi-Lab B Worksheet".

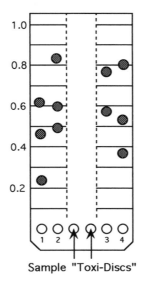

Sample "Toxi-Discs"

Toxi-Lab B Drug Standards and Order of Migration:

Position	Name	Use	Approx R_f
	Secobarbital	Hypnotic	0.62
B-1	Phenytoin	Anti-convulsant	0.47
	Phenobarbital	Hypnotic	0.24
	Glutethimide	Hypnotic	0.83
B-2	Pentobarbital	"	0.60
	Aprobarbital	"	0.49
B-3	Ethinamate	Hypnotic	0.78
	Amobarbital	"	0.57
	Diazepam	Tranquilliser	0.80
B-4	Butabarbital	Hypnotic	0.53
	Barbital	"	0.37

Pink dye markers are included with the reference samples. These migrate just behind the solvent front. The "Toxi-Grams" can be photocopied as a permanent record.

Additional standards are available as drug impregnated "Toxi-Discs". Photographs of the chromatograms, obtained from the urine of patients on various drugs, are included with the kit ("Photo-Grams"). These include the metabolite patterns (most metabolites have a lower R_f value than the parent drug). A table of R_f values, for the two developing systems, is also supplied for a large number of drugs; together with their staining characteristics, etc.

Using 5 ml of urine, the threshold of detection for most drugs is approximately 1.0 µg or less/ml of urine. Some substances, however, are not detected with Toxi-Lab and a negative result does not mean that all drugs are absent. Questionable findings should be confirmed by employing a different methodology.

REAGENTS AND EQUIPMENT:

1. Toxi-Lab A Kit.
    This includes: Toxi-Grams A. These have a V printed on top to indicate that they contain vanadium.
    Toxi-Discs A. For basic extract concentration.
    Toxi-Tubes A. Extraction tubes.
    Toxi-Dip A. Modified Dragendorff's reagent.
    Toxi-Dip A jars.
    Toxi-Lab A Worksheets.
    Toxi-Grams B.
    Toxi-Discs B.
    Toxi-Tubes B
    Toxi-Dips B. Diphenylcarbazone, silver nitrate and mercuric nitrate reagents.
    Toxi-Dip B jars.
    Toxi-Lab B Worksheets.
    Chromatography jars.
    Toxi-Grams. Photographs of run chromatograms.
2. Developing Solution A.
    Mix 3 ml of methanol, 1.5 ml of distilled water and 87 ml of Analar ethyl acetate. Shake for 20 - 30 seconds.
3. Developing Solution B.
    Mix 60 ml of dichloromethane and 40 ml of Analar ethyl acetate. Shake for 20 - 30 seconds.
4. Toxi-Dip A-1.
    Add 10 - 12 ml of formaldehyde solution ($\approx$ 37% HCHO) to the bottom of the A-1 jar. Blot dry the base plate and cap tightly.
5. Toxi-Dip A-2.
    Fill the jar to about 5 mm from the top with concentrated sulphuric acid. Place this jar in the base of a large Petri dish.
6. Beaker or Jar Filled with Distilled Water.
7. Toxi-Dip A-3.
    Add the contents of the A-3 vial and 10 ml of glacial acetic acid to the A-3 jar and mix. Add distilled water to bring the level to about 5 mm from the top. Cap tightly and mix well.
8. Toxi-Dip B-1.
    Add the contents of the diphenylcarbazone vial to the B-1 jar. Add dichloromethane to bring the level to about 5 mm from the top. Cap tightly and mix well.
9. Toxi-Dip B-2.
    Add the contents of the silver nitrate vial to the B-2 jar. Add distilled water to bring the level to about 5 mm from the top. Cap tightly and mix well.

10. Toxi-Dip B-3.
>Add the contents of the mercuric nitrate vial to the B-3 jar. Add distilled water to bring the level to about 3/4 full. Slowly, with constant stirring, add 10 ml of concentrated sulphuric acid. Add distilled water to bring the level to about 5 mm from the top. Cap tightly and mix well.
11. Ammonium Hydroxide (SG 0.880).
12. Urines.

A.	10 mg of codeine	+	1 mg of secobarbital.
B.	2 mg of diazepam	+	2 mg of phenobarbital.
C.	10 mg of paracetamol	+	2 mg of pentobarbital.
D.	5 mg of chlorpromazine	+	2 mg of barbital.
E.	2 mg of amitriptyline	+	2 mg butabarbital.
F.	1 mg of imipramine	+	2 mg of amobarbital.

>Dissolve in 1 ml of methanol. Add 100 ml of distilled water and mix well. Add 100 ml of urine and mix well.
>Distribute 10 ml aliquots into numbered McCartney bottles.

366 nm UV Lamp in a Darkened Room.
Electric Warmer (Food Warmer).
Positive Displacement Sampler (1 - 50 µl).
Ceramic Spotting Tiles (Evaporation Plates).
Pins.
Bench Centrifuges to hold the "Toxi-Tubes".
Plastic Forceps.
Pasteur Pipettes with Tips.
White Card Pieces ($\approx$ 30 x 30 mm).
Paper Towels.
Bench Covered with Brown Paper.

PROCEDURE:

The Toxi-Grams should only be handled with the plastic forceps provided.

A "spiked" urine sample is provided for each student. Each urine has had 2 drugs added to it. Carry out the following procedure to identify the drugs:

Preliminary Steps:

1. Plug in the electric warmer.

2. Into separate wells of the cool evaporation plate, place 2 "A" Blank Discs and 2 "B" Blank Discs. (Use a pin to transfer the discs).

Extraction:

1. Shake a Toxi-Tube A to dislodge the salt.

2. To the Toxi-Tube A and Toxi-Tube B, add urine to the "pour" line (5 ml arrow for Toxi-Tube A and 4.5 ml arrow for Toxi-Tube B). Cap and mix both tubes by inversion for approximately 2 minutes. (Do not shake).

3. Centrifuge the tubes for 2-5 minutes. (The coloured aqueous layer should be on the bottom, after centrifugation).

Concentration:

1. Place the evaporation plate on the warmer.

2. With a Pasteur pipette, transfer 5 drops of the organic layer (top) from Toxi-Tube A to one well containing an A disc. Transfer the remaining organic layer to the other well containing an A disc. Allow the solvent to evaporate between adding aliquots of the extract to the well. Carry out a similar transfer for the Toxi-Tube B. (This gives a ratio of approximately 1:20 for each sample).

3. Immediately after the extracts have completely evaporated, remove the discs from the wells with a pin.

Inoculation:

1. Remove a Toxi-Gram A and a Toxi-Gram B from their jars and transfer to a firm flat surface, which is covered with a sheet of clean paper towelling.

2. Using a pin, insert the "A" Discs into the centre holes of the Toxi-Gram A and the "B" discs into the centre holes of the Toxi-Gram B. Be careful not to damage either the Toxi-Discs or the Toxi-Grams. Use a piece of card to press home the discs.

3. Place the "loaded" Toxi-Grams on the warmer with the disc ends slightly off the edge.

Development:

(Do not proceed to this step until inoculation is completed).

1. Transfer 3 ml of "A" and 3 ml of "B" developing solutions into the respective chromatography jars. With a positive displacement sampler, transfer the volume of ammonium hydroxide, recommended on the Toxi-Gram label, directly to the fluid in the specific chromatography jar. Cap immediately and swirl each jar vigorously for a few seconds.

2. Remove the Toxi-Grams from the warmer and lower each, disc ends first, into the specific chromatography jar. The side edges of the Toxi-Gram should not touch the jars. Cover and do not disturb during migration.

3. Remove the Toxi-Grams when the dye spots reach the level of 0.95 (13-18 minutes) and place on the warmer **face down** for 20-30 seconds, until the fumes have vanished.

Detection and Identification:

Toxi-Lab A

Preliminary Step:

Place the Toxi-Gram into Toxi-Dip A-1 (formaldehyde vapour) for approximately two minutes. (Several Toxi-Grams may be placed in the A-1 jar and removed at intervals). Remove the Toxi-Gram and place on the warmer for 5-10 seconds to remove **some** of the formaldehyde fumes.

Stage I:

Dip the Toxi-Gram slowly in and out of Toxi-Dip A-2 (concentrated H_2SO_4), lean against the jar until the yellow meperidine spot becomes visible. Attempt to match the position and colour of any spot in the unknown zone with a standard drug having the same characteristics. Record your observations on the Toxi-Lab A Worksheet.

Stage II:

Dip the Toxi-Gram briefly in and out of a jar or beaker of distilled water. Wait about 5 seconds and then dip again. Some drugs will become visible for the first time, others will fade, and still others will change colour. Attempt to match the position and colour of any spot detected in the central unknown zone with a standard drug having the same characteristics. Record your observations on the worksheet. Carry out one further dip in the water and note any changes in the spots.

Stage III:

Re-dip the Toxi-Gram several times in the same jar of water and then view, in the dark, under long-wave ultraviolet light (366 nm). Use both transmitted and reflected UV light. Attempt to match the position and fluorescence of any spot detected in the central unknown zone with a standard drug having the same characteristics. Record your observations on the worksheet.

Stage IV:

Place the Toxi-Gram into Toxi-Dip A-3 (modified Dragendorff's reagent) and agitate for a few seconds. Remove and compare the position and colour of any spot detected in the central unknown zone with a standard drug having the same characteristics. Record your observations on the worksheet. Blot dry the Toxi-Gram and photocopy.

Toxi-Lab B

Preliminary Step:

Dip the Toxi-Gram smoothly in and out of Toxi-Dip B-1 (diphenyl- carbazone solution) and lean it against the jar to "dry". All traces of dichloromethane should disappear before continuing. However, do not heat.

Stage I:

Remove the caps from Toxi-Dips B-2 (silver nitrate) and B-3 (mercuric sulphate). Dip the Toxi-Gram into Toxi-Dip B-2, waiting momentarily for a golden-brown background colour to develop. Then **immediately** transfer into Toxi-Dip B-3, agitating the Toxi-Gram up and down several times until the background clears. Record your observations on the Toxi-Lab B Worksheet.

Stage II:

View the Toxi-Gram in the dark under long-wave UV light (366 nm). Record your observations on the worksheet. Blot dry the Toxi-Gram and photocopy.

Note:

Urine from patients, to whom the drugs have been administered, will contain metabolites of the drugs. Look up the Photo-Grams of the drugs that you have identified, and compare the metabolite pattern with chromatogram that you obtained.

EXERCISE 8-3

THE SEMIQUANTITATIVE DETERMINATION OF METHYLMALONIC ACID IN URINE BY THIN LAYER CHROMATOGRAPHY

This exercise can be done in conjunction with Exercise 9-6, as it uses the same sample. This is urine from a 14 week old girl with methylmalonic acidaemia. Details of this condition are given in Exercise 9-6.

The organic acids are extracted from the urine by the same method as used in Exercise 9-6, and taken to dryness.

The organic acids are then redissolved in ethanol and 1 and 5 µl samples spotted onto a 0.2 mm thick silica gel TLC plate. Standards are also applied to the plate, which is then developed with an amyl acetate : acetic acid : H_2O (30 : 10 : 3) mixture.

After drying the plate, the methylmalonic acid is localised with Fast Blue B Salt reagent. The sizes of the methylmalonic acid spots in the unknown are compared with those of the standards.

Running the extract on the TLC plate allows the methylmalonic acid to be separated from other components in the extract that may interfere with the colour reaction. It also allows the methyl malonic acid to diffuse in the gel to give spot sizes that are proportional to the amount present.

REAGENTS AND EQUIPMENT:

1. Analar Sodium Chloride.
2. Analar Anhydrous Sodium Sulphate.
      Also weigh out 0.5 gram, as an indication of the required amount.
3. 6 M HCl.
4. Ethyl Acetate.
5. Diethyl Ether.
6. Patient's Urine.
      Dissolve 1 g of MMA and 0.7 g of glycine in 300 ml of normal urine.
7. Ethanol.

8.	Developing Solution.
	Mix well:
		300 ml of amyl acetate,
		100 ml of glacial acetic acid,
		30 ml of distilled water.
9.	100 mmol/l Methylmalonic Acid Standard.
	Dissolve 1.18 g of methylmalonic acid in distilled water and make to 100 ml.
	Store at 4°C.
10.	Fast Blue B Reagent.
	Dissolve 500 mg of Fast Blue B Salt (o-dianisidine tetrazotized - Sigma Cat. No. D3502) in 100 ml of 75% ethanol. Add 4 ml of glacial acetic acid and mix well.
	Prepare fresh.

6 ml Screw-Top Culture Tubes.
Spatulas.
Bench Centrifuges.
Centrifuge Tubes.
Pasteur Pipettes.
50°C Heating Block with Nitrogen Cylinder and Fine Tubes from a Manifold above the Block.
Test Tubes.
Wassermann Tubes.
2 and 10 ml Graduated Pipettes.
100, 200 and 500 µl Pipettors with Tips.
TLC Plates.
	Merck Cat No 5553 - Silica Gel G 60 on aluminium foil, 200 x 200 mm.
TLC Tanks with Lids.
100 ml Measuring Cylinders.
1 and 5 µl Capillary Pipettors with Capillaries.
Brown Paper Covered Bench.
Pencils and Rulers.
Hair Dryers.
Spray Gun.
	In fume cupboard, containing the Fast Blue B reagent. Connected to a compressed air source.
Oven at 80°C.

PROCEDURE:

1.	To a 6 ml screw-top culture tube, add:

		0.5 g NaCl,
		2 drops of 6 M HCl,
		2 ml of ethyl acetate,
		1.0 ml of patient's urine.

2.	Cap the tube and shake vigorously for 2 minutes.

3.	Centrifuge for about 2 minutes at moderate speed.

4.	Transfer the top layer to a test tube, with a Pasteur pipette.

5.	Re-extract the aqueous phase with another 2 ml of ethyl acetate, as in steps 2 and 3, and transfer the top layer to the test tube.

6.	Re-extract the aqueous layer with 2 ml of diethyl ether, as above, and add the ether extract to its test tube.

7.   Add about 0.5 g of anhydrous sodium sulphate to the test tube and swirl the tube to adsorb any aqueous material.

8.   Carefully decant the liquid into a centrifuge tube.

9.   Evaporate to dryness under a stream of nitrogen, at 50°C in a heating block.

10.   Redissolve the organic acids in 2.0 ml of ethanol.

11.   Mark out a TLC plate as shown below:

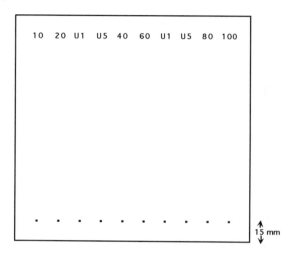

12.   Apply 1 and 5 µl of the extract to the U1 and U5 spots, using a 1 µl capillary pipettor. (Use a hair dryer to dry the 5 µl spot after each µl application).

13.   Prepare the following standards, in Wassermann tubes, using the 100 mmol/l (100 nmol/µl) methyl malonic acid standard:

	10	20	40	60	80
µl of distilled H$_2$O	900	800	600	400	200
µl of MMA standard	100	200	400	600	800

14.   Apply 1 µl of the 10, 20, 40, 60, 80 and 100 nmol/µl standards to their appropriate spots.

15.   Place 100 ml of the developing solvent in a chromatography tank.

16.   Add the plates (2 per tank) and cover with the lid.

17.   Run the plates for 35 minutes.

18.   Remove the plates and dry them in an 80°C oven.

19.   Spray the plates with the Fast Blue B reagent.

20. Dry the plates with warm air from the hair dryer (methylmalonic acid should appear as purple spots).

21. Compare the spot size and intensity of the unknowns with the standards.

Calculate the approximate amount of methylmalonic acid excreted in 24 hours. Assume that the sample is from a 450 ml, 24 hour collection.

Reference Values:

| Normal children and adults | : | $< 40\ \mu mol/24$ hours. |
| Children with methylmalonic acidaemia | : | 2-50 mmol/24 hours. |

GENERAL REFERENCES:

Widdop, B., (1986), "Hospital Toxicology and Drug Abuse Screening". In "Clarke's Isolation and Identification of Drugs", Ed. Moffat, A.C., p 6-14, Pharmaceutical Press.

Thin Layer Chromatography.

Curry, A. S. (1969), "Thin-Layer Chromatography" in "Isolation and Identification of Drugs", Vol. 1, Ed. Clarke, E.G.C., p 43-58, Pharmaceutical Press.

Bastos, M.L., Kananen, G.E., Monforte, J.R. and Sunshine, I., (1975), "TLC of Basic Organic Drugs", in "Methodology for Analytical Toxicology", Ed. Sunshine, I., p 434-442, CRC Press.

Stead, A.H., Gill, R., Wright, T., Gibbs, J.P. and Moffat, A.C., (1982), Standardised Thin-Layer Chromatographic Systems for the Identification of Drugs and Poisons, Analyst, 107, 1106-1168.

Extraction Techniques.

Higgins, G. and Leach, M., (1975), "Extraction Procedures", part of "Screening Tests for Common Drugs" in "Isolation and Identification of Drugs", Vol. 2, Ed. Clarke, E.G.C., p 878-880, Pharmaceutical Press.

"Extrelut" Information Booklet, published by Merck.

Toxi-Lab.

Toxi-Lab Drug Detection System Instruction Manual, published by ANSYS Inc.

Methylmalonic Acid by TLC.

Gutteridge, J. M. C. and Wright, E. B., (1970), A Simple and Rapid Thin-Layer Chromatographic Technique for the Detection of Methylmalonic Acid in Urine, Clin. Chim. Acta 27, 289-291.

CHAPTER 9

GAS CHROMATOGRAPHY

The basic components of a gas chromatography system are illustrated below:

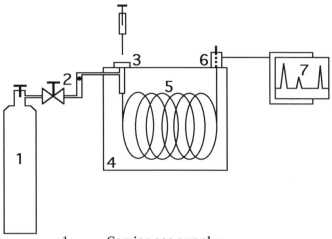

1. Carrier gas supply.
2. Gas flow regulator (and flow meter).
3. Sample injection system.
4. Thermostatically controlled oven.
5. Column.
6. Detector.
7. Recorder or computing integrator.

The inert carrier gas flows continuously through the injection system, column and detector. The flow rate is carefully controlled to ensure reproducible retention times and to minimise detector drift.

The sample is injected into a heated injection port, where it is vaporised and carried into the column. The column can be a packed column or a capillary (open tubular) column. The packed column is usually made of glass or stainless steel and packed with solid particles. These particles may be coated with a high boiling point liquid (Gas-Liquid Chromatography, GLC) or be used as adsorbents (Gas-Solid Chromatography, GSC). Capillary columns are made of fused silica and coated on the inside with a thin layer of high boiling point liquid.

The sample partitions itself between the stationary phase (adsorbent or high boiling point liquid) and the carrier gas. Different components of the sample are adsorbed onto, or dissolved in, the stationary phase to different degrees. Hence, the components of the mixture are separated on the column.

After the column, the carrier gas, together with the sample components as they are eluted from the column, pass through the detector. An electrical signal is generated by the detector as the sample components pass through. This signal can be fed to a recorder or computing integrator. The components of a mixture can be identified by their retention times (ie. the time in the column) and quantitated by the amount of signal generated by the detector (eg. the peak height on a recorder trace).

The various components will now be examined in more detail:

1.   The Carrier Gas.

The purpose of the carrier gas is to carry the sample through the column. It must be inert, ie. it must not react with the stationary phase or with the sample components.

It also provides a suitable matrix for the detector to measure the sample components. Helium is a good gas for thermal conductivity detectors. Nitrogen will provide the best efficiency with a flame ionisation detector, but only over a narrow gas flow rate. Very dry nitrogen is desirable for electron capture detectors.

The gases should be of high purity to avoid interaction of impurities with the column or sample. Oxygen removal is important, especially with capillary columns, as most stationary phases can de oxidised at high temperatures.

2.   Flow Control.

Accurate control of the carrier gas flow is essential for reliable analysis. There is a particular gas flow rate, dependent upon the column packing and type of carrier gas, which gives optimum efficiency for a column. Flow rates above or below this reduce the efficiency of the column.

Obviously, the gas flow rate also effects the retention time of components of the sample.

If a flow meter is not included in the system before the injector, monitoring the gas flow leaving the detector, with a bubble flow meter, can be used to measure flow through the column. Gas leaks can also be detected by this method. Air and hydrogen flow rates, for flame ionisation detectors, can also be measured as they leave the detector.

Water vapour and then oxygen adsorbers can be included in the carrier gas flow line.

3.   Sample Injection Systems.

Most samples are liquids and these are usually injected into the injection port, via a silicone rubber septum, by a microsyringe. Volumes injected are usually in the order of 1 to 10 µl.

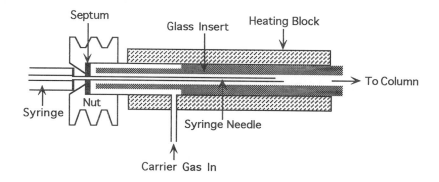

The block is heated to a temperature that rapidly vaporises the sample, but not hot enough to produce decomposition of the sample. Therefore, once injected, the sample is immediately vaporised and swept into the column.

The sample volumes injected into capillary columns are much smaller (can be as low as 1 nl). Hence the sample is usually split, after vaporisation and mixing with carrier gas, and only a small proportion enters the column.

4.   Thermostatically Controlled Oven.

The degree of retention of compounds on a column is dependent upon temperature, as well as upon the type of stationary phase and carrier gas flow rate. The higher the temperature, the shorter will be the retention time due to increased volatility of the sample components. However, as the temperature increases, the resolution (ability to separate sample components) decreases. Therefore, a compromise is reached between analysis time and resolution. Retention time is approximately halved for each 30°C rise in temperature.

Where the temperature of the column is kept constant, it is known as isothermal analysis. Temperature programming incorporates a rise in temperature during the analytical run. This is usually essential where the sample contains components with a wide range of boiling points.

The diagram below illustrates the use of isothermal analysis and temperature programming to separate a hydrocarbon mixture:

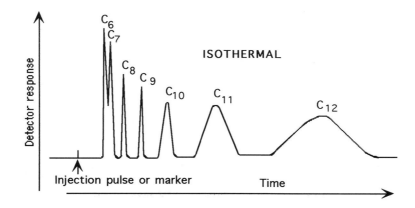

TEMPERATURE PROGRAMME

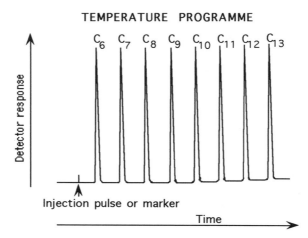

Detector response

C_6 C_7 C_8 C_9 C_{10} C_{11} C_{12} C_{13}

↑ Injection pulse or marker

Time →

At constant temperature, the early peaks (low boiling point components) emerge very rapidly and tend to overlap. The higher boiling point components take a long time to emerge and appear as short, broad peaks. Some high boiling point components remain on the column at the end of the run. These may appear as baseline drift or "ghost" peaks in the next chromatographic run.

These problems are overcome with temperature programming. The run is started at a lower temperature than that used for isothermal analysis. This gives good separation of the low boiling point components. The temperature is increased at a constant rate and reaches a final temperature in excess of that used in the isothermal run. This brings the higher boiling point components off the column in a reasonable time. They also appear as higher and narrower peaks than those seen in the isothermal analysis. This makes it easier to quantitate these components by peak area measurement.

5.   Column.

There are 2 classes of columns, capillary (open tubular) and packed.

Capillary Columns

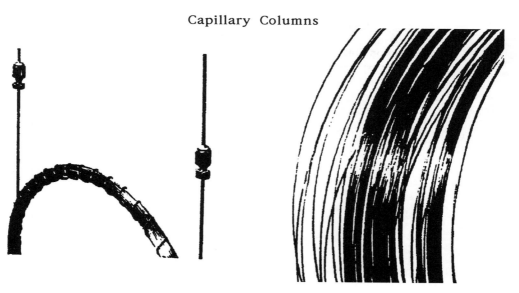

Most capillary columns have a thin film (usually 0.3-0.5 μm thick) of stationary phase coated on the inside of a fused silica capillary. The capillary has a polyimide coating, on the outside, which gives it strength and allows the capillary to be bent without breaking. These columns can be referred to as "wall-coated open tubular" (WCOT).

In "support-coated open tubular" (SCOT) columns, the liquid stationary phase is coated onto a relatively thick layer of support material that lines the column. Similar columns used for adsorption or molecular sieving are known as "porous-layer open tubular" (PLOT) columns. Typical examples are illustrated below:

	WCOT Small bore	SCOT Wide bore
Internal diameter (mm)	0.2	0.5
External diameter (mm)	0.35	1.0

The SCOT columns have a higher capacity (amount of material that the column can handle) than the WCOT. However, these have been largely replaced by 530 μ ("Megabore") columns. These are fused silica columns with an internal diameter of 0.53 mm and a coating with a film thickness of about 1-3 μm. These columns are designed to replace packed columns and can take up to about 5 μl without the need for sample splitting.

Most modern capillary columns have their stationary phase bonded onto the silica by covalent bonds and cross-linking of the phase material. These columns can be washed with solvent to remove impurities that build up on the column.

Columns of 10 m length can separate about 15 components; whereas, those of 25-30 m can separate up to 50 components. 50 m, or longer, columns should be used where more than 50 components are present.

Packed Columns

These are made of glass or stainless steel and are usually 2 m long with an internal diameter of about 3 mm. Glass columns have a number of advantages over stainless steel. With glass, the packing can be observed for changes in composition or tightness of packing. Glass is also more inert than stainless steel. However, it is less robust and material can be adsorbed onto the glass surface. The latter problem can be overcome by "silanising" the glass, prior to packing.

The packing material in GSC is an adsorbent such as charcoal, alumina or silica gel. Very large surface areas can be obtained by using porous particles. Large molecular weight materials tend to be adsorbed more strongly than the lower molecular weight ones. Hence, lower molecular weight compounds tend to be eluted first.

GLC columns are packed with an inert support material that has been coated with a high boiling point liquid. The most commonly used support is diatomaceous earth (eg. Kieselguhr). This is purified by acid washing and then "silanised" to reduce the number of -OH groups and hence, the adsorptive effects of the support. Reducing the number of adsorptive -OH groups improves the chromatic resolution by reducing "tailing". The support has a large surface area per unit mass of material.

About 400 high boiling point liquids are commercially available as stationary phases for GLC. However, only about 6 are necessary to separate any mixture of compounds that can be separated by GLC.

Stationary phases can be divided into "non-selective" and "selective". "Non-selective" phases separate components largely on their molecular weight and "selective" phases have a high affinity for a particular chemical group. For example, phenyl groups in a stationary phase have an affinity for unsaturated and aromatic compounds.

Silicone greases are the most common stationary phase. They have the following basic structure:

$$
CH_3 - \underset{\underset{CH_3}{|}}{\overset{\overset{CH_3}{|}}{Si}} - O \left[\underset{\underset{CH_3}{|}}{\overset{\overset{CH_3}{|}}{Si}} - O \right]_n - \underset{\underset{CH_3}{|}}{\overset{\overset{CH_3}{|}}{Si}} - CH_3
$$

The greater the value of n, the greater the viscosity. These Compounds are "non-selective". Replacement of some of the methyl groups can produce "selective" stationary phases (eg. phenyl groups will retain aromatics and fluorinated groups will retain polar compounds).

Apiezon L greases are mixtures of high molecular weight hydrocarbons. These act as "non-selective" liquid phases. Polyethylene glycols, on the other hand, show selectivity for polar groups.

To coat the column, the stationary phase is dissolved in a volatile solvent and then mixed with the support. The mixture is then taken to dryness in a rotary evaporator.

Column efficiency increases as the thickness of the stationary phase coating decreases. This is due to the rapid attainment of equilibrium of the distribution of the analyte between the stationary phase and the gas. However, the capacity of the column is decreased as the stationary phase is reduced. A compromise is usually reached between the two.

Due to the nature of the stationary phase, there is a maximum temperature at which the column can be operated. As the temperature is increased, the stationary phase will start to evaporate and be eluted from the column. This not only depletes the column of stationary phase, but will also give a high background response from the detector. This loss of stationary phase is known as column "bleed". Columns also have a minimum operating temperature, below which the stationary phase solidifies. Below this temperature, column efficiency drops drastically and peaks broaden significantly.

Once the column has been packed, it has to be conditioned. The system is run overnight at a temperature of 20-30°C above that which will be used in the assay. This elutes impurities and short chain polymers from the column that would produce a noisy base-line response from the detector. Obviously the conditioning temperature must be below the maximum operating temperature of the column.

Selection of a stationary phase.

It is desirable to select a stationary phase that will separate all the components in the mixture under consideration. A number of classification systems have been put forward for stationary phase material. The most commonly used is that of McReynolds.

The system is based on retention indices (Kovats indices). This uses the n-paraffins as a reference base. By definition, the retention index of each paraffin is 100 times the number of carbon molecules in the molecule (ie. the retention index of hexane is 600). This value is independent of the column and conditions used. However, the retention index of other compounds will depend upon the column and upon the conditions used and hence these must be quoted.

To obtain the retention index of a compound, the retention time of the compound and at least 3 n-paraffins are determined. The paraffins used should have retention times that bracket the compound under investigation. The retention time of a non-retained compound is also obtained. This is subtracted from the other retention times to give "adjusted retention times". The adjusted retention times are then plotted (on a log scale) against the retention index of the n-paraffins (on a linear scale). The retention index of the compound under investigation is read from the graph, using its adjusted retention time. Eg. the determination of the retention index of benzene:

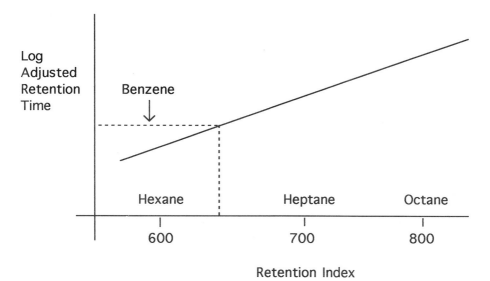

As mentioned before, the retention index of a compound will be different on different columns. "Non-polar" phases give low retention indices; whereas, a "polar" phase (with respect to the compound under consideration) will increase the retention, in comparison with the n-paraffins, and hence increase the retention index. A squalene stationary phase is the most "non-polar" and will give the lowest retention indices. Other stationary phases will give higher retention indices.

McReynolds constants are the difference between the corrected retention index, of particular compounds, on squalene and on the stationary phase under investigation. The corrected retention indices are obtained using 20% stationary phase on acid washed Chromosorb W (a diatomite support) at a temperature of 120°C. There are 5 compounds used to give the major McReynolds constants. Each of them is a specific type of compound and will be retained by a particular type, or types, of stationary phase. These are tabulated below:

Assigned letter	x'	y'	z'	u'	s'
Compound	Benzene	n-butanol	2-pentanone	Nitropropane	Pyridine
Interaction with stationary phase	Inter- molecular forces	Electron attraction	Electron repellers	Complex	Complex
Type of compound retained by column	Aromatics Olefines	Alcohols Nitriles Acids -CHCl$_2$ -CCl$_3$ -CH$_2$Cl -NO$_2$	Ketones Ethers Aldehydes -N(Me)$_2$ Esters Epoxides	Nitromethane	Pyridine Dioxane

Some examples are illustrated below:

Name	Type	x'	y'	z'	u'	s'
OV-101	Dimethyl fluid	17	57	45	67	43
OV-1	Dimethyl gum	16	55	44	65	42
OV-17	Methyl- phenyl	119	158	162	243	202
OV-210	Methyl- fluoro alkyl	146	238	355	468	310

The McReynolds constants are used in selecting a stationary phase to separate different classes of compounds, when their boiling points are similar. For example, the separation of n-butyl ethyl ether (B Pt : 91°C) and n-propanol (B Pt : 97°C). Columns with a high y' value, with respect to the z' value, will retain the n-propanol and the ether will come off the column first. However, columns with a high z' value, compared with the y' value, will retain the ether and the alcohol will come off first.

For separation within a class of compounds, columns with a stationary phase similar in composition to the sample, should be used (eg. alcohols on carbowax, hydrocarbons on Apiezon etc). This gives symmetric peaks; ie. avoids "tailing". Compounds are separated in order of their boiling points.

Where one or more of a class differ from other members by the addition of a double bond or a second functional group, the column should be selected on the basis of this additional group. For example, columns with high x' values will retard unsaturated fatty acids, when separating fatty acid esters.

Derivatives.

Where compounds are unstable, have high boiling points or give tailing peaks, the formation of a chemical derivative will often improve the chromatography. For example, methyl esters of fatty acids give better chromatograms than their parent acids (the acids tend to produce tailing peaks) as they are not adsorbed by the free -OH groups on the support.

Derivatives can also be produced to improve detector response. For example, halogenated derivatives for use with an ECD.

Derivatisation can be carried out "pre-column". Eg. the formation of trimethylsilyl derivatives of urinary organic acids for "inborn errors of metabolism" investigation. These derivatives are formed in heated tubes, prior to injection. "On-column" methylation can be carried out in a hot injector port. For example, anticonvulsant drugs can be methylated when extracted into tetramethyl ammonium hydroxide and this extract injected into the GC.

6.   Detectors.

The detector generates an electrical signal proportional to the amount of sample leaving the column. Detectors may be classified into two groups, concentration-dependant and mass rate-dependant detectors.

In concentration-dependent detectors (eg. thermal conductivity [TCD] and electron capture [ECD] detectors, the signal is proportional to the concentration of the compound in the carrier gas. In mass rate-dependant detectors (eg. flame ionisation [FID] and nitrogen-phosphorus [NPD] detectors), the signal is dependant upon the mass of compound passing through the detector in unit time.

Detectors can also be classified into universal response and selective response. The former respond to all types of compounds (eg. TCD) and the latter respond to specific types of compound (eg. FIDs respond to organic compounds).

Detectors are heated to avoid condensation of the sample components or column "bleed" in the detector.

Thermal Conductivity Detector.

The TCD works on the principle that a hot body will loose heat at a rate which depends upon the composition of the surrounding gas. Hence, the rate of heat loss can be used as a measure of gas composition.

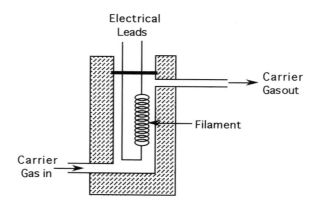

When carrier gas molecules strike the heated filament, heat is transferred to the gas molecules. The smaller the carrier gas molecule, the higher its mobility, and hence, the greater the chance of contacting the filament. Therefore, smaller molecules have greater thermal conductivity than larger ones. Hydrogen and helium, therefore, have high thermal conductivity. Sample components, in the carrier gas (H_2 or He), reduce the thermal conductivity to en extent dependent upon their concentration and molecular weight.

A constant current is applied to the filament and, when the carrier gas only is flowing, a constant temperature is reached. When the carrier gas contains a sample component, the thermal conductivity is reduced and the temperature of the filament rises. This increases the resistance of the filament, which is measured by a Wheatstone bridge circuit. This change in filament resistance is, therefore, recorded on a chart recorder or recording integrator.

It should be noted that these detectors are sensitive to changes in carrier gas flow rate. More molecules strike the filament, per second, as the flow rate increases.

Flame Ionisation Detector.

This detector works on the principle that the electrical conductivity of a gas is directly proportional to the concentration of charged particles within the gas.

The ionising source in a FID is a hydrogen flame. Organic molecules, in the carrier gas, ionise during combustion to produce a mixture of anions, cations and electrons.

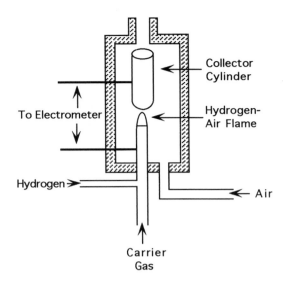

A potential (300-400 volts) is applied between the burner jet and the collector cylinder. The presence of charged particles in the flame allows a current to flow. This produces a voltage drop across a resistor in the circuit. This voltage drop is amplified by an electrometer and fed to a recorder or recording integrator.

The FID responds only to organic compounds and hence does not respond to water vapour etc.

FID performance depends upon the flow rate of all three gases. A flow ratio of 1:1:10 (H_2 : carrier gas : air) generally produces a good response.

Nitrogen-Phosphorus Detector.

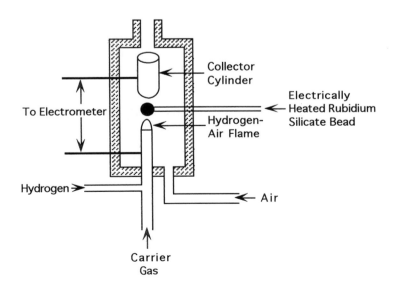

This detector is very similar to a FID. However, the flame uses less hydrogen. This suppresses ionisation produced by burning organic compounds. Above the flame is an electrically heated rubidium or caesium bead. Burning compounds containing nitrogen and/or phosphorus (and to a lesser extent halogens) generate ions when in contact with the bead.

Compared with a FID, this detector is about 50 times more sensitive to nitrogen and 500 times more sensitive to phosphorus. Even though it is very selective for nitrogen and phosphorus, it will respond to some halogenated compounds. One of its main uses is in drug and pesticide analysis.

Electron Capture Detector.

The ECD measures a loss in signal, rather than an increase. As the carrier gas flows through the detector, it is ionised by a β radiation source (usually ^3H or ^{63}Ni). During ionisation, slow electrons are formed and these migrate to a central anode.

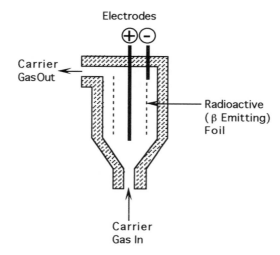

About 90 volts is applied across the electrodes and this produces a current of about 10 nA. Electron capturing molecules (eg. halogenated compounds) capture the slow electrons and, hence, reduce the current flow between the electrodes. This drop in current is then amplified and fed to a recorder or integrator.

Modern ECDs have pulse-voltage, variable frequency, power supplies, which maintain a constant current. When no sample is present, the pulse rate is low. When the sample enters the detector, the pulse rate increases to offset the current loss. The pulse frequency is proportional to the sample concentration and can, therefore, be used for quantitation.

The ECD is very sensitive to compounds containing halogens, conjugated carbonyls, nitriles, nitrates and organometallics. It is insensitive to hydrocarbons, alcohols and ketones.

^{63}Ni is a better radiation source than ^3H, as it can be used with detector temperatures up to 350°C. ^3H tends to be lost from the source foil at temperatures above 220°C. Hence, a tritium foil can become contaminated due to condensation of sample or column "bleed", as it is operated at lower temperatures than ^{63}Ni.

The characteristics of these detectors are listed below:

	TCD	FID	NPD	ECD
Minimum detectable quantity	10^{-8} g/sec	5×10^{-12} g/sec	10^{-11} g/sec	10^{-12} g/sec
Response	Universal	Only organic compounds	Mainly nitrogen and phosphorus	Mainly halogens and organo-metallics
Linear range	10^4	10^6	10^4	500-10,000
Stability to temperature and gas flow changes etc	Good	Excellent	Fair	Fair
Carrier gas	He or H_2	N_2, He or H_2	N_2 or He	N_2 or Ar + 5% CH_4
Temperature limit	400°C	400°C	300°C	220°C (^{3}H) 350°C (^{63}Ni)

Gas Chromatography - Mass Spectrometry (GC-MS).

The ultimate detector for a GC is a mass spectrometer (MS). These are commercially available as dedicated GC-MS instruments. Mass spectrometry requires samples to be pure. This is achieved by gas chromatography which presents compounds, one at a time, to the mass spectrometer. As the mass spectrometer operates at high vacuum, the instruments invariably use capillary columns to reduce the amount of carrier gas presented to the MS.

The instrument layout is illustrated below:

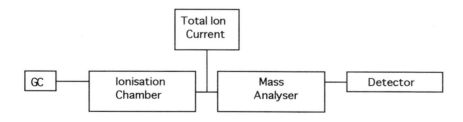

The sample components enter the ionisation chamber where they are bombarded by an electron beam of about 70 electron volts. This is enough to break chemical bonds and dislodge electrons from the fragments. Hence, a number of positively charged fragments are produced.

The total ion current, measured at this point, produces a chromatogram similar to that of a universal detector.

The positively charged fragments then pass into the mass analyser which allows only those with a specific mass/charge (m/z) ratio to pass to the detector. Quadrupole mass analysers are used in GC-MS. This type of analyser consists of a quadrant of four rods energised with a specific DC voltage and variable radio frequency AC voltages. These are applied to opposing rods and cause the ions to be alternatively attracted to, and then be repelled by, the rods. Hence, the ions start to oscillate. Only those of a specific m/z will undergo stable oscillation and pass to the detector. All other ions strike the rods. By altering the radio frequency, ions of different m/z pass to the detector. Hence, a wide range of m/z can be rapidly scanned to give a mass spectrum. These spectra are stored in the computer and linked to the total ion current peaks. Hence, each eluted compound will have one or more mass spectra. These spectra are then compared with spectra in the instrument's memory in order to identify the chemical present.

This type of mass spectrometry is known as Electron Impact Ionisation MS. As most of the molecules are fragmented, few survive intact to give an indication of the molecular weight of the compound. A gentler form of mass spectrometry is Chemical Ionisation MS. Methane or isobutane is mixed with the sample in the ionisation chamber. The ionisation of methane and its interaction with the compound (M) eluted from the GC is illustrated below:

$$CH_4 + e^- \quad \rightarrow \quad CH_4^+ + 2\,e^-$$

$$CH_4^+ + CH_4 \rightarrow \quad CH_5^+ + CH_3$$

$$CH_5^+ + M \quad \rightarrow \quad MH^+ + CH_4$$

MH^+ has the molecular weight of the compound plus 1. Hence, this method can be used to determine the molecular weight of compounds eluting from the GC column. An example of the two types of mass spectrometry is illustrated below for N-(3-chloropropyl) amphetamine (M Wt : 211). This has the structure:

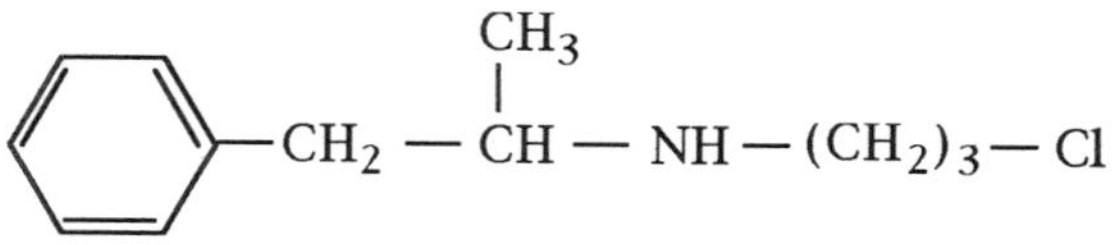

Two of the fragments generated in the electron impact ionisation are:

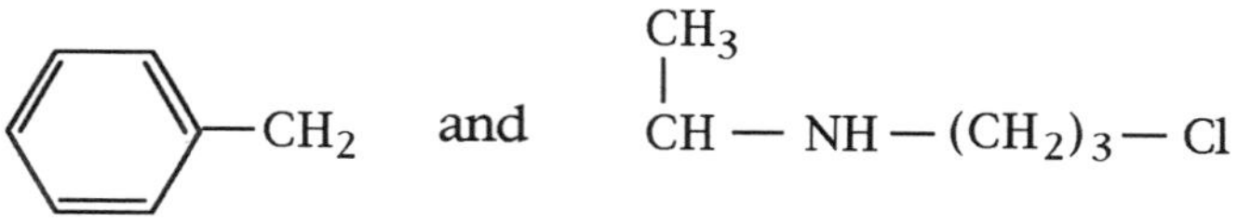

M Wt : 91   M Wt : 120

These fragments can be clearly seen in the electron impact spectrum. The peak at 122 is the same fragment as that at 120, except that it contains the chlorine isotope 37, instead of 35. Naturally occurring chlorine is made up of about 75% isotope 35 and 25% isotope 37. This ratio can be seen in the peak heights of the 120 and 122 fragments. The chemical ionisation spectrum also shows this distribution with the unfragmented molecule (M+1) at 212 and 214 mass units.

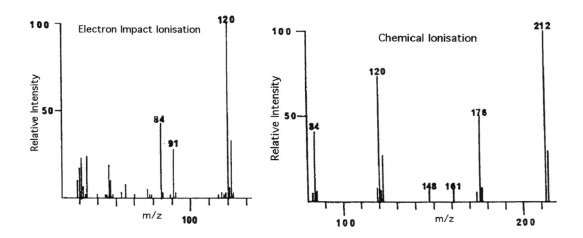

When GC-MS is used for quantitative work, and the fragment pattern of the compound of interest is know, specific fragments only can be examined by the instrument. This is known Selective Ion Monitoring (SIM). About 3 ions are usually chosen. One is usually the base peak (maximum intensity peak) and another is a fragment not likely to be generated by other components of the original mixture. This technique greatly increases the sensitivity of system as the instrument can do many SIM scans on a GC peak and combine the results.

7.   Recorder or Computing Integrator.

The output from the detector is fed into a recorder or computing integrator.

From the recorder trace, compounds can be identified by their retention time (distance on the chart from the peak to the injection point) and/or quantitated by measuring the peak height, or area, and comparing this with standards.

A computing integrator can measure the area of peaks much more accurately than can be done from chart paper. They can also accurately record retention times and do a number of calculations to obtain a final value for the concentration of the compound in the original sample.

The capabilities of computing integrators include:

1.   Peak and baseline detection.
2.   Peak area integration.
3.   Baseline correction of peaks.
4.   Allocation of the areas of fused peaks.
5.   Measurement of peak retention time.
6.   Elimination of data not of interest, such as solvent peaks and minor component peaks.
7.   Calculation of the results in terms of concentration of each sample component.

The computing integrator integrates continuously and transfers this data to a microprocessor every 100 ms. The microprocessor collects these data samples into data bunches; eg. 1 Data Bunch = 5 Data Samples. The memory looks at the 5 preceding data bunches when a peak is detected. Therefore, a "window" passes across the chromatogram.

The integrator decides that a peak is being eluted when the slope exceeds a set threshold value. The slope is calculated at the end of each data bunch.

When a peak is detected, the integrator refers back 5 data bunches in the peak and retains the value of the baseline in the memory.

The slope sensitivity (threshold value) that is set on the integrator depends on the noise level of the input. One has to have a high threshold if the noise level is high.

The integrator has a baseline offset. This area is included in the original assessment and the baseline is drawn at the end of the peak. The area below the baseline is subtracted from the total area. This offset allows for an increasing or decreasing baseline.

The types of baseline correction that the integrator can carry out are shown below:

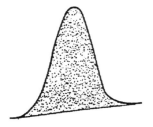

Correction for baseline drift.

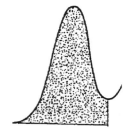

Baseline-valley selection.

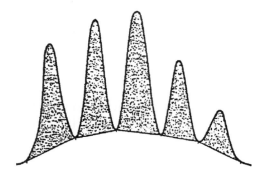

Valley-valley selection.

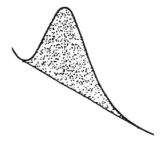

Tangent drawn for rider peak.

The method used is selected by a single baseline testing parameter and the relationship of the chromatographic signal level to the beginning baseline level. The end of the peak is detected when the slope of the data bunches falls below the threshold value. A baseline test is then applied and, if this is the set value, a baseline is drawn.

A rider peak on a peak tail can be detected if the integrator counts more negative data bunch slopes than positive ones. These rider peaks will then be given a tangent baseline (see above). The remaining area will be included in the major peak, if desired.

Chromatogram interpretation can be carried out automatically by the integrator. The type of interpretations are shown below:

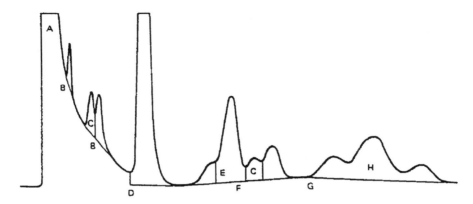

A : Elimination of solvent peak.   B : Tangents drawn for rider peaks.
C : Fused peaks split.   D : Reset to original baseline.
E : Shoulder peak split.   F : Trapezoidal baseline correction.
G : Automatic increased sensitivity for broader peaks.
H : Integration of several peaks as a single area.

As an example of the calculations that the computing integrator can carry out, internal standardisation is illustrated. The system is calibrated with an internal standard of known concentration and the analyte of known concentration. These concentration values for each peak (located by retention times) are entered into the integrator. When a known amount of internal standard is added to an unknown mixture and run through the column, the integrator can calculate the concentration of the analyte from the area of the internal standard peak and the unknown peak area.

The peak area = K x the sample concentration.

K is the response factor of the detector to the analyte.

Therefore, during calculation, we have

$$\text{Peak Area}_A = K_A \times \text{conc}_A \text{ for a Peak A.}$$

and $\quad \text{Peak Area}_{IS} = K_{IS} \times \text{conc}_{IS}$ for the internal standard peak.

$$\frac{\text{Peak area}_A}{\text{Peak area}_{IS}} = \frac{K_A \times \text{Conc}_A}{K_{IS} \times \text{Conc}_{IS}}$$

$$\frac{K_A}{K_{IS}} = \frac{\text{Peak area}_A \times \text{Conc}_{IS}}{\text{Peak area}_{IS} \times \text{Conc}_A}$$

$\dfrac{K_A}{K_{IS}}$ is the response factor of the detector to compound A, relative to the internal standard.

Therefore, during an analytical run, the concentration of sample A can be calculated from:

$$\text{Conc}_A \;=\; \frac{\text{Peak area}_A}{\text{Peak area}_{IS}} \;\times\; \frac{\text{Conc}_{IS}}{K_A/K_{IS}}$$

With this method of quantitation, the precision of the analysis is not dependent on the injection of an accurately known amount of sample, as the ratio of peak areas is used in the calculation. The system can only be used where there is a linear response from the detector, with changes in the concentration of the analyte.

Most GC and HPLC detectors have a 1 volt and a 10 mV output. Either of these can be fed to the integrator.

Column Selectivity and Efficiency.

A typical chromatogram of a mixture of two substances (1 and 2) and a small amount of a non retained substance (m) is shown below:

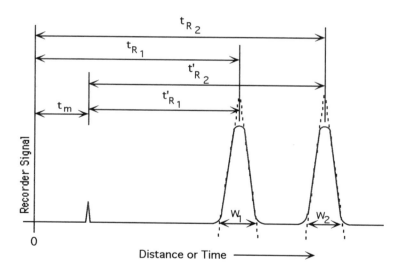

$$
\begin{array}{lll}
t_m & = & \text{Retention time of a non-retained compound (Eg. methane).} \\
t_{R1} \text{ and } t_{R2} & = & \text{Retention times of substances 1 and 2 respectively.} \\
W_1 \text{ and } W_2 & = & \text{The base widths of peaks 1 and 2 respectively (from tangents} \\
& & \text{drawn at the inflection points).}
\end{array}
$$

The adjusted retention times are $t_{R1} - t_m$ and $t_{R2} - t_m$ (ie. t'_{R1} and t'_{R2}).

The distance between the two peaks (Δt_R) is a measure of the column's selectivity for these 2 compounds. This can be expressed as the separation factor:

$$\text{Separation factor} = \frac{t_{R2} - t_m}{t_{R1} - t_m}$$

When the compound emerges from the column it will be present in a certain volume of carrier gas. With an efficient column, a fixed amount of compound will be in a small amount of carrier gas. The detector response will give a sharp, narrow peak. If the column is not very efficient, the compound will appear in a large volume of carrier gas. This appears as a blunt, broad peak. However, if the compound is strongly retained by the column, it will still appear, after a long time, as a blunt, broad peak from an efficient column. Therefore, both peak broadening and retention time are taken into consideration when examining the efficiency of a column.

The concepts of selectivity and efficiency are illustrated below:

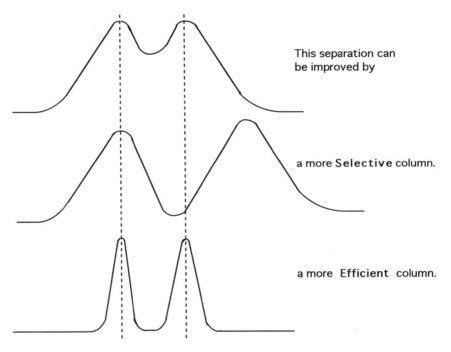

This separation can be improved by

a more **Selective** column.

a more **Efficient** column.

The efficiency of a column can be expressed on terms of the number of theoretical plates (N) it contains. This concept compares chromatography with distillation. N can be calculated from the following formula:

$$N = \left(\frac{\text{Adjusted retention time}}{\text{Standard deviation of peak}} \right)^2$$

$$\text{or} \quad \left(\frac{t'_R}{\sigma} \right)^2$$

For a symmetric peak (Gaussian), the base width (W) = 4 x the standard deviation.

$$\therefore N = \left(\frac{t_R - t_m}{\frac{1}{4}\,W} \right)2$$

$$\therefore N = 16 \left(\frac{t_R - t_m}{W} \right)2$$

Efficiency can also be expressed as the height equivalent to a theoretical plate (HETP):

$$HETP = \frac{L}{N}, \text{ where L is the length of the column bed, usually in mm.}$$

The effectiveness of a column is expressed in terms of its resolution (R_S). This combines both selectivity and efficiency. Resolution is defined as the distance between the two peak centres, divided by the average base width.

Hence:

$$R_S = 2 \left(\frac{t_{R2} - t_{R1}}{W_1 + W_2} \right)$$

Where $R_S = 1.5$, separation is almost complete. Where $R_S = 1$, the peaks overlap, but separation is still acceptable. Where R_S is below 1, the separation is poor. The resolution of the two peaks in the top trace in the diagram on the previous page would be about 0.7 and the resolution in the bottom trace would be about 2.

Factors that contribute to column efficiency can be simply expressed by the following equation, where A, B and C are constants and $\bar{\mu}$ is the average carrier gas flow rate:

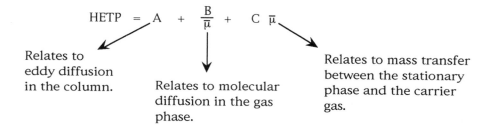

$$HETP = A + \frac{B}{\bar{\mu}} + C\,\bar{\mu}$$

Relates to eddy diffusion in the column.

Relates to molecular diffusion in the gas phase.

Relates to mass transfer between the stationary phase and the carrier gas.

As can be seen, the efficiency is dependent upon the carrier gas flow rate. Hence, there is an optimum flow rate for any particular column.

Support particle size effects A, and efficiency improves as particle size is reduced.

As regards B, efficiency is improved if uniform particle size and shape are used. Efficiency is increased at lower temperatures due to reduced rate of sample diffusion in the gas phase. The type of carrier gas used also effects the rate of sample diffusion in the gas phase.

C is effected by the thickness of the stationary phase. Efficiency increases with decreasing phase thickness, due to the rapid attainment of equilibrium between the stationary and mobile phases. However, with decreasing stationary phase thickness, the capacity of the column is reduced and the column will more easily become overloaded with the resultant decrease in efficiency.

As mentioned above, the efficiency is dependent on the type of carrier gas used. This is illustrated in the "Van Deemter" plot, where HETP is plotted against the average linear carrier gas flow rate for nitrogen, helium and hydrogen:

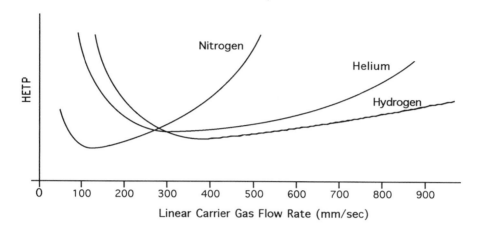

The actual position and shape of the curves depends upon the type of column used, the shape and size of the support and upon the type and thickness of the stationary phase.

Optimal practical gas flow rates are usually selected at values that correspond to 1.5-2 times the flow rate that gives the minimum HETP. This compensates for decreasing flow rates that are seen with rising temperatures in temperature programming.

For most columns, optimum practical efficiency is obtained with a nitrogen flow of 150-350 mm/sec, a helium flow of 300-450 mm/sec or a hydrogen flow of 450-600 mm/sec. Due to the rapid decrease in efficiency at higher flow rates, nitrogen is not recommended as a carrier gas for capillary columns.

Qualitative Analysis.

The most common method of qualitative analysis is to run a series of likely compounds, together with the unknown, and compare their retention times.

If the standard mixture contains the compound present in the unknown, then a mixture of standard plus unknown will give an increase in peak height of the compound present in the unknown, when compared with the other standard peak heights.

The retention index (Kovats Index) can be calculated, using the method previously described, and then compared with those in published tables for the appropriate stationary phase and temperature.

However, retention data does not absolutely confirm the peak identity, as several compounds may have identical retention times. The retention times on different stationary phase types can improve identification, but a GC-MS is the best method of confirming the identity of the compound.

Quantitative Analysis.

Due to the fact that detectors respond differently to different compounds, the instrument should be calibrated separately for each component in a mixture. However, published data is available to interrelate detector response to components of a mixture. Hence, a series of standards are run for one of the components and correction factors are used for the others.

The amount of compound, present in the sample, is proportional to the area of the peak produced on the chromatogram. The measurement of area is easily obtained with a computing integrator connected to the detector. However if area measurements are done by hand, it is time consuming and not very accurate.

A less accurate method of obtaining the amount of compound present is to measure the peak height and to compare this with standards. This is obviously easier than measuring area, if one has to do it by hand.

With the small samples that used in GC, obtaining good volume accuracy is difficult. Hence, a system of internal standards is usually used in quantitative GC work.

The internal standard is a component that is not likely to be present in the sample. It should have a retention time similar to the components being analysed, but not identical.

The internal standard (a constant amount per unit volume) plus the compound being analysed (in increasing amounts per unit volume) are run through the GC. A standard curve of [analyte] is plotted against the ratio of analyte peak area (or peak height) to internal standard peak area (or peak height). As the ratio is independent of volume, the volume injected into the GC does not have to be accurate.

A known amount of internal standard is added to the unknown and a proportion of this mixture is injected into the GC. The peak area (or peak height) ratio is then calculated. If the concentration of the internal standard is the same in the unknown mixture as it is in the standard mixtures, then the unknown concentration can be read directly from the standard curve. However, if the concentration of the internal standard, in the unknown mixture, is different from that in the standard mixtures, then a correction factor must be applied to the concentration read from the graph:

$$\text{Analyte concentration} = \text{Conc from graph} \times \frac{\text{Conc of IS in Unknown}}{\text{Conc of IS in Standards}}$$

EXERCISES

These exercises are written for a Packard gas chromatograph fitted with a flame ionisation detector.

Other instruments will probably require different attenuation settings etc.

EXERCISE 9-1

THE DETERMINATION OF THE RETENTION INDEX OF BENZENE ON AN OV-17 COLUMN

OV-17 is a liquid phase used in Gas-Liquid Chromatography. It is a silicone grease with about 50% of the methyl side chains replaced by phenyl groups. The column used is a 3% coating of OV-17 on Chromosorb W (a purified diatomaceous earth). Unsaturated compounds are retained by this column to a greater extent than saturated compounds.

The retention index of benzene will be determined using this column. This will then be used to calculate the first McReynolds constant (x') for OV-17. Due to the high volatility of the lower n-paraffins, a lower temperature (60°C) than that used by McReynolds (120°C) will be employed. Temperature changes do not produce a marked variation in retention indices, and so the McReynolds constant will be similar to the quoted value of 119.

A mixture in the n-paraffins (C_5 to C_9) will be run, with and without, added benzene. T_m will be determined by the injection of a sample of natural gas.

REAGENTS AND EQUIPMENT:

Mixture A.
    Equal volumes of n-pentane (C_5), n-hexane (C_6), n-heptane (C_7), n-octane (C_8) and n-nonane (C_9).
        Keep in a small tightly sealed vial.
Mixture B.
    A + benzene (equal volume to a n-paraffin).
        Keep in a small tightly sealed vial.

1 μl GC Syringe.
100 μl GC Syringe.
GC Parameters:
    Column:
        3% OV-17 on Chromosorb W, 100-120 mesh, 2 metre x 4 mm (internal diameter) Pyrex glass.
    Instrument Settings:

Carrier gas (N_2) flow rate	:	20 ml/min.
Detector gas flow	:	30 ml H_2/min.
		300 ml air/min.
Oven temperature controls	:	Isothermal.
Setpoint °C	:	60°C.
Attenuator	:	64.
Range	:	10^3.
Injection port temperature	:	200° C.
Detector Temperature	:	200°C.
Output lead	:	Recorder socket at the rear of the instrument.

Recorder:

Input	:	10 mV.
Atten/Cal	:	Cal.
Chart Speed	:	50 mm (5 cm)/min.
Chart drive	:	Off.
Pen	:	Use the recorder "Zero" to set the pen to the 10/90 line on the right of the chart.

PROCEDURE:

1.   Take 1 μl of solution A into a 1 μl GC syringe (It has a needle with a sharp tip) and carefully insert the needle through the septum on the injection port until the needle is fully inserted. (Be careful not to damage the syringe needle or plunger.)

2.   Start the recorder chart drive. Rapidly inject the 1 μl sample and mark the start point on the recorder at the same time by rapidly turning the zero control on the recorder clockwise and back again.

3.   After the n-nonane has come off the column, stop the recorder chart drive.

4.   Repeat the above with solution B and identify the benzene peak.

      The recorder sensitivity can be adjusted by altering the attenuator on the GC (increasing the value decreases the sensitivity), or the input voltage on the recorder.

5.   Determine the t_m value as follows:

      (a)   Using a 100 μl GC syringe and draw in 100 μl of gas from an open gas tap.

      (b)   Inject the 100 μl of natural gas and mark the chart, as before.

      (c)   Switch off the chart drive after the peak has been recorded.

Calculations:

1.   Using the traces for solution B and natural gas, draw in the baseline and a line from the apex of each peak to the baseline.

2.   Measure the distance from the injection point to these lines, for each peak and for t_m.

3.   Subtract the t_m value from the others, to give adjusted retention times.

4.   Plot the adjusted retention times on the log axis and the retention indices of the n-paraffins (pentane = 500, hexane = 600 etc) on the linear axis of semi-log paper.

5.   Read off the retention index of benzene, from its adjusted retention time.

6.   Calculate the first McReynolds constant (x') for OV-17, as described on page 9-7. The retention index of benzene, on a squalene column, is 649.

      The x' McReynolds constant should be 119.

EXERCISE 9-2

THE EFFECT OF TEMPERATURE ON RETENTION

It will have been noted that pentane and hexane come off the column as sharp peaks that are close together; whereas, octane and nonane are well separated, but emerge as blunt broad peaks. The effect of increased column temperature on the retention of these peaks will be examined in this exercise. This will be done by using the temperature programmer.

REAGENTS AND EQUIPMENT:

As for Exercise 9-1, except:
GC Parameters:
    Oven temperature controls:

Select temperature programming	:	Prog.
Isothermal pre-period	:	0 minutes.
Isothermal final period	:	1 minute.
Cool-down period	:	3 minutes.
Rate	:	40°/min.
Initial temperature	:	60°C.
Final Temperature	:	160°C.

PROCEDURE:

1.   Use 1 μl of solution A and inject as for Exercise 9-1, but also press the "start" button on the temperature programmer, 10 seconds before you inject the sample.

Compare the trace with that obtained in Exercise 9-1.

The state of the temperature programme is indicated by the lights on the programmer panel. The Setpoint °C control indicates the current temperature selected by the programme. At the end of the run, the detectors (and columns) are lifted and vents below the oven are opened. An air stream then rapidly cools the oven and columns. When the system is ready for the next injection, the light on the programmer will glow dimly. It will be fully alight if the oven is still heating.

Appendix.

The boiling points of the n-paraffin are listed below:

Pentane	:	35°C
Hexane	:	67°C
Heptane	:	94°C
Octane	:	126°C
Nonane	:	151°C

The boiling point of benzene is 80°C.

EXERCISE 9-3

THE DETERMINATION OF BLOOD ALCOHOL

Ethanol is one of the commonest xenobiotics (substances foreign to the body) that affect cerebral function. There are few controls on its availability to the general public. Although death from acute ethanol poisoning, by itself, is uncommon, it potentiates the effects of many other xenobiotics and, therefore, can lead to death when ethanol is consumed in conjunction with other drugs.

Ethanol ingestion is also important, at sub-lethal levels, from a medico-legal point of view, especially in relation to the Road Traffic Acts.

Ingested ethanol is rapidly absorbed into the body, with about 20% being absorbed via the gastric mucosa and the rest through the intestine. The rate of absorption is reduced by the presence of food in the stomach and intestines. Fatty food, in particular, delays absorption by reducing stomach motility, delaying the emptying of the stomach and forming a lining on the stomach mucosa. Under most conditions, the absorption of a single dose of ethanol is rapid enough to produce a peak level in the blood after about one hour. Ethanol freely diffuses across plasma membranes and so all body fluids contain an almost equal concentration of ethanol.

The majority of ethanol, ingested by the body, is metabolised to carbon dioxide and water. The remainder, only about 3%, is lost in the urine and in the lungs. Because of its easy movement across plasma membranes, ethanol is only about 20% more concentrated in urine than in the blood. The concentration of ethanol in alveolar air is directly proportional to the blood concentration. The level in blood is about 2,100 times greater than that in alveolar air.

The primary site of metabolism of ethanol is the liver, which contains most of the body's alcohol dehydrogenase. The alcohol is oxidised to acetaldehyde and then to acetyl CoA.

$$CH_3\text{-}CH_2\text{-}OH \; + \; NAD \; \xrightarrow{\;Alcohol\,Dehydrogenase\;} \; NADH \; + \; CH_3\text{-}CHO$$

$$CH_3\text{-}CHO + NAD + CoA\text{-}SH \; \xrightarrow{\;Acetaldehyde\,Dehydrogenase\;} \; NADH + CoA\text{-}S\text{-}CO\text{-}CH_3$$

Acetyl CoA can then be further oxidised in the Krebs cycle, or converted into fatty acids or cholesterol. The alcohol dehydrogenase system is saturated with substrate at low ethanol concentrations (about 0.1 g/l). Above this level, alcohol is metabolised at a constant rate for any one individual (about 10 g/hour). Pyruvate greatly increases this rate and large amounts of ingested carbohydrates facilitate the oxidation of ethanol.

Blood ethanol levels below 0.5 g/l (the Australian limit for driving) produce very little effect on the subject. A sense of exhilaration and euphoria may be produced. 1.0 g/l is the legal limit for driving in most states of the USA. At the level of 1.5 g/l most drinkers will show some decline in normal judgement and muscular co-ordination. 50% will also show some signs of drunkenness. At 2 g/l ethanol, 80% of the drinkers will be showing signs of drunkenness. (This is the limit at which a person is considered drunk in the USA.) At 3 g/l, even the hardened drinker will be showing signs of irresponsibility or even stupor. At 4 g/l, coma will have set in and 5 g/l is fatal in almost all cases. The effects can be humorously summarised in the following table:

Blood Level (g/l)	Effect
<0.5	Dry and Decent
0.5-1.5	Delighted and Devilish
1.5-2.5	Delinquent and Disgusting
2.5-3.5	Dizzy and Delirious
3.5-4.5	Dazed and Dejected
>4.5	Dead Drunk

Case History.

The sample is a blood sample from an 18 year old man who had been apprehended by the police because of erratic driving.

In this method, the internal standard is n-propanol. The samples and standards are diluted with the internal standard and then protein is removed by precipitation with tungstic acid and centrifugation. 1 µl aliquots are then injected onto the column. The ratios of ethanol peak heights to n-propanol peak heights are calculated and plotted against the ethanol standard concentrations. The unknown is then read from this graph.

REAGENTS AND EQUIPMENT;

1.   50 g/l Ethanol Standard.
           Accurately weigh 2.5 g (3.17 ml) of Analar ethanol and make to 50 ml with distilled water. Mix well.
2.   100 mg/l n-Propanol.
           Add 100 mg (125 µl) of Analar n-propanol to 1 litre of distilled water and mix.
3.   100 g/l Sodium Tungstate.
           Dissolve 5g sodium tungstate in distilled water and make to 50 ml.
4.   0.33 M Sulphuric Acid.
           Add 16.6 ml of 1 M H_2SO_4 to distilled water and make to 50 ml.
5.   Suspects Sample.
           Add 150 µl of Analar ethanol to 100 ml of blood and mix well.

Centrifuge Tubes.
Wassermann Tubes.

Bench Centrifuge.
100, 200 and 250 µl Pipettors and Tips.
1, 2, 5 and 10 ml Graduated Pipettes.
1 µl GC syringe

GC parameters:

Column: 2 metre x 2 mm Pyrex glass, packed with Carbopack (a carbon support) loaded with 0.2% Carbowax 1,500 (a paraffin wax).

Injector port temperature	:	200°C
Column temperature	:	100°C
Detector temperature	:	200°C
Nitrogen (Carrier gas) flow rate	:	15 ml/min
Detector gas flow	:	30 ml H_2/min.
		300 ml air/min.
Attenuator	:	2 x 10
GC output lead	:	Recorder

Recorder:

Recorder input voltage	:	10 mV
Chart speed	:	10 mm (1 cm)/min

PROCEDURE:

1. Using the 50 g/l ethanol standard, set up the following standards:

    A :   1.5 g/l   (300 μl plus 9.7 ml of distilled H_2O).
    B :   1.0 g/l   (200 μl plus 9.8 ml of distilled H_2O).
    C :   0.5 g/l   (2.0 ml of B + 2.0 ml of distilled H_2O).

2. Add 3.0 ml of 100 mg/l n-propanol to 2 centrifuge tubes, followed by 200 μl of the suspect's blood.

3. Add 3.0 ml of the 100 mg/l n-propanol to 3 Wassermann tubes (A, B & C), followed by 200 μl of each standard to its appropriate tube.

4. Add 250 μl of 100 g/l sodium tungstate and 250 μl of 0.33 M H_2SO_4 to all 5 tubes. Mix well after each addition.

5. Stand for 5 minutes and then centrifuge the blood tube for 2 minutes.

6. Decant the supernatant into Wassermann tubes.

7. Run 1.0 μl samples of water, each standard and the unknown through the GC. Wait for both peaks (ethanol and n-propanol) to be eluted, before you inject the next sample. (Wash the syringe with at least 5 rinses of distilled or deionised water after each injection).

Draw a baseline for the ethanol and n-propanol peaks. Measure the peak heights for both n-propanol and ethanol, from their baselines, in mm. Divide the ethanol peak height by the n-propanol peak height for each sample. Plot a standard curve of this ratio against the ethanol concentration and read off the unknowns. Average these values.

Compare this value with the legal limit for driving a motor vehicle.

EXERCISE 9-4

HEAD-SPACE GAS ANALYSIS

Head-space gas analysis is a convenient method of identifying volatile compounds in complex samples, such as gastric contents or blood.

In this method, some n-propanol (internal standard) is added to the sample and standards, which are then warmed in sealed containers. An aliquot of the head-space gas (ie. the vapour above the sample) is then analysed by gas chromatography. Volatile compounds can then be identified by comparing their relative retention times (to n-propanol) with that of the known standards.

If no peak appears within 20 minutes, there are probably no volatile compounds present.

Case Study.

The sample is blood from a 25 year old man. He was found unconscious at 2.00 am this morning and is a known diabetic. A head-space gas analysis has been requested to assist in identifying the cause of his coma.

REAGENTS AND EQUIPMENT:

1.   Methanol standard.
            $100\,\mu l$ in 50 ml of distilled water.
2.   Ethanol Standard.
            $100\,\mu l$ in 50 ml of distilled water.
3.   Acetone Standard.
            $25\,\mu l$ in 50 ml of distilled water.
4.   Chloroform Standard.
            $10\,\mu l$ in 50 ml of distilled water and shake well.
5.   20 g/l n-propanol.
            Make 2.5 ml of n-propanol to 100 ml with distilled water and mix well.
6.   Patient's Blood Sample.
            Add $100\,\mu l$ of ethanol and $25\,\mu l$ of acetone to 50 ml of blood and mix well.

5 ml McCartney bottles fitted with perforated caps and silicone rubber sealing pads.
Water bath at 56°C.
1 ml tuberculin syringes fitted with 25 gauge needles.
    1 for each standard and 1 for the unknown blood (Labelled).
GC Parameters:
As for Exercise 9-3 except:
     Column temperature   : 90°C.
     Nitrogen (Carrier gas) flow rate)   :   30 ml/min.
     Attenuator   : 64×10^{2}.
                                       (This may have to be altered. The higher the value, the smaller the peak height.)

Chart recorder:
    Speed :   20 mm (2 cm)/min.

PROCEDURE:

The following samples are provided:

Standards { 2 ml/l methanol in distilled water.
2 ml/l ethanol in distilled water.
0.5 ml/l acetone in distilled water.
0.2 ml/l chloroform in distilled water.

The patient's blood sample.

1. Add 0.5 ml of each sample to a 5 ml McCartney bottle fitted with a perforated cap and silicone rubber sealing pad.

2. Add 100 µl of 20 g/l n-propanol (in distilled water) to each bottle and gently mix.

3. Tightly cap the bottles and place them in a 56°C water bath.

4. After about 3 minutes, inject a 1 ml aliquot of the vapour above each sample into the GC, using a 1 ml tuberculin syringe fitted with a 25 gauge needle. (Keep a finger on the syringe plunger so that it is not blown out when the syringe needle penetrates the injection septum).

    Do not allow the vapour to cool before injection and mark the chart at the injection point.

    Use a separate syringe for each standard and for the unknown.

    Make sure all the volatile components are off the column before you inject the next sample.

The unknown sample does not contain components with a longer retention time than chloroform.

Calculations:

1. Record the distance between the injection point and the peak.

2. Identify the n-propanol peak, for each standard and the unknown.

3. Divide the standard (or unknown) peak retention time(s) by the n-propanol peak retention time for each standard and for the unknown. This gives the relative retention times.

4. From the standard values, identify the volatile compound(s) in the unknown.

5. Compare these values with the table below:

RETENTION TIMES FOR SOME VOLATILE SUBSTANCES

Relative Retention Time	Xenobiotic	Relative Retention Time	Xenobiotic
1.0	n-Propanol (Reference)	1.6	Paraldehyde
0.22	Methanol	2.1	Ethyl acetate
0.44	Ethanol	2.3	Trichloroethane
0.7	Acetone	2.8	Acetaldehyde
0.8	Isopropanol	2.9	Benzene
1.0	Ether	3.1	Trichloroethylene
1.45	Chloroform	14.0	Tetrachloroethane

EXERCISE 9-5

THE DETERMINATION OF THE CONCENTRATION OF ANTICONVULSANT DRUGS IN PLASMA

An epileptic fit is the result of an unphysiological synchronous discharge of neurones. This can be localised to a group of neurones or can rapidly spread to large areas of the brain. Each of these 2 groups can be further subdivided, depending on the characteristics of the seizure. The EEG pattern can also be used to help to identify the type of epilepsy.

The aim of treatment is to suppress fits without impairing mental or motor function. Even if this is achieved, there is little reversion of the EEG pattern to normal. Drugs that are effective in controlling *Grand Mal* (major convulsions) may be harmful in *Petit Mal* (a brief loss of consciousness with some muscle contraction and relaxation).

The commonly used anticonvulsant drugs are tabulated below, with the usefulness of monitoring their plasma levels:

Drug	Therapeutic Range μmol/l (mg/l)	Usefulness of TDM
Phenytoin	40-80 (10-20)	Essential
Phenobarbital	40-170 (10-40)	Moderately
Primidone	Monitored as Phenobarbital	Not very
Carbamazepine	20-50 (4-12)	Yes
Valproate	350-700 (50-100)	Not very
Ethosuximide	300-700 (40-100)	May be
Clonazepam	40-200 (15-60)	Not known

These anticonvulsants act to prevent the spread of neuronal excitation by mechanisms that are not fully understood. However, they can be divided into those that stabilise excitable membranes and those that enhance the functional activity of neurotransmitters that block synaptic transmission.

All the anticonvulsants have fairly low therapeutic indices. It is, therefore, important to tailor the dose to the individual. This is especially important in the case of phenytoin (Dilantin) where the enzyme system that metabolises the drug is easily saturated.

Where one drug does not fully control the fits, a second drug is used. Hence, epileptic patients are often on more than one anticonvulsant. These drugs can interact to alter the blood level of the other drug(s). For example, carbamazepine (Tegretol) is a potent enzyme inducer and will lower the blood level of phenytoin if both drugs are used together.

Therefore, in order to individualise drug therapy, plasma or serum drug concentrations should be monitored once steady-state levels have been reached, and the dose adjusted to keep the plasma level in the therapeutic range.

The dose-related toxic effects of anticonvulsants are sedation, nystagmus (involuntary movement of the eyeballs), ataxia and psychological changes such as confusion, memory loss and depression.

Some patients can be slowly withdrawn from anticonvulsants after being fit-free for 2 years. However, the majority of patients have to remain on medication for the rest of their lives.

A number of methods have been used to monitor the level of anticonvulsants in plasma or serum. Gas chromatographic methods can be used to assay all the drugs simultaneously. However, derivatisation is required prior to GC separation. Immunoassays will only monitor one drug at a time. With the exception of valproate, all the drugs can be monitored using HPLC.

Case History.

The plasma sample is from a 25 year old man being treated with phenytoin and phenobarbital. The sample has been collected for routine therapeutic drug monitoring (TDM).

The anticonvulsant drugs are extracted from an acidified sample into toluene containing an internal standard (ethyl-tolylbarbital).

The drugs are then re-extracted into aqueous tetramethyl ammonium hydroxide (alkali). One μl aliquots of this extract are then injected onto the GC column.

Using a hot injection port, the tetramethyl ammonium hydroxide rapidly methylates the nitrogens in the drugs. These methylated derivatives are then separated on the OV-101 column and detected by a flame ionisation detector.

The ratio of the standard peak areas, to that of the internal standard, are plotted and the unknown read from this standard curve.

REAGENTS AND EQUIPMENT:

1.   Drug standard.

        10 mg phenobarbital
        10 mg phenytoin   }   Dissolve in 10 ml of methanol.
        10 mg primidone

2.   Drug Free Plasma.
3.   Patient's Plasma.

        Dissolve 10 mg of phenobarbital and 5 mg of phenytoin in 3.0 ml of methanol.
        Add 200 µl of this solution to 20 ml of drug free plasma.

4.   0.2 M Phosphoric acid.
5.   Internal Standard.

        Dissolve 4 mg of 5-ethyl-5-tolylbarbituric acid in 200 ml of toluene.

6.   20% TMAH in methanol.

        Dissolve 1.0 g of tetramethyl ammonium hydroxide in methanol and make to 5.0 ml.

7.   Ethanol.

        In a beaker for syringe rinsing.

8.   Distilled Water.

        In a beaker for syringe rinsing.

50, 100 and 200 µl Pipettors and Tips.
1, 2 and 5 ml Graduated Pipettes.
Centrifuge Tubes with a Narrow Tip at their Base.

Bench Centrifuges.
Glad-wrap.
16 x 100 mm Screw Top Culture Tubes.
1 µl GC Syringe.
GC Parameters:

Column	:	3% OV-101 on Chromosorb W, 100-120 mesh, 2 metre x 2 mm Pyrex.
Column gas flow	:	30 ml nitrogen/min.
Detector gas flow	:	30 ml H_2/min.
		300 ml air/min.
Temperatures:		
Injection port	:	350°C
Oven	:	215°C
Detector	:	280°C
Attenuation	:	1×10^3
Output lead	:	Recorder

Integrator:

Attenuation	:	0
Threshold	:	0
Zero	:	0
Chart speed	:	1 cm/min.

PROCEDURE:

1. Prepare the plasma standards using the stock drug standard (this contains 1.0 g of phenobarbital, 1.0 g of phenytoin and 1.0 g of primidone per litre, in methanol) as follows:

    (i)   Add 4.0 ml of drug free plasma to 4 tube, labelled A to D.

    (ii) Add the stock drug standard, as shown below, and mix well:

A	:	50 µl.
B	:	100 µl.
C	:	200 µl.
D	:	400 µl.

2. Add 2.0 ml of sample (standards and unknowns) to 16 x 100 mm screw-top culture tubes.

3. Add 1.0 ml of 0.2 M phosphoric acid and 5.0 ml of internal standard solution (20 mg/l 5-ethyl-5-tolylbarbituric acid in toluene).

4. Shake vigorously for 2 minutes.

5. Centrifuge for 3 minutes to separate the phases.

6. Transfer about 4 ml of the upper (toluene) layer to a centrifuge tube with a narrow tip at its base.

7. Add 100 µl of 20% tetramethyl ammonium hydroxide in methanol.

8. Place 2 layers of "Glad-wrap" between your thumb and the top of the tube. Extract gently by inverting the tube 30 times.

9. Centrifuge for 5 minutes.

10. There should be a small aqueous layer at the bottom ($\cong$ 50 µl). Remove and discard as much of the toluene layer as possible.

11. Slowly, over 2 - 3 seconds, inject 1 µl samples of these aqueous layers onto the column and start the integrator. Inject the sample soon after the syringe needle has been fully inserted through the injection port septum.

    Wash the syringe with water (3 times) and then ethanol (3 times) and air dry between each injection.

Identify the peaks (the drugs come off the column in the order: phenobarbital, ethyl-tolylbarbital (IS), primidone and then phenytoin).

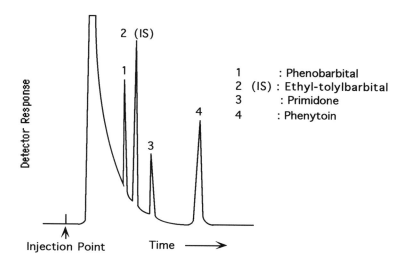

1 : Phenobarbital
2 (IS) : Ethyl-tolylbarbital
3 : Primidone
4 : Phenytoin

Calculate the area ratios of the phenobarbital and phenytoin peaks to the tolylbarbital peak.

Calculate the concentration of the drugs in the standard plasmas.

Plot a standard curve of the area ratios against the standard concentrations and read off the unknown.

Compare the patient's plasma drug levels with the therapeutic ranges.

EXERCISE 9-6

URINARY CARBOXYLIC ACID PROFILE

A number of inborn errors of metabolism give rise to organic acidurias. These disorders are associated with the urinary excretion of large amounts of normal metabolic intermediates or unusual metabolites, due to the accumulation of large amounts of intermediates that are metabolised by alternative pathways.

The identification of the organic acids in the urine assists in the localisation of the metabolic defect.

Even though most organic acidurias are due to enzyme defects, some can be due to enzyme inhibition by environmental toxins or due to nutritional deficiencies.

The criteria used by the Royal Children's Hospital in Melbourne for a urinary carboxylic acid profile are:

1. Persistent vomiting.
2. Family history of mental retardation.
3. Urine with abnormal smells.
4. Amino acid profile suggestive of organic aciduria.
5. Other clinical data suggesting an organic aciduria.

Case History.

The sample is urine from a fourteen week old girl with methylmalonic acidaemia.

This disorder results from a defect in the conversion of methylmalonyl-CoA to succinyl-CoA.

The precursors and enzymatic steps are illustrated below:

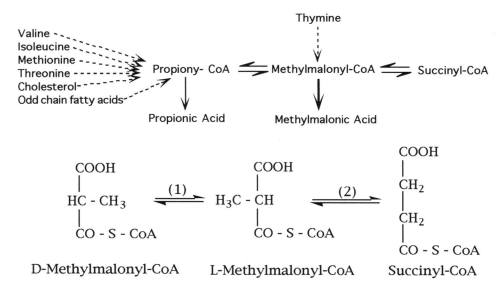

(1)   Methylmalonyl-CoA Racemase.

(2)   Methylmalonyl-CoA Mutase.
       Requires adenosylcobalamin as a co-enzyme.

The enzymatic abnormalities can be due to defective methylmalonyl-CoA mutase or to defects in the synthesis of adenosylcobalamin from vitamin B_{12}.

The disease presents with lethargy, failure to thrive, recurrent vomiting, dehydration, respiratory distress and muscular hypotonia. The biochemical abnormalities include metabolic acidosis, hyperammonaemia, hypoglycaemia and hyperglycinaemia. Most of the patients with a complete absence of methylmalonyl-CoA mutase die in infancy.

Treatment is to restrict the protein intake of all patients in order to reduce the amount of methylmalonyl precursors. Vitamin B_{12} administration also improves the situation where the defect is in the synthesis of adenosylcobalamin.

The urinary output of methylmalonic acid, in children with methylmalonic acidaemia is in the range 240 - 5,700 mg/24 hours. The normal output in children and adults is less than 5 mg/24 hours.

Organic acids do not chromotograph well on GC, due to their polar nature. The formation of derivatives results in less polar compounds that produce sharper GC peaks.

In this method, trimethylsilyl (-O-Si(CH$_3$)$_3$) derivatives are formed. These are ethers of hydroxyl groups and esters of carbonyl groups.

The carboxylic acids are rendered non-ionised with HCl and then extracted into ethyl acetate (and diethyl ether). The extract is then dried with anhydrous sodium sulphate and taken to dryness under nitrogen. The derivatives are then formed using N, O-Bis (trimethylsilyl)-trifluoro acetamide (BSTFA) with 1% trimethylchlorosilane (TMCS) at 60°C. The TMCS enhances the silylating action of BSTFA. These derivatives are then injected onto the GC.

The retention times of the derivatives are compared with those of n-paraffins for the calculation of methylene units. Normal urine and 10 mM methylmalonic acid will be treated in the same way as the patient's sample.

REAGENTS AND EQUIPMENT:

1.   Analar Sodium Chloride.
2.   Analar Anhydrous Sodium Sulphate.
3.   6 M HCl.
4.   Ethyl Acetate.
5.   Diethyl Ether.
6.   10 mmol/l Methylmalonic Acid.
         Dissolve 59 mg of MMA in 50 ml of distilled water.
7.   Normal Urine.
8.   Patient's Urine.
         Dissolve 1 g of MMA and 0.7 g of glycine in 300 ml of normal urine.
9.   BSTFA + 1% TMCS.
         1 ml Ampoules.
              Alltech Cat. No. 18089.
10.   Hydrocarbon Standard C$_{10}$-C$_{18}$.
         Add 500 µl of n-decane to a 2 ml vial of Alltech Cat. No. NP-MIX-G (C$_{12}$-C$_{18}$).
         Add 200 µl of this mixture to 5 ml of AR n-hexane and mix well.
         Store in a tightly sealed vial.
11.   HPLC Grade Methanol in a beaker.
         For GC syringe rinsing.

6 ml culture tubes with caps.
Heating Block (45°C) with fine tubes connected by a manifold to a nitrogen cylinder.
Heating Block at 60°C.
Bench Centrifuges.
Top Loading Balance.
1 and 2 ml Graduated Pipettes.
1 µl GC Syringe.
Pasteur Pipettes.
Variable Pipettor (set to 125 µl) with Tips.
Spatulas.
GC Parameters:
         Column : 2 meter glass, packed with 3% OV-101 on gas chrom Q.
         Instrument settings:

Injection port temperature	:	255°C.
Detector temperature	:	270°C.
Temperature programme	:	80 - 210°C.
Temperature rise rate	:	7.5°C per minute.
Final hold time	:	3 minutes.
Cool down time	:	3 minutes.
Attenuator	:	$10^2 \times 64$.
Nitrogen (Carrier gas) flow rate	:	≈ 25 ml/minute.

Detector gas flow	:	30 ml H_2/minute. 300 ml air/minute.
Output lead	:	Recorder.

Integrator:

Chart speed	:	0.5 cm per minute.
Attenuation	:	2.
Threshold	:	3.

PROCEDURE:

1. To three 6 ml screw-top culture tubes, add:

    0.5 g NaCl.
    2 drops of 6 M HCl.
    2 ml of ethyl acetate.

Tube (i)	:	1 ml of patient's urine.
Tube (ii)	:	1 ml of normal urine.
Tube (iii)	:	1 ml of 10 mM methylmalonic acid.

2. Cap the tubes and shake vigorously for 2 minutes.

3. Centrifuge for about 2 minutes at moderate speed.

4. Transfer the top layers to labelled test tubes, with Pasteur pipettes.

5. Re-extract the aqueous phase with another 2 ml of ethyl acetate, as in steps 2 and 3, and transfer the top layer to the appropriate test tube.

6. Re-extract the aqueous layer with 2 ml of diethyl ether, as above, and add the ether extract to its test tube.

7. Add about 1 g of anhydrous sodium sulphate to each test tube and swirl the tubes to absorb any aqueous material.

8. Carefully decant the liquid into 6 ml screw-top culture tubes.

9. Evaporate to dryness under a stream of nitrogen, at 45°C in a heating block.

10. Add 125 µl of BSTFA/TMCS reagent, cap the tubes tightly and mix briefly.

11. Incubate at 60°C for 30 minutes in the heating block.

12. Allow the tubes to cool.

13. Inject 1 µl of the normal urine derivative onto the column and then start the temperature programme. Start the integrator as you inject the sample.

14. After the completion of the programme, inject 1 µl of the patient's urine derivative when the oven temperature light dims and start the programme again. Start the integrator as you inject the sample.

15. Repeat the above with 1 µl of the methylmalonic acid derivative.

16. Repeat the above with 1 μl of the hydrocarbon standard (C_{10} - C_{18}) in hexane.

17. Mix together equal volumes of the patient's derivative and the hydrocarbon standard. Inject 1 μl of this mixture into the GC, as above.

18. Mix together equal volumes of the methylmalonic acid derivative and the hydrocarbon standard. Inject 1 μl of this mixture into the GC, as above.

Notes:

1. Peak heights can be increased by decreasing the attenuation on either the GC or on the integrator.

2. Wash out the GC syringe, a number of times with methanol, after each injection.

The type of chromatogram obtained is illustrated below:

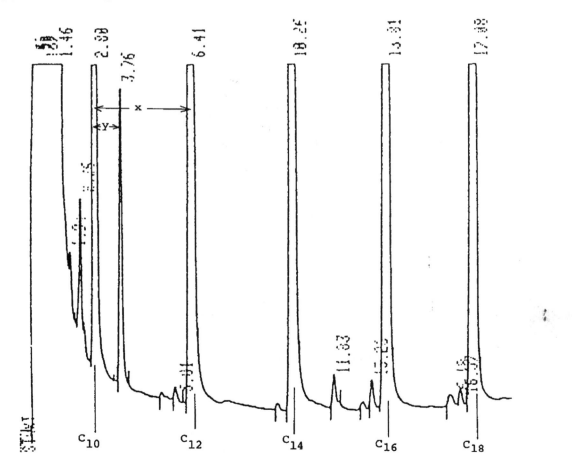

Calculate the methylene units (MU) of any significant peaks, as follows:

$$MU = n + (2y/x)$$

Where n = the number of carbons in the paraffin,
                 x = the time between C_n and C_{n+2} peaks,
                 y = the time difference between C_n and peak of interest.

For example, in the chromatogram above, the peak at 3.76 minutes is probably lactic acid.

$$x = 6.41 - 2.88 = 3.53$$

$$y = 3.76 - 2.88 = 0.88$$

$$\therefore MU = 10 + (2 \times 0.88 \div 3.53) = 10.50$$

Other peaks can be identified from tables of MU values. Some of those used by the Royal Children's Hospital are shown below:

MU	Acid	Implication
10.54	Lactic	Lactic Acidosis
10.83	Pyruvic	Lactic Acidosis
11.25	2-Hydroxybutyric	Lactic Acidosis
11.35	3-Hydroxypropionic	Propionic Acidaemia
11.60	3-Hydroxybutyric	Ketosis
12.00	Acetoacetic	Ketosis
12.09	Methylmalonic	MMA Aciduria
12.38	3-Hydroxyvaleric	Propionic Acidaemia
12.81	3-Ketovaleric	Propionic Acidaemia
15.05	Salicylic	Aspirin
15.86	Phenyl Lactic	PKU
17.07	Phenyl Pyruvic	PKU
18.40	Homogentisic	Alkaptonuria
19.15	4-Hydroxyphenyl Acetic	Tyrosine Metabolite
20.53	4-Hydroxyphenly Pyruvic	Tyrosine Metabolite

Note that some variation in the second decimal value can be expected.

GENERAL REFERENCES:

Cramers, C. A. and McNair, H. M., (1983). "Gas Chromatography" in J. Chromat. Library Vol. 22A, Chromatography - Fundamentals and Applications, A. Fundamentals and Techniques. Ed. Heftmann, E., p A195-A224. Published by Elsevier.

Poklis, A., (1989), "Gas Chromatography" in Clinical Chemistry - Theory, analysis and correlation, 2nd Edition. Eds. Kaplan, L. A. and Pesce, A. J., p 110-125. Published by The C.V. Mosby Company.

Willard, H. H., Merritt, L. L., Dean, J. A. and Settle, F. A., (1981), Instrumental Methods of Analysis, 6th Edition, p 454-494. Published by Wadsworth Publishing Co.

Supina, W. R., (1974), "The Packed Column in Gas Chromatography". Published by Supelco Inc.

Geddes, A. J., (1980), "Mass Spectrometry", Chapter 9 in An Introduction to Spectroscopy for Biochemists, Ed. Brown, S. B., p 336-391. Published by Academic Press.

Blood Alcohol.

Cooper, J. D. H., (1971), Clin. Chim. Acta, $\underline{33}$, 483.

Head-Space Gas Analysis.

Leach, H., (1969), Isolation and Identification of Drugs, Volume 1. Ed.: Clarke, E. G. C., p 73.

Anticonvulsants.

Solow, E. B., Metaxas, J. M. and Summers, T. R., (1974), Antiepileptic Drugs: A Current Assessment of Simultaneous Determination of Multiple Drug Therapy by Gas Liquid Chromatography - On Column Methylation, J. Chromat. Sci., $\underline{12}$, 256-260.

Carboxylic Acid Profile.

Tanaka, K., Hine, D. G., West-Dull, A. and Lynn, T. B., (1980), Gas-Chromatographic Method of Analysis of Urinary Organic Acids. I. Retention Indices of 155 Metabolically Important Compounds, Clin. Chem., $\underline{26}$, 1839-1846.

Tanaka, K., West-Dull, A., Hine, D. G., Lynn T. B. and Lowe, T., (1980), Gas-Chromatographic Method of Analysis of Urinary Organic Acids. II. Description of the Procedure, and its Application to Diagnosis of Patients with Organic Acidurias, Clin. Chem., $\underline{26}$, 1847-1853.

CHAPTER 10

HIGH PERFORMANCE LIQUID CHROMATOGRAPHY

Modern chromatography can be traced back to the work of Martin and Synge, published in 1941. Their paper predicted the use of very small particle sizes and high pressure in liquid chromatography. They also predicted gas chromatography. Ten years passed before the development of gas chromatography and 30 before the use of very small particle sizes in liquid chromatography.

High Performance Liquid Chromatography (HPLC) has, in the last 20 years, become a major method in separating organic compounds. It has the advantage over gas chromatography (GC) in that non-volatile and heat labile compounds can be separated. It also has greater selectivity than GC in that both the mobile and stationary phases can be varied. Larger volumes of sample can be used with HPLC than with GC. This is because GC samples have to be vaporised immediately on injection.

It has been estimated that GC can only be used to separate 15-20% of all known organic compounds; whereas, HPLC can separate 80-85%.

Separation is achieved by a competitive distribution of the sample between two phases. One is a mobile liquid and the other is a stationary liquid or solid.

The high separating efficiency of HPLC has been achieved by the optimisation of column parameters and, in particular, of the particle size of the column packing. Liquids have viscosities 20-100 times that of gases and so greater pressures have to be applied to liquid chromatography columns to produce desirable flow rates. The diffusion rate of the sample in a liquid is 3,000 - 30,000 times less than in a gas, and so much smaller particles allow rapid equilibrium with the stationary phase.

Smaller particles tend to reduce the flow rate through the column and so short columns are also used to produce an optimum flow rate. High efficiency can be obtained with very small particles and short columns.

HPLC therefore uses very small particles, short columns and high pressure delivery systems.

The components of a HPLC system are shown below:

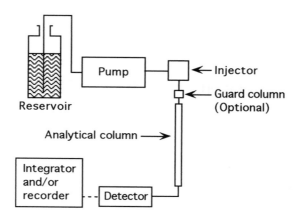

The pump aspirates mobile phase from the reservoir and then forces it through the injector, the column and thence through the detector. The detector response is recorded as a chromatogram such as that below:

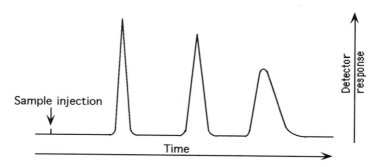

Useful data can be obtained from the chromatogram, such as retention time and peak area. The retention time is the time, after injection, for the sample peak to be detected. This is constant under fixed conditions and so retention time can be used for qualitative analysis.

The area under each peak is proportional to the concentration of that component of the sample. Peak area can, therefore, be used for quantitative analysis.

In order to obtain sharp peaks (high efficiency), the dispersion of the sample and its separated components must be kept to a minimum. The dispersion of a solute band in the connector tubes is dependent upon the internal diameter of the tube, the length of the tube and the flow rate within the tube. The degree of band spreading is dependent on the fourth power of the radius of the tube. Hence, connecting tube diameter is an important factor in maintaining high efficiency in the HPLC system.

Connecting tubing between the injector, guard column, analytical column and detector should have a diameter of no more than 0.25 mm and be no longer than 200 mm. The couplings between the injector, the column, the connector tubes and the detector should have a butt fit, so that no additional liquid spaces are included.

The components of a HPLC system will now be looked at in more detail.

Equipment Components (a).

The Mobile Phase Reservoir.

This should hold about 1 litre and must be chemically resistant. If chlorinated hydrocarbons are used in the mobile phase, light should be excluded to avoid photo-decomposition.

A facility for solvent degassing should be included. Solvents must be degassed to remove the possibility of air bubbles forming in the check valves of the pump. If this happens the efficiency of the pump is considerably reduced. Air bubbles passing through the detector can also produce spikes on the chromatogram. Oxygen can also react with some solutes and stationary phases.

Air is soluble in all solvents used for the mobile phase. Polar solvents, water in particular, have a high solubility for air; whereas, non-polar solvents, such as hexane, have a lower air solubility.

A number of methods for degassing are available. A convenient method is to bubble helium through the solvent from a 15 μm glass sinter for about 5 minutes. The helium flow is then reduced to maintain an atmosphere of helium above the solvent. Helium displaces dissolved air and has a very low solubility itself. The degassed solvent is filtered before entering the pump to remove any small particular material, which could damage the check valves of the pump.

Another convenient method for degassing (and filtering) is to filter the mobile phase through a 0.45 μm filter, under suction. Teflon filters can be used for organic mobile phases and cellulose ester filters for aqueous ones. However, "Durapore" filters, made of a polymeric material, can be used for either type of solvent.

Solvents must be of high purity to avoid adsorption of contaminants onto the column, reducing the efficiency of the column. Impure solvents also tend to have higher far-UV absorption, than pure solvents.

The Pump.

The pump should have the following:

1. A flow rate of 0.01 - 10 ml/minute.

2. Capable of producing pressures of up to 70 MPa (10,000 psi).

3. Pulse free delivery. This is necessary to reduce detector flow cell "noise".

4. An adjustable safety pressure cut-off.

Pumps fall into two types: (1) mechanical, producing constant flow, and (2) pneumatic, producing constant pressure. Constant pressure pumps are best for column packing and constant flow pumps are best for chromatographic analysis.

Most modern chromatographic pumps use a dual piston reciprocating system as shown below:

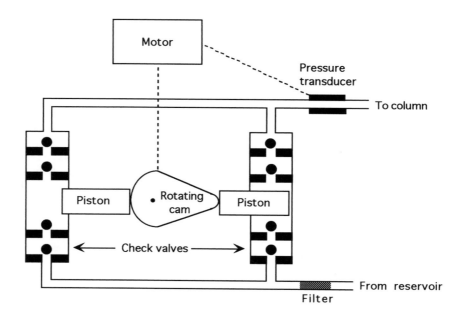

The two pistons are 180° out of phase so that one side is pumping while the other is refilling. Even with this system there are still pulse variations due to compliance of the piston chambers and seals and so the pump is fitted with a feedback system. A pressure transducer is located downstream of the pump. Pressure at this point is proportional to the flow rate. Output from the transducer is fed to the pump speed control via a feedback mechanism. As the pressure falls at piston changeover, the pump is speeded up to maintain original pressure. In this way the pump output is maintained at a constant flow rate.

The Waters (A division of the Millipore Company and major HPLC producer) pump operates slightly differently. It consists of 2 pistons, one having twice the capacity of the other. When the larger one is filling, the smaller one is emptying. On completion of the stroke, the larger piston now pumps half its contents into injector and the other half into the smaller piston, which is now refilling.

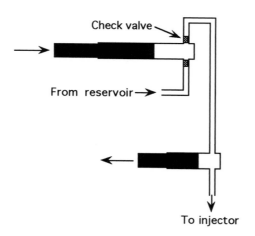

For most separations the pump is run at 1-3 ml/minute.

Gradient Elution.

In a number of cases, the use of a single solvent may not be able to separate all the components of a mixture, if all the components are to be eluted from the column in a reasonable time. The use of a single solvent is known as isocratic HPLC.

In gradient elution HPLC, the composition of the mobile phase is changed, in a controlled manner, during analysis. One method of gradient elution is to have two solvent reservoirs and a programmable control over the switching valve that feeds the two to a mixer before entry to the pump.

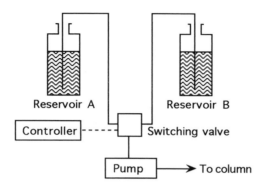

Other systems, such as that used by Waters, use mixing on the high pressure side of 2 or 3 pumps. The composition of the mobile phase is controlled by varying the pump flow rates. This is done by a gradient controller. The outputs from the pumps are fed around a cone, which facilitates mixing.

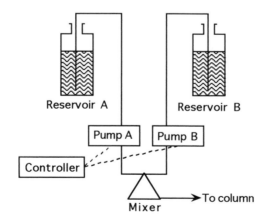

Some of the types of gradient profiles that can be generated by modern microprocessor gradient programs are shown below:

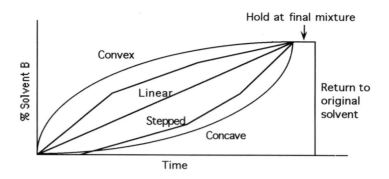

Sample Injection.

To obtain high efficiency from the column, the sample should be injected as a very narrow band into the centre of the column.

Syringe injection systems (such as those used in gas chromatography) cannot be used with pressures above 15 MPa, even when stop flow systems are used. Syringe injection also has the disadvantage that small pieces of septum, dislodged by the syringe needle, can contaminate the column head.

Therefore, most HPLC systems use rotary injection valves. A diagram of a 7 port valve is shown below:

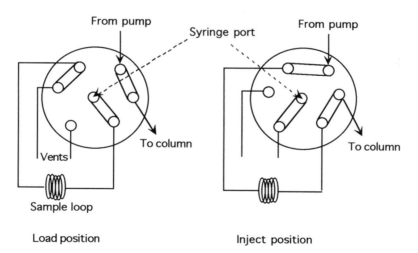

In the "load" position the sample is injected from a microlitre syringe into the needle port. The sample enters the sample loop. The loop full of sample will hold 20 µl (or other stated volume). Smaller volumes can be added to the loop, but 0.5 µl (or similar volume) is left in the needle port when the valve is rotated. The solvent flows directly from the pump to the column when the valve is in this position.

When the valve is rotated to the "inject" position, the solvent flows via the sample loop. The sample is thus added to the centre of the top of the column. The needle port can now be washed out ready for the next sample.

Rotary injection valves will operate in pressures up to 60 MPa (9,000 psi).

The Waters injector (U6K) operates on a different principle. This involves the mobile phase flowing through a narrow tube (restrictor loop, 0.23 mm internal diameter) while the sample is loaded into the sample loop, which has an internal diameter of 1.02 mm. In the inject position, the mobile phase has access to both loops. As there is much more resistance to flow in the restrictor loop, most of the mobile phase flows via the sample loop.

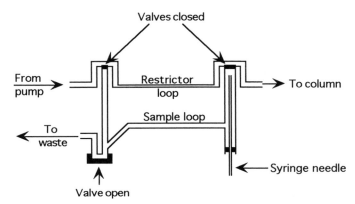

Sample loading.

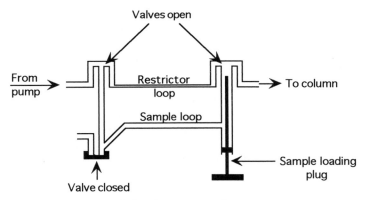

Injection onto column.

Guard Columns.

The purpose of a guard column is to protect the analytical column from contaminants that may be in the samples, as analytical columns are quite expensive.

They are usually very short columns that are often packed with pellicular material. Pellicular material consists of glass spheres of about 40 μm diameter, coated with the stationary phase to a depth of 1-2 μm. The dry material is very easy to pack into columns by a "tap and fill" technique.

Once this material has become contaminated, it can be discarded and the guard column repacked.

Other guard columns consist of cartridge holders. The cartridges contain the same filling as that used in the analytical column. These cartridges can be easily disposed of when they become contaminated.

Analytical Columns.

Most analytical columns are packed with microparticulate material. It consists of very small particles of silica or alumina with diameters of 2 to 10 μm. These particles can be spherical or irregular. The stationary phase can be chemically bonded to the surface of the silica particles. These packing materials were first introduced in 1971.

Open column chromatography uses particles with a diameter of 100-150 μm. These particles contain deep pores where the mobile phase tends to stagnate. Mass transfer between the phases is poor due to this stagnation and to the large interparticle channels and so the efficiency of these columns is very low.

The surface area of pellicular particles range from 5 to 15 m^2/g. Compared with porous material this is low and so the sample capacity is low. However, mass transfer is greatly improved and so, therefore, is column efficiency.

Column efficiencies of microparticulate material are greater than those with the previous two types due to the, higher rates of solute mass transfer between the phases. This is due to the small pore depths and the reduction in interparticle channels.

The active surface area of these particles range from 80 m^2/g using alumina to over 500 m^2/g using silica gel. These small particles tend to agglomerate in the dry state due to the electrostatic charge on their surface. As a result particles below 20 μm in diameter cannot be packed by the "tap and fill" method. The use of a suitable liquid, usually polar, will nullify this aggregation effect.

These microparticulate columns are, therefore, packed by a slurry packing technique. Slurries are prepared in liquids of high density or high viscosity to reduce settling out. These are degassed and pumped under high pressure rapidly into the column. The more rapidly the column is packed the higher the efficiency. A pre-column is usually included as the density of the packing falls off sharply at the top of the column. The main column is used for chromatography.

Most microparticulate columns are designed so that none of the sample touches the side of the column. Column efficiency is considerably reduced if part of the sample touches the side, due to the fact that relatively large spaces can exist between the column wall and the particles, when compared with the space between particles.

The degree of radial spread of the injected sample is dependent on five factors:

    (1)   The initial band diameter.
    (2)   Stream splitting, which is dependent upon the particle diameter,
    (3)   The diffusion rate of the solute in the mobile phase,
    (4)   The flow rate of the mobile phase and
    (5)   The distance travelled in the column.

In most cases, where 5 or 10 μm microparticulate material is used, a column of 5 mm internal diameter can be used for columns up to 300 mm in length. 7 mm internal diameter columns should be used for 300-700 mm long columns. It is essential that the columns be very evenly packed so that no solute molecules reach the wall.

Wall effects are eliminated in the Waters Radial Compression columns. These are plastic columns of 100 x 8 (internal diameter) mm. The tube is compressed by a mechanical or hydraulic device. This presses the plastic into the space between the packing material adjacent to the wall. Radial pressure on the packing also reduces the void space between the particles and, hence, increases column efficiency.

Column Theory.

Linear elution chromatography is where the concentration of the solute in the stationary phase is directly proportional to the concentration of the solute in the mobile phase. The elution peaks will be Gaussian in shape and most of the information on column selectivity and efficiency, in chapter 9 (Pages 9-18 to 9-21) will apply. Hence, efficiency (N) is given by the formula:

$$N = 16 \left(\frac{t_R}{W} \right)^2$$

Where: N   =   The number of theoretical plates.

t_R =   The retention time of the compound used to assess the efficiency.

W   =   The base width of the peak, from tangents drawn at the inflection points.

Hence, the efficiency of a column is related to the degree of peak broadening. The narrower the peaks, the greater the efficiency. The plate number (N) measures the efficiency of the entire chromatographic bed, which includes extra-column contributions. Correction can be made for the void volume of the system. The corrected efficiency (N_{eff}) is calculated from the formula:

$$N_{eff} = 16 \left(\frac{t_R - t_m}{W} \right)^2$$

Where: t_m =   The retention time of a non-retained solute.

Another measure of column efficiency is plate height, H (or HETP, Height Equivalent to a Theoretical Plate).

$$H = \frac{L}{N}$$

Where: L   =   The length of the column bed.

H has the dimension of length, usually mm.

It is sometimes more useful to express the plate height in terms of the particle diameter. h, the "Reduced Plate Height", is given by the formula:

$$h = \frac{H}{dp}$$

Where: dp   =   The particle diameter.

For a highly efficient column, h is around 2-3.

The fine structure of the column packing material is shown below:

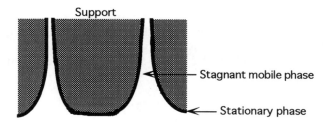

A finite time is required for a solute molecule to migrate from the flowing zone of the mobile phase to the stagnant zone and thence to the stationary phase. Thus, equilibrium takes a finite time to be reached. As a result, the concentration profile (concentration of the compound of interest at different points in the column), in the flowing mobile phase, runs ahead of that in the stationary phase. This is illustrated below:

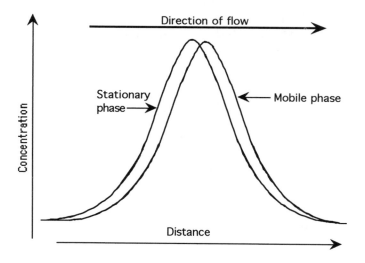

The rate of movement of solute is faster upstream than downstream. This effect leads to peak spreading and hence to a reduction in efficiency. The faster the mobile phase moves, the greater is this effect.

The uneven flow around the packing material also causes dispersion due to velocity variations across interparticle streams. These two effects are reduced by slower flow rates.

However, lateral dispersion of the sample is time dependent. With a slower flow rate the sample is in the column for longer and will, therefore, disperse further. There is, therefore, an optimum flow rate for each column. The plot of HETP against the mobile phase flow rate gives a similar shape to the Van Deemter plots seen in gas chromatography.

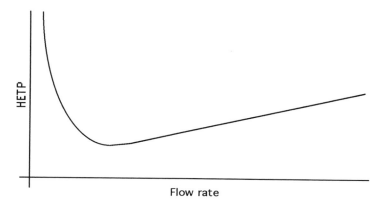

The Peak Capacity of the column is the number of equispaced peaks that the column can separate. Calculation of this value includes the square root of the Number of Theoretical Plates. An average HPLC column (with about 10,000 Theoretical Plates) will have a Peak Capacity of around 50; whereas, a classical liquid chromatography column (with about 100 Theoretical Plates) will have a Peak Capacity of about 5.

As discussed in chapter 9 (Page 9-20), the separation of two adjacent peaks is called the resolution (R_S) and is defined as the distance between the two peak centres divided by the average base width (in the same units).

From the elution diagram on page 9-18:

$$R_S = 2\left(\frac{t_{R2} - t_{R1}}{W_1 + W_2} \right)$$

Where the peaks are closely separated, W_1 is approximately equal to W_2.

$$R_S = \frac{t_{R2} - t_{R1}}{W}$$

and W= 4 standard deviations (SD) for a Gaussian curve.

Therefore, when R_S = 1.5, separation is almost complete with less than 1% overlap (6 SD separation).

When R_S = 1, separation = 4 SD and two peaks can be seen. However, as R_S falls below 1, it becomes more difficult to distinguish the two peaks especially if they are of unequal intensity.

Three factors contribute to the resolution obtained. These are the selectivity of the column, the retention of the column and the column efficiency.

The resolution equation can be expressed as follows:

$$R_S = \frac{1}{4} \left(\frac{\alpha - 1}{1 + \alpha} \right) \left(\frac{\bar{k}'}{1 + \bar{k}'} \right) \sqrt{N}$$

Selectivity (Relative Partition) Retention Column efficiency

The separation factor $(\alpha) = \frac{K_2}{K_1}$, where K_1 and K_2 are the distribution coefficients of the two solutes in the stationary and mobile phase.

The capacity factor, $\bar{k}'$, is the average column capacity factor of the two solutes:

$$\frac{k'_1 + k'_2}{2}$$

Where k' = $K \frac{V_S}{V_m}$

V_S = The volume of the stationary phase.

V_m = The volume of the mobile phase.

N is altered by changing the length of the chromatography column bed, the particle size or the velocity of the mobile phase. α can be changed by using different phase systems and k' by altering the ratio of mobile to stationary phase as well as by using different phase systems.

The effect of reducing each of N, α and $\bar{k}'$, is shown below:

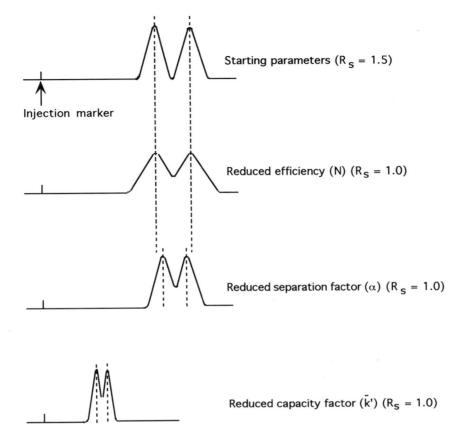

High k' values lead to long retention times with broad dilute peaks that are hard to detect. Optimum k' values are in the range 1.5 to 5 with the best at 2.

k' is related to retention time by the formula:

$$k' = \frac{t_R - t_m}{t_m}$$

Therefore, when k' = 2, the retention time of the compound of interest will be three times the retention time of an unretained solute.

A typical isocratic chromatogram of a mixture containing many components is shown below:

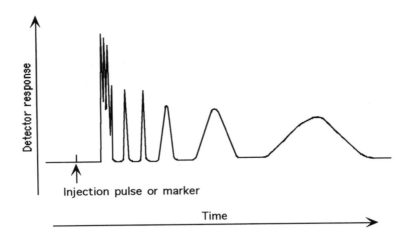

Injection pulse or marker

Time

As can be seen, the front of the chromatogram shows a series of bunched peaks that are poorly resolved. These components have low k' values. Components in the middle portion have k' values in the optimum range (1.5 - 5) and are well resolved. The k' values of the final peak is greater than 5. Peaks with k' values greater than 5 can be hard to detect and quantify.

This situation can be improved by using gradient elution. k' values are changed during the elution. The disadvantage of gradient elution is that time is required to return the column to its original state, after the completion of analysis, and before the next run can be carried out. The type of result obtained with gradient elution is shown below:

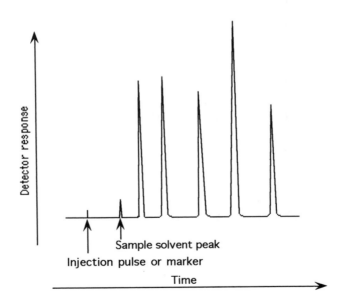

Sample solvent peak

Injection pulse or marker

Time

Other peaks seen on the chromatogram.

The injection point may be seen as a pressure pulse when the injection valve is rotated. This appears as a very sharp spike on the recorder trace. Most injectors have a system that sends a short electrical pulse to the integrating recorder.

The sample solvent may also show a peak, this can be positive or negative or both. This peak depends, to an extent, on the solvent used to dissolve the sample.

Injection pressure pulses and sample solvent peaks are not always present.

Non-Gaussian Peaks.

In well-behaved chromatography, the peaks will be symmetrical, ie. Gaussian in shape. For this to take place, the distribution coefficient must not alter with changes in sample concentration. However, a number of factors associated with chromatography can lead to a change in the distribution coefficient with changing concentration. If this happens, the peaks will be skewed and the retention times will be altered. This is illustrated in the figures below:

Distribution Coefficient	Constant	Decreases	Increases
Detector Response			
Retention Time	Constant	Decreases	Increases

The left hand recorder trace shows the well-behaved system with no change in K. The centre trace shows the response when K decreases with increase in sample concentration and the right hand trace shows the response when K increases with increasing sample concentration.

Practical Aspects of non-Gaussian Peaks.

A tailing peak, as in the centre above, would suggest that an incorrect column is being used. A reverse tailing peak, as in the right-hand trace above, suggests that the column is overloaded.

CHROMATOGRAPHIC TECHNIQUES

The various types of chromatography discussed in chapters 7 and 8 can also be used in HPLC. Hence, material covered in these chapters will only be briefly mentioned.

Adsorption Chromatography.

The adsorbents used in HPLC are mainly silica gel and alumina. Due to the high porosity of these materials, a large surface area is exposed to the mobile phase. Solute molecules are mainly retained by hydrogen bonds and Van der Waals forces. It is mainly the polar functional groups of the solute that interact with the stationary phase. This type of chromatography is best suited to non-ionic solutes that are soluble in organic solvents.

Solutes are eluted form the column by the mobile phase, which competes for the binding sites on the stationary phase. More polar solvents have a greater affinity for the binding site than the less polar ones. Therefore, the more polar the solvent is, the greater its eluting power.

Mixtures of solvents can be used. The more polar will be adsorbed more readily onto the column than the weaker. Time must, therefore, be allowed for the column to reach equilibrium with a solvent mixture, before the sample is applied. Solvent mixtures can often produce better resolution than a pure solvent. Greatest improvements in selectivity are seen when the concentration of the stronger component in the mixture is less than 5% by volume or greater than 50% by volume.

The water content of the stationary phase is very important as this determines the strength of the binding sites. A highly active surface is required for separating non-polar solutes with a non-polar solvent. The water content of the particles must be very low in this case. Separating polar solutes with a non-polar solvent requires a deactivated surface with a fair amount of water in the particle. Water breaks the hydrogen bonds between hydroxyl groups on the particle. These groups are the strongest adsorbers. Equilibrium between the water content of the solvent and the column must be reached; otherwise the retention time of samples will change. Isopropanol can be used instead of water. This has the advantage of being miscible with most of the non-polar solvents.

Gradient elution can be used so that the elutropic strength is increased during the run. A linear change in elutropic strength, with time, is best. This is achieved by a concave gradient profile. After the elution, the adsorbent must be regenerated back to its original condition. It is better to reverse the gradient rather than a direct return to the original solvent.

Partition Chromatography.

In partition chromatography, the solute molecules distribute themselves between two immiscible liquid phases according to their relative solubilities.

In early forms of partition chromatography, the stationary phase was held on the support, such as silica gel, by adsorption forces only. In this system, the stationary phase must be completely insoluble in the mobile phase. In practice this is virtually impossible and so the stationary phase was slowly eluted from the column. This phenomenon is considerably retarded if the mobile phase is saturated with stationary phase, prior to application to the column.

To prevent elution of the stationary phase, it can be chemically bonded to the support material. This type of chromatography is termed "bonded phase chromatography", as the chromatographic characteristics can be different from adsorbed stationary phase.

The surface of silica gel particles contains a large number of reactive silanol groups ($\equiv$Si-OH). Two stable types of compounds can be prepared.

1. Siloxanes:

$$\text{Particle} \equiv \text{Si} - \text{OH} \; + \; R_2SiCl_2$$
$$\downarrow$$
$$\text{Particle} \equiv \text{Si–O–Si–Cl}$$

with R groups attached above and below the Si–O–Si–Cl.

2. Silica - carbon bond:

    (a) Chlorination.

$$\text{Particle} \equiv \text{Si–OH} \; + \; SOCl_2$$
$$\downarrow$$
$$\text{Particle} \equiv \text{Si–Cl}$$

    (b) Grignard or similar reaction.

$$\text{Particle} \equiv \text{Si} - \text{Cl} \; + \; RLi$$
$$\downarrow$$
$$\text{Particle} \equiv \text{Si} - R$$

The Siloxanes are attacked by strong acids (pH <2) and by alkali (pH >9). Most bonded stationary phases are made by this type of reaction and so can only be used in the pH range 2-9.

The Si–C bond is more stable, but the silica gel structure contains $\equiv$ Si–O–Si $\equiv$ bonds. These bonds are attacked at pHs greater than 9. Therefore, separation techniques using silica as the support material should not be used outside the pH range 2-9.

Some commercially available chemically bonded materials are listed below:

Stationary phase	Type
Octadecyl (C_{18})	Non polar (aliphatic)
Phenyl	Non polar (aromatic)
Alkyl nitrate (cyano)	Medium polarity
Ether	Medium polarity
Alkyl amine (-NH_2)	Medium polarity
Diol	High polarity

Reverse phase partition chromatography (using C_{18} columns) is probably the most common form used in HPLC. Here the stationary phase is non-polar and the mobile phase is polar.

Polar solutes, therefore, readily dissolve in the mobile phase and are eluted first from the column. Samples are eluted in decreasing order of polarity.

Water (or an aqueous buffer) alone, or in a mixture with water miscible organic solvents, is used for the mobile phase. Solutes are retained by their hydrophobic groups and separated by their relative solubility in the mobile phase.

The medium polarity group stationary phases are usually used in the normal phase mode, where the solute is retained by its polar groups. It is eluted with a non-polar solvent. The high polarity diol stationary phase is used for normal phase chromatography.

The addition of small amounts of phosphoric acid (or perchloric acid) to the mobile phase can often improve peak shapes in reverse phase chromatography.

Ion Exchange Chromatography.

In gravity fed ion-exchange columns the functional groups are bonded to a polystyrene, or similar, matrix. This matrix structure tends to collapse when high pressures are applied to the column. These materials cannot, therefore, be used in HPLC at pressures above 7 MPa.

Satisfactory ion exchangers, for HPLC, have been prepared by attaching ionic functional groups to organic molecules chemically bonded to silica particles. For example, an anionic and cationic exchange are shown below:

Anionic:

$$Particle \equiv Si - (CH_2)_n - N^+ - R \qquad Cl^-$$

with R groups above and below the N^+ center.

Cationic:

$$Particle \equiv Si - (CH_2)_n - SO_3^- \qquad Na^+$$

Aqueous solutions are most commonly used as the mobile phase, due to the good solvating and ionising properties of water.

Retention is governed by the total concentration of the ions in solution (displacement of the exchange equilibrium), and by the pH (ionisation of the substances to be analysed).

The mobile phase is usually a buffer, the ionic strength of which can be varied. Increasing the ionic strength decreases the retention time. The retention also depends on the nature of the eluting ion. The greater the affinity of the ion exchanger for this ion, the shorter the retention time.

Increases in temperature improve column efficiency with the resultant reduction in peak base width.

In addition to true ion exchange phenomena, other effects (such as adsorption and partition) play some part in the separation achieved by ion exchange columns. Therefore, it is often more difficult to predict the behaviour of ion exchange columns for a particular analysis than it is for other forms of liquid chromatography.

Ion Suppression - Reverse Phase Chromatography.

Weakly ionic compounds can be analysed by reverse phase chromatography, if they are in the unionised form.

In this method of chromatography, the pH of the mobile phase is adjusted so that ionisation is suppressed. Thus acid materials can be separated, in a mobile phase of low pH, by reverse phase chromatography.

$$HA \quad \xleftarrow{\text{Low pH}} \quad H^+ + A^-$$

As columns cannot be used at pHs greater than 9, this makes ion suppression of basic materials rather limited.

Ion-pair - Reverse Phase Chromatography.

Ion-Pair Chromatography offers an alternative to ion-exchange in the separation of strongly ionic materials.

In this technique, ionisation of the sample is encouraged by the use of a mobile phase with an appropriate pH.

ie.
(a)
$$HA \quad \xrightarrow{\text{High pH}} \quad A^-$$

(b)
$$B \quad \xrightarrow{\text{Low pH}} \quad BH^+$$

An ionic species, of opposite charge to that of the ion of interest is added to the mobile phase to form an "Ion Pair" with the sample ion.

ie.
(a) $\quad A^- + X^+ \quad \longrightarrow \quad AX$

(b) $\quad BH^+ + Y^- \quad \longrightarrow \quad BHY$

The solubility characteristics of the ion-pair are substantially different to the original ionic sample and it is more soluble in non-polar solvents. The ion-pair can be analysed using reverse phase techniques.

Quaternary ammonium salts can be used as a counter ion with acids and alkyl sulphonic acids used with bases.

Illustrated below is the ion-pair interaction of octane sulphonate with a primary amine, and a tetrabutyl ammonium ion with a carboxylic acid, plus their interaction with the C_{18} stationary phase:

Particle Si—O—Si(CH$_3$)(CH$_3$)—(alkyl chain)—SO$_3^-$ R-NH$_3^+$

Particle Si—O—Si(CH$_3$)(CH$_3$)—(alkyl chain)—N$^+$(C$_4$H$_9$)(C$_4$H$_9$)(C$_4$H$_9$) R-COO$^-$

This technique has the advantage over ion-exchange in that the chromatographic behaviour can be predicted better, and both ionic and non-ionic material in the same mixture can be analysed.

Steric Exclusion Chromatography.

In this type of chromatography, components of a mixture are separated in order of their molecular size.

The stationary phase is a porous matrix with the pores closely controlled in size. Molecules that are too large to enter any of the pores are excluded from the matrix and pass through the column with the solvent front. Molecules of intermediate size can enter some pores and not others. These are retarded, to an extent dependant on their size, as they flow down the column. Small molecules can enter all the pores and are eluted last from the column.

Material that can be used at high pressure includes porous glass and porous silica. Porous glass is available in a number of different pore diameters. It can be used to separate molecules ranging from 10^3 to 10^{10} Daltons. Porous silica can separate molecules from 4 x 10^4 to 4 x 10^6 Daltons and is available in a smaller range of pore sizes than the porous glass.

Both these materials have polar surfaces and are sensitive to polar interactions with large molecules. This is especially important if non-polar solvents are used. The surface can, however, be treated with alkyl groups and this leads to a marked reduction in adsorption effects.

Highly cross-linked polystyrene is also used as the porous material. This can be used with pressures up to 40 MPa and with organic solvents. These gels allow the range from 100 to 5 x 10^8 Daltons to be covered. This material does not show adsorption effects.

Chromatography Type Selection in HPLC.

The figure below illustrates a selection scheme:

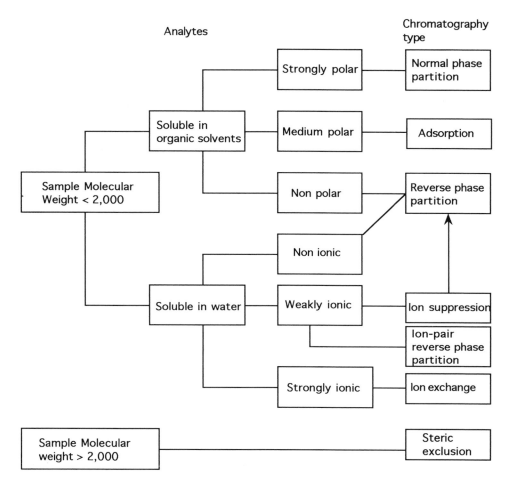

Equipment Components (b).

The Detector.

The ideal detector should:

1. Respond to all solutes,
2. Have high sensitivity and predictable response,
3. Respond independently of the mobile phase,
4. Be unaffected by changes in temperature and mobile phase flow rate,
5. Have a response, which increases linearly with solute concentration,
6. Be non-destructive of solute,
7. Provide qualitative information,
8. Introduce no peak spreading.

In practice, however, no detector is universally applicable to all liquid chromatography problems. However, it is usually possible to find a satisfactory detector for a particular problem. The major criterion is finding a detector with the desired sensitivity.

Detectors for HPLC can be divided into two categories, universal and selective.

Universal detectors are ones that respond to all solutes. These have low sensitivity. An example is the refractive index detector.

Selective detectors measure some characteristic unique to the sample component. These are more sensitive than universal detectors. For example, ultra violet absorption, fluorescence emission and electrochemical detectors fit into this category.

Refractive index detectors.

These are very versatile column monitoring devices, but they suffer from a number of disadvantages. The main one is that RI measurement is very temperature sensitive. It is also pressure sensitive and cannot be used with gradient elution. The extent of the temperature effect is reduced in modern instruments by comparing the RI of the eluate with a reference flow of pure mobile phase. However, very fine temperature control is still critical for the accurate functioning of the instrument.

Three types of differential refractometer are commercially available. The Fresnel type measures the change in percentage of reflected light at a glass-liquid interphase as the RI of the liquid changes. The volume of these cells is about 3 µl and so is suitable for high efficiency columns.

The deflection type measures the deflection of a beam of monochromatic light by a double prism as the eluate is passed through one half of it and the pure mobile phase passed through the other half. The cell volume is larger in this type.

In the interference type, a beam of monochromatic light is split and passes through the two cells. It is then recombined and directed onto a photomultiplier. When a solute passes through the sample cell the light velocity of the beam passing through the cell is reduced. The recombined beams are, therefore, out of phase and interfere. The resultant light intensity detected by the photomultiplier is reduced. The linear response of detector output with concentration is limited with this method.

The detection limit of RI detectors ranges from 10 µg to 1 mg/l. They have been used extensively to monitor steric exclusion column eluates.

Ultraviolet absorption detectors.

These are the most widely used detectors in liquid chromatography. These fall into 2 groups: fixed wavelength and variable wavelength.

Fixed wavelength detectors use the emission lines from excited metal sources. Eg. the 254 nm line from mercury and the 214 nm line from zinc.

The use of the deuterium lamp and a diffraction grating means that wavelengths between 185 and 380 can be used. These variable wavelength detectors have two advantages compared with fixed wavelength detectors. These are:

1.   The wavelength can be adjusted to that of the maximum absorbance of the substance, to provide maximum sensitivity.

2.   In some applications, they can provide useful selectivity. For example, in trace analysis it may be possible to set the detector to a wavelength at which only the trace component absorbs, thus eliminating interference problems for the major components of the sample.

The disadvantage is that the emission lines, from fixed wavelength sources, are much more intense than the same wavelength from a variable wavelength source. Hence, the sensitivity of fixed wavelength detectors is greater at these wavelengths.

The optical layout of a typical variable wavelength UV detector is shown below:

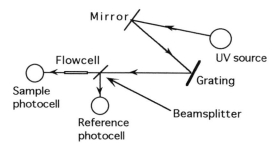

Compensation for variations in the intensity of the lamp output, at different wavelengths and with power fluctuations, is achieved by using a beam splitter to direct a small percentage of the light onto a reference photocell. The ratio of the two photocells is then fed into the electronics.

In order to avoid diffusion of the sample, leading to peak spreading and resultant decrease in efficiency, it is essential that the detector cell should have as small a volume as possible while retaining a reasonable light path length. The standard cell holds 8 μl, being 10 mm long and 1 mm in diameter.

The reduction in efficiency of the system is greater with the less retained (low k' values) components. With high efficiency columns (HETP values of less than 0.05 mm) there is appreciable peak broadening, using the standard 8 μl cell, for components with k' values of less than 2. 1 μl cells should, therefore, be used with these columns.

A typical flow cell is shown below:

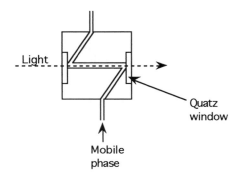

The detection limit of UV detectors is about 1 µg/l. The detectors are subject to pressure variation effects and so a pulse free pump system should be used. Temperature changes can lead to a base line drift. Solvents that absorb in the UV cannot be used in the mobile phase. However, most solvents used in HPLC do not absorb at wavelengths above about 210 nm.

The Waters fixed wavelength detector uses two flowcells and two photodiode detectors. A common filter serves both photodiodes and is used to select the desired emission line from a mercury or zinc lamp. One flowcell is for the sample and the other is for the reference cell. The reference cell is usually left empty, but can be filled with mobile phase solvent. The outputs from both photodiodes are fed to the amplifier where the absorbance of the reference is subtracted from the absorbance of the sample.

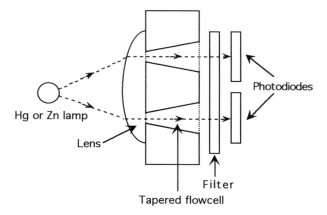

The flowcells are of a taper design. This allows the light, dispersed by the "liquid lens effect", to reach the photodiode.

The "liquid lens effect" is due to 2 phenomena. (1) The liquid in the flow cell tends to flow faster in the centre of the cell than on the edges. (2) The light source tends to heat up the stainless steel support of the flow cell. As the liquid near the wall is flowing more slowly than that in the centre, it tends to heat up more than the centre. This produces a change in refractive index between the centre and sides of the flowcell. This change in refractive index acts as a lens to diverge the light rays.

In recent years, **diode array** detectors have become a popular means of monitoring HPLC eluates. These detectors provide UV spectra of compounds as they are eluted from the column. The layout of these detectors is illustrated below:

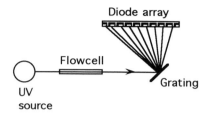

There is usually 1 diode per nm (or less) with 500 or more diodes in the detector. These are scanned in 0.1 of a second to produce a spectrum. 10 scans are made in 1 second. These are averaged to produce the spectrum that is stored on disk and displayed on the screen. The advantages of this type of detector are:

1. Rapid spectra production.
2. Wavelength accuracy (no moving monochromator grating).
3. Improved sensitivity (average of 10 scans).
4. Less stray light (less optics).

Fluorescence detectors.

Both filter and grating fluorescence detectors are used in HPLC. They can be used with gradient elution.

Below is the layout of a typical grating fluorescence detector:

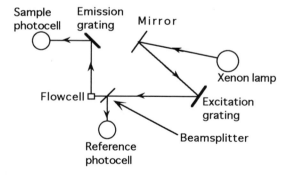

Fluorescence detectors are of use in the analysis of porphyrins, aflatoxins, riboflavins and oestrogens etc. They can also be used with induced fluorescence. For example, dansyl chloride derivatives of amino acids fluoresce strongly.

Electrochemical detectors.

These can be used to detect compounds that can be oxidised or reduced by an electrode. A cathode provides electrons for reduction and an anode removes electrons for oxidation.

Energy has to be supplied to the electron to carry out the reaction. This is provided by the voltage applied to the electrode (-2 [Red] to +2 [Oxid] volts).

The electrode potential must exceed the reaction potential for the reaction to take place. Therefore, the electrode potential can be set to detect some compounds (low red-ox potentials) and not others (higher red-ox potentials).

The mobile phase must be electrically conductive. Therefore, non-polar solvents can not be used. Electrolytes are added to water, methanol or acetonitrile (Eg. Potassium nitrate added to methanol). Buffer strengths of 0.05 M have adequate conductivity.

An oxidation reaction is illustrated below:

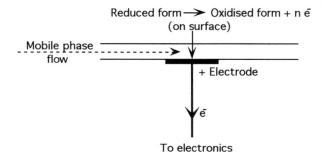

Electrons are removed at the + electrode surface as the components in the HPLC eluate are oxidised. The current flow is amplified and recorded.

The layout of a typical electrochemical detector is illustrated below:

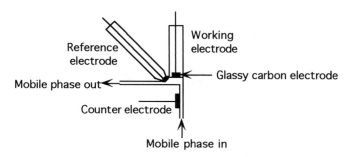

Electrochemical detectors are used to detect catecholamines, vitamins and drugs etc.

Mass spectrometer detectors.

As with gas chromatography, the ultimate detector is a mass spectrometer. As the mass spectrometer operates at high vacuum, the instruments invariably use capillary columns to reduce the amount of mobile phase presented to the mass spectrometer (MS).

Two approaches have been taken to interfacing an HPLC with a MS. These involve the removal of the mobile phase on a heated moving belt before the sample enters the MS, or using the mobile phase as part of the ionisation process. Thermosprays and electrosprays are the most commonly used interfaces in the second group.

Thermosprays rely on the use of a volatile buffer such as ammonium acetate. The column eluate passes through a heated capillary into a heated chamber. The liquid emerges as a high velocity mist. The solvent and volatile buffer vaporise, leaving sample molecules that are protonated $(M+1^+)$ or deprotonated $(M-1^-)$. Some of these are then fed into the quadrupole mass analyser.

In an electrospray, the eluate enters a chamber via a fine capillary. At the other end of the chamber is a glass capillary that leads to the quadrupole mass analyser. Between 3,000 and 5,000 volts are applied between the 2 capillaries. This produces a spray of highly charged droplets. Solvent is volatilised under the vacuum system to leave protonated analyte molecules.

These techniques are known as "soft ionisation" processes and generate molecular ions. Hence, the molecular weight of the analytes can be obtained from the mass spectra. However, electrosprays and thermosprays can be modified to produce ion fragments that can yield more structural information about the analytes.

As an example, HPLC - electrospray MS has been used to identify haemoglobin variants. Either a straight haemolysate, or globin chains separated on HPLC, can be introduced into the electrospray MS. Molecular ions of the globins with 11 to 21 protons are generated. These give a series of peaks on the MS. Each globin chain gives a separate peak for each level of protonation (ie for each m/z). From these, the mass of these ions is then determined and the molecular weight of the globin calculated. These are then compared with normal haemoglobin. The mass difference can suggest the amino acid change.

The actual site of the change is identified by digesting the globin with trypsin, and running the digest through the HPLC-MS. This gives the mass of the digest fragments. Again, a different mass from the normal fragment can suggest the amino acid change. Some mass changes can correspond to one of two or more possible amino acid changes. In this case the fragment would have to be sequenced.

Recorder and/or Computing Integrator.

Most detectors have outputs of 10 mVolts and 1, 2 and/or 10 volts. Most recorders can be used with the 10 mV output.

Quantitative analysis requires an accurate measurement of the area of the peaks. The modern computing integrator is best suited to this purpose. Any of the outputs can be connected to the computing integrator. Computing integrators are covered in chapter 9 (pages 9-15 to 9-18).

EXERCISES

These exercises are written for a Waters HPLC system using 2 pumps, Model 680 gradient controller (for Exercise 10-3 only), a U6K injector, Guard-PAK, Nova-Pak C_{18} radially compressed column (C_{18} on 4 μm diameter spherical silica particles), Model 440 fixed wavelength detector (254 nm), extended wavelength detector (214 nm), a recorder and/or a QA-1 (Hewlett Packard 3390A) integrator. The recorder is connected to the 10 mV output of the lower channel of the Model 440 and the integrator to the 10 mV output of the upper channel.

Modifications to the HPLC system and conditions may have to be made with other instruments.

The electrical connections used in the system are illustrated below. These can be modified to suit the available equipment. The mobile phase flow is included as the dotted lines.

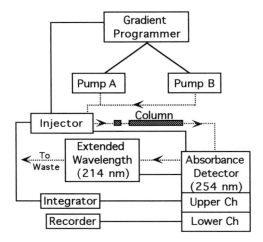

Injection should produce a 1 second pulse of 10 - 15% of full scale on the recorder chart or integrator trace (if already running). The solvent programme and integrator should also start.

If neat methanol is pumped through the system after a phosphate buffer has been used, phosphate will precipitate. This will block mobile phase flow and can cause damage to the pump and column etc. Hence, it is essential that all phosphate has been removed from the system before neat methanol is used. The following is a suggested shutdown procedure:

1.   Run both pumps at 2 ml per minute with filtered distilled water, for 5 minutes.

2.   Run both pumps at 2 ml per minute with filtered 50:50, (methanol : water), for five minutes.

3.   Run both pumps at 2 ml per minute with filtered HPLC methanol, for 5 minutes. Then turn off all components and at the mains.

Following are suggested operating instructions for the U6K injector:

Load the HPLC syringe. (NB. Do not use a gas chromatography syringe, which has a sharp tip) as follows:

(a)   Place the tip of the syringe in the sample and draw up a syringe full. Expel this to waste.

(b)   Repeat (a) twice.

(c)   With the tip of the syringe in the sample mixture, pump the syringe (to about the halfway mark) about three times, to remove air bubbles.

(d)   Slowly draw up a syringe full of liquid.

(e)   Expel the solution until the required amount remains. Wipe the syringe needle with a paper tissue.

Inject the sample, as follows:

(a)   On the U6K injector, move the top lever to "Load".

(b)   Move the bottom lever to the vertical position.

(c)   Remove the "Sample Loading Plug" and place it in the hole in the top lever.

(d)   Fully insert the needle of the syringe, inject the sample and remove the syringe.

(e)   Replace the "Sample Loading Plug" and return the lower lever to the horizontal position.

(f)   Move the top lever to the "Inject" position. (The sample now enters the system.) If the injection pulse is not recorded on the chart, mark the chart (or start the integrator) at the same time as you complete the movement of the lever.

EXERCISE 10-1

THE EFFECT OF SOLVENT COMPOSITION ON THE SEPARATION OF THREE PARABENS, USING A C_{18} COLUMN

Parabens are esters of parahydroxybenzoic acid:

$$HO-\langle\text{benzene ring}\rangle-COOR$$

The R groups used in this exercise are:

- CH_3	(Methyl).
- CH_2-CH_2-CH_3	(Propyl).
- CH_2-CH_2-CH_2-CH_3	(Butyl).

Increasing the hydrocarbon chain decreases the polarity of the molecule and, hence, it will be held more tightly by the C_{18} chains on the stationary phase of the column. A strongly polar mobile phase, such as water, will not elute these parabens from the column. However, a less polar solvent, such as methanol, will elute them.

In this exercise, the effect of varying the methanol : water composition of the mobile phase, on the elution of the parabens, will be examined.

The upper pump (B) will be used to provide HPLC grade methanol and the lower pump (A), distilled water. A total flow rate of 2.0 ml per minute will be used and the pumps adjusted to give mixtures from 100% methanol to 75% methanol.

A recorder will be used to detect the peaks. This is connected to the 10 mV output of the lower amplifier channel.

REAGENTS AND EQUIPMENT:

1.   Paraben Mixture.
            Dissolve 1.0 mg of p-hydroxybenzoic acid methyl ester, 1.0 mg of p-hydroxybenzoic acid propyl ester and 1.0 mg of p-hydroxybenzoic acid butyl ester in 100 ml of HPLC grade methanol.
2.   Uracil Solution.
            Dissolve 2.0 mg of uracil in 100 ml of 50% HPLC grade methanol, 50% distilled water.
3.   Mobile Phase 1.
            Distilled water.
            Vacuum filter, just before use, using a Millipore 0.45 μm filter.
4.   Mobile Phase 2.
            HPLC grade methanol.
            Vacuum filter, just before use, using a Millipore 0.45 μm filter.

HPLC System and Recorder.
            Set up as in the procedure below, with mobile phase 1 (H_2O) connected to the lower pump (A) and mobile phase 2 (methanol) connected to the upper pump (B). The pumps can be controlled individually or with a gradient programmer.
            The suggested column is a Waters Nova-Pak C_{18} radial compression column. However, similar C_{18} columns can be used.
            The recorder should have a variable input so that the peak heights can be adjusted to give peaks that occupy most of the chart.
10 or 25 μl HPLC syringe.
2 x 100 ml Beakers with Distilled Water and HPLC Methanol.
            For syringe washing.
Paper Tissues.

PROCEDURE:

1.   Check the following on the instrument:

        Model 440 Absorbance Detector:
            Display   :   Lower Channel Absorbance.
            Lower Channel Amplifier:
                        0.2 Absorbance Units Full Scale (AUFS).
                        λ_1 (254 nm).
            Recorder:
                        Zero   :   Out.
                        Polarity   :   In.

Upper Pump (B):
          2.0 ml/minute.
          Pressure Limit   :   2,000 p.s.i. (The maximum allowed for radial compression columns.)

Lower Pump (A):
          0.0 ml/min.
          Pressure Limit   :   2,000 p.s.i.

Recorder:
          Input   :   10 mV.
          Atten/Cal   :   Cal.
          Speed   :   1 cm/min.

2. Adjust the lower channel "Offset" control so that the digital display reads 0.000.

3. Adjust the recorder pen to 10/90 on the right hand side of the paper (chart flow towards you), with the recorder "Zero".

4. Load 10 µl of the paraben mixture (10 mg of each paraben, per litre, in methanol) into a 10 or 25 µl HPLC syringe by the procedure described above.

5. Inject the sample, by the procedure described above, and mark the chart (if no injection pulse is recorded) at the same time as you complete the movement of the top lever, by rapidly turning the recorder "zero" clockwise and then bringing it back to the 10/90 position on the chart.

6. When the combined peak is off the column, set the pumps as follows:

          Upper Pump (B)   :   1.9 ml/minute.
          Lower Pump (A)   :   0.1 ml/minute.

7. Wait 2-3 minutes for the column to equilibrate and then inject another 10 µl of sample, as in steps 4 and 5.

8. Repeat step 7 with the following pump settings:

    (a)   Upper pump (B)   :   1.8 ml min.
          Lower Pump (A)   :   0.2 ml/min.

    (b)   Upper pump (B)   :   1.7 ml/min.
          Lower pump (A)   :   0.3 ml/min.

    (c)   Upper pump (B)   :   1.6 ml/min.
          Lower pump (A)   :   0.4 ml/min.   } Select an AUFS of 0.1 for these runs.

    (d)   Upper pump (B)   :   1.5 ml/min.
          Lower pump (A)   :   0.5 ml/min.

9. Wash the syringe by drawing in, and expelling to waste, HPLC methanol. Repeat this procedure 5 times.

10.   Wash the syringe, three times, with distilled water.

11.   Inject 10 µl of the 20 mg/l uracil in 50% methanol, using the above procedure and the flow rates of step 8 (d). Mark the point of injection on the chart.

12.   Wash the syringe with distilled water (5 times) and then methanol (3 times).

Uracil is not retained by the C_{18} column and is eluted at the end of the void volume (t_m). t_m is, therefore, obtained from the uracil run.

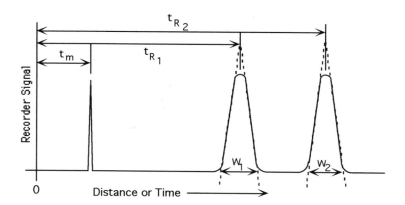

The column Capacity Factor (k') is a useful retention parameter. It indicates the degree of retention of a compound in a particular chromatographic system. It is defined as the ratio of the amount of solute in the stationary phase, to the amount in the mobile phase. It can be calculated by determining the number of void volumes required to elute the solute. As k' is a ratio, there is no need to calculate the actual volume, as the same value can be obtained using retention times or by measuring retention distances on the chart-paper.

$$k' = \frac{t_R - t_m}{t_m}$$

Calculate the k' values for the 3 parabens from the recorder traces obtained in Steps 11 and 8(d), measuring the retention distances in mm.

The Selectivity (α) of a system is the relative retention of two compounds. It is calculated from the ratio of the capacity factor (k') of each compound.

$$\alpha = \frac{k'_2}{k'_1}$$

Using the k' values, calculated above, calculate α for the following pairs of parabens:

(a)   Methyl paraben and propyl paraben.
(b)   Methyl paraben and butyl paraben.

The actual separation of the compounds is indicated by the Resolution (R_s) of the system. This includes the peak spreading (efficiency) as well as the retention of the compounds.

R_S is defined as the distance between the two peak centres divided by the average peak width:

$$R_S = 2 \left(\frac{t_{R2} - t_{R1}}{W_1 + W_2} \right)$$

Calculate the R_S values for the separation of the methyl and propyl parabens and for the propyl and butyl parabens.

EXERCISE 10-2

THE DETERMINATION OF THE EFFICIENCY OF A C_{18} COLUMN.

An efficient column is one in which peak spreading is minimal, ie. compounds are eluted in small volumes. This gives sharp narrow peaks on the recorder chart.

Efficiency is expressed as the Number of Theoretical Plates (N) of the system. The system includes the entire chromatographic bed from the injector to the detector. Poor connections, between components in a system, will reduce the efficiency. With good connections, the efficiency will depend on the packing of the column.

Efficiency can also be expressed as the Height Equivalent to a Theoretical Plate (HETP of H) or as reduced Plate Height (h). HETP is the length of column occupied by one plate and h is the number of packing particles per plate.

Acenaphthene (ethylene naphthalene) is a commonly used compound for the determination of the efficiency of C_{18} columns. It is retained relatively strongly by the C_{18} stationary phase. As mentioned in Exercise 10-1, uracil is not retained by the C_{18} stationary phase. Hence, t_m can be measured from the uracil peak. Acenaphthene and uracil have the structures shown below:

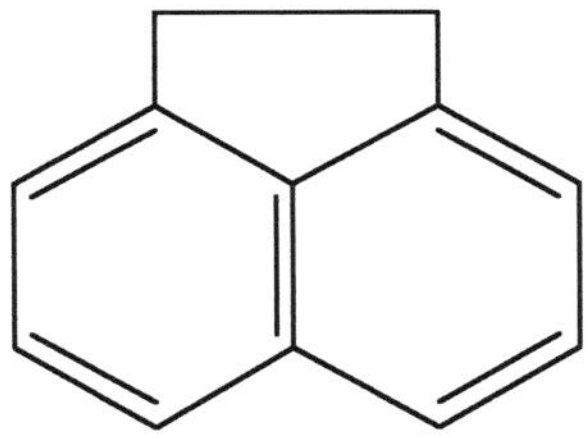

Acenaphthene 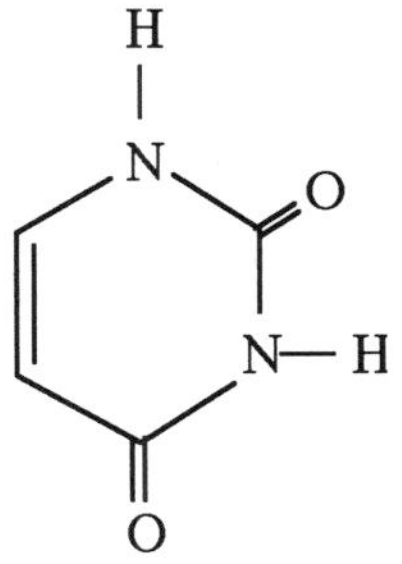   Uracil

The number of theoretical plates (N) is given by the formula:

$$N = \left(\frac{t_R}{\sigma} \right)^2$$

σ is the standard of deviation of the Gaussian peak produced in an ideal chromatogram. σ can be calculated from any of the parameters shown below:

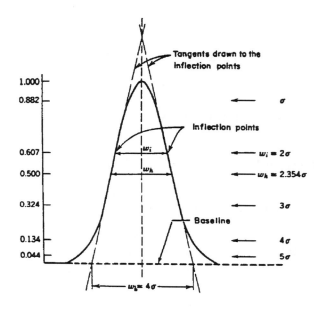

N will be calculated using the 4 and 5 σ methods. However, N does not include a correction for the void volume of the column. The Corrected Efficiency (N_{eff}) can be calculated from the formula:

$$N_{eff} = \left(\frac{t_R - t_m}{\sigma} \right)^2$$

N_{eff} = The effective number of plates.

The corrected efficiency will also be calculated.

REAGENTS AND EQUIPMENT:

1.   Uracil/Acenaphthene Mixture.
            Dissolve 2.0 mg of uracil and 90 mg of acenaphthene in 100 ml of 90% HPLC methanol, 10% distilled water.
2.   Mobile Phase 1.
            Distilled water.
            Vacuum filter, just before use, using a Millipore 0.45 μm filter.
3.   Mobile Phase 2.
            HPLC grade methanol.
            Vacuum filter, just before use, using a Millipore 0.45 μm filter.

HPLC System and Recorder.
        Set up as in the procedure below, with mobile phase 1 (H$_2$O) connected to the lower pump (A) and mobile phase 2 (methanol) connected to the upper pump (B). The pumps can be controlled individually or with a gradient programmer.
        The suggested column is a Waters Nova-Pak C$_{18}$ radial compression column. However, similar columns can be used.
        The recorder should have a variable input so that the peak heights can be adjusted to give peaks that occupy most of the chart.

10 or 25 µl HPLC syringe.
3 x 100 ml Beakers with Distilled Water 90% HPLC Methanol and HPLC Methanol.
    For syringe washing.
Paper Tissues.

PROCEDURE:

1. Check the following on the instrument:

    Model 440 Absorbance Detector:
        Display   :   Lower channel Absorbance.
        Lower Channel Amplifier:
                     0.2 Absorbance Units Full Scale (AUFS).
                     λ_1 (254 nm).
        Recorder:

                Zero   :   Out.
                Polarity   :   In.

    Upper Pump (B):
        1.6 ml/minute.
        Pressure Limit   :   2,000 p.s.i.

    Lower Pump (A):
        0.4 ml/min.
        Pressure Limit   :   2,000 p.s.i.

    Recorder:
        Input   :   10 mV.
        Atten/Cal   :   Cal.
        Speed   :   2 cm/min.

2. Adjust the lower channel "Offset" control so that the digital display reads 0.000.

3. Adjust the recorder pen to 10/90 on the right hand side of the paper (chart flow towards you), with the recorder "Zero".

4. Load 10 µl of the uracil/acenaphthene mixture (20 mg/l uracil and 900 mg/l acenaphthene in 90% HPLC methanol) into a 10 or 25 µl HPLC syringe.

5. Inject the sample and mark the chart (if no injection pulse is recorded) at the same time as you complete the movement of the top lever, by rapidly turning the recorder "zero" clockwise and then bringing it back to the 10/90 position on the chart.

6. Wash the syringe 5 times with 90% methanol, 3 times with distilled water and then 3 times with 100% methanol.

7. After the acenaphthene peak has been eluted, switch the pumps to:

        Upper pump (B)   :   2.0 ml/minute.
        Lower pump (A)   :   0 ml/minute.

8. Run for 5 minutes and then turn off the pumps.

10-35

From the trace, calculate N and N_{eff} using the 4 σ method. For the acenaphthene peak, draw the baseline and draw tangents at the inflection points. Measure the distance between these lines at the baseline, to obtain 4 σ. Measure the t_R and t_m.

$$N = 16 \left(\frac{t_R}{4\sigma} \right)^2 \qquad N_{eff} = 16 \left(\frac{t_R - t_m}{4\sigma} \right)^2$$

The 4 σ method does not take into account tailing peaks that can be produced by overloading or by a column with poor efficiency. The 5 σ method includes peak tailing.

Measure the peak height (from the baseline) and calculate 4.4% of this value. Draw a line, 4.4% of the peak height, above the baseline and measure the distance occupied by the peak on this line. This gives the 5 σ value.

$$N = 25 \left(\frac{t_R}{5\sigma} \right)^2$$

$$N_{eff} = 25 \left(\frac{t_R - t_m}{5\sigma} \right)^2$$

Calculate the HETP and corrected HETP (H_{eff} = Effective plate height) from the following formulae:

$$HETP = \frac{L}{N}$$

$$H_{eff} = \frac{L}{N_{eff}}$$

L is the length of the column bed, which is 100 mm for the column used.

Calculate the reduced plate height (h) (the number of particles per plate) and the effective reduced plate height (h_{eff}) from the formula:

$$h = \frac{HETP}{dp}$$

$$h_{eff} = \frac{H_{eff}}{dp}$$

dp is the average diameter of the packing particles. This is 0.004 mm for the column used. h for a highly efficient column is around 2 - 3, and 10 - 20 for a moderately efficient one.

EXERCISE 10-3

THE IDENTIFICATION AND SEMIQUANTITATION OF DRUGS IN PLASMA

In the method used in this exercise, plasma proteins are precipitated with acetonitrile and glacial acetic acid. After centrifugation, an aliquot of the supernatant is injected into the HPLC instrument.

The drugs are separated on the column by their relative distribution in the C_{18} stationary phase and in the aqueous mobile phase. The more polar drugs will be eluted first, as they a greater solubility in the mobile phase, compared with the stationary C_{18}. Due to the fact that drugs have quite different distribution coefficients, some drugs would take a very long time to be eluted from the column if a single (isocratic) mobile phase was used. Other drugs would be bunched up at the start of the chromatogram. Hence, a solvent program is used to elute the drugs.

The two solvents used for elution are acetonitrile and 50 mM phosphate buffer, pH 3.2. These will be used with a linear gradient from 5% acetonitrile to 50%, over a period of 10 minutes.

The column to be used is a radially compressed, 100 x 8 mm, column containing C_{18} on 4 μm diameter silica particles.

The 214 nm zinc emission line, from the Extended Wavelength module, will be used to detect the drugs. Most drugs absorb this wavelength of light and will, therefore, give rise to peaks on the recording integrator as they pass through the flow cell.

Case History.

The sample is plasma from a 38 year old man. He was found unconscious in his hotel room at 7 am by a maid delivering breakfast. He is a known epileptic and was being treated with phenobarbital (phenobarbitone). An empty bottle of Serepax (oxazepam) was also found on the bedside table.

An HPLC drug screen has been requested to see if the above drugs were present, or not, and to assist in identifying any other drug that might be present.

A sample of drug-free plasma will also be treated in the same way as the patient's sample to indicate HPLC peaks that are present in drug-free plasma.

7 drugs are included in the standard mixture. This mixture includes the two drugs that are likely to be in the patient's plasma.

The retention time of the drugs will be used in confirming the presence of these drugs. Any additional peaks in the patient's sample, that do not correspond to peaks in drug-free plasma, will indicate extra drugs. These could be identified by including drugs with similar retention times to the extra peaks, in an additional standard drug mixture.

The standard drug mixture, that will be used in this exercise, includes 250 mg/l (in methanol) of the following:

Paracetamol	-	Analgesic
Codeine	-	Analgesic
Salicylate	-	Analgesic, anti-inflammatory (Active metabolite of aspirin)
Phenobarbital	-	Anticonvulsant, sedative
Pentobarbital	-	Sedative, hypnotic
Oxazepam	-	Anti-anxiety
Imipramine	-	Antidepressant

The drugs are eluted from the column in this order, as illustrated below:

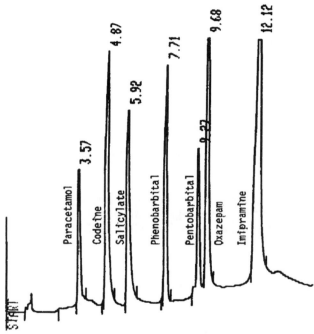

The area of the drug peaks will be used to obtain semiquantitative values of the drug concentrations in the patient's plasma. For quantitative analysis, an internal standard should be added to the patient's plasma and taken through all the steps of the analysis (See Exercise 10-4).

REAGENTS AND EQUIPMENT:

1.   Standard Drug Mixture.
            Add 5 mg of each of the following drugs to 20 ml of HPLC grade methanol:

            Paracetamol.   Codeine.
            Salicylic acid.   Pentobarbital.
            Phenobarbital.   Oxazepam.
            Imipramine.
        Shake well to dissolve.
        Store at 4ºC.

2.   Drug Free Plasma.
            Or Serum.
            Store at 4ºC.

3. **Patient's Plasma.**
    Weigh out 32 mg of phenobarbital and 10 mg of oxazepam and dissolve in the minimum amount of HPLC grade methanol. Add distilled water to give a final volume of 40 ml and mix well. Mix 4.0 ml of this solution with 36 ml of drug free plasma (or serum). Store at $4^{o}C$.

4. **Mobile Phase 1.**
    HPLC grade acetonitrile.
    Vacuum filter, just before use, using a Millipore 0.45 μm filter.

5. **Mobile Phase 2.**
    0.05 M phosphate buffer, pH 3.2.
    Dissolve 7.5 g of NaH_2PO_4 in 1 l of distilled water. Adjust the pH to 3.2 with phosphoric acid.
    Vacuum filter, just before use, using a Millipore 0.45 μm filter.

6. **Glacial Acetic Acid.**
    Analar.

7. **Acetonitrile.**
    HPLC grade.

HPLC System and Integrator.
    Set up as in the procedure below; with mobile phase 1 (acetonitrile) connected to the upper pump (B) and mobile phase 2 (phosphate buffer) connected to the lower pump (A). The pumps are controlled with a gradient programmer (Eg. Waters Model 680).
    The suggested column is a Waters Nova-Pak C_{18} radial compression (C_{18} on 4 μm diameter spherical silica particles) column. However, similar columns can be used.
    The integrator is connected to the 10 mV output of the upper channel of the absorbance detector (Waters Model 440). The pulse marker, from the injector, connects to the integrator and to the gradient programmer so that both start as the sample is injected.
10 or 25 μl and 100 μl HPLC syringes.
2 x 100 ml Beakers with Distilled Water and HPLC Methanol.
    For syringe washing.
Paper Tissues.
125, 500 and 1,500 μl Pipettors and Tips.
Centrifuge Tubes.
Bench Centrifuges.
Vortex Mixers.
Wassermann Tubes.

PROCEDURE:

1. Add 500 μl of the patient's plasma to a centrifuge tube and 500 μl of drug-free plasma to a second centrifuge tube.

2. Add 1.5 ml of acetonitrile to each tube.

3. Add 125 μl of glacial acetic acid to each tube and mix on the vortex mixer for 20 seconds.

4. Centrifuge the tubes for 3 minutes.

5. Decant the supernatants into labelled Wassermann tubes.

6. Check the following on the Waters HPLC instrument:

    Mobile Phases:
        Acetonitrile (HPLC grade) to upper pump (B).
        50 mM phosphate buffer (pH 3.2) to lower pump (A).
    Gradient Programmer (Model 680):

Programme No. 1.

Time	Flow	%A	%B	Curve
Initial	3.00	95	5	* (Hold)
10.00	3.00	50	50	6 (Linear)
15.00	3.00	95	5	11(Return to initial conditions)

Mode   :   Operate Gradient.
Function   :   Initial.
Pumps:
  Power   :   On.
  Pressure limit   :   2,000 p.s.i.
Absorbance Detector:
  Display   :   Upper Channel Absorbance.
  Upper Channel Amplifier   :   1.0 AUFS.
                                       λ_2 (214 nm).
                                       Recorder zero   : Out.
                                       Recorder polarity : In.
Extended Wavelength Module:
  Power   :   On-1.
Integrator:
  Speed   :   5 mm/min (0.5 cm/min)
  Attenuation   :   1   (Increasing the number by 1, halves the peak height).
  Threshold   :   2.
  Zero   :   0.

7.   Make sure that the "Initial Conditions" have been running through the column for at least five minutes before injecting a sample.

8.   Adjust the upper channel "Offset" control so that the digital display on the absorbance detector reads 0.000.

9.   Press "List" and then "Zero" on the integrator. Repeat this until the second number on the print out in close to 0.0. (The position of the baseline can be changed by pressing "Zero", "a value from -6 to 100" and then "Enter". (-6 is the left margin and 100 is the right margin).

10.   Load a 10 or 25 µl HPLC syringe with 10 µl of the standard drug mixture.

11.   Inject the sample. (The integrator and gradient flow should start automatically. Start them, if they don't.)

12.   Wash the syringe with 3 lots of HPLC methanol.

13.   After the completion of the solvent programme, press "Stop" on the integrator.

14. Maintain the instrument on "Initial Conditions" for at least 5 minutes and then inject 100 μl of one of the supernatants from step 5, using a 100 μl HPLC syringe. Wash out the syringe with HPLC methanol, 3 times.

15. Repeat steps 13 and 14 with the other supernatant.

16. After completion, wash out the syringe with 3 lots of HPLC methanol, 3 lots of distilled or deionised water and then another 3 lots of HPLC methanol.

Identify the peaks due to endogenous compounds; ie. those present in the drug-free plasma.

Compare the retention times of the drugs in the standard mixture with those of the drugs in the patient's plasma. Use this information to identify the drugs present in the patient's plasma.

By comparing the area of the drug peaks, calculate the approximate concentration of the drugs in the patient's plasma.

Reference Values:

Phenobarbital.

Therapeutic range	:	15 - 40 mg/l.
Toxic above	:	50 mg/l.
Reported lethal levels	:	78 - 116 mg/l.

Oxazepam.

Therapeutic range	:	0.2 - 1.4 mg/l.
Toxic above	:	About 5 mg/l.
Lethal levels	:	Oxazepam, on its own, is unlikely to be lethal.

Benzodiazepines, such as oxazepam, have synergistic effects when taken with alcohol or other CNS depressants, such as phenobarbital. These combinations can have serious effects, such as severe CNS and respiratory depression.

EXERCISE 10-4

THE QUANTITATIVE DETERMINATION OF PHENOBARBITAL AND OXAZEPAM IN A PLASMA SAMPLE

Case History.

The sample is the same as that used in Exercise 10-3. It is plasma from a 38 year old man. He was found unconscious in his hotel room at 7 am by a maid delivering breakfast. He is a known epileptic and was being treated with phenobarbital (phenobarbitone). An empty bottle of Serepax (oxazepam) was also found on the bedside table. The presence of these two drugs has been suggested by HPLC screening. A quantitative determination has now been requested.

The drugs will be extracted, together with internal standards (barbital [barbitone] for the barbiturate and diazepam for the benzodiazepine) into ether by adjusting the pH to produce the non-ionised form of the drug. Phenobarbital and oxazepam standards will be treated in a similar manner. The drugs are further purified by the steps outlined below.

Barbiturates are weak acids. Hence, they are non-ionised at pH 2-3 and will be extracted into ether. However, they are ionised at pH 13 (0.45 M NaOH) and will move into the aqueous layer. They can then be re-extracted back into ether when the pH of the aqueous phase is re-adjusted to 2-3. This process isolates the barbiturates from interfering material in the sample.

Benzodiazepines are weak bases and are non-ionised at pH 7.0. However, they form ionised hydrochlorides with 6 M HCl. Hence, they will move from an original ether extract into 6 M HCl. This can then be re-adjusted back to pH 7.0 and the drugs re-extracted back into ether. This process isolates the benzodiazepines from interfering material in the sample.

The ether extracts will be taken to dryness and redissolved in HPLC methanol for injection into the HPLC.

The column used is a Waters radial compression (C_{18} on 4 μm diameter silica particles) column.

The mobile phase for the barbiturates is 0.1 M NaH_2PO_4 (Adjusted to pH 7.5) : methanol (60:40). For the benzodiazepines, it is acetonitrile : 0.01 M sodium acetate buffer, pH 4.6 (50:50).

The 214 nm zinc emission line will be used for the detection of the barbiturates and the 254 nm mercury line for the benzodiazepines.

REAGENTS AND EQUIPMENT:

1. Drug Standards.
   Stock:
   Dissolve 50 mg of phenobarbital and 10 mg of oxazepam in 5 ml of HPLC methanol. Add 245 ml of distilled water and mix well.
   Standards:

Standard	1	2	3	4
ml of stock	12.5	25	50	75
ml of dist H_2O	87.5	75	50	25

   Mix well and store at 4°C.
2. Patient's Plasma.
   Weigh out 32 mg of phenobarbital and 10 mg of oxazepam and dissolve in the minimum amount of HPLC grade methanol. Add distilled water to give a final volume of 40 ml and mix well. Mix 4.0 ml of this solution with 36 ml of drug free plasma (or serum). Store at 4°C.
3. Internal Standard.
   Dissolve 100 mg of barbital and 10 mg of diazepam in 10 ml of HPLC methanol.
4. Concentrated Phosphoric Acid.
5. Diethyl Ether.
   Analar grade.

6. 0.45 M NaOH.
    Dissolve 1.8 g of NaOH in distilled water and make to 100 ml.
7. Anhydrous Sodium Sulphate.
8. Mobile Phase 1.
    Dissolve 15 g of NaH_2PO_4 in about 800 ml of distilled water. Adjust the pH to 7.5, with a strong NaOH solution, and make up to 1 l with distilled water.
    Mix 600 ml of the above buffer with 400 ml of HPLC grade methanol.
    Vacuum filter, just before use, using a Millipore 0.45 μm filter.
9. Mobile Phase 2.
    Dissolve 820 mg of sodium acetate in about 800 ml of distilled water. Adjust the pH to 4.6 with acetic acid and make up to 1 l with distilled water.
    Mix 500 ml of the above buffer with 500 ml of HPLC grade acetonitrile.
    Vacuum filter, just before use, using a Millipore 0.45 μm filter.
10. 6 M HCl.
    Mix 50 ml of concentrated HCl (12 M) with 50 ml of distilled water.
11. 1 M Phosphate Buffer, pH 7.0.
    (A) Dissolve 13.6 g of anhydrous KH_2PO_4 in distilled water and make to 100 ml.
    (B) Dissolve 17.4 g of anhydrous K_2HPO_4 in distilled water and make to 100 ml.
    Add B to all of A, until a pH of 7.0 is reached.
12. HPLC Grade Methanol.

HPLC System and Integrator.
    Set up as in the procedure below, with mobile phase 1 (phosphate buffer/methanol) connected to the lower pump (A) and mobile phase 2 (acetate buffer/acetonitrile) connected to the upper pump (B). The pumps are controlled manually.
    The suggested column is a Waters Nova-Pak C_{18} radial compression (C_{18} on 4 μm diameter spherical silica particles) column. However, similar columns can be used.
    The integrator is connected to the 10 mV output of the upper channel of the absorbance detector. The pulse marker, from the injector, connects to the integrator so that it starts as the sample is injected.
10 or 25 μl HPLC syringe.
2 x 100 ml Beakers with Distilled Water and HPLC Methanol.
    For syringe washing.
Paper Tissues.
pH Paper (2-8) and Glass Rods.
100 μl Pipettors and Tips.
1, 5 and 10 ml Graduated Pipettes.
50 ml Conical Flasks.
50 ml Measuring Cylinders.
100 ml Separating Funnels.
100 ml Beakers.
Steam Bath with Adjacent Compressed Air Connected to Rubber Tubes and Pasteur Pipettes.
Stands, Bosses and Rings.
2 or 5 ml Screw Capped Vials.

PROCEDURE:

Work in pairs for this exercise.

2 pairs should use the patient's plasma and the remaining pairs should use one of the four standards, each. One member of each pair should do the barbiturate extraction and the other the benzodiazepine extraction. The standards are:

1.	25	mg/l phenobarbital,	5	mg/l oxazepam.
2.	50	mg/l "	10	mg/l "
3.	100	mg/l "	20	mg/l "
4.	150	mg/l "	30	mg/l "

1.   Add 100 μl of the internal standard (10 g/l barbital and 1.0 g/l diazepam, in ethanol) to 10.0 ml of sample or standard, in a conical flask, and mix well.

Barbiturate Assay:

2.   Add 4.5 ml of the solution from step 1 to a 100 ml separating funnel. Adjust the pH to 2-3, using indicator paper and a glass rod, with concentrated H_3PO_4.

3.   Add 30 ml of ether and carry out the following extraction scheme (Each extraction involves vigorous shaking for about 2 minutes. Make sure that you hold the stopper in place. Release the pressure after a few shakes by inverting the funnel and opening the tap.):

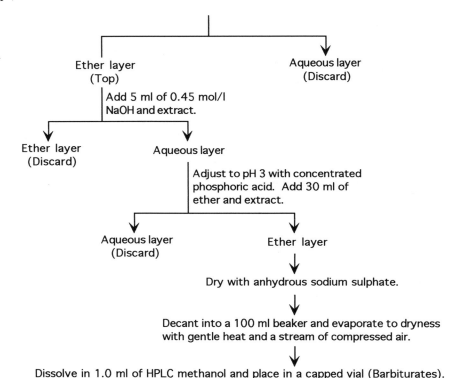

Ether layer
(Top)

Aqueous layer
(Discard)

Add 5 ml of 0.45 mol/l NaOH and extract.

Ether layer
(Discard)

Aqueous layer

Adjust to pH 3 with concentrated phosphoric acid. Add 30 ml of ether and extract.

Aqueous layer
(Discard)

Ether layer

Dry with anhydrous sodium sulphate.

Decant into a 100 ml beaker and evaporate to dryness with gentle heat and a stream of compressed air.

Dissolve in 1.0 ml of HPLC methanol and place in a capped vial (Barbiturates).

4.   **HPLC Assay:**

Check the following on the HPLC system:

       Mobile Phase: 0.1 M NaH_2PO_4 (adjusted to pH 7.5) : methanol (60:40). Fed to the bottom pump (A).

Top Pump (B)	:	Power	:	Off.
Bottom Pump (A)	:	Power	:	On.
		Flow rate	:	2.5 ml/min.
		Pressure limit	:	2,000 p.s.i.
Extended Wavelength Module	:	On-1.		

Absorbance Detector:
    Top panel
    (Display module) :   Upper channel absorbance.
    Central panel   : λ_2 (214 nm) and 0.5 (AUFS) depressed.
    Lower panel   :   Not used.

Integrator:
    Speed   :   5 mm/min.
    Attenuation   :   2.
    Threshold   :   4.

Injection Volume   :   10 µl.

5.   Make sure that the mobile phase has been pumped through the column for at least 5 minutes.

6.   Adjust the upper channel "Offset" control so that the digital display on the absorbance detector reads 0.000.

7.   Press "List" and then "Zero" on the integrator. Repeat this until the second number on the printout is as close to 0.0 as possible. (The position of the baseline can be changed by pressing "Zero", "a value from -6 to 100" and then "Enter". [-6 is the left hand margin and 100 is the right hand margin.]).

8.   Make sure that the HPLC syringe has been washed out with HPLC methanol, 3 or 4 times, and then air dried.

9.   Load a 10 or 25 µl HPLC syringe with 10 µl of sample and inject into the instrument. (The integrator should start automatically when the top lever is switched to "Inject". If not, press "Start" on the integrator.)

10.   Wash the syringe with 3 lots of HPLC methanol and air dry.

11.   After the last peak has returned to the baseline, press "Stop" on the integrator.

Identify the drug peaks. Barbital elutes at about 2 minutes and phenobarbital at about 3 minutes.

Benzodiazepine Assay:

12.   Add 4.5 ml of the solution from step 1 to a 100 ml separating funnel.

13.   Add 4.5 ml of 1 M phosphate buffer, pH 7.0.

14.   Add 30 ml of ether and carry out the following extraction scheme (Each extraction involves vigorous shaking for about 2 minutes. Make sure that you hold the stopper in place. Release the pressure after a few shakes by inverting the funnel and opening the tap.):

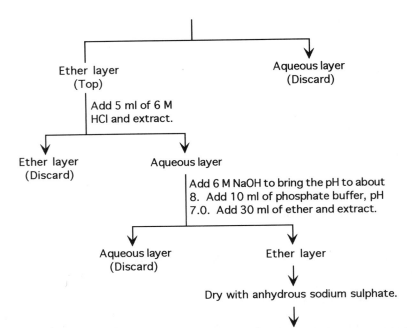

Ether layer
(Top)

Aqueous layer
(Discard)

Add 5 ml of 6 M
HCl and extract.

Ether layer
(Discard)

Aqueous layer

Add 6 M NaOH to bring the pH to about
8. Add 10 ml of phosphate buffer, pH
7.0. Add 30 ml of ether and extract.

Aqueous layer
(Discard)

Ether layer

Dry with anhydrous sodium sulphate.

Decant into a 100 ml beaker and evaporate to dryness
with gentle heat and a stream of compressed air.

Dissolve in 1.0 ml of HPLC methanol and place in a capped vial (Benzodiazepines).

15.   HPLC Assay.

Check, or set, the following on the HPLC system:

Mobile phase: Acetonitrile : 0.01 M sodium acetate buffer, pH 4.6 (50:50).
                  Fed to top pump (B).

Bottom pump (A)   :   Power   :   Off.
Top pump (B)   :   Power   :   On.
                        Flow rate   :   3.0 ml/min.
                        Pressure limit: 2 000 p.s.i.
Extended wavelength module   :   Off-O.
Absorbance Detector:
      Top panel   :   Upper channel absorbance.
      Central panel : λ_1 (254 nm) and 0.1 (AUFS) depressed.
      Lower panel   :   Not used.

Integrator:
      Speed   :   5 mm/min.
      Attention   :   2.
      Threshold   :   4.

Injection volume   :   10 µl.

16. Make sure that the mobile phase has been pumped through the column for at least 5 minutes.

17. Adjust the upper channel "Offset" control so that the digital display on the absorbance detector reads 0.000.

18. Press "List" and then "Zero" on the integrator. Repeat this until the second number on the printout is as close to 0.0 as possible. (The position of the baseline can be changed by pressing "Zero", "a value from -6 to 100" and then "Enter". [-6 is the left hand margin and 100 is the right hand margin.]).

19. Make sure that the HPLC syringe has been washed out with HPLC methanol, 3 or 4 times, and then air dried.

20. Load 10 or 25 μl HPLC syringe with 10 μl of sample and inject into the instrument. The integrator will start automatically when the top lever is switched to "Inject".

21. Wash the syringe with 3 lots of HPLC methanol and air dry.

22. After the last peak has returned to the baseline, press "Stop" on the integrator.

Identify the drug peaks. Oxazepam elutes at about 2 minutes and diazepam at about 4.5 minutes.

Calculations:

1. Plot the ratio of the peak area of the standards to the peak area of the internal standard (barbital or diazepam) against the standard concentration.

2. Calculate the peak area ratios for the unknown and read the concentration off the graph.

Reference Values:

Phenobarbital.

Therapeutic range	:	15 - 40 mg/l.
Toxic above	:	50 mg/l.
Reported lethal levels	:	78 - 116 mg/l.

Oxazepam.

Therapeutic range	:	0.2 - 1.4 mg/l.
Toxic above	:	About 5 mg/l.
Lethal levels	:	Oxazepam, on its own, is unlikely to be lethal.

Benzodiazepines, such as oxazepam, have synergistic effects when taken with alcohol or other CNS depressants, such as phenobarbital. These combinations can have serious effects, such as severe CNS and respiratory depression.

EXERCISE 10-5

THE QUANTITATIVE DETERMINATION OF ANTICONVULSANT DRUGS IN PLASMA

As discussed in Exercise 9-5, number of methods have been used to monitor the level of anticonvulsants in plasma or serum. Gas chromatographic methods can be used to assay all the drugs simultaneously. However, derivatisation is required prior to GC separation. Immunoassays will only monitor one drug at a time. With the exception of valproate, all the drugs can be monitored using HPLC.

In this exercise, an HPLC method will be used. The anticonvulsants, together with an internal standard (hexobarbital) will be adsorbed onto activated neutral charcoal. The drugs are then eluted from the charcoal with a mixture of dichloromethane, isopropanol and diethyl ether (65:10:25). This removes most of the endogenous interferents present in plasma. After the removal of the solvent, the drugs are dissolved in methanol for injection into the HPLC.

A radially compressed C_{18} column is used together with a methanol, acetonitrile, phosphate buffer (20 mM, pH 4.0) mixture (25:17:58) as the mobile phase. The anticonvulsants are detected by their absorption at 214 nm. Ethosuximide absorbs poorly at this wavelength and can not be quantitated in the levels found in the plasma. 195 nm should be used if this drug is present in the patient.

Case History.

The sample is the same as that used in exercise 9-5, which is plasma from a 25 year old man being treated with phenytoin and phenobarbital. The sample has been collected for routine therapeutic drug monitoring (TDM). This sample will be treated as above, as will a series of standards made from adding anticonvulsants to drug free plasma.

REAGENTS AND EQUIPMENT:

1. Drug standard.
    10 mg phenobarbital
    10 mg phenytoin   } Dissolve in 10 ml of methanol.
    10 mg primidone
2. Drug Free Plasma.
3. Patient's Plasma.
    Dissolve 10 mg of phenobarbital and 5 mg of phenytoin in 3.0 ml of methanol.
    Add 200 μl of this solution to 20 ml of drug free plasma.
4. Internal Standard.
    Dissolve 10 mg of hexobarbital in 10 ml of HPLC grade methanol.
5. Eluting Solvent.
    Mix together:
        65 ml of dichloromethane,
        10 ml of isopropyl alcohol
        and 25 ml of diethyl ether.
6. Phosphate Buffer, pH 4.0, 20 mM.
    Dissolve 2.78 g of $(NH_4)_2HPO_4$ in about 1 l of distilled water. Adjust the pH to 4.0 with phosphoric acid and make to 2 l.
    For reagent 7.

7.   Mobile Phase.
            Mix together:
                250 ml of methanol,
                170 ml of acetonitrile
                and 580 ml of reagent 6.
            Vacuum filter, just before use, using a Millipore 0.45 μm filter.
8.   Activated Neutral Charcoal.
            Sigma Cat. No. C 5385.
9.   HPLC Grade Methanol.

HPLC System and Integrator.
    Set up as in the procedure below, with the mobile phase (phosphate buffer/methanol/
    acetonitrile) connected to either pump.
    The suggested column is a Waters Nova-Pak C_{18} radial compression (C_{18} on 4 μm diameter
    spherical silica particles) column. However, similar columns can be used.
    The integrator is connected to the 10 mV output of the upper channel of the absorbance
    detector. The pulse marker, from the injector, connects to the integrator so that it starts as
    the sample is injected.
25 μl HPLC syringe.
2 x 100 ml Beakers with Distilled Water and HPLC Methanol.
    For syringe washing.
Paper Tissues.
50, 100 and 200 μl Pipettors and Tips.
5 ml Graduated Pipettes.
Wassermann Tubes.
Centrifuge Tubes.
Bench Centrifuges.
Spatulas.
    Weigh out 20 mg of activated charcoal to indicate the required amount.
Vortex Mixers.
50°C Heating Block with Fine Tubes Connected to a Nitrogen Cylinder or Compressed Air connected
to Pasteur Pipettes.

PROCEDURE:

1.   Prepare the plasma standards using the stock drug standard (this contains 1.0 g of
        phenobarbital, 1.0 g of phenytoin and 1.0 g of primidone per litre, in methanol) as
        follows:

        (i)   Add 4.0 ml of drug free plasma to 4 tubes, labelled A to D.

        (ii)   Add the stock drug standard, as shown below, and mix well:

                A   :   50 μl.
                B   :   100 μl.
                C   :   200 μl.
                D   :   400 μl.

2.   Add 1.0 ml of each of the above plasma standards and duplicate 1.0 ml aliquots of the
        patient's plasma to centrifuge tubes.

3.   Add 4 ml of distilled water to each tube and mix well.

4.   Add 100 μl of the internal standard (1.0 g/1 hexobarbital in methanol) to each tube and
        mix well.

5. Add about 20 mg of activated neutral charcoal to each tube.

6. Mix each tube for 30 seconds on a vortex mixer.

7. Centrifuge at about 2,000 g for 2 minutes. Decant and discard the supernatant. Keep the tubes inverted and wipe the excess liquid from the mouth of the tubes with paper tissues.

8. Add 500 µl of the eluting solvent and vortex each tube for 30 seconds.

9. Centrifuge for 2 minutes and decant the supernatant into labelled centrifuge tubes.

10. Repeat steps 8 and 9, combining both supernatants. Decant into another centrifuge tube, if any charcoal is in the combined supernatants.

11. Gently evaporate the solvent, using compressed air or nitrogen, at 50°C, in a heating block.

12. Dissolve the residue in 200 µl of HPLC methanol.

13. Check the following on the HPLC system:

    Mobile phase fed to the appropriate pump.

    Pump (Top or Bottom):
Power	:	On
Flow rate	:	3.0 ml/min.
Pressure limit	:	2,000 psi.
Extended Wavelength Module	:	On - 1.

    Absorbance Detector:
Top panel (Display module)	:	Upper channel absorbance.
Central panel	:	λ_2 (214 nm) and 0.5 (AUFS) depressed.
Lower panel	:	Not used.

    Integrator:
Chart speed	:	5 mm (0.5 cm)/min.
Attenuation	:	1
Threshold	:	2

14. Make sure that the mobile phase has been pumped through the column for at least 5 minutes.

15. Adjust the upper channel "Offset" control so that the digital display on the absorbance detector reads 0.000.

16. Press "List" and then "Zero" on the integrator. Repeat this until the second number on the printout is as close to 0.0 as possible. (The position of the baseline can be changed by pressing "Zero", "a value from -6 to 100" and then "Enter". [-6 is the left hand margin and 100 is the right hand margin.]).

17. Make sure that the HPLC syringe has been washed out with HPLC methanol, 3 or 4 times, and then air dried.

18. Load the 25 μl HPLC syringe with 20 μl of sample and inject into the instrument. (The integrator will start automatically when the top lever is switched to "Inject". If not, press "Start" on the integrator.)

19. Wash the syringe with 3 lots of HPLC methanol and air dry.

20. After the last peak has returned to the baseline, press "Stop" on the integrator.

21. Identify the anticonvulsant drug peaks by comparison with the trace below:

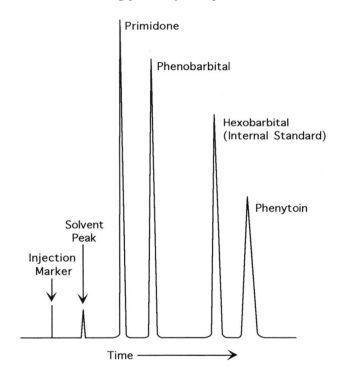

Ethosuximide would elute just before primidone and carbamazepine would elute just after phenytoin.

Calculations:

1. Calculate the concentration of the anticonvulsants in each of the standards.

2. Divide the anticonvulsant peak areas by the hexobarbital peak area, for each standard and for the unknowns.

3. Plot these values, for the standards, against the standard concentration.

4. Read off the anticonvulsant concentrations of the unknowns and compare these with the quoted therapeutic ranges.

Reference Values:

Phenobarbital.
    Therapeutic range   :   15 - 40 mg/l.
Phenytoin.
    Therapeutic range   :   10 - 20 mg/l.

GENERAL REFERENCES:

Martin, A. J. P. and Synge, R. L. M., (1941), Biochem. J., **35**, 91.

Ridgeon, P. A., (1978), Liquid Chromatography Separations, a booklet published by Pye Unicam Ltd.

Simpson, C. F., (1978), Practical High Performance Liquid Chromatography, Heyden and Son, Ltd.

Bristow, P. A., (1976), Liquid Chromatography in Practice, HETP.

Yost, R. W., Ettre, L. S. and Conlon, R. D., (1980), Practical Liquid Chromatography - An Introduction, Perkin-Elmer.

Bowers, L. D., (1989), Liquid Chromatography, in Clinical Chemistry - Theory, Analysis and Correlations, (2nd Edition) Eds. Kaplan, L. A. and Pesce, A. J., 94-109, The C. V. Mosby Co.

HPLC-MS.

Voyksner, R. D., (1994), Atmospheric Pressure Ionisation LC/MS, Environ. Sci. Technol., **28**, 118A-127A.

Shackleton, C. H. L., Falick, A. M., Green, B. N. and Witkowska, H. E., (1991), Electrospray Mass Spectrometry in the Diagnosis of Variant Hemoglobins, J. Chromat., **563**, 175-190.

The identification and semiquantitation of drugs in plasma.

Kabra, P. M., Stafford, B. E. and Marton, L. J., (1981), Rapid Method for Screening Toxic Drugs in Serum with Liquid Chromatography, J. Anal. Tox., **5**, 177-182.

The quantitative determination of phenobarbital and oxazepam in a plasma sample.

Gill, R., Lopes, A. A. T. and Moffat, A. C., (1981), Analysis of Barbiturates in Blood by High Performance Liquid Chromatography, J. Chromatography, **226**, 117-123.

Kabra, P. M., Stevens, G. L. and Marton, L. J., (1978), High Pressure Liquid Chromatographic Analysis of Diazepam, Oxazepam and N-dimethyldiazepam in Human Blood, J. Chromatography, **150**, 355-360.

The quantitative determination of anticonvulsant drugs in plasma.

Waters and Perkin-Elmer method sheets.

CHAPTER 11

ELECTROPHORESIS

The principle of electrophoresis is that ions will move when exposed to an electric field. Ions can be separated by their rate of migration in that electric field. The movement of ions, relative to each other, in an electric field will depend upon the charge on the ion and upon the ability of the ion to move in the supporting medium.

When the supporting medium offers no resistance to the movement of ions, the rate of movement of an ion is dependent upon its charge/cross sectional area ratio. The velocity of the ion is due to the balance of forces acting on it. The electromotive force (Voltage x Charge on the ion) is equal and opposite to the frictional force. The frictional force is proportional to the viscosity of the liquid and to the cross sectional area of the hydrated ion. The viscosity is usually constant, if the temperature is kept constant. Hence, a large molecule with many charges will have the same rate of migration as a small molecule with the same charge/area ratio.

Therefore, the rate of migration of a particle (molecule) in a buffer solution is dependent upon:

1.   The net charge on the molecule (z).
2.   The size and shape of the molecule.
3.   The electric field strength (volts/cm) (E).
4.   Properties of the support medium and buffer.
5.   The temperature ($\cong 2.4\%$ increase / $^{\circ}C$ rise, due to a decrease in viscosity).

The force acting to accelerate the molecule is Ez. This is countered by the resistance of the buffer. Where the molecule is spherical this resistance is equal to $6\ \pi r \eta v$, where:

r   =   The radius of the hydrated molecule.
η =   The viscosity of the buffer.
v   =   The velocity of the molecule.

Very rapidly, these forces counter each other and the molecule attains a constant velocity.

Moving Boundary Electrophoresis.

Moving boundary electrophoresis was the first form of electrophoresis that was used for protein separation. It was introduced in 1937 by Arne Tiselius, a Swedish biochemist.

In this technique, the sample (in buffer) is placed in a "U" tube. Buffer is layered above the solution, in each arm of the U. Electrodes are inserted into the buffer and a potential applied. Ions will move towards the electrode with the opposite charge. This is illustrated below for a solution containing two different proteins, where A has a greater negative charge, to cross sectional area, than B:

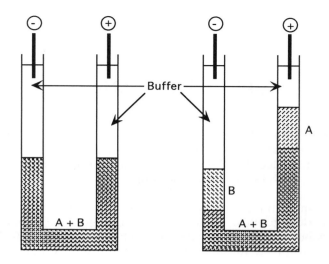

The amount of protein present can be quantitated by the refractive index change at the various boundaries (ie. Buffer to A and A to A + B). This is achieved using a Schlieren optical system:

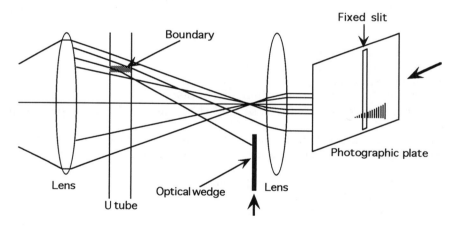

As the optical wedge is moved up, the photographic plate is moved passed the slit. The boundary acts as a lens, due to the change in refractive index. This deflects the light downwards where it is occluded by the optical wedge. As a result, a shadow is cast on the photographic plate. As the wedge moves higher, this shadow increases. This is illustrated for the electrophoretic separation of serum proteins:

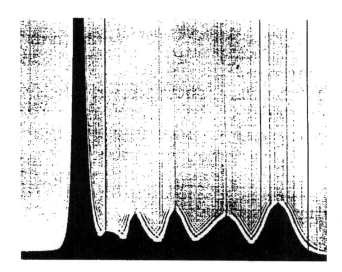

The large band on the left is albumin. This is followed by α_1, α_2, β and γ globulins. The band on the right is due to the boundary with straight buffer.

Moving boundary electrophoresis is not suitable for processing many samples; nor is it suitable for small molecules, due to the diffusion of these in solution. Convection currents in the U tubes are also a problem, even when the tubes are cooled. Hence, for most applications, zone electrophoresis has replaced moving boundary.

Zone Electrophoresis.

Zone electrophoresis was introduced in 1949. In this technique, the sample is applied as a spot, or as a narrow band, to a support medium. The support medium can be paper, cellulose acetate, starch, agarose or polyacrylamide gels. The structure of paper and cellulose acetate is such that it offers little impediment to the movement of most ions within the liquid supported by the paper or cellulose acetate. On the other hand, starch and polyacrylamide gels have a tighter structure that impedes the movement of large molecules. Agarose gels restrict the movement of very large, or long (such as DNA), molecules. Therefore, these latter supports also act as molecular sieves with the ions being separated due to their molecular size differences as well as due to their charge differences.

While the sample is in solution, it will diffuse outwards. The rate of diffusion is faster for small molecules than for large ones. Hence electrophoresis of small molecules should be conducted rapidly and the sample should be fixed to the support medium as soon as possible after the electrophoretic run, or transferred to a membrane for fixation.

The speed of electrophoretic separation is increased by increasing the voltage gradient across the support medium. Hence, high voltage electrophoresis is used for the separation of small ionised molecules.

For example, amino acids can be separated on a paper support medium, using high voltage ($\cong$ 4,000 volts) electrophoresis and a 8% formic acid buffer, pH 1.6. The paper is rapidly dried after the electrophoretic run and the amino acids located with a stain reaction:

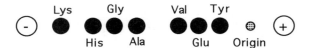

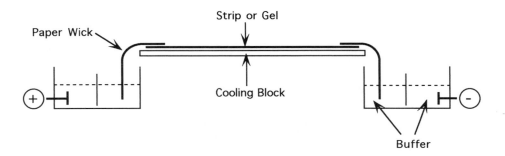

The basic electrophoretic apparatus is illustrated below:

With some support media, such as cellulose acetate, the wick connection to the buffer is unnecessary, as the support medium can dip directly into the buffer.

The use of double buffer chambers avoids changes in the buffer composition in contact with the support. Hence, electrolysis at the electrodes is isolated from the support.

The charge on an ionisable molecule will depend upon the pH of the buffer solution in which it is dissolved. Hence, the pH of the buffer used plays a very important role in electrophoresis of biological molecules. For example, all plasma proteins carry net negative charges at pH 8.8 and will migrate towards the anode (+ electrode).

Increasing buffer concentration causes the migration rate of the sample components to be reduced. This is because the buffer ions are now carrying most of the current, and their movement also retards oppositely charged sample ions by the attraction between the ions. However, increasing buffer strength tends to give sharper bands. But, high strength buffers are also likely to denature some proteins. If the strength is too low, the buffering is poor and poor resolution is obtained. The strength of buffer, used in electrophoresis, is usually expressed in terms of ionic strength, μ. This can be calculated using the formula:

$$\mu = \tfrac{1}{2} \Sigma \, Mz^2$$

Where: Σ = the sum of,
M = the molar strength of each ion,
z = the charge on each ion.

Most buffers used in electrophoresis have an ionic strength of 0.05 to 0.1.

The electrophoretic support medium acts as a resistance to current flow, so that the actual current flow will depend upon the type of medium, the buffer used, and its strength, and upon the voltage applied. As the medium is acting as a resistance, heat is generated in the medium and its temperature will rise. This results in a decrease in resistance (hence, an increase in current flow, if a constant voltage is applied) and evaporation of the buffer solvent.

The buffer solvent loss is not usually a problem with low voltage electrophoresis using paper or cellulose acetate support medium. Problems with changing current flow can be overcome by using constant current power packs. With some low voltage gel electrophoresis or with high voltage paper electrophoresis, some cooling mechanism has to be used.

Electroendosmosis (Endosmosis, Electro-osmosis).

If plasma is electrophoresed, at pH 8.8, on a paper, cellulose acetate or agarose support, some of the γ-globulins will migrate towards the cathode (- electrode), even though all the plasma proteins carry a negative charge. This is due to the phenomenon of electroendosmosis.

Hydroxyl ions, from water, tend to be adsorbed, to varying degrees, by most electrophoretic support media to produce a surface negative charge. Paper, cellulose acetate and agarose also contain acid groups (Eg. agarose contains pyruvate and sulphate). These media produce a greater surface negative charge in neutral or alkaline buffers. Positively charged buffer ions are attracted to this negative surface. Those close to the negative charge are held tightly, but those further away are free to move. When an electric potential is applied to the system, these loosely held cations will migrate towards the cathode.

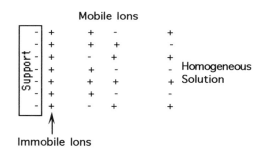

As these ions are surrounded by a cloud of water (ie. they are hydrated), they carry water towards the cathode. Hence, there is a net flow of fluid towards the cathode. This effect is minimal with starch and polyacrylamide gels, due to a lack of surface charge on these latter two supports.

Electrophoretic Support Media.

Paper (usually thick chromatography paper) and cellulose acetate are commercially available ready to use. Starch gel is not used very often these days. This leaves agarose and polyacrylamide as gels that are prepared in the laboratory. These will be examined in more detail.

Agarose.

Agarose is a linear polymer of D-galactose and 3,6-anhydro-L-galactose. It is purified from agar, which is extracted from red algae (seaweed). When a hot solution of agarose cools, double helix chains are formed. These aggregate, on standing, to form the gel that is used as an electrophoretic support medium.

The agarose strength controls the pore size in the gel. 8.5 g/l is the usual strength for plasma protein separation. However, large lipoproteins and immune complexes are retarded in the gel. DNA fragments are separated in gels ranging between 3 and 15 g/l. The gel restricts the movement of the linear DNA, so that the smallest fragments move fastest in the gel.

Highly purified agarose still contains some sulphate and pyruvate groups. These groups give the gel a negative charge. As a result of this, electroendosmosis will take place when a voltage is applied across the gel. This is usually not a problem with most types of electrophoresis. It is, however, a problem in isoelectric focusing where one needs the system to be static once the pH gradient has been established.

The agarose used in isoelectric focusing is "Agarose IEF". Positive charges, equivalent to the net negative charges on the purified agarose, are incorporated into the polymers. These positive charges on the gel introduce an electroendosmosis in the opposite direction to that induced by the negative charges. Hence, there is no net movement of the liquid.

Polyacrylamide.

Polyacrylamide gels are made by the polymerisation of acrylamide ($CH_2=CH.CO.NH_2$) and bisacrylamide (methylene bisacrylamide, $(CH_2=CH.CO.NH)_2.CH_2$). Polymerisation of acrylamide forms long chain molecules and these are cross-linked by the inclusion of bisacrylamide in the mixture. Gel formation occurs when the bisacrylamide content is greater than 2% of the total acrylamide.

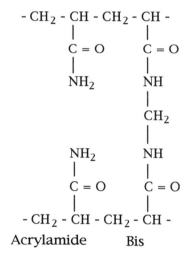

Acrylamide and bisacrylamide are polymerised only by the presence of free radicals. These can be produced by the photodecomposition of riboflavin in the presence of a trace of oxygen. Tetramethylethylenediamine (TEMED) is also used as a chain initiator. Alternatively, TEMED and ammonium persulphate can be used to generate the free radicals to initiate polymerisation.

Increasing the bisacrylamide percentage increases the number of cross links. This increases the chain thickness to a maximum of about 6 nm diameter. The cross linking leaves pores in the structure, the diameter of which depends upon the total acrylamide concentration. Pore size also depends upon the percentage of bisacrylamide, and the smallest pores are produced when the bisacrylamide makes up 5% of the total acrylamide.

Hence, by altering the total amount of acrylamide and the percentage of bisacrylamide, gels of different pore sizes can be produced. These range from 0.6 to 4 nm in diameter and are useful sizes for the separation of proteins. However, any particular gel will contain pores of a range of sizes.

Gels are often referred to in terms of the total percentage of acrylamide (monomer and bis) and the bis (cross linker) percentage (of total acrylamide). ie. a T5 C3 gel consists of 4.85% acrylamide and 0.15% bis. This is a large pore gel, suitable for separating plasma proteins. Most proteins can be separated using gels in the range 5-15%. Low percentage gels (large pore size) are used for separating high molecular weight proteins and high percentage gels (small pore size) are used for low molecular weight proteins. Small DNA fragments (10-1,000 base pairs) can be separated on gels that range from 20% (10-100 bp) to 3.5% 100-1,000 bp).

Polyacrylamide does not adsorb proteins, nor act as an ion-exchanger. Hence, when it is used as the support for electrophoresis, the proteins are separated on the basis of molecular size, as well as their charge/surface area ratio. The gel acts as a molecular sieve. The ability of molecules to move through the gel depends upon the size and shape of the pores in the gel and upon the size and shape of the molecules.

As polyacrylamide contains no charged groups, no electroendosmosis takes place. However, hydrostatic pressure may cause the movement of buffer through the gel. Hence, this must be taken into account, if measuring the mobility of compounds in polyacrylamide gel electrophoresis (PAGE).

The temperature of the gel is important as regards the mobility of the components being separated. Mobility is inversely proportional to the viscosity of the mixture in which the components are being separated.

Viscosity decreases as the temperature increases and, hence, the mobility will increase. There is an approximate increase of 2.4% in mobility per degree rise in temperature. Hence, cooling baths are usually used for PAGE and the gels are kept reasonably thin. Dilute buffers can also used to reduce the heat generated during the run.

In order to obtain sharp bands in PAGE, the proteins in the sample must be concentrated prior to separation. This can be done in two ways:

1.   Buffer Strength.

     The sample, in a dilute buffer, is applied to the top of gel. As the buffer has low conductivity, a large voltage drop is generated across sample. Hence, sample proteins rapidly migrate to form a layer on the top of the gel.

2.   Discontinuous Electrophoresis.

     The sample is applied to the top of a large pore stacking gel (pH 6.7) that is on top of the separating gel (pH 8.5). On top of the sample is the electrode buffer (pH 8.5) which contains glycine. The negatively charged glycine and proteins migrate into the stacking gel.

     At the pH of the stacking gel (6.7), glycine carries little charge and migrates slowly. Chloride, being fully charged, migrates rapidly. As a result a large voltage drop is generated which concentrates the proteins between the glycine and the chloride.

On entering the separating gel, the glycine becomes negatively charged and now migrates ahead of the proteins. The proteins are now separated on their charge to cross sectional area ratio and their ability to move in the polyacrylamide.

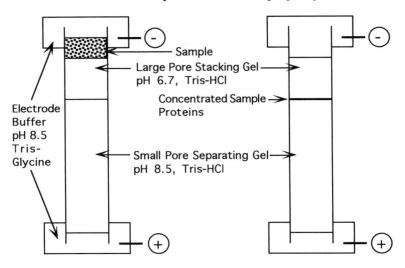

Plasma (or Serum) Electrophoresis.

Cellulose acetate and agarose gels are the most commonly used supports. Agarose gel gives better resolution. Whereas cellulose acetate electrophoresis of serum gives 5 or 6 bands, agarose gel can yield 15.

A buffer of pH 8.6 to 8.8 is usually used. At this pH, all plasma proteins carry a negative charge. However, some of the γ globulins move towards the cathode. This is due to electroendosmosis. The major bands are illustrated below:

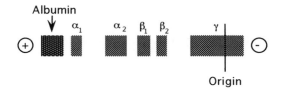

Fibrinogen, present in plasma, gives a band between β_2 and γ.

A commonly used buffer is Gelman High Resolution buffer of pH 8.8 and ionic strength of 0.05. This is a barbital-Tris buffer. 0.4 g/l of calcium lactate is added to this buffer to resolve the β globulins into two distinct bands (β_1 and β_2), seen above. α_1-antitrypsin is the major component of the α_1 band. α_2-macroglobulin and haptoglobin are the major components of the α_2 band. Transferrin is the major component of β_1 and complement protein C_3 is the major component of β_2. The immunoglobulins comprised most of the γ band.

After separation of the protein fractions, they are fixed onto the support by precipitation and then stained with a protein stain. The major stains are amido schwarz, Commassie blue and silver nitrate; in order of increasing sensitivity. After staining, the excess stain is removed and the support "cleared". The amount of protein bound stain is then quantitated by scanning the bands with a densitometer and measuring the area of each peak.

The quantitation is not an absolute measure of the protein concentration because albumin absorbs more dye than the globulins, on a weight-for-weight basis. Linearity between amounts of dye adsorbed and weight of protein exists only over low protein levels and falls off with increasing protein content. As a result, the albumin band tends to understain. These two factors tend to balance each other, but even so, results should be expressed as % of bound dye rather than as % of total protein.

SDS-PAGE.

In this technique, proteins are separated on the basis of their molecular weights.

Sodium dodecyl (C_{12}) sulphate (SDS) is an ionic detergent that strongly binds to, and denatures, proteins. One SDS molecule binds to two amino acid residues. This gives a constant negative charge per unit mass of the protein. Therefore, all SDS-protein complexes migrate towards the anode on electrophoresis. The smaller molecular weight complexes migrate faster, as they can move freely in the gel.

Therefore, in SDS-PAGE, SDS in incorporated into the polyacrylamide gel and the sample is given a prior treatment by heating it with SDS. Reduction of the protein can also be included in the initial treatment of the sample. This breaks the disulphide (-S-S-) links between protein chains. The final protein-SDS complexes are rod-like structures.

Illustrated below is the separation of some proteins, that have not been subject to reduction of disulphide links, on a T4-22.5, C4 gradient gel:

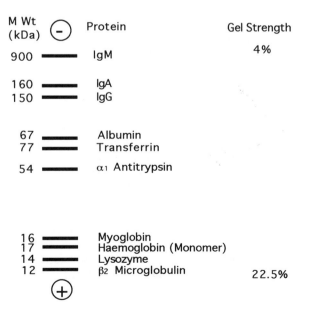

The presence of -S-S- bonds in a protein can be identified using a gel that has a reducing agent in part of it. This is illustrated below:

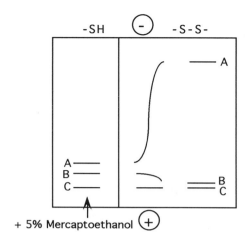

Mercaptoethanol (a reducing reagent) is incorporated in the left hand side of the gel. Some of it diffuses to the right. Hence, the left hand lane has broken -S-S- links and the right hand lane has them intact. The left side of the centre lane has broken -S-S- bonds and right side has them intact.

Protein A contains inter-chain -S-S- bonds that are broken to produce a number of equisized chains that migrate faster than the original protein. Protein B contains intra-chain -S-S- bonds that are broken to produce a longer chain that migrates more slowly. Protein C contains no -S-S- bonds.

Isoelectric Focusing.

In this technique, proteins are separated on the basis of their isoelectric points (pI). At these pHs the proteins have an equal number of positive and negative charges. Hence, they have no net charge.

Even though isoelectric separation was demonstrated in 1912, it was not until the synthesis of carrier ampholytes, in 1969, that the technique was fully utilised.

The system relies on the production of a stable pH gradient. Amphoteric molecules, when exposed to an electric potential, will migrate to their isoelectric pH regions in this pH gradient. If the molecule is not at its isoelectric pH region it will carry a net charge and will, therefore, move due to the electric potential across the system. For example, if the molecule is at a pH greater than its isoelectric point, it will carry a negative charge. It will, therefore, move towards the anode. In doing so, it moves into a region of lower pH and, hence, its negative charge is reduced. When it reaches the region of its isoelectric point, it will carry an equal amount of positive and negative charge and will, therefore, stop.

As will be realised, the proteins will be concentrated at their isoelectric regions, even though they may have been applied over a broad area. The technique is, therefore, a very efficient means of separating and concentrating individual proteins. Proteins with isoelectric points that differ by only 0.01 of a pH unit can be separated by this technique. An occasional drawback of the technique is that the proteins may precipitate, as proteins are least soluble at their isoelectric points.

Early pH gradients were established by layering an acidic buffer above a basic buffer an allowing them to diffuse into each other. However, the pH gradient was short lived, due to the rapid movement of the buffer ions in the electric field.

Protein hydrolysates and synthetic peptides were used as ampholytes, but these only covered a small pH range. Ampholytes are now prepared by reacting acrylic acid with different polyethylene polyamines to give the following structures:

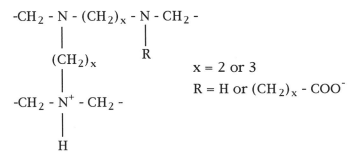

$$x = 2 \text{ or } 3$$
$$R = H \text{ or } (CH_2)_x - COO^-$$

These ampholytes are of low molecular weight (300-1,000) and have a whole range of pI values. Starting with a neutral pH, those with low pI values will be negatively charged and will move towards the anode. Those with high pI values will be positively charged and will move towards the cathode. Hence, the acid ampholytes move towards the anode and the basic ones move toward the cathode. One of the requirements of the ampholytes is that they have a high buffering capacity. Hence they start to set up a pH gradient, as illustrated below:

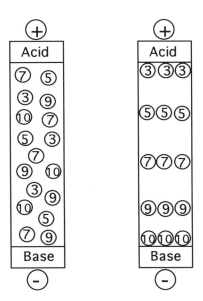

To establish a pH gradient, convection currents must be avoided and the rate of ampholyte diffusion reduced. A density gradient was used in the early days, but an electrophoresis matrix is now used. This can be a polyacrylamide gel, a cross-linked dextran gel or an agarose gel. An agarose gel, containing non-crosslinked polyacrylamide (this reduces electroendosmosis and diffusion within the gel) is commonly used.

Electroendosmosis is a problem when using purified agarose, as the voltages used are usually higher than in most types of electrophoresis and purified agarose contains negative groups. Therefore, the agarose used in isoelectric focusing is "Agarose IEF", where positive charges equivalent to the net negative charges on the agarose are incorporated into the polymers. As a result there is no net electroendosmosis.

The conductivity of the gel is relatively high at the beginning of the isoelectric focusing (IEF) run. As the pH gradient is established, the conductivity decreases. Hence, a constant-power source is usually used.

Very thin gels (0.4 mm) are often used for IEF. These gels have a cost saving advantage, as well as not requiring cooling during the run. These gels can be placed directly on the electrodes, rather than having an acid and a base in contact with the electrodes.

An example of a monoclonal IgG IEF serum is illustrated below. The multi bands of the monoclonal IgG are due to post transcriptional modifications. These are mainly the deamination of basic amino acids and modifications in the carbohydrate part of the molecule.

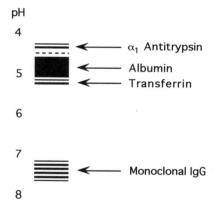

Two Dimensional Electrophoresis.

In this technique, sample components are initially separated by isoelectric focusing. This is then run, at right angles, in SDS-PAGE. Hence, proteins are separated by their isoelectric point in one direction and by molecular weight in the other. Plasma yields more than three hundred spots, six of which are illustrated below:

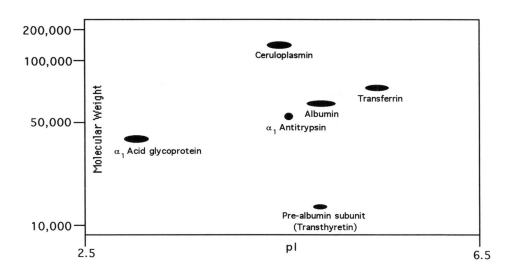

Immunofixation.

To a large extent, this technique has replaced immunoelectrophoresis (See chapter 13). Plasma proteins are separated by normal electrophoresis on cellulose acetate or agarose gels. The protein of interested is then precipitated with a specific antibody. After washing to remove all the other proteins (still soluble), the protein of interest is stained. This can then be quantitated by densitometry.

Eg.   The identification of the immunoglobulin type of a paraprotein:

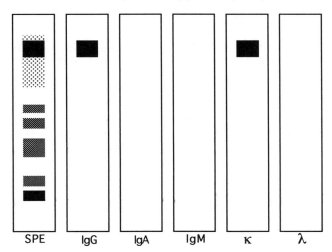

SPE (Serum Protein Electrophoresis).
    Proteins are chemically fixed and then stained.

IgG etc.
    Specific antisera is used to fix the protein, which is then stained.

Western Blotting (Immunoblotting).

Individual proteins can be localised with labelled antibodies. With certain media, such as polyacrylamide gels, it is difficult to get the antibody into the gel. Hence, the electrophoresed proteins have to be transferred and fixed to a membrane (eg. nitrocellulose or positively charged nylon) and then probed with specific antibodies. Transfer is usually carried out in an electrical blotting cell. The first antibody (Antibody 1) is directed against the protein of interest and an enzyme labelled Antibody (2) is directed against Antibody 1, as illustrated below:

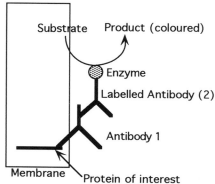

This procedure also increases the sensitivity of the technique by a factor of 10 - 100, when compared with direct staining techniques.

Other Electrophoretic Techniques.

Nucleic acid electrophoresis is covered in chapter 12.

Immunoelectophoresis, crossed immunoelectrophoresis (2-dimensional immunoelectro-phoresis) and electroimmunoassay (Laurell's rockets) are covered in chapter 13.

CAPILLARY ELECTROPHORESIS

Capillary electrophoresis combines standard electrophoretic techniques with the speed and ease of HPLC. The sample components, in a capillary, are separated using a high voltage. These components are detected as they pass the detector:

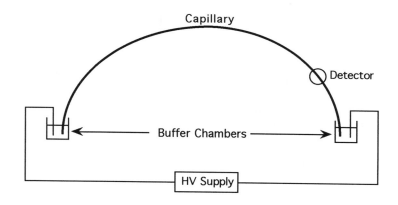

The technique was first used for amino acid separation in 1981. This was followed by protein separation in 1986. The first commercial instruments were produced in 1988.

The technique has the advantages that very small (1-2 nl) volumes of sample are required and that many electrophoretic techniques can be used with a single capillary.

The capillaries range in length from a few centimetres to one metre. They have internal diameters between 25 and 100 μm. These are usually made of fused silica and are coated, on the outside, with a polymer to increase the mechanical strength. The inside can be uncoated or coated:

Uncoated.

At pHs above 3, the silica surface has a negative charge:

Surface

$$
\begin{array}{ccc}
O^- & OH & O^- \\
| & | & | \\
- Si - O - & Si - O - & Si - O - \\
| & | & | \\
O & O & O \\
| & | & |
\end{array}
$$

Therefore, there is a net electroendosmotic flow of buffer towards the cathode. This flow increases as the pH of the buffer increases.

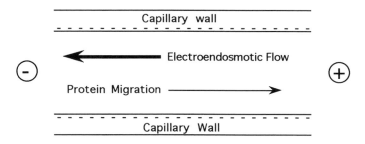

Coated.

The capillary can be coated with a polymer to abolish the electroendosmotic flow and/or to eliminate the binding of the analyte (eg. protein) to the capillary wall.

Detectors.

Ultra violet or visual light absorption, by the compounds of interest, is the most commonly used method of detection. The polymer coating of the capillary is removed, in the detector, to allow light to pass through the capillary.

Wider bore capillaries give better sensitivity, due to the longer light path length, but separation efficiency, due to band broadening, decreases due to diffusion and heat generation. 25 - 75 μm ID capillaries are, therefore, used as a compromise.

As the analytes move at different speeds, peak areas are a function of migration speed as well as concentration.

Photodiode array detectors (512 diodes, 190-600 nm) are available for use with Beckman instruments and high-speed wavelength scan detectors are used by Bio-Rad.

Beckman also provides laser fluorescence and mass spectrometry detectors.

HV Power.

Usually 200 - 400 volts/cm.

Sample Loading.

The sample is moved into the capillary by either an electrical or displacement method. Electrical methods involve either the electrokinetic movement of the sample components by normal electrophoresis or by being swept into the capillary by buffer movement in electroendosmosis.

Displacement loading involves the use of pressure, a partial vacuum or gravity to move the sample into the capillary.

SEPARATION METHODS.

Free Zone.

This can be carried out in uncoated or coated capillaries and involves normal electrophoresis. Hence, negatively charged molecules will migrate towards the anode. However, electroendosmosis is the dominant movement, with uncoated capillaries, sweeping the buffer and the analytes back towards the cathode.

Additives can be included in the buffer to reduce binding to the capillary wall (which is a problem with protein separation) or to enhance separation. For example, chiral additives can be included in the buffer to resolve stereo isomers.

Due to the strong affinity of the silanol groups in the capillary wall for proteins, coated capillaries are often used for protein separation.

An example of serum capillary electrophoresis is shown below. The diagram also shows the technique of immunosubtraction, a method for identifying a specific protein in serum or plasma. A portion of the diluted sample is passed through a column containing antibodies to the protein of interest, attached to a solid support. The treated and untreated samples are then run on capillary electrophoresis equipment. The diagram, below, shows the immunosubtraction of α_1-antitrypsin (arrow):

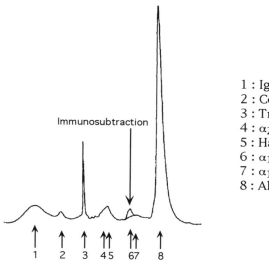

1 : IgG.
2 : Complement C3.
3 : Transferrin.
4 : α_2-Macroglobulin.
5 : Haptoglobin.
6 : α_1-Antitrypsin.
7 : α_1-Acid glycoprotein.
8 : Albumin.

Micellar Electrokinetic Capillary Chromatography (MECC).

This is a type of free zone separation. However, a surfactant (Eg. SDS) is used as a carrier of the analytes. The carrier has a surface charge and, therefore, migrates towards an electrode. Analyte molecules bind to the carrier by partitioning with the hydrophobic part of the micelle (C_{12} chains). In the illustration below, the micelle has a large number of negatively charged sulphate groups on the surface and will migrate towards the anode. However, in an uncoated capillary, the electroendosmotic flow will sweep it back towards the cathode and detection:

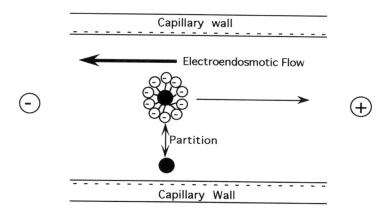

This technique has been successfully used to separate drugs, as illustrated for anticonvulsants, below:

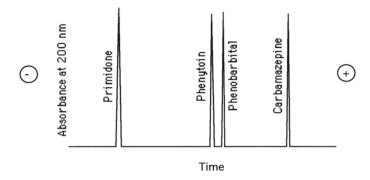

Isoelectric Focusing.

IEF can only be carried out in coated capillaries, as electroendosmosis must be avoided during the run. The sample is mixed with ampholytes (and often methyl cellulose) and introduced into the capillary by displacement loading. The anode chamber usually contains dilute phosphoric acid and the cathode chamber, dilute NaOH. The focusing voltage is then applied until the current reaches a low value.

After the focusing is complete, the sample components must be mobilised passed the detector. In one method, NaCl is added to the NaOH at the cathode chamber. Cl⁻ enters the capillary and causes a drop in the pH gradient. The focused proteins now move past the detector. Alternatively, pressure can be applied to force the contents of the capillary passed the detector.

Illustrated below is the IEF separation of deoxygenated haemoglobin variants in normal blood:

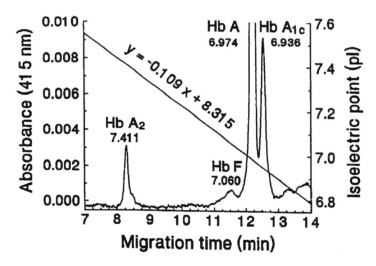

The absorbance scale has been adjusted to highlight the minor variants. The straight line illustrates the corresponding pH of the ampholytes as they extrude from the capillary.

Molecular Sieving.

Even though gels can be cast in capillaries, this is difficult. These gel filled capillaries can only be used for a limited number of runs, which limits their use. However, linear polymers can easily be introduced into the capillary to form a matrix that retards large molecules. Therefore, molecules with similar electrophoretic migration in a free zone system can then be separated on the basis of their size.

This technique has successfully been used to separate DNA fragments from about 100 to 2,000 base pairs. Plasma lipoproteins have been separated on the basis of their size, by this technique, as have SDS-protein complexes.

EXERCISES

EXERCISE 11-1

HIGH VOLTAGE ELECTROPHORESIS OF URINE AMINO ACIDS

The pattern of amino acids in urine, separated by high voltage electrophoresis (HVE), can be used to assist in the diagnosis of many inborn errors of metabolism.

HVE is a rapid method for separating small ionisable compounds such as amino acids. However, not all commonly encountered amino acids can be separated from each other, as some amino acids migrate at the same rate in a specific buffer. In this case, a 2-dimensional system can be used where the HVE is used in one direction and then normal phase chromatography is used at right angles to the original run.

In this exercise, a buffer of pH 1.9 will be used for the electrophoresis. At this pH, the basic amino acids travel rapidly towards the cathode, whilst the acidic amino acids remain near the origin.

Case History.

The sample is urine from a fourteen week old girl with methylmalonic acidaemia. This disorder results from a defect in the conversion of methylmalonyl-CoA to succinyl-CoA and details can be found in chapter 9, exercise 9-6.

Hyperglycinaemia, seen in children with methylmalonic acidaemia, is due to the inhibitory action of high levels of methylmalonyl-CoA and Propionyl-CoA on mitochondrial glycine cleavage enzyme (The reversal of glycine synthase to yield CO_2 and NH_3). Hence, glycine accumulates in the mitochondria. This diffuses into the cytosol and thence into the extracellular fluid. From there it enters the blood and then passes into the urine.

As part of the investigation in this case, high voltage electrophoresis of urine amino acids has been requested. A sample of normal urine will also be run, with the patient's urine, in order to compare the amino acid patterns. A selection of amino acid standards (including glycine) will also be run in order to identify any amino acid(s) present in abnormally large amounts.

After the run, the paper will be dried and then dipped into a ninhydrin solution and heated to 105°C. (Details of the reaction are given in exercise 7-2). The ninhydrin positive spots (purple) will then be fixed by dipping the paper into a copper nitrate solution. This converts the purple spots to orange.

In the method used at the Melbourne Royal Children's Hospital (RCH), the amount of urine applied to the paper is inversely proportional to the urine creatinine concentration. The volume of urine containing 0.11 µmol of creatinine is used in the run. This makes adjustment for variations in the urine concentration.

As well as using the ninhydrin reagent, the RCH also uses a fast blue B stain to localise methylmalonic acid (see exercise 8-3) and iodoplatinate to localise the sulphur containing amino acids that do not stain well with ninhydrin.

A typical high voltage electrophoresis apparatus is illustrated below:

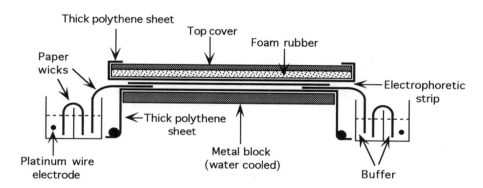

REAGENTS AND EQUIPMENT:

1. pH 1.9 Buffer.
        Mix together:
                41.6 ml of 90% formic acid,
                150 ml of glacial acetic acid,
                and make to 2 l with distilled water.
2. D/A/R Standard.
        Dissolve 20 mg of each of aspartic acid, alanine and arginine in 10 ml of reagent 1.
3. E/G/K Standard.
        Dissolve 20 mg of each of glutamic acid, glycine and lysine in 10 ml of reagent 1.
4. S/H Standard.
        Dissolve 20 mg of each of serine and histidine in 10 ml of reagent 1.
5. Normal Urine.
6. Patient's Urine.
        Dissolve 1 g of methylmalonic acid and 0.7 g of glycine in 300 ml of normal urine.
7. 0.2% Ninhydrin in Acetone.
8. 2% Pyridine in Distilled Water.
9. Copper Nitrate Solution.
        Add 200 µl of 10% nitric acid to 100 ml of ethanol.
        To this solution, add 500 µl of a saturated copper nitrate solution and mix well.

High Voltage Electrophoresis Apparatus and Power Pack.
        This exercise was written for the Shandon single cooled-plate apparatus (11 x 24"). However, the setting up techniques may be slightly different if other equipment is used.

Electrophoretic Strips.
        Whatman 3 MM paper cut to 500 x 100 mm.
Wicks.
        Whatman No. 1 paper cut to 220 x 400 mm and 220 x 200 mm.
Pencils and Rulers.
Filter Paper.
Glass Rectangles.
        Made from glass tubing or rod:

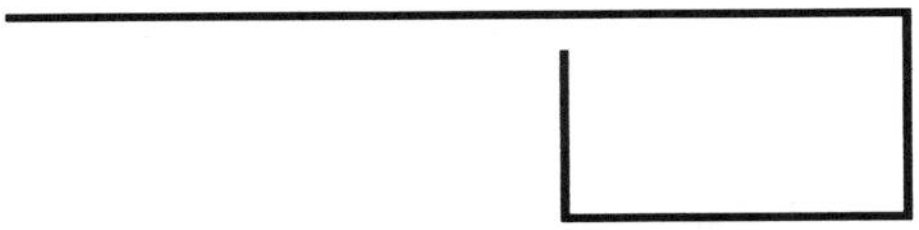

```
 _____
|                     _____  |
|                    |          | |
|                    |          | |
|                    |_____| |
```

2 µl Capillary Samplers and Capillaries.
Dish with reagent 1, for soaking strips and wicks.
Chromatogram Dip Trays.
        Shandon Scientific Co. or equivalent.
Oven at 105°C with Clean Paper on the Shelves.
String with Clothes Pegs under Fume Hood.
Forceps.
Disposable gloves.
Paper Tissues.
Cover the bench in clean brown paper.

PROCEDURE:

This exercise can be done in conjunction with exercises 8-3 and 9-1, as it uses the same sample.

The ninhydrin reaction is very sensitive and so the electrophoretic strips should be handled with forceps, or disposable gloves should be used.

1.   Place a strip of the Whatman 3 MM paper (500 x 100 mm) on a clean piece of brown paper. With a pencil, mark the paper as shown below:

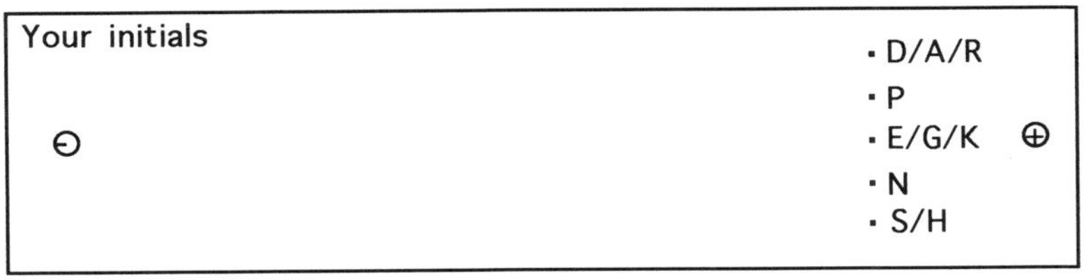

```
 _____
| Your initials                                           |
|                                              · D/A/R     |
|                                              · P         |
|                                              · E/G/K   ⊕ |
|          ⊖                                   · N         |
|                                              · S/H       |
|_____|
```

2.   Dip the strip in the formic acid - acetic acid buffer, pH 1.9.

3.   Place the strip on a clean piece of brown paper and blot off the excess buffer with filter paper.

4.   Place a glass rectangle under the origin line, so that the application sites are in the centre of the rectangle.

5.   Apply 2 µl samples of the appropriate solution to its application site using a capillary sampler:

        P   :   Patient's urine.

11-21

N	:	Normal urine.
D/A/R	:	2 g/l of aspartic acid, alanine and arginine in the buffer.
E/G/K	:	2 g/l of glutamic acid, glycine and lysine in the buffer.
S/H	:	2 g/l of serine and histidine in the buffer.

Wash out the capillary with 2 or 3 water washes between each sample. Blot dry the capillary with paper tissues before use.

6. Clean and dry the surface of the thick polythene sheets in the electrophoresis apparatus.

7. Fill the electrode chambers to about 3/4 full, with the pH 1.9 buffer.

8. Use the Whatman No. 1 chromatography paper for the wicks; dampen with buffer before use.

> 270 x 200 mm (folded to 270 x 100 mm) as the electrode buffer wicks.

> 270 x 400 mm (folded to 270 x 200 mm) as the wicks from the buffer to the plate. Lay the wick on the plastic so about 100 mm of wick is over the cooling plate.

9. Place the strips on the plate, so that the ends overlap the wicks.

10. Clamp down the top plate.

11. Close the protective Perspex cover and make sure that the cooling water is flowing.

12. Run the electrophoresis for 30 minutes at 4,000 volts.

13. Turn the power off, open the cover and then remove the strips.

14. Dry the strips, on clean paper, in a 105°C oven for 10 minutes.

15. Run the strips through a dip tray containing 0.2% ninhydrin in acetone to which 2 to 3 ml of 2% pyridine has been added just before dipping.

16. Hang the strips up to dry under the fume hood for 2 to 3 minutes.

17. Place them in the 105°C oven, on clean paper, until the purple spots are fully developed (2 to 3 minutes).

18. Dip the strips in a dip tray containing the copper nitrate reagent and hang up to dry, under the fume hood.

Identify any abnormal amino acid in the patient's sample.

EXERCISE 11-2

CELLULOSE ACETATE ELECTROPHORESIS OF SERUM OR PLASMA PROTEINS

As discussed earlier in this chapter, plasma proteins (at a pH of 8.6 or above) are all negatively charged and, if exposed to an electric potential, will migrate towards the anode. The rate of migration will depend on the potential applied, the charge on the molecule, the diameter of the hydrated molecule, the viscosity of the buffer and the ability of the molecule to move in the supporting medium.

Cellulose acetate strips are used as the supporting medium in this exercise. They are soaked in a buffer solution of the required pH, the serum or plasma is added and an electrical potential applied across the length of the strip.

The buffer used in this exercise is a commercial high resolution buffer of pH 8.8 and ionic strength of 0.05. This is a barbital-Tris buffer. Calcium lactate is added to this buffer to resolve the β globulins into two distinct bands (β_1 and β_2).

The fine texture of the cellulose acetate reduces migration due to diffusion and there is no adsorption of protein to the cellulose acetate. Both these factors produce good resolution of serum or plasma into a number of distinct bands.

As previously mentioned, cellulose acetate exhibits electroendosmosis. Hence, some of the γ globulins appear on the cathodic side of the application point, even though all the proteins carry a net negative charge at pH 8.8.

After separation of the protein fractions, they are fixed onto the strip by precipitation with trichloroacetic acid and stained with Amido Schwarz. The amount of protein bound stain is then quantitated by scanning the strip and measuring the area of each peak.

Case History.

A 61 year old male patient presented with severe back pain, loss of weight and an increased susceptibility to infection. Plasma electrophoresis has been requested to assist in the possible diagnosis of myelomatosis (multiple myeloma).

Myelomatosis is associated with the proliferation of monoclonal plasma cell that invade bone. These cells mobilise the calcium from the bone. Punched out areas of bone can be seen on X-ray. The released calcium can produce renal damage. Patients are usually anaemic.

The plasma cells produce a monoclonal immunoglobulin known as a paraprotein. This produces an intensely staining band in the β or γ regions on electrophoresis. A reduction in normal immunoglobulin production occurs and this leads to the increased susceptibility to infections. Plasma albumin levels are usually reduced.

Fortunately, myelomatosis is rare with about 3 cases per 100,000 being diagnosed per year. Most patients will die within a few years of diagnosis.

REAGENTS AND EQUIPMENT:

1. High Resolution Buffer, pH 8.8, + Calcium Lactate.
        Dissolve the contents of 1 sachet (Gelman Cat No. 51104 or Helena Cat No. 5805) in distilled water, add 0.4 g of calcium lactate and make to 1,000 ml.
        Store at 4°C.
2. Stain Fixative.
        Dissolve 1.25 g of Amido Schwarz (Naphthol Blue Black) and 15 g of trichloroacetic acid in 500 ml of methanol.
        Store in a dark bottle and put out a funnel for reagent return after use.
3. Clearing Fluid.
        Mix 300 ml of glacial acetic acid with 2 l of methanol.
4. Normal Serum or Plasma.
        Keep on ice.
5. Myelomatosis Serum or Plasma.
        Obtained from the local cancer hospital. Freeze in vials at -20°C. The frozen vials can be kept for at least 5 years.
        Keep on ice.

Electrophoretic Tank and Magnets or Clips.
        Fill with buffer and level the liquid by tilting the tank.

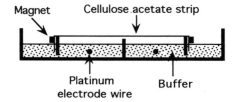

Constant Current Power Supply.
Cellulose Acetate Strips.
        Sepraphore III, 6 x 1" (Gelman Cat No. 51003).
Sample Applicators.
        Gelman Cat No. 51225.
2 μl Capillary Samplers and Capillaries.
Staining and Clearing Tanks (5).
        Gelman Cat No. 51457.
Large Glass Sheet.
Test Tube (without lip).
Forceps.
Filter Paper (18.5 cm diameter).
Dish Containing Buffer.
Large Chromatography Sheets.
Water Insoluble Textacolour Pen.
        Eg. Gelman Cat No. 51224.
Scissors.
Densitometer, or Manual Scanner and Planimeter (+ Thick Cardboard Sheet, Drawing Pins and Magnifying Glass).

PROCEDURE:

The cellulose acetate strips should only be handled by the ends, using forceps. All work should be carried out on a clean sheet of chromatography paper.

1. Write your name (or initials) on one end of a cellulose acetate strip with a water insoluble textacolour pen and allow the ink to dry.

2. Float each strip on the surface of the buffer until all the white areas have disappeared and then immerse in the buffer for one minute. This prevents the trapping of air bubbles in the strip.

3. Remove the strip from the buffer and lay it on a sheet of filter paper that has been soaked in the buffer. Very lightly blot the strip with filter paper to remove any surface buffer.

4. Apply 2 µl of serum or plasma (using a 2 µl capillary sampler and wiping the capillary before application) along the wires of the Gelman applicator in a strip about 10 mm long. Leave a few mm clear at each end of the wires.

Use the serum or plasma from a healthy subject for one strip and the serum or plasma from a myelomatosis patient for a second strip. Use separate samplers and applicators, or wash and dry them between samples.

5. Touch the applicator on the strip so that the serum is transferred to a point about 25 mm from the middle of the strip.

6. Transfer the strip to the electrophoresis tank, so that the sample is on the cathode side, and apply tension with the magnets (magnet arrows point towards the electrode terminals) or clips. Make sure that the cellulose acetate strip contacts the buffer on both sides of the tank. Replace the lid.

7. When the tank is full, connect the power supply. Make sure that the cathode is on the side of the strip to which the sample was applied.

8. Switch on the current and adjust for a constant current of one mA per strip.

9. Run for 75 minutes.

10. After this period, turn off the power. Remove the strips using two pairs of forceps applied at the ends of the horizontal part of the strip; <u>hold horizontally</u> and immerse in the stain-fixative (0.25% Amido Schwarz and 3% trichloroacetic acid in methanol) for 5 minutes. (After staining return the stain to its bottle).

11. Wash the strips in the clearing fluid (15% glacial acetic acid in methanol) until the background is white. (4 washes with about 2 minutes in each wash tank).

12. Carefully and quickly transfer the strips to a sheet of clear glass and spread out to dry. Roll a test tube on the strip to exclude trapped air bubbles. The strips will clear and dry on the glass.

(N.B. Do not touch the strips while they are clearing, as they partially dissolve in the acetic acid.)

13.   Compare the electrophoretic patterns of the two samples.

14.   After about 2 hours, the cleared dry strips can be removed from the glass.

DENSITOMETRY.

Quantitation of the amount of bound dye per protein band can be carried out by densitometry. The techniques for manual scanning, using an "EEL" scanner, and densitometry, using a Helena Auto Scanner, are described below.

"EEL" Scanner.

In this instrument, the strip is manually moved past a narrow light beam. The absorbance is displayed on a galvanometer on top of the instrument. This is then marked on a sheet of paper held below a ruled scale attached to the strip holder. The area of each peak, corresponding to an electrophoretic band, can then be determined using a planimeter.

1.   Trim the ends off the cellulose acetate strip.

2.   Turn on the scanner and allow a 5 minute warm up period.

3.   Place your electrophoretic strips between the two glass plates of the strip holder.

4.   Place the glass plates in the holder and insert into the scanner.

5.   Place a sheet of paper in the paper clamp on top of the instrument.

6.   Draw a base line by placing a pen in the notch at zero on the scale and move the holder up the paper.

7.   Open the photocell compartment on the left of the instrument and observe the position of the protein bands in relation to the light beam.

8.   Move the holder so that all bands are located towards the back of the instrument, in relation to the light beam, and that the light beam passes through a clear section of the strip.

9.   Close the photocell compartment and zero the galvanometer with the "zero" control on the right-hand side.

10.   Move the holder towards you in very small steps and plot the absorbance reading on the paper, using the absorbance scale. Make sure you pick each peak and each trough. (The more points you have, the better).

11.   After you have scanned all the bands, remove the holder and remove your strip.

12.   Join up all the points on the scan that you have just produced. If the cellulose acetate strip is placed over the scan, the peaks should exactly match the bands.

13.   Draw a vertical line from each trough to the baseline of the scan.

14. Measure the area under each peak using the planimeter as follows (This is proportional to the amount of dye bound by the protein fraction under consideration):

15. Pin the scan to the cardboard sheet provided and position the planimeter so that the arms are almost at right angles, with the pointer over the scan, the scale at the left and the pin above the left hand side of the scan:

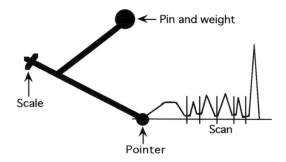

16. Place the pointer of the planimeter over the left-hand end of the baseline of the first peak.

17. Read the vernier scale with a magnifying glass.

18. With the pointer, follow the perimeter of the peak around clockwise, 5 times, and end up back at the left-hand end of the baseline.

19. Again read the vernier scale and record the difference in the two readings.

20. Repeat steps 16 to 19 with all the other peaks.

21. Express each band as a percentage of the total bound dye. Compare the two sets of results and calculate the A/G ratio for each.

Helena Auto Scanner.

This scanner generates 2 traces. The top trace displays the absorbance on the vertical axis and the strip position on the horizontal axis. This generates a series of peaks, each corresponding to an electrophoretic band. Below this trace is the integrator trace which consists of 10 horizontal lines. The integrator pen moves up and down over these lines. The greater the absorbance, the faster the pen moves. Hence, the number of ruled lines crossed, below a peak, is proportional to the area of the peak.

1. Trim the ends off the cellulose acetate strip.

2. Turn on the instrument and wait one minute for it to warm up.

3. Select: Auto Gain, % Scale and Auto Zero.

4. Open the lid and select: Vis, lamp "On" and the small slit. Check that the 525 nm (green) filter is in position.

5. Place the cellulose acetate strip on the holder, over the central slot, and clamp with the bar magnets.

6. Slide the carriage back so that the light beam is in the centre of the strip.

7. Adjust the thumbscrews so that all the electrophoretic bands (+ about 5 mm on each end of the strip) will be scanned. This is done by moving the carriage from left to right and back.

8. Slide the carriage fully to the left and close the lid.

9. Press the "Scan" button.

10. After the scan has been completed and the trace drawn, remove your piece of chart paper and strip from the instrument.

11. Draw vertical lines from each trough to the bottom of the integrator trace on the chart paper.

12. For each peak, count the number of horizontal lines crossed by the integrator pen (one full movement of the pen, from the bottom to the top and back again, is 20 lines).

13. Express each band as a percentage of the total bound dye. Compare the two sets of results and calculate the A/G ratio for each.

Reference Ranges:

Total serum protein		:	59 - 82 g/l
Albumin		:	35 - 55 g/l
Globulins	α_1	:	2 - 4 g/l
	α_2	:	4 - 9 g/l
	β_{1+2}	:	6 - 10 g/l
	γ	:	7 - 15 g/l
A/G ratio		:	1.1 - 2.2

Notes:

i. If plasma is used instead of serum, the fibrinogen will be seen as a band between the β_2 and γ bands on the electrophoretic strip. The reference range for plasma fibrinogen is 2-4 g/l.

ii. A decrease in total plasma proteins is always due to a reduced albumin level, with the globulin remaining normal or slightly decreased. This produces a lowering of the A/G ratio. An increase in total plasma proteins, other than dehydration, also produces a lowered A/G ratio due to the increase being in one or more of the globulin components. Albumin is usually reduced as well.

EXERCISE 11-3

AGAROSE GEL ELECTROPHORESIS OF SERUM OR PLASMA PROTEINS

Greater resolution of plasma proteins can be obtained with agarose gels, when compared with cellulose acetate electrophoresis. In this exercise, the agarose gel is poured on GelBond. This is a plastic sheet, one side of which has agarose incorporated into its structure. This side in very hydrophilic and the gel is poured on this side.

With the 8.5 g/l of agarose gel, used in this exercise, most plasma proteins are free to move within the gel. However, the large proteins, such as the chylomicrons, very-low-density lipoproteins and immune complexes will be retarded or may not move at all.

The buffer used is a barbital one of pH 8.6. At this pH, plasma proteins carry a negative charge. Calcium is included in the buffer solution to resolve some of the β band proteins.

Running a number of samples in the same medium makes it easier to detect minor mobility and/or concentration changes between samples, when compared with separate cellulose acetate strips used in exercise 11-2.

Samples of normal and myelomatosis plasma or serum will be run. An electrophoretic marker, consisting of bromphenol blue bound to bovine serum albumin, will also be run.

After electrophoresis, the gels will be fixed with picric acid-acetic acid, dried, stained with Coomassie blue, destained and photocopied.

Case History.

This is the same sample as used in exercise 11-2 and is from a 61 year old male patient who presented with severe back pain, loss of weight and an increased susceptibility to infection. Plasma electrophoresis has been requested to assist in the possible diagnosis of myelomatosis (multiple myeloma).

REAGENTS AND EQUIPMENT:

1. Barbital Buffer, pH 8.6, 75 mmol/l + Calcium Lactate.
        Dissolve 4.14 g of barbital (diethylbarbituric acid) in about 1 l of boiling distilled water. Add 1.15 g of calcium lactate and 26.28 g of sodium barbital. When dissolved, cool and make to 2 l with distilled water.
        Store at 4°C.
2. Fixing Solution.
        Add 1.5 g of picric acid to 100 ml of distilled water and stir overnight at room temperature. Filter and add 20 ml of glacial acetic acid.
3. Stain.
        Dissolve 200 mg of Coomassie Brilliant Blue R-250 in 100 ml of reagent 4 and filter, using Whatman No. 1 paper.
4. Destaining Solution.
        Mix together 500 ml of methanol, 100 ml of glacial acetic acid and 500 ml of distilled water.
5. Normal Serum or Plasma.
        Keep on ice.

6. Myelomatosis Serum or Plasma.

Obtained from the local cancer hospital. Freeze in vials at -20°C. The frozen vials can be kept for at least 5 years.

Keep on ice.

7. Electrophoretic Marker.

Dissolve 10 mg of bromphenol blue in 10 ml of 0.9% NaCl.

Dissolve 1.0 g of bovine serum albumin in 1.0 ml of the above solution.

Store at 4°C.

8. Agarose Gels.

Suspend 850 mg of agarose in 100 ml of reagent 1. Heat, with stirring, to boiling so that the agarose dissolves. Let cool to about 50°C.

Cut GelBond (Pharmacia Biotech Cat No. 80-1129-32) into 110 x 140 mm rectangles and level on a levelling table, with the hydrophilic surface upwards. Pipette 16 ml of the agarose solution onto each rectangle. Allow to set for 15 minutes. Store the gels on moist filter paper in sealed (food storage) boxes at 4°C.

Use the gels the following day.

9. 10% Glycerol.

Electrophoresis Apparatus with Water Cooled Plate and Power Pack.

Eg. LKB system (2117 Multiphor tank and 3371 E power pack).

Sample Application Strip.

Pharmacia Cat No. 80-1129-47.

1 and 2 μl Capillary Samplers with Capillaries.

Wicks.

Whatman No. 1 chromatography paper, 14 cm wide and long enough, when folded double, to reach from the buffer to the gel. Interleaf 3 of these doubled sheets.

Glass Plates.

TLC plates are suitable.

1 l Plastic Bottle filled with Water. (As a 1 kg weight).

Filter Paper (18.5 cm diameter).

Paper Towels.

Dish for Staining and Destaining.

Hair Dryer.

Bench Covered with Clean Brown Paper.

PROCEDURE:

1. Place the application strip on the gel so that the slits are about 35 mm from the cathode edge (ie. the edge of the strip is about 15 mm from the edge of the gel) of the 140 mm dimension of the gel. The strip has one 7 mm slot per cm and, hence, 13 slots are available for samples (leaving spaces at the edges of the gel).

2. Apply 2 μl of the plasma or serum samples along the slits with 2 μl capillary samplers and 1 μl samples of the two plasmas or sera with 1 μl capillary samplers. Also apply 1 μl of the marker albumin with a 1 μl capillary sampler.

3. Allow 5 minutes for the samples to enter the gel and then peel off the application strip (Do not discard the strip. Wash it with distilled or deionised water, dry it and return it to its packet).

4. Make sure that the electrode chambers are filled with buffer (ie. the buffer must be above the well dividers for both electrodes) and cooling water is flowing through the unit.

5. Spread a few drops of 10% glycerol solution on the cooling plate.

6. Place one edge of the gel on the cooling plate and lower it so that air bubbles are not trapped between the gel base and the plate.

7. Soak the wicks (6 thicknesses of Whatman No. 1 chromatography paper) in buffer and lay them on the gel so that about 10 mm of gel is covered and the wicks dip onto the electrode buffers.

8. Place a glass plate on top of the wicks to hold them in place.

9. Close the lid and apply 250 volts (constant voltage).

10. When the marker spots have migrated 50-55 mm (about 70 minutes) turn off the power.

11. Place the gel in the fixative (saturated picric acid in 85% acetic acid) for 15 minutes.

12. Remove the gel and place on a glass plate. Place a sheet of wet chromatography paper on the gel and then a wad of paper towels on top of this. Add another glass plate and then a 1 kg weight.

13. Leave for 15 minutes and then dry the gel with a hair dryer.

14. Place the dried gel in a dish and cover with Coomassie Brilliant Blue R-250 (0.2 % in destaining solution) for 5 minutes.

15. Pour the stain back into its bottle and add destain (methanol : acetic acid : water [5 : 1 : 5]) to the dish. Swirl to elute the stain and then discard the wash.

16. Repeat the destaining until the background is clear.

17. Stand the gel upright on a wad of paper towels and allow to dry.

18. Observe the band patterns of the two samples.

19. The gel can now be scanned in a densitometer and/or photocopied.

The reference values for the protein bands is given in exercise 11-2.

EXERCISE 11-4

CELLULOSE ACETATE ELECTROPHORESIS OF LACTATE DEHYDROGENASE ISOENZYMES

Plasma proteins are separated as in exercise 11-2. The lactate dehydrogenase isoenzymes are then located by the following set of reactions:

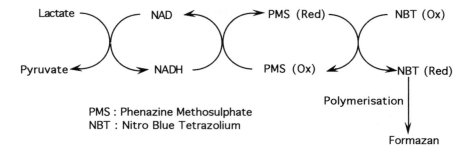

PMS : Phenazine Methosulphate
NBT : Nitro Blue Tetrazolium

The first reaction requires the catalytic activity of lactate dehydrogenase; whereas, the other reactions are spontaneous.

The location reagent is used by soaking a cellulose acetate strip in the reagent and placing this in contact with the electrophoretically run strip. These are sealed between glass slides and incubated at 37°C. Clearing the strip, after colour development, is not very satisfactory, as the violet formazan tends to elute. However, reflectance densitometry can be used to quantitate the bands and the strip can be photocopied as a record.

Lactate dehydrogenase isoenzymes, and their diagnostic use, were discussed in chapter 3 (pages 3-13, 3-14 and 3-17).

REAGENTS AND EQUIPMENT:

1.   High Resolution Buffer, pH 8.8.
            Dissolve the contents of 1 sachet (Gelman Cat No. 51104 or Helena Cat No. 5805) in distilled water and make to 1,000 ml.
            Store at 4°C.
2.   Location Reagent.
            0.1 M Tris-HCl buffer, pH 9.2:
                Dissolve 1.21 g of Tris base in 50 ml of H_2O
                Add 0.2 M HCl until pH 9.2 is reached.
                Dilute with 2% bovine serum albumin to 100 ml.
            Add: 50 mg of NAD,
                    5 mg of nitroblue tetrazolium and
                    1.0 g of lithium lactate.
            Mix well.
            Just before class, add 0.5 mg of phenazine methosulphate and mix well.
            Keep on ice in a dark bottle.
3.   Fresh Normal Serum or Plasma.
            Keep on ice.
4.   Abnormal Serum or Plasma.
            If you have ready access to sera or plasma from patients that have abnormal LD isoenzyme patterns (Eg. myocardial infarction, liver disease etc.); these samples can be run.
            Keep on ice.

Electrophoretic Tank and Magnets or Clips.
        Fill with buffer and level the liquid by tilting the tank.
Constant Current Power Supply.
Cellulose Acetate Strips.
        Sepraphore III 6 x 1" (Gelman Cat No. 51003).
Substrate Strips.
        Sepraphore III strips cut in half (ie. 3 x 1").
Sample Applicators.

Gelman Cat No. 51225.
2 μl Capillary Samplers and Capillaries.
500 μl Samplers and Tips.
Glass Sheet.
      Eg. TLC plate.
Forceps.
Filter Paper (18.5 cm diameter).
Dish Containing Buffer.
Large Chromatography Sheets.
Water Insoluble Textacolour Pen.
      Eg. Gelman Cat No. 51224.
3 x 1" Microscope Slides.
Clingfilm or other Food Wrap.
(Reflectance Densitometer).

PROCEDURE:

The cellulose acetate strips, and the substrate strips, should only be handled by the ends, using forceps. All work should be carried out on a clean sheet of chromatography paper.

1. Carry out steps 1 to 9, as in exercise 11-2, but use 4 applications of the sample to the same origin line.

2. Place a 3 x 1" microscope slide on a piece of Clingfilm about 8 x 15 cm.

3. Just before the electrophoresis run is complete, pipette 500 μl of the location reagent onto a clean glass sheet.

4. Saturate a substrate strip by floating it on the reagent.

5. Place the substrate strip on the microscope slide.

6. After the electrophoretic run, turn off the power. Remove the strip with forceps, **hold horizontally** and carefully place it on the substrate strip so that the application site is about 20 mm from one end, with the anodic side covering the rest of the slide. Do not realign the strips, once they make contact.

7. Place another microscope slide on top and fold over the protruding strip.

8. Wrap the Clingfilm around the slides and place in a dark 37°C incubator oven.

9. When the bands are well developed (15-30 min), the strip can be scanned in a reflectance densitometer and/or photocopied.

With normal plasma or serum, LD_1 and LD_2 stand out as clear bands with LD_3 as a fainter band. LD_4 and LD_5 are usually rather indistinct.

Reference Ranges (as a percentage of total LD):

LD_1 :   18 - 33
LD_2 :   28 - 40
LD_3 :   18 - 30
LD_4 :   6 - 16
LD_5 :   2 - 13

EXERCISE 11-5

MONOCLONAL GAMMOPATHY TYPING BY IMMUNOFIXATION

The presence of a paraprotein in the electrophoretic pattern of serum or plasma from a routine investigation of a patient would suggest myelomatosis, but does not confirm it. Typing this band, as monoclonal, helps in confirming the diagnosis.

In this exercise, the plasma proteins are separated by agarose gel electrophoresis, as in exercise 11-3. However, six aliquots of the sample are run. One of these is fixed and stained as in exercise 11-3. The remaining 5 samples are treated with specific antisera to precipitate immunoglobulin types A, G and M (α, γ and μ heavy chains) as well as the κ and λ light chains. These segments are then washed, stained and dried. The type of immunoglobulin can be obtained from the results.

REAGENTS AND EQUIPMENT:

1. Barbital Buffer, pH 8.6, 75 mmol/l + Calcium Lactate.
    Dissolve 4.14 g of barbital (diethylbarbituric acid) in about 1 l of boiling distilled water. Add 1.15 g of calcium lactate and 26.28 g of sodium barbital. When dissolved, cool and make to 2 l with distilled water.
    Store at 4°C.
2. Fixing Solution.
    Add 1.5 g of picric acid to 100 ml of distilled water and stir overnight at room temperature. Filter and add 20 ml of glacial acetic acid.
3. Stain.
    Dissolve 200 mg of Coomassie Brilliant Blue R-250 in 100 ml of reagent 4 and filter, using Whatman No. 1 paper.
4. Destaining Solution.
    Mix together 500 ml of methanol, 100 ml of glacial acetic acid and 500 ml of distilled water.
5. Myelomatosis Serum or Plasma.
    Obtained from the local cancer hospital. Freeze in vials at -20°C. The frozen vials can be kept for at least 5 years.
    Keep on ice.
6. Electrophoretic Marker.
    Dissolve 10 mg of bromphenol blue in 10 ml of 0.9% NaCl.
    Dissolve 1.0 g of bovine serum albumin in 1.0 ml of the above solution.
    Store at 4°C.
7. Agarose Gels.
    Suspend 850 mg of agarose in 100 ml of reagent 1. Heat, with stirring, to boiling so that the agarose dissolves. Let cool to about 50°C.
    Cut GelBond (Pharmacia Biotech Cat No. 80-1129-32) into 110 x 140 mm rectangles and level on a levelling table, with the hydrophilic surface upwards. Pipette 16 ml of the agarose solution onto each rectangle. Allow to set for 15 minutes. Store the gels on moist filter paper in sealed (food storage) boxes at 4°C.
    Use the gels the following day.
8. 10% Glycerol.
9. Antisera.
    Anti-human IgA, IgG, IgM, κ and λ.
    Keep on ice.
10. Antiserum Application Strips.
    Cellulose acetate strips cut to 5 x 50 mm.
11. 0.9% NaCl.

Electrophoresis Apparatus with Water Cooled Plate and Power Pack.
 Eg. LKB system (2117 Multiphor tank and 3371 E power pack).
Sample Application Strip.
 Pharmacia Cat No. 80-1129-47.
1 μl Capillary Samplers with Capillaries.
Wicks.
 Whatman No. 1 chromatography paper, 14 cm wide and long enough, when folded double, to reach from the buffer to the gel. Interleaf 3 of these doubled sheets.
Glass Plates.
 TLC plates are suitable.
1 l Plastic Bottle filled with Water. (As a 1 kg weight).
Filter Paper (18.5 cm diameter).
Paper Towels.
Dish for Staining and Destaining.
Hair Dryer.
Bench Covered with Clean Brown Paper.
Moist Box.
 Food container box (with lid) with a layer of damp filter or chromatography paper lining the box.
Forceps.

PROCEDURE:

1. Place the application strip on the gel so that the slits are about 35 mm from the cathode edge (ie. the edge of the strip is about 15 mm from the edge of the gel).

2. Apply 6 x 1 μl of the plasma or serum sample (1 - 6) along the slits with a 1 μl capillary sampler, leaving gaps for Bl slots. Also apply 1 μl of the marker albumin to the M slots, with a 1 μl capillary sampler:

Bl	1	Bl	M	Bl	2	3	4	5	6	Bl	M	Bl

Note that samples with very intense paraprotein bands may need diluting 1/10 with 0.9% NaCl for the immunofixation (sample lots 2 - 6).

3. Carry out steps 3 to 10, as for exercise 11-3.

4. Cut the GelBond between samples M and 2.

5. Process the sample 1 gel section as in steps 11 to 17 of exercise 11-3.

6. Saturate the antiserum application strips in antiserum by floating them on the surface. Lightly blot them with paper tissues to remove excess antiserum. Apply the strips to the β and γ regions of the other section of the gel, as follows:

2	IgA	3	IgG	4	IgM	5	κ	6	λ

Make sure that no air bubbles are trapped between the gel and the cellulose acetate.

7. Place the gel in the moist box and stand at room temperature for 30 minutes.

8. Remove the cellulose acetate strips and place the gel on a glass plate. Place a sheet of wet chromatography paper on the gel and then a wad of paper towels on top of this. Add another glass plate and then a 1 kg weight. Leave for 15 minutes.

9. Soak the gel in 0.9% NaCl overnight.

10. Soak the gel in distilled water for about 30 minutes.

11. Place a sheet of wet chromatography paper on the gel and then a wad of paper towels on top of this. Add another glass plate and then a 1 kg weight.

12. Leave for 15 minutes and then dry the gel with a hair dryer.

13. Place the dried gel in a dish and cover with Coomassie Brilliant Blue R-250 (0.2 % in destaining solution) for 5 minutes.

14. Pour the stain back into its bottle and add destain (methanol : acetic acid : water [5 : 1 : 5]) to the dish. Swirl to elute the stain and then discard the wash.

15. Repeat the destaining until the background is clear.

16. Stand the gel upright on a wad of paper towels and allow to dry.

Identify the immunoglobulin type.

EXERCISE 11-6

SODIUM DODECYL SULPHATE-POLYACRYLAMIDE GEL ELECTROPHORESIS (SDS-PAGE) OF URINE PROTEINS

SDS-PAGE is a convenient technique for classifying proteinurias (more than 150 mg lost in the urine per day). These can be classified into glomerular, tubular and overflow. However, there is often overlap between the groups.

Glomerular proteinuria (usually 5-50 g/day) is due to increased permeability of the glomerulus and the loss of plasma proteins in the urine. With minor damage to the glomeruli, proteins with molecular weights of less than 100,000 (ie. albumin and transferrin) are found in the urine. This is referred to as selective proteinuria. With more severe glomerular damage, non-selective proteinuria is seen. In this case, haptoglobins and IgG appear in the urine. In highly non-selective proteinuria, IgA, complement C3 and α_2 macroglobulins appear.

Tubular proteinurias are due to tubular damage or defects. This results in the reduced ability to re-absorb filtered low molecular weight proteins. These include α_1 microglobulin, retinol binding protein and β_2 microglobulin. Protein loss is usually less than 1 g/day.

Overflow, or pre-renal, proteinuria is due to high levels of low molecular weight proteins in the blood. These then saturate the tubular re-absorptive mechanisms. Bence-Jones proteins, haemoglobin and myoglobin are examples of such proteins.

For a complete separation of urinary proteins, based on their molecular weight, a SDS - polyacrylamide pore gradient is required where the polyacrylamide strength is varied between 4 and 22.5%.

With the 12% separating gel used in this exercise, the very low molecular weight proteins will not be separated and very high molecular weight proteins will not enter the gel. The gels will have been prepared as 0.75 mm thick slabs. These will take up to 10 samples. A stacking gel of 4% polyacrylamide is layered above the separating gel.

One normal and 2 abnormal urine samples (diluted to about 0.5 g of protein per litre) will be run, together with selected proteins of known molecular weight. The samples will not be treated with a reducing agent and so disulphide links will not be broken.

After the running of the gels, the protein bands will be fixed to the gel and stained with Coomassie Brilliant Blue R-250. The excess stain is then removed with a destaining solution and the gels can be photographed.

REAGENTS AND EQUIPMENT:

1. 10% SDS.
   Dissolve 10 g of sodium dodecyl sulphate in distilled water and make to 100 ml.
   For reagents 2, 5, 8 and 9.
2. Electrode Buffer.
   Dissolve 3.0 g of Tris base and 14.4 g of glycine in distilled water. Add 10 ml of reagent 1 and make to 1 l with distilled water.
3. Stacking Gel Buffer (0.5 M Tris-HCl, pH 6.8).
   Dissolve 3.0 g of Tris base in distilled water and adjust to pH 6.8 with concentrated HCl. Make to 50 ml with distilled water.
   For reagent 9.
4. Separating Gel Buffer (1.5 M Tris-HCl, pH 8.8).
   Dissolve 9.15 g of Tris base in distilled water and adjust to pH 8.8 with concentrated HCl. Make to 50 ml with distilled water.
   For reagent 8.
5. Sample Buffer.
   Mix 10 ml of reagent 1 and 12.5 ml of reagent 3. Add 10 g of sucrose and 50 mg of bromphenol blue and stir to dissolve. Make to 50 ml with distilled water.
6. Acrylamide Stock Solution (30%).
   Dissolve 7.2 g of acrylamide (handle with care, as it is a neurotoxin) and 300 mg of bisacrylamide in distilled water and make to 25 ml.
   For reagents 8 and 9.
7. 10% Ammonium Persulphate.
   Dissolve 100 mg of ammonium persulphate in 1 ml of distilled water.
   For reagents 8 and 9.
8. Separating Gel (T12, C4).
   Mix together the following:
       8.0 ml of reagent 6,
       5.0 ml of reagent 4,
       200 μl of reagent 1 and
       6.7 ml of distilled water.
   Degas for about 10 minutes and then add:
       100 μl of reagent 7 and
       10 μl of TEMED (N,N,N',N'-Tetramethylethylenediamine).
   Immediately use this solution to pour the gel, as described below.
9. Stacking Gel (T4, C4).
   Mix together the following:
       0.67 ml of reagent 6,
       1.25 ml of reagent 3,
       50 μl of reagent 1 and
       3.0 ml of distilled water.
   Degas for about 10 minutes and then add:
       25 μl of reagent 7 and
       2 μl of TEMED.
   Immediately use this solution to pour the gel, as described below.

10. Water Saturated n-Butanol.
   Shake about 100 ml of n-butanol with 100 ml of distilled water, in a separating funnel. Discard the aqueous layer (bottom).
   For the pouring of the gel.

11. Samples.
   A : Normal urine (neat).
   B : Dissolve 10 mg of human albumin in 20 ml of 1/20 urine.
   C : Add 400 µl of blood to 20 ml of 1/20 urine.

12. Standards.
   10 mg of the following dissolved in 20 ml of distilled water, each:
      Cytochrome C.
      Lysozyme.
      Ovalbumin.
      Bovine serum albumin.

13. Stain.
   Add 0.5 g of Coomassie Brilliant Blue R-250 to 250 ml of methanol. Swirl to dissolve and add 50 ml of glacial acetic acid. Dilute to 500 ml with distilled water. Filter, using Whatman No. 1 paper.

14. Destaining Solution.
   Mix 100 ml of methanol, 70 ml of glacial acetic acid and 830 ml of distilled water.

15. 10% Glacial Acetic Acid.

Vertical Electrophoresis Set Up.
   Hoefer Scientific Instruments SE 250 Mighty Small II vertical electrophoresis units, or similar.
   The procedure for pouring the gel in a Hoefer system is described below:
   The system is designed to run 2 gels at the same time. These gels can be poured in situ. However, this requires sealing the bottom of the gel space with agarose. It is more convenient to use an accessory caster (Eg. The Hoefer SE 245).
   Assemble on the caster: a glass plate, an alumina plate (with notched side upwards), one 0.75 mm spacer on each side and a second glass plate. Clamp these in place. Fill the gel space to about 3/4 full with reagent 8. Overlay the top of the gel with water saturated n-butanol and allow one hour for the gel to polymerise. Pour off the n-butanol, rinse with distilled water and drain. Fill the gel space, almost to the notch on the alumina plate, with reagent 9, and insert the 10 well comb. Allow the stacking gel to polymerise. Gently remove the comb and wash the wells with reagent 2. Lightly grease the silicon rubber gasket on the inner core of the electrophoresis apparatus. Transfer the alumina plate, gel and front glass plate to the electrophoresis apparatus and clamp in place with the red clamps.

Power Pack.
   One that can produce a constant current of 20 mA.

500 µl Pipettors and Tips.

Wassermann Tubes.

Heating Block at 95°C.

25 µl GC Syringe.

100 and 250 ml Measuring Cylinders.

Large Petri Dishes.
   Or similar covered container for staining and destaining the gels.

(Camera and Light Box or Filter Paper).

PROCEDURE:

1. Add 500 µl of the sample buffer (2% SDS, 20% sucrose and 0.1% bromphenol blue in Tris-HCl buffer, pH 6.8) to 7 Wassermann tubes.

2. Add 500 µl of each urine sample and 500 µl of the standard protein solutions (0.05% cytochrome C [M.Wt.: 12,300], 0.05% lysozyme [M.Wt.: 14,300], 0.05% ovalbumin [M.Wt.: 45,000] and 0.05% bovine serum albumin [M.Wt. 66,000] to the sample buffer.

3. Mix well and heat in a heating block at 95°C for 5 minutes.

11-38

4.   Place the electrophoresis unit in the lower buffer chamber. Rinse the sample wells with distilled or deionised water.

5.   Invert the unit over the sink to drain the wells.

6.   Wet the sample template and place it against the glass plate so that the 10 well segment corresponds to the sample wells.

7.   Fill the sample wells and upper and lower buffer chambers with the electrode buffer. (Tris-glycine and 1% SDS). (150 ml for the lower and 75 ml for each upper chamber.)

8.   Use a fine needle syringe (Eg. a GC syringe) to apply 15 µl of each sample to selected wells so that the sample underlayers the buffer. Wash out the syringe with distilled water, between samples.

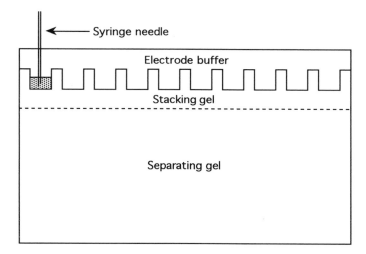

9.   Place the lid on the unit and connect the leads to the power supply.

10. Run the gels at 20 mA (constant current) per gel.

11. When the tracking dye (bromphenol blue) reaches the bottom of the gel, turn off the power.

12. Remove the unit from the lower buffer chamber and pour off the upper chamber buffer.

13. Remove the side clamps and lift off the gel sandwich.

14. Pry open the gel sandwich from the bottom. The gel usually sticks to the alumina plate.

15. Remove the spacers and peel off the gel.

16. Place the gel in the stain (0.1% Coomassie Brilliant Blue R-250 in 50% methanol, 10% glacial acetic acid).

17. Allow the gels to stain overnight and then destain in 10% methanol, 7% glacial acetic acid over the next 3-4 days (The destain solution being changed twice a day.)

The gels can then be photographed on a sheet of filter paper or when placed on a light box.

The gels can be stored in 10% glacial acetic acid.

EXERCISE 11-7

ISOELECTRIC FOCUSING OF BLOOD PROTEINS

Isoelectric focusing can be used in the diagnosis of haemoglobinopathies and plasma cell malignancies. The technique will be demonstrated in this exercise.

The gels contain 10 mg of Agarose IEF, 5 mg of non-crosslinked polyacrylamide, 18 μl of ampholyte solution (3-10) per ml of gel. The ampholytes (3-10) cover the pH range 3 to 10.

The gel is formed in an electrophoresis template and in contact with GelBond. This is a plastic sheet, one side of which has agarose incorporated into its structure. This side of the sheet is very hydrophilic and binds to the gel.

The gel thickness is only about 0.4 mm and so the gel does not have to be cooled during electrofocusing.

The conductivity of the gel is relatively high at the beginning of the IEF. As the pH gradient is established, the conductivity decreases. Optimum IEF is obtained with a constant power input; hence, a constant power source will be used.

After the run, the proteins will be fixed to the gel with sulphosalicylic acid and TCA. The gel is then washed to remove most of the ampholytes, which tend to stain with Coomassie Blue.

The gels are then squashed to remove most of the water and are dried with a hair dryer. After a rinse in destain, the gels are stained with Coomassie Brilliant Blue R-250. The background is then destained and the gels are dried.

Samples of normal serum (or plasma), haemolysed blood and the Pharmacia Broad pI kit calibration proteins will be run.

Albumin focuses in the pH range 4.7-5.0, α_1-antitrypsin around 4.2-4.7, transferrin at 5.2 and fibrinogen around 6.1-6.3. IgG extends in a diffuse zone between 5 and 9.5 (mainly in the range 7-9.5). The majority of other plasma proteins have isoelectric points in the range 4.5-5, but are hard to identify.

The pI of HbA is about 7.0. HbA_2 has a higher pI ($\cong 7.4$), as the δ chain has more basic (pK $\cong$ 10) and less acidic (pK $\cong$ 4)) amino acids than the β chain. The pI HbA_{1C} is slightly more than HbA due to the glycation of the terminal amino groups on the β chains (pK $\cong$ 8) and will probably not be resolved from HbA.

The 6.55 pI calibration standard is human carbonate dehydratase (carbonic anhydrase). This band should also be seen in the haemolysed blood sample.

GEL PREPARATION.

Reagents and Equipment:

1. Non-Crosslinked Polyacrylamide.
    A. Dissolve 2.0 g of acrylamide (handle with care, as it is a neurotoxin) in distilled water and make to 10 ml. Add 10 µl of TEMED (N,N,N',N'-Tetramethylethylenediamine) and gently mix, avoiding aeration.
    B   Dissolve 14 mg of ammonium persulphate in 10 ml of distilled water.
    Carefully mix A and B, avoiding aeration, and immediately transfer to dialysis tubing, tied at the bottom. Stand at room temperature for 2 hours. Then add a glass marble and tie the top of the tubing. Dialyse against 4 - 6 litres of distilled water, overnight, with gentle stirring. Transfer the polyacrylamide solution to a 50 ml measuring cylinder and measure its volume. Calculate the polyacrylamide concentration.
2. IEF Grade Agarose.
    Eg. Pharmacia Biotech Cat No. 17-0468-01.
3. Ampholyte Solution (pH Range 3 - 10).
    Eg. Pharmacia Biotech Cat No. 17-0456-01.

Template.
70 x 115 x 0.4 mm. Corning templates can be shortened to 70 mm by applying 2 strips of Dymo tape (2 x 0.2 mm) parallel to the electrodes, so that the wells are about 20 mm from the anode.
GelBond Film.
Pharmacia Biotech Cat No. 80-1129-32, cut to 75 x 124 mm.
Dialysis Tubing.
For reagent 1. M Wt cut off: 6,000 - 8,000.
Water bath at 60°C.
200 µl Pipettors and Tips.
5 and 10 ml Graduated Pipettes.
Heating Block.
100 ml Conical Flask.
60°C Water Bath.
60°C Oven.

Procedure:

For 10 plates.

1. Calculate the volume of reagent 1 that contains 200 mg of polyacrylamide. Add this, 400 mg of Agarose IEF and distilled water to make a final volume of 38 ml, to a 100 ml conical flask.
2. Boil gently until the agarose is dissolved.
3. Cool to 60°C in a water bath.
4. Prewarm 2.0 ml of reagent 3, the templates and a 5 ml graduated pipette to 60°C.
5. Add the warmed reagent 3 to the gel mixture.
6. Avoiding bubbles, pipette 4.0 ml of the warm gel mixture into the warm template.
7. Layer the GelBond, hydrophilic side down, onto the molten gel and allow to spread evenly across the template.
8. Cool on a level surface.
9. Store the gels in airtight bags, at 4°C, overnight. Use within 1 month.

OTHER REAGENTS AND EQUIPMENT:

1. Fixative.
    Dissolve 17.5 g of sulphosalicylic acid and 25 g of trichloroacetic acid in 165 ml of methanol. Make to 500 ml with distilled water.
2. Stain.
    Dissolve 1.5 g of Coomassie Brilliant Blue R-250 in 220 ml of methanol. Add 220 ml of distilled water and 55 ml of glacial acetic acid. Filter, using Whatman No. 1 paper.

3.   Destain.
        Mix 400 ml of methanol, 400 ml of distilled water and 100 ml of glacial acetic acid.
4.   Samples.
        Normal serum or plasma.
        1/5 haemolysed blood.
        Keep on ice.
5.   pI Calibration Standards.
        Pharmacia Biotech broad pI kit, pH 3.5 - 9.3, Cat No. 17-0471-01. Dissolve the contents
        of one vial in 25 µl of deionised water.
        Keep on ice.
6.   Ethanol.

IEF Tank.
    Bio-Rad Model 111 Mini IEF Cell or homemade Perspex box, with lid, containing graphite rod
    electrodes, 60 mm apart.
Power Pack.
    Providing constant power. Eg. Bio-Rad Model 3000 Xi.
1 ml GC Syringe.
Stain and Wash Containers.
    Eg. Food containers.
Glass Plates.
    Eg. TLC plates.
500 ml Plastic Bottle.
    Fill with water for the 500 g weight.
Filter Paper Rectangles 80 x 130 mm.
Paper Towelling.
Hair Dryer.
(Camera).

PROCEDURE:

1.   Carefully peel the GelBond film away from the template and place it on a dark
     background.

2.   Apply the samples (0.5 µl of serum, 1.0 µl of 1/5 haemolysed blood and 1.0 µl of pI
     calibration standard) to the wells, without touching the gel, using a 1 µl syringe. Wash
     out the syringe with distilled water, between samples.

3.   Moisten the graphite electrodes of the IEF tank with distilled or deionised water.

4.   Invert the gel and place it directly onto the electrodes, with the sample wells nearest the
     anode. Replace the tank lid.

5.   Set the integrator on the Bio-Rad 3000 Xi power pack for 210 volt hours, 1 watt constant
     power, for the home made IEF tank; or 175 volt hours, 1 watt constant power, for the
     Bio-Rad 111.

6.   After completion of the focusing ($\approx$ 40 min), identify the coloured bands in the blood
     and pI standard. The gel can be photographed at this stage.

7.   Place the gel in the fixative for 10 minutes.

8.   Place the gel in a container of distilled or deionised water for at least 20 minutes. (Make
     sure that the gel does not float).

9.   Place the gel on a glass plate. Moisten two of the chromatography paper rectangles with ethanol and place on the gel.

10. Place 2 or 3 sheets of folded paper towelling on the chromatography paper and then cover with another glass plate.

11. Place a $^1/_2$ kg weight on the glass plate and leave for about 5 minutes.

12. Dry the gel with a hair dryer.

13. Immerse the gel in destain for 5 minutes.

14. Stain with Coomassie Blue for 2 minutes.

15. Rinse in destain until the background is clear.

16. Allow the gel to air dry and photograph or photocopy it.

By comparison with the position of pI calibration standard bands, identify the main proteins in the samples from their pI values.

The proteins used in the Pharmacia Broad pI Calibration kit have the following pI values:

3.50	4.55	5.20	5.85	6.55	
6.85	7.35	8.18	8.45	8.65	9.30
(Red)	(Red)				

The kit also contains methyl red (pI : 3.75). However this is lost in the staining procedure.

GENERAL REFERENCES:

Epstein, E. and Karcher, R. E., (1994), Electrophoresis. Chapter 7 in Tietz Textbook of Clinical Chemistry (2nd Edition), Eds. Burtis, C. A. and Ashwood, E. R., p 191-205, W. B. Saunders Company.

Brewer, J. M., (1989), Electrophoresis. Chapter 8 in Clinical Chemistry - Theory, Analysis and Correlation (2nd Edition), Eds. Kaplan, L. A. and Pesce, A. J., p 140-152, The C. V. Mosby Company.

Varley, H., Gowenlock, A. H. and Bell, M., (1980), Separative Procedures - Electrophoresis. Chapter 4 in Practical Clinical Biochemistry, Volume 1 (5th Edition), p 76-102, William Heinemann Medical Books Ltd.

Pharmacia Fine Chemicals publish the following useful booklets:

    Polyacrylamide Gel Electrophoresis - Laboratory Techniques.
    Isoelectric Focusing - Principles and Methods.

Capillary Electrophoresis.

An Introduction to Capillary Electrophoresis, (1991). A booklet published by Bio-Rad Life Science Group.

Introduction to Capillary Electrophoresis, (1991). A booklet published by Beckman Instruments Inc.

Evenson, M. A. and Wiktorowicz, J. E., (1992), Automated Capillary Electrophoresis Applied to Therapeutic Drug Monitoring, Clin. Chem., **38**, 1847-1852.

Hempe, J. M. and Craver, R. D., (1994), Quantitation of Hemoglobin Variants by Capillary Isoelectric Focusing, Clin. Chem., **40**. 2288-2295.

High voltage electrophoresis of urine amino acids.

Efron, M. L., (1968), High Voltage Paper Electrophoresis, Chapter 5 in Chromatographic and Electrophoretic Techniques, Volume II - Zone Electrophoresis (2nd Edition), Ed. Smith, I., p 166-193, William Heinemann Medical Books Ltd.

Pitt, J., (1989), Personal communication and RCH method sheets.

Cellulose-acetate electrophoresis of plasma proteins.

Gelman Manual No. 70176-B.

Agarose gel electrophoresis of plasma proteins.

Jeppsson, J.-O., Laurell, C.-B. and Franzen, B., (1979), Clin. Chem., **25**, 629-638.

Cellulose acetate electrophoresis of lactate dehydrogenase isoenzymes.

Procedures, Techniques and Apparatus for Electrophoresis. A booklet published by Gelman Sciences Inc.

Monoclonal gammopathy typing by immunofixation.

Ritchie, R. F. and Smith, R., (1976), Immunofixation I - General Principles and Application to Agarose Gel Electrophoresis, Clin. Chem., **22**, 497-499.

Ritchie, R. F. and Smith, R., (1976), Immunofixation III - Applications to the Study of Monoclonal Proteins, Clin. Chem., **22**, 1982-1985.

Attaelmannan, M. and Levinson, S. S., (2000), Understanding and Identifying Monoclonal Gammopathies, Clin. Chem., **46**, 1230-1238.

Sodium dodecyl sulphate polyacrylamide gel electrophoresis of urine proteins.

Skerritt, J. H., (1985), Clin. Biochem. Monograph, Advanced Electrophoretic Techniques for Protein Investigation in Clinical Diagnosis, p 64-71. Published by the Australian Association of Clinical Biochemists Inc. (AACB).

Biegler, B. and Dennis, P. M., (1989), Clin. Biochem. Monograph, Advanced Electrophoretic Techniques in Clinical Diagnosis, p 38-43 and 72-74. Published by the AACB.

Beetham, R. and Cattell, W. R., (1993), Proteinuria: pathophysiology, significance and recommendations for measurement in clinical practice, Ann. Clin. Biochem., 30, 425-434.

Isoelectric focusing of blood proteins.

Biegler, B. and Sykes, S., (1989), Advanced Electrophoretic Techniques in Clinical Diagnosis, p 64-66. Published by the AACB.

CHAPTER 12

MOLECULAR DIAGNOSTICS

To quote from the newsletter of the Molecular Pathology Division of the AACC (Vol. 5, No. 3, June 1993) "Molecular diagnostics may be defined as the application of the techniques of molecular biology to laboratory medicine". It is a relatively new science. It is used in the diagnosis and screening of genetic disease, cancer screening, detection and monitoring and in microbiology. It also plays an important role in forensic science. This involves evidence (such as bloodstains) identification and paternity testing.

This chapter will mainly concentrate on genetic disease and oncologic applications with some forensic applications. It will also only examine DNA techniques and not those associated with RNA.

DNA in human cells is located in the nucleus and in mitochondria. The distribution in somatic cells (all cells except the germ cells) is illustrated in the table below:

	NUCLEAR	MITOCHONDRIAL
Number of bases:	2 x 3,000 megabases.	16.6 kilobases.
Number of chromosomes:	2 x 22 autosomes + 2 sex chromosomes (XX or XY).	Several circular chromosomes per mitochondrion. 500 - 3,000 mitochondria per cell.
Number of genes:	2 x 65,000 - 80,000.	37.
Percentage of DNA that codes for m, t and rRNA:	$\cong$ 3.	$\cong$ 93.
DNA mass per cell:	$\cong$ 6.4 pg.	Variable.
Inheritance:	Mendelian for autosomes. X-linked for X. Paternal for Y.	Maternal.

The germ cells (gametes - ovum and sperm cells) have half the number of nuclear chromosomes. Somatic cells are "Diploid" and germ cells are "Haploid".

POLYMORPHISM.

Polymorphisms are variations in the nucleotide sequence of the DNA, due to mutations, insertions or deletions and variations in the number of repeat sequences.

It is estimated that 1 in 100 to 200 base pairs in the human genome are polymorphic.

Most of these are in the non-coding regions of the DNA.

Changes in the exon regions of structural genes can lead to the loss of a protein or the production of an abnormal protein. However, mutations in the 3rd nucleotide of a codon may give rise to the normal amino acid being incorporated into the protein. Hence, no abnormality is produced.

Restriction Fragment Length Polymorphisms (RFLPs).

In 1980, it was noted that bacterial endonucleases (enzymes that split DNA at specific sites) split DNA into different sized fragments, when the DNA was obtained from different people.

The appearance, or disappearance, of specific fragments can be associated with genetic diseases.

Variable Number of Tandem Repeats (VNTRs - Minisatellites).

VNTRs are a type of RFLP. In 1985, a repeating sequence of 33 base pairs was found in an intron of the human myoglobin gene. This contained a 16 base pair core that was common in many other minisatellites and present in a variable number of repeats.

Other repeats (9 - 24 base pairs [bp]) have been found in all chromosomes, producing segments of DNA that are 100 - 20,000 bp in length. It is estimated that VNTRs are found at more than 10,000 loci.

These can be used as linked markers of genetic diseases in families, and in suspect identification (DNA fingerprinting).

Short Tandem Repeats (STRs - Microsatellites).

In 1989, di-, tri- and tetra-nucleotide repeats were observed. It is estimated that STRs are found at more than 10,000 loci.

They are used in forensic identification, early tumour detection (changes in the size of the STR or deletion of a STR) and are used as linked genetic disease markers.

DNA SOURCES.

Genetic Diseases.

Leucocytes from the blood and buccal epithelium cells from inside the mouth are the most convenient source of DNA for screening and diagnosis in children and adults.

Guthrie blood spots provide a source of DNA from diseased infants. These are blood samples that are collected on filter paper, about 2 weeks after birth. These samples are used for the screening of phenylketonuria, cystic fibrosis and hypothyroidism in newborn infants.

Chorionic villi and fibroblasts from amniocentesis are the sources of prenatal DNA. Chorionic villi are usually sampled between the 9th and 12th week after conception. This can be done via the cervix or through the abdomen. The procedure carries a 2 - 3% risk of abortion. Amniocentesis is usually carried out between weeks 15 and 20. A needle is inserted through the abdomen into the amniotic fluid and a sample withdrawn. The risk of abortion is 0.5 - 1%.

A single cell from a 4 or 8 cell blastocyst can be removed for DNA studies, prior to implantation. The remaining cells will grow and differentiate normally.

The first polar body, generated during the first meiotic division of the primary oocyte, can be examined, prior to fertilisation. If this carries 2 copies of the defective gene, the secondary oocyte will have 2 normal copies and can then be taken on for fertilisation.

Oncology.

As an aid to diagnosis, DNA from suspected tumour cells can be extracted and examined for abnormalities. Abnormal DNA found in a large collection of "normal" cells can be used to monitor the effectiveness of treatment.

"Normal" cells can also be examined for polymorphisms that predispose the subject to a specific type of cancer. For example, about 10% of breast cancers are inherited. Most of these are due to mutations of the BRCA1 and BRCA2 genes. Women who inherit a mutated gene have a high probability of developing breast cancer before the age of 40.

Forensic.

Blood, blood stains, semen, semen stains, tissues (fresh or preserved), bone marrow and hair follicles are all sources of DNA that can be used in forensic work.

DNA EXTRACTION.

DNA in the nucleus is associated with histone proteins. Hence, DNA has to be separated from protein during the extraction process. This can be done by using an extraction with phenol and chloroform. This leaves the denatured protein at the junction of the aqueous and organic layers, with the DNA in the aqueous layer.

Alternatively, a more gentle procedure can be used. The plasma and nuclear membranes are lysed and the proteins denatured with SDS. This mixture is treated with proteases and the protein fragments salted out.

With both procedures, further treatment with proteases is usually required. RNase is used during the extraction to remove the RNA. The DNA is then precipitated with ethanol. It can be redissolved in buffer.

Nuclear DNA can be separated from mitochondrial DNA by initially lysing the cells and collecting the nuclei by centrifugation. DNA is then extracted from the nuclei.

The shearing forces, generated during extraction, will break the DNA. This, together with the action of endogenous nucleases (before inactivation) yields DNA fragments in the order of 100 - 200 kilobases (kb).

DNA DIGESTION.

For a number of techniques, such as RFLP studies, the DNA must be digested by bacterial restriction endonucleases. These enzymes are very specific in their recognition sites. Most are palindromic; that is they recognise nucleotide sequences that are the same on both strands of the DNA. As the DNA strands run in the opposite direction to each other, the nucleotide sequences run in opposite directions. For instance, the endonuclease *Mst* II recognises:

$$5' - C C T N A G G - 3'$$
$$3' - G G A N T C C - 5'$$

N can be any of the 4 nucleotides. The enzyme splits between the C and T to yield:

$$5' - C C \qquad\qquad T N A G G - 3'$$
$$3' - G G A N T \qquad\qquad C C - 5'$$

Restriction endonucleases yield $10^5 - 10^7$ fragments from the human genome. These are of varying size.

DNA ELECTROPHORESIS.

DNA fragments can be separated by electrophoresis on the basis of their size, very much like SDS-protein complexes. DNA carries one negative charge per nucleotide (from the phosphate) and hence will move towards the anode. The speed of migration will depend on the size and form of the DNA.

For DNA with the same number of bases, closed circular (supercoiled) DNA will migrate the fastest, followed by a linear duplex and then the open circular form. Restriction fragments and amplified segments of a chromosome are linear duplexes, and so they are separated only on the number of their base pairs. The rate of migration in the gel is inversely proportional to the number of base pairs in the fragment. Hence, the smallest fragment migrates the fastest.

Horizontal, submerged (submarine) agarose gels are used for most separations. A thin layer of buffer covers the gel. The conductivity of the gel and buffer are about the same, so most current flows through the gel.

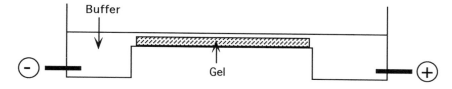

The pore size of agarose gels is larger than polyacrylamide gels; hence, polyacrylamide is used to separate small fragments:

Resolution of DNA fragments

Agarose

% Composition	Size (bp)
0.3	5,000 - 40,000
0.6	1,000 - 20,000
0.9	500 - 7,000
1.2	400 - 6,000
1.5	200 - 4,000
2-3	100 - 2,000

Polyacrylamide

% Composition	Size (bp)
3.5	100 - 1,000
5.0	80 - 500
8.0	60 - 400
12.0	40 - 200
20.0	10 - 100

Recently, new types of agarose have been developed that can separate small DNA segments, with good resolution. For example, FMC BioProducts NuSieve agarose, in 2 - 6% gels, can separate 10 - 1,000 base pairs. This avoids the use of the neurotoxin acrylamide.

Best resolution, in agarose gels, is obtained with a tris-borate buffer, 1 volt per cm of gel and a temperature of 4°C.

The DNA fragments can be located by incorporating ethidium bromide into the gel. Ethidium bromide interacts with adjacent purines and pyrimidines in the double helix. This greatly increases the ethidium bromide fluorescence. The complex absorbs UV light and emits orange light. Hence, DNA bands can be located by shining UV light through the gel.

Bromphenol blue can be included with the sample. This migrates ahead of the DNA fragments and can be used to follow the progress of the electrophoresis.

Pulsed-Field Gel Electrophoresis.

Very large DNA fragments have difficulty in entering even the lowest concentration agarose gels. Hence, a different type of technique has to be used to separate these large fragments.

DNA tends to break during normal extraction. Hence, for the production of very large fragments or for the isolation of whole chromosomes, cells are embedded in agarose for the extraction process. Large fragments are produced by endonucleases that recognise an 8 base pair sequence. These will produce only a few fragments per chromosome.

The technique used to separate these fragments is pulsed-field gel electrophoresis. In this technique, the electric field is alternatively applied at different angles (ie. alternating between the corners of the gel) for defined time periods (5 seconds to 2 minutes). Each time the field is changed, the DNA molecules have to reorientate themselves before they can move in the gel. This takes longer for larger molecules. Hence, the smallest fragments still migrate the fastest.

Ultra-low electroendosmosis agarose is used for the electrophoretic run. The buffer is circulated, during the run, to keep the temperature constant (usually 15°C). After the electrophoresis, the DNA can be isolated from the agarose, by using the enzyme agarase to digest the gel.

For example, yeast's 16 chromosomes can be separated on a 1% agarose gel over a 48 hour period, using 4.5 V/cm. These chromosomes range from 225 to 1,900 kilobases in size.

MEMBRANE TRANSFER.

As mentioned, restriction endonucleases yield somewhere in the order of one million fragments from the human genome. Hence, ethidium bromide will produce a broad fluorescent streak in the gel. To localise a specific fragment, a probe must be used that will bind with the nucleotide sequence. It is usually difficult to get these probes into the gel. Hence, the separated DNA fragments must be transferred to a membrane. Nitrocellulose membranes were used, but these have been mainly replaced by positively charged nylon, as nitrocellulose is very flammable and quite brittle. The transfer process is known as Southern blotting, after Ed Southern who devised the technique in 1975.

The gel is first treated with acid to break large DNA fragments. It is then treated with alkali to produce single stranded DNA. The DNA is then transferred to the membrane by a vacuum, electric potential or by capillary action. The latter technique was used by Ed Southern and is illustrated below:

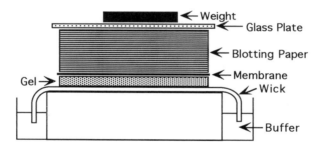

The DNA is then fixed to nitrocellulose membranes by heating or to nylon by UV treatment.

PROBES.

The DNA fragment of interest is located with a probe. Probes are labelled segments of DNA or RNA that hybridise with specific nucleotide sequences. Hybridisation is the base pairing between the probe and the DNA of interest.

Eg.

    Probe: G T G C A T G C C G A T G A C A
    Fragment: - T G T C A C G T A C G G C T A C T G T A T C C G -

Cloned DNA up to 25 kilobase (kb) can be used as a probe. This must be made single stranded before it can be used.

cDNA (Complementary DNA) is made from RNA, using retroviral reverse transcriptase. This enzyme uses RNA as the template and synthesises the complementary DNA. Messenger RNA is usually used as the template. The complementary DNA is the same as one of the strands of the membrane-fixed DNA, except that the introns are missing. The RNA is then removed by alkali hydrolysis and double stranded DNA produced. This is then cloned to produce large amounts of DNA. This has to be converted to single stranded DNA before it can be used as a probe.

Synthetic oligonucleotides (10 - 100 bases) are single stranded and can be used directly. Synthetic oligonucleotide, of about 20 nucleotides, are known as allele specific oligonucleotides (ASOs). These can be used to detect single base mutations, which can not be detected with the larger probes.

Probe Labels.

These are used to localise the hybridised probe.

Radioisotopes, such as ^{32}P, ^{33}P, ^{35}S or ^{3}H, can easily be incorporated into the probe during synthesis. These isotopes are β (electron) emitters, with ^{32}P producing the highest energy and ^{3}H the least energy. The high energy emitters are located by placing a photographic film on the membrane. ^{3}H labelled probes are used to localise *in situ* DNA sequences. In this case a photographic emulsion is poured directly onto the specimen.

A variety of non-isotopic labels are available for use with probes. For example, biotin or digoxigenin can be attached to the DNA. After hybridisation of the biotin labelled probe, an avidin-enzyme conjugate is added. Avidin binds very strongly to the biotin. A colourless substrate is then added and a coloured product generated. In the case of the digoxigenin labelled probe, a fluorophore-antidigoxigenin antibody complex can be added. The fluorescence is then observed.

MEMBRANE WASHING.

Prior to localisation, the excess probe must be removed. This is achieved by the use of wash buffers. The "stringency" of the wash buffer can be adjusted to remove partly bound probes. The table below compares low and high stringency buffers:

Low Stringency	High Stringency
High ionic strength (> 0.3 M NaCl)	Low ionic strength (< 0.3 M NaCl)
Low temperature (25-40°C)	High temperature (65-95°C)
Neutral pH	Alkaline pH
Base-pair mismatches are allowed.	No base-pair mismatches are allowed with ASO probes.

EXAMPLES:

Muscular Dystrophy.

Duchenne and Becker muscular dystrophies are serious diseases seen in boys and men. The diseases are associated with progressive muscle weakness leading to respiratory impairment, infections and death. Death is usually in the early 20s with the Duchenne form of the disease; whereas, patients with the Becker form can survive into their 60s.

Most cases of the disease are associated with deletions of parts of the dystrophin gene. Frame shifts are associated with the deletions seen in the Duchenne form of the disease. In this case, the number of deleted nucleotides can not be divided by 3. Hence, the remaining (downstream) DNA will code for incorrect amino acids. As a result, the Duchenne form of the disease is worse than the Becker form.

The dystrophin gene is on the X chromosome and, as males only have one X chromosome, the disease is seen in males. Females, with two X chromosomes, may be carriers with one normal X and one abnormal. The gene is very large (2.4 Mb). Due to its size, it is prone to spontaneous mutations. Hence, new cases often appear without a family history. The gene consists of 79 exons and these only occupy 14 kb in the mRNA.

In the diagnosis, using a Southern blotting technique, cDNA is prepared from the m-RNA and split into 6 fragments. These are labelled with ^{32}P and used as probes. The patient's DNA is digested with the endonuclease *Hind* III and the fragments electrophoresed in 1% agarose. The separated fragments are transferred to a nylon membrane, fixed and then hybridised with the probes. The results, using one these probes, are shown below:

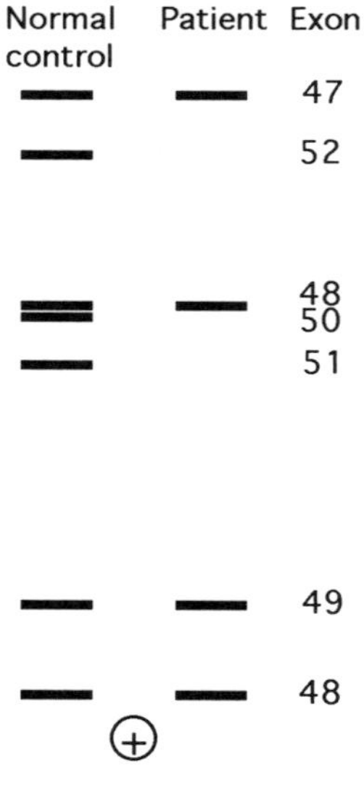

As can be seen, the patient lacks exons 50, 51 and 52.

Monoclonal Lymphocyte Detection in Lymphoma.

Using an antibody J region DNA probe, antigen receptor genes from polyclonal B lymphocytes do not give any specific bands on Southern blotting, due to the variety of different sized DNA fragments. When 1 - 5% or more lymphocytes are monoclonal, a specific band will be seen.

DNA "Fingerprinting".

DNA "Fingerprinting" can be used to identify individuals, due to the high degree of polymorphism between individuals.

Multilocus probes can be used to locate specific VNTRs in the genome. On electrophoresis, 20 - 50 bands are produced. These range in size between 1 and 20 kb. The larger fragments have a unique distribution in individuals.

Single locus probes can be used with highly polymorphic segments of the genome. With this type of probe there is a higher probability of 2 individuals having the same pattern than with multilocus probes. Therefore, a series of single locus probes are usually used.

Below is an example of DNA fingerprinting used to compare evidence (E) with that from the victim (V) and 2 suspects (S1 and S2):

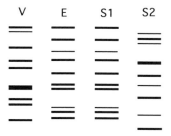

As can be seen, the evidence does not come from the victim or suspect 2. However, there is a high probability that it came from suspect 1.

Below is an example of the use of DNA fingerprinting in paternity testing:

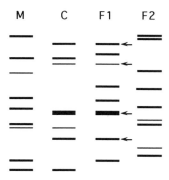

The child (C) inherits the fingerprint bands from both parents. If one deletes the corresponding maternal bands (M), the remainder (unless there is a new mutant band) should come from the father (F1 or F2). There are no corresponding bands with F2 and so he can be ruled out as the father. With F1, the arrowed bands match those of the child, so there is a strong probability that he is the father.

GENE TRACKING

Gene tracking is following mutant genes as they segregate in families. This can be direct or indirect. Direct gene tracking is the detection of a specific mutant allele at a genetic locus. With indirect gene tracking, polymorphic genetic markers (that are close to the mutant gene) are followed through a series of generations in a family.

Sickle cell anaemia is due to a single mutation. It is an autosomal recessive disease. This means that the patient has to inherit a defective gene from both parents. Hence, it can be followed in families by direct gene tracking.

The defective gene carries a point mutation (A → T) in the 6th codon of the β globin gene on chromosome 11. This means that thymine replaces adenine and results in valine replacing glutamic acid at position 6 of the β chains of haemoglobin. As a result, haemoglobin molecules aggregate, and distort the shape of erythrocytes, when the haemoglobin becomes deoxygenated. This leads to increased red cell fragility. Most patients die in childhood.

The mutation deletes a *Mst* II endonuclease site:

$$5' - C C \lfloor T N A G G - 3'$$
$$3' - G G A N T \rceil C C - 5'$$

Normal DNA contains this site:

```
  5      6
  Pro   Glu
        - C C T  G A G  G -
        - G G A  C T C  C -
```

However, it is lost in sickle cell DNA:

```
  5      6
  Pro   Val
        - C C T  G T G  G -
        - G G A  C A C  C -
```

As a result, one *Mst* II restriction site is lost in the HbS β DNA. This is illustrated below, where the restriction sites are indicated by the arrows:

Normal (HbA β chain):

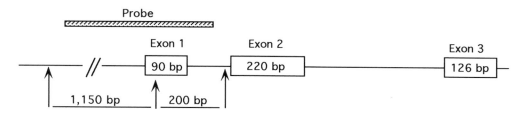

Sickle cell (HbS β chain):

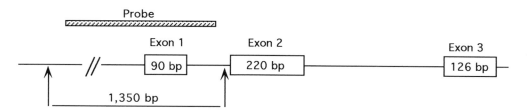

Hence, the probe will detect the 1,150 bp fragment from HbA and the 1,350 fragment from HbS. The small 200 bp fragment runs off the end of the gel in the system used.

DNA fragment electrophoresis patterns from a family are shown below. Squares represent males, circles represent females and the diamond is a foetus of unknown sex. White areas represent normal genotype and black the mutated genotype.

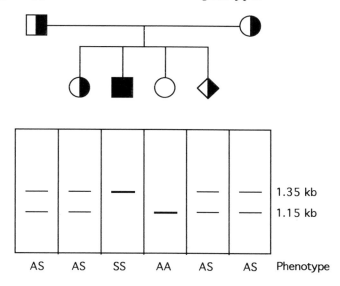

As can be seen, both parents, the first daughter and the foetus are carriers (1.15 and 1.35 kb fragments). The son has sickle cell disease (two 1.35 kb fragments) and the second daughter is normal (two 1.15 kb fragments).

Indirect gene tracking can be illustrated by following a genetic marker, associated with cystic fibrosis, through members of a family.

Cystic fibrosis is a defect in ion transport across epithelial membranes as a result of mutations in the Cystic Fibrosis Transmembrane Conductance Regulator (CFTR) gene. This results in decreased secretion of chloride and increased sodium absorption. This leads to dehydration of the secreted fluid in the lungs and pancreas (and other secretory tissues) and the formation of thick mucus plugs in the lungs and pancreas. The mucus in the lungs is a site for bacterial infection. Only about 70% of patients survive into adult life.

It is an autosomal recessive disease and so defective genes have to be inherited from both parents. The CFTR gene is located on chromosome 7 and is 250 kb in size with 27 exons. Over 500 mutations have been observed. Hence, if the mutation is not known, indirect gene tracking has to be used follow the defective gene in the family.

With most indirect gene tracking, polymorphic genetic markers on either side of the gene of interest are followed in family members. This reduces the likelihood of both markers being separated from the gene of interest during recombination (cross over) in meiosis.

A short tandem repeat (STR) exists in one of the introns of the CFTR gene. This intragenic STR is a highly polymorphic TA and CA repeat site between exons 17b and 18. Hence, loss of the marker during recombination is avoided. An example is shown below, where the number of repeats is examined:

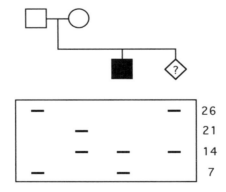

The son presents with cystic fibrosis and the parents want to know the status of the foetus. The defective genes are associated with the 7 repeats from the father and the 14 repeats from the mother. The foetus has inherited only one of these (the 14 repeats from the mother) and will, therefore, be a carrier.

DOT (SLOT) BLOTTING

For specific, known mutations, it is often not necessary to separate DNA fragments by electrophoresis before hybridisation with a probe. In dot (or slot) blotting, digested and denatured DNA can be directly applied to a nylon or nitrocellulose membrane and fixed.

These are then probed with oligonucleotide probes (ASOs) to detect single base substitutions. High stringency washes are required to remove the ASO if there is not a perfect match.

EXAMPLES:

Sickle Cell Anaemia.

2 probes are used for the diagnosis of sickle cell anaemia, one for the normal sequence (β^A) and the other for the sickle cell sequence (β^S):

β^A ASO   :   TG ACT CCT G$\underline{A}$G GAG AAG TC

β^S ASO   :   TG ACT CCT G$\underline{T}$G GAG AAG TC

The normal subject's DNA hybridises with the normal probe (β^A) only. The sickle cell patient's DNA hybridises with the mutant probe (β^S) only. The carrier's DNA hybridises with both probes indicating that he or she possesses both types of the gene.

Oncogene Activation.

The first step in the development of a number of tumours is the activation of an oncogene (Proto-oncogene → Oncogene) by point mutations.

These mutations can be investigated using a series of ASO probes on dot-blots and using high stringency washes:

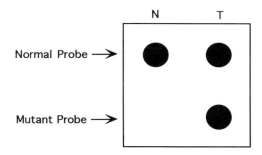

The normal tissue DNA (N) hybridises with the normal probe only. The tumour tissue DNA (T) hybridises with both probes, indicating that both normal and mutated cells are present.

FLUORESCENCE *IN SITU* HYBRIDISATION (FISH)

FISH is a technique for looking at the chromosomes of single cells for gene deletions, amplifications and translocations of parts of the chromosomes.

In the standard technique, cells are grown in culture and then colchicine is added. This inhibits microtubule formation and so cell division is halted at metaphase, as spindle filaments can not form to pull the divided chromosomes apart. These condensed chromosomes can then be probed with fluorescent probes.

The cells are placed on microscope slide and the chromosomes denatured (converted to single strands) with formamide. The probe is added and allowed to hybridise. Non-hybridised probe is washed off and the slide examined under UV light.

Probes for specific genes can be used, or a collection of probes can be added to cover the whole chromosome. This latter technique is known as chromosome painting.

Fiber-FISH is a technique where the probes are hybridised to interphase chromosomes. Plasma and nuclear membranes are lysed with detergent at one end of a glass slide. Tipping the slide allows the DNA to extend down the slide. The binding of the large fluorescent probe produces a "beads on a string" effect, when viewed under the microscope. This produces much higher resolution than with standard FISH. For example, deleted segments of the dystrophin gene can be detected in muscular dystrophy.

Standard FISH techniques can be used in the diagnosis of Prader-Willi and Angelman syndromes. The genes associated with these syndromes lie close together on chromosome 15.

Expression of the genes is inhibited by an imprinting centre some distance from the genes. This involves methylation of some of the bases. Different methylation takes place in males and females. Normal expression is illustrated below:

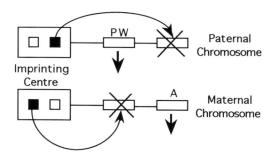

Most cases of Prader-Willi or Angelman syndromes involve the deletion of both genes on one chromosome. Hence, the deletion of the paternal genes gives rise to Prader-Willi syndrome (the Angelman gene is still expressed) and the deletion of the maternal genes gives rise to Angelman syndrome (The Prader-Willi gene is still expressed).

Some of the symptoms of these diseases are tabulated below:

Prader-Willi	Angelman
Hypotonia (Floppy baby). Moderate mental retardation. Gross obesity in childhood.	Severe mental retardation. Growth retardation. Jerky movements. Fair colouration.

The FISH results of the two chromosome 15s, from a patient with Prader-Willi or Angelman syndrome, is illustrated below, where the Prader-Willi / Angelman gene site is missing from one of the divided chromosome 15s:

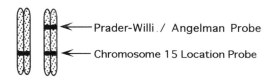

Oncogene amplification is seen in a number of tumours. This means that a large number of copies of an oncogene are produced. Oncogenes code for protein kinases, growth factor receptor proteins and for other proteins that facilitate the action of growth factors. Hence, the rate of cell division will be accelerated when oncogenes are amplified.

Chromosome painting FISH can be used to demonstrate oncogene amplification. Tumour cells are hybridised with green fluorescent probes and normal cells hybridised with red fluorescent probes. These chromosomes are scanned and a computer generates the green/red fluorescence ratio. The part of the chromosome exhibiting oncogene amplification will produce a rise in this ratio.

In about 90% of cases of Burkitt's lymphoma, the *myc* oncogene is translocated from chromosome 8 into the immunoglobulin heavy chain gene in chromosome 14. This leads to excessive expression of the *myc* gene as it is now located in a very active part of the human genome. This translocation can be detected by using chromosome 8 paint probes.

TARGET AMPLIFICATION

For the techniques that have been described, other than FISH, about 10 µg of DNA is required. This can be obtained from about 50 mg of wet tissue. Therefore, when only a small amount of material is present, it is essential to produce many copies of the DNA segment of interest.

One advantage of the amplification process is that there is no need to purify DNA prior to amplification.

Another advantage of target amplification is that enough DNA can be produced so that probes are not needed. A single band will be seen on an electrophoretic gel, when stained with ethidium bromide.

POLYMERASE CHAIN REACTION (PCR)

The first PCR publication appeared in 1985, but it was not until 1988 that the technique started to become widely used. This was due to the use of a heat stable DNA polymerase. Prior to that, fresh polymerase had to be added to each amplification cycle.

The first step in PCR is to separate the DNA chains. This is achieved by heat. The target is selected by 2 nucleotide primers, which have unique sequences that hybridise with DNA on each side of the target. The DNA polymerase then extends the nucleotide chain. After the target sequence has been replicated, the sample is heated to stop the reaction and separate the new chains.

This cycle is then repeated, doubling the number of chains with each cycle. Initially, DNA synthesis progresses past the ends of the target. As the primers define the 5' end of the new chain and the replicated template finishes at the 3' end defined by the other primer, target length products dominate after a few cycles. Targets are usually of the order of 100 - 1,000 base pairs. 25 - 30 cycles will yield about a million copies, after which the efficiency of the process declines markedly, mainly due to the inactivation of the polymerase. The process is illustrated below:

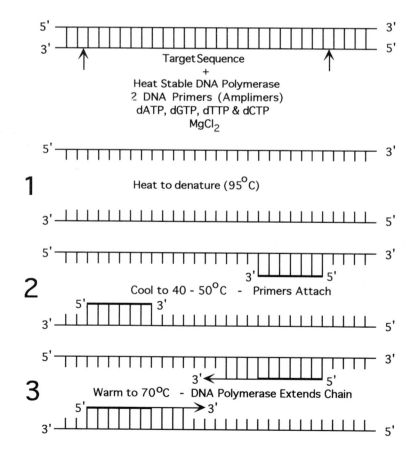

EXAMPLES:

Cystic Fibrosis.

Even though over 500 mutations have been observed in the CFTR gene, approximately 70% of Caucasians have the ΔF508 mutation in Exon 10. This is the deletion of phenylalanine from amino acid position 508. It involves the deletion of the nucleotides C from one codon (507) and TT from the next (508). This is illustrated below:

DNA sequence ⟶ ...ATC A⎾TC TT⏋T GGT GTT...
Amino acid sequence⟶ ...ILe ILe (Phe) Gly Val...
Amino acid number ⟶ ...506 507 (508) 509 510...

As both ATC and ATT code for isoleucine, the ΔF508 mutation sequence reads: ... ILe ILe Gly Val ...

In the examination of patients for this mutation, the mutation region is amplified by PCR and the products electrophoresed. The DNA is located with ethidium bromide.

The normal PCR product is 3 nucleotides longer than the ΔF508 product; hence, it does not migrate so far on the gel. The heteroduplexes, in the heterozygous carriers, are made up of one chain of each product. As they do not hybridise around the mutation site, there is a bulge in the DNA duplex. This migrates more slowly in the polyacrylamide or MDE (Mutation Detection Enhancement - a polyacrylamide like matrix) gel, as illustrated below:

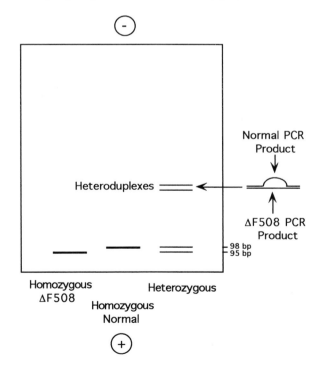

12-17

Huntington's Disease.

Huntington's disease (HD) is an autosomal dominant disease. This means that patients only have to inherit one copy of the defective gene to suffer from the disease.

It is a degenerative disorder of the brain which mainly attacks people in their 40s and 50s.

In 1993, the gene was isolated on chromosome 4. The genetic defect relates to the number of CAG repeats near the 5' end of the gene. Unaffected people have 8 - 35 repeats. HD patients have more than 36 repeats. The more repeats, the earlier the likely onset of HD.

The number of CAG repeats tends to increase in sperm, but not in other tissues. Hence, an earlier onset can often be seen in children from an affected father.

In the diagnosis of HD, the section of the gene containing the CAG repeats is amplified by PCR and the products subjected to electrophoresis. The products are located with ethidium bromide. In the example below, lane 1 is from a normal subject, lane 2 is from a HD patient and lanes 3 and 4 are from a father (3) and daughter (4) with HD.

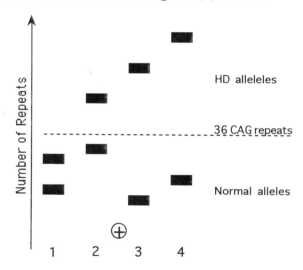

DNA "Fingerprinting".

The amount of evidence material available is often not enough for VNTR Southern blotting techniques. Hence, PCR provides a useful technique for amplifying what is available. Multiplex PCR is commonly used to amplify many loci. That is, many targets are amplified at the same time.

Primers with different coloured fluorescent labels can be used. For example, 5 loci in each of 3 colours can be amplified to give 15 loci. Size markers in a 4th colour can be included to provide a pattern that is almost unique for an individual.

PCR AND DOT BLOTTING

PCR can be used to amplify a specific target and this amplified product is then denatured and attached to a nylon membrane. This can then be probed with ASOs for specific nucleotide changes.

The thalassaemias are the most common single gene disorders in the human population. These are due to mutations in the α or β globin genes. PCR and dot blotting are commonly used for screening for thalassaemia mutations.

β thalassaemia is characterised by the absence (β^0), or reduced production (β$^+$), of the β chain of haemoglobin. The β-globin gene is on chromosome 11 and is illustrated below:

Exon 1	Exon 2	Exon 3
90 bp	222 bp	126 bp
(20 codons)	(74 codons)	(42 codons)

One of the most common mutations, in the Mediterranean population, is a C → T nonsense mutation in codon 39, in exon 2. The normal codon 39 is CAG, and is TAG in the mutant form. TAG is transcribed to the mRNA UAG, which is the code for stop. Therefore, protein synthesis stops at this point. This truncated protein has no physiological function. This results in a β^0 phenotype.

β thalassaemia, like α thalassaemia, is an autosomal recessive disease and patients have to inherit 2 defective genes to suffer from β thalassaemia major. These patients suffer from anaemia, which can be treated with blood transfusions. However, this leads to an iron overload and an early death.

Patients and suspected carriers can be screened for a variety of mutations, including the one above. Here, the nucleotide sequence around the suspected mutation is amplified. The product is denatured and applied as 2 spots to a nylon membrane. After fixation, they are probed with a normal and a mutant ASO. Labelled ASO are used for easy detection after stringent washing:

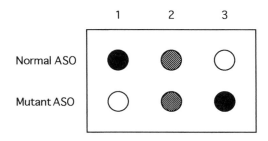

1 is a person with 2 normal DNA sequences at this site, 2 is a carrier and 3 has both β globin genes carrying the mutation.

PCR AND REVERSE DOT BLOTTING

With reverse dot blotting, the probe is attached to a solid support, rather than the PCR product being attached. The PCR products are labelled, usually by using labelled primers. The Roche Amplicor system for the detection of specific DNA sequences (mainly microbial) is an example of reverse dot blotting. In this system, the probe is attached to microtiter wells.

Biotin is attached to the 5' end of the primers and PCR is carried out in the usual way. This PCR product is then added to the wells. If there is base pair matching, the PCR product hybridises with the probe. After washing, an avidin-horseradish peroxidase conjugate is added. The avidin binds strongly to the biotin. After washing, the substrate (o-tolidine) and hydrogen peroxide are added. The production of tolidine blue is a positive result:

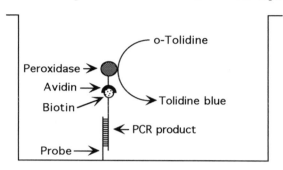

PCR-MEDIATED SITE-DIRECTED MUTAGENESIS

Where a genetic disease is associated with a single base change, normal PCR will produce equal sized products from both the normal and mutant gene and, hence, these can not be differentiated on electrophoresis. However, if the mutation introduces or deletes a restriction site, the restriction enzyme can be used to produce 2 fragments of the PCR product that contains the restriction site. Hence, faster moving bands will be seen on electrophoresis. If a restriction site is not introduced or deleted by the mutation, one can be introduced into the PCR product by using a primer with a base mismatch. An example of this technique is the screening for α_1-antitrypsin deficiency mutations.

α_1-antitrypsin is a 52,000 Dalton glycoprotein, produced in the liver and released into the blood. It is a protease inhibitor and its main role is to inhibit elastase released from leukocytes broken down in the lung. Low levels, or an absence, of α_1-antitrypsin are associated with destruction of elastic tissue in the lungs and a resultant emphysema.

α_1-Antitrypsin deficiency is an autosomal recessive disease; hence the patient has to inherit a defective gene from both parents. The gene has 5 exons and is located on chromosome 14. There are at least 30 known mutations of the gene. The 2 most common are the Z and S variants.

With 2 Z mutations (ZZ), the altered protein is unable to leave the liver. Liver damage results, which leads to cirrhosis and possible death. SS and M (normal gene) S individuals are unaffected. However, SZ individuals may be affected. The Z mutation is G → A in codon 342 (exon 5) and lysine replaces glutamic acid. The S mutation is A → T in codon 264 (exon 3) and valine replaces glutamic acid.

These mutations do not create or remove a restriction site and so normal and mutant PCR products can not be differentiated. However, a restriction site can be introduced during PCR amplification. This is illustrated, below, for Z mutation screening:

```
Normal sequence          : ...T C G A C G A G A...
Z allele                 : ...T C G A C A A G A...
PCR primer               : ...T C G T C  3' end

Normal product           : ...T C G T C G A G A...
Z allele product         : ...T C G T C A A G A...
```

This deletes a *Taq* I restriction site (...T↓C G A...) from both products, but introduces a new *Taq* I site into the normal PCR product, but not into the Z allele product.

Both PCR products are 179 base pairs in length. However, after treatment with *Taq* I, the normal product yields 157 and 22 bp fragments. Hence, the 157 and 179 bp products can be distinguished after electrophoresis and ethidium bromide treatment:

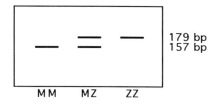

The 22 bp fragment runs off the bottom of the gel.

ALLELE SPECIFIC AMPLIFICATION

AMPLIFICATION REFRACTORY MUTATION SYSTEM
(ARMS)

This type of PCR is designed to detect specific single nucleotide changes. 2 sets of primers are used. Each set will amplify either the normal or the mutant target.

Primers that have a 3' end nucleotide mismatch will not amplify the target. A common primer is used at one end and normal, or mutant, primer is used at the other end. Only where the 3' end is base paired will there be amplification:

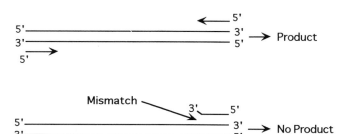

An example of this technique is the Zeneca CF(4)m-PCR kit, which is used to detect 4 common mutations in cystic fibrosis. These mutations are:

621 +1(G→T).
   Guanine is replaced by thymine in the 1st position in the intron after codon 621.

G551D.
   Glycine is replaced by aspartic acid in codon 551,

G542X.
   Glycine is replaced by a stop codon in position 542.

ΔF508.
   Phenylalanine (codon 508) is deleted.

Two sets of tubes are used. Tube A will amplify the normal 508 + 621 and mutant 551 + 542 sites. Hence, normal DNA will yield 508 (160 bp) and 621 (367 bp) products.

Tube B will amplify mutant 508 + 621 and normal 551 + 542 sites. Hence, normal DNA will yield 542 (255 bp) and 551 (283 bp) products:

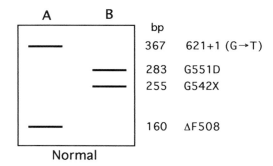

Normal

Hence, for one site, two normal genes will give a band in the normal tube only. Heterozygotes will give bands in both tubes and two mutant genes will give a band in the mutant tube only. For example, The ΔF508 mutation (160 bp product) will give the bands shown below:

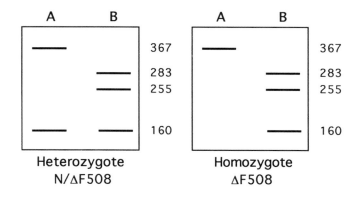

12-22

HETERODUPLEX GENERATION

Heteroduplexes are the result of hybridisation of 2 DNA strands where there is not an exact base pair match. These duplexes have bulges in them and this restricts their movement on electrophoresis. An example was seen in the PCR diagnosis of the ΔF508 mutation in cystic fibrosis, where 3 nucleotides are missing from one strand of the duplex. The heteroduplex is made up of one strand of normal DNA and the other strand with the ΔF508 mutation.

Heteroduplexes can be generated as a means of identifying specific mutations. In one of these techniques, amplified patient DNA is mixed with synthetic DNA to generate heteroduplexes that migrate differently on electrophoresis.

Some phenylketonuria (PKU) mutations can be identified by this technique. Most cases of PKU are associated with mutations to the phenylalanine hydroxylase gene. Severe mental retardation develops, if the disease is not treated by reducing the phenylalanine intake.

The phenylalanine hydroxylase gene consists of 13 exons. 40% of PKU cases are associated with mutations to exon 12.

In this technique, a synthetic DNA of modified exon 12 plus parts of the flanking introns, is used:

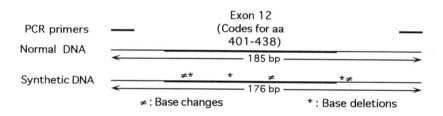

Patient and synthetic DNA are treated separately for 35 PCR cycles. Then they are mixed and given a further 4 cycles. The products are then electrophoresed on polyacrylamide gels. The heteroduplexes produce bands that migrate a distance equivalent to homoduplexes (base pair matched) of 354 to 446 base pairs:

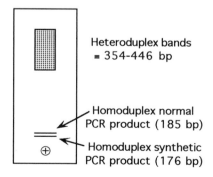

5 common mutations can be recognised by their specific heteroduplex band pattern. For example, the R408W (arginine replaced by tryptophan at position 408) mutation gives the pattern below:

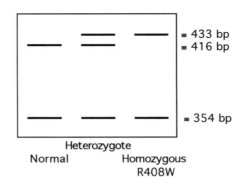

Normal | Heterozygote / Homozygous R408W

OTHER PCR TECHNIQUES

SPECIFIC MUTATIONS.

Oligonucleotide Ligation Assay.

In this assay, normal PCR is carried out for the region around the mutation site. 2 probes that join at the mutation site are added, together with DNA ligase. One probe is tagged with biotin (B) and the other with digoxigenin (D). Where there is a mismatch at the ligation site, the probes will not be joined:

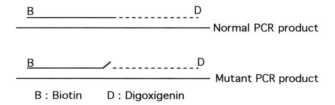

B : Biotin   D : Digoxigenin

This mixture is then added to avidin coated microtiter wells. After denaturation and washing, labelled anti-digoxigenin antibody is added. The unbound antibody is removed by washing and the presence of the label determined:

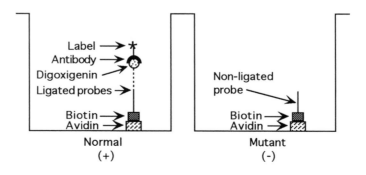

MUTATION SCREENING.

Mutations that cause the activation of oncogenes, or the deactivation of tumour suppressor genes, can occur at many sites. When screening for a mutation at an unknown site, a variety of tests can be used. Some of these are briefly discussed below:

1.   Heteroduplex Mobility.

When the PCR reaction generates normal and mutant products, heteroduplexes will form. These are slow moving on electrophoresis. This technique is only suitable for products of less than 200 base pairs.

2.   Single Strand Conformational Polymorphism.

PCR product are denatured and run into a non-denaturing gel. In the gel, the DNA chains fold and form base pairs within the chain. The configuration of the folded chain depends on the nucleotide sequence. Hence, a different sequence will give a different folded structure. These structures migrate differently in the gel, on electrophoresis. Again, this technique is only suitable for products of less than 200 base pairs.

3.   Chemical or Enzymatic Cleavage of Mismatches.

The test PCR product is mixed with normal DNA PCR product, heated and cooled to form heteroduplexes. This is then treated with chemicals or enzymes that will split the heteroduplexes at the mismatch site. Heteroduplexes are not formed with a normal test. After treatment, the sample is examined by electrophoresis. Only one band is seen with the normal (N) sample. Mutant samples (M) will show low molecular weight fragments as well as the full length PCR products from homoduplexes:

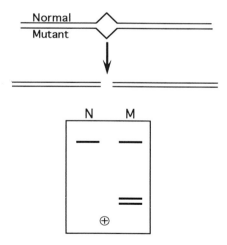

This technique can be used with kilobase-length samples.

4.   Denaturing Gradient Gel Electrophoresis.

Denaturing gels are made by increasing the urea concentration in the gel, as one moves towards the anode. PCR products migrate to the point where the DNA strands separate and then the migration slows markedly. Heteroduplexes denature earlier and, hence, do not move so far into the gel.

5.   Sequencing.

In this method, the concentration of one primer is about 100 times that of the other. After about 15 PCR cycles, the lowest concentration primer is used up and then linear amplification takes place. This yields a single strand product.

This product is used in the Sanger sequencing method. Four tubes are set up with the deoxyribonucleotide triphosphates (dNTPs), a primer and DNA polymerase. Each tube also contains a separate dideoxyribonucleotide triphosphate (ddCTP etc):

C	T	A	G
ddCTP	ddTTP	ddATP	ddGTP
dATP	dATP	dATP	dATP
dCTP	dCTP	dCTP	dCTP
dGTP	dGTP	dGTP	dGTP
dTTP	dTTP	dTTP	dTTP
Primer	Primer	Primer	Primer
DNA polymerase	DNA polymerase	DNA polymerase	DNA polymerase

When the dideoxyribonucleotide is incorporated into the chain, synthesis stops. This is due to the lack of an -OH group in position 3' of the ribose; and hence there is no interaction with the incoming nucleotide triphosphate. Therefore, each tube generates a variety of different length products, as there is competition between the dideoxyribonucleotide and the normal deoxyribonucleotide for incorporation in the growing chain. These different sized products can be separated by electrophoresis on a vertical polyacrylamide denaturing gel:

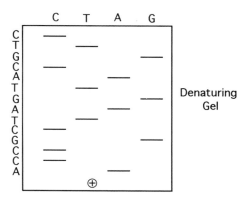

Hence, the nucleotide sequence is read from the bottom of the gel: ACCG etc. This is the complementary sequence to the original PCR product.

The products are located by including a marker in the reaction mixture. This can be a radioisotope, usually ^{33}P or ^{35}S, incorporated into the dNTPs. Fluorescent labelled primers can be used. Where different coloured fluorescent labels are used in each tube, the products can be run in a single well for electrophoresis. If different coloured fluorescent labels are attached to the ddNTPs, only a single synthesis tube is needed.

LIGASE CHAIN REACTION (LCR)

LCR is an alternative target amplification technique to PCR. Two pairs of amplimers cover the target region. In the standard method, the members of the pair butt onto each other. The double stranded DNA is denatured by heating. The mixture is then cooled to about 55°C to allow the amplimers to hybridise with the target. Heat stable DNA ligase then joins the members of each pair. The cycle is then repeated a number of times. The technique is illustrated below:

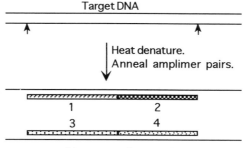

The technique has been used for the diagnosis of sickle cell anaemia. 22 base amplimers are used; one member of the pair has a 21 base capture "tail" and its partner is tagged with biotin:

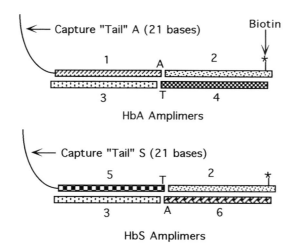

Amplification is carried out with 2 cycles of 94°C for 1.5 minutes and 55°C for 6 minutes. This is then followed by 25 cycles of 91°C for 0.5 minutes and 55°C for 6 minutes. The amplified products are then captured by the specific complementary oligonucleotides attached to microtiter wells. Alkaline phosphatase - streptavidin conjugates are then added. After washing, the substrate is added and the plate incubated:

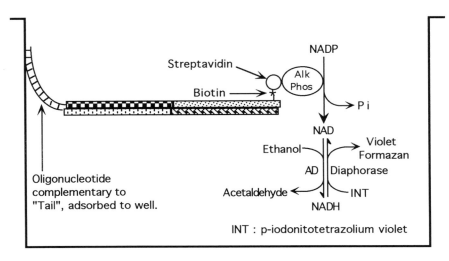

The rate of absorbance increase is measured in the A and S wells and the A/S ratio determined. The following results illustrate the genotype:

Genotype	A/S
AA	38
AS	$\cong 1$
SS	0.03

A commercial LCR system (the LC_X) was introduced by Abbott Diagnostics Division in 1994. In their system, the amplimer pairs do not butt onto each other. Hence, the gaps have to be filled by DNA polymerase before ligation. This system avoids the ligation of amplimer pairs that are not hybridised to the target or amplified target.

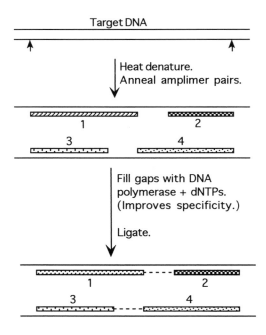

Detection of the product is similar to the system used in the Abbott IM$_X$. Antigens A and B are attached to members of a pair of amplimers. After amplification, the product is captured by anti-A antibody coated microparticles. These are then held by fine glass fibres in the fluorescence detector. Alkaline phosphatase labelled anti-B antibody is added. After washing, the substrate (methyl umbelliferyl phosphate) is added:

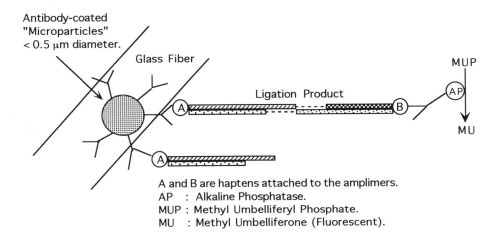

The enzyme removes the phosphate to produce methyl umbelliferone, which is fluorescent. The rate of increase in fluorescence is measured.

CONTAMINATION CONTROL

In theory, the amplification procedures only require one copy of the target DNA. Hence, contamination of new samples with product is a problem. It is advisable to do all sample preparation in an area away from the amplification and detection. Aerosols, that can be generated in a number of laboratory procedures, must be avoided. Positive displacement pipettors or aerosol-prevention tips should be used for pipetting.

The presence of contamination in the laboratory can be checked by running reagent only (no target DNA) blanks. These should yield no amplification product.

Work areas can be decontaminated by using ceiling mounted 254 nm mercury lamps that are turned on when the laboratory is not in use. Thymines are cross linked by this process and the product will not act as a template for amplification.

The Roche Amplicor system uses dUTP instead of dTTP in the PCR. Prior to amplification, the samples are treated with uracil-N-glycosylase. This enzyme removes uracils from any contaminating PCR product. As a result, the contaminant can not be amplified.

DNA MICROARRAYS AND MICROCHIPS

The recent advances in microtechnology are having a profound effect on molecular diagnostics. For example, rather than spending weeks using different probes to identify a cystic fibrosis mutation, now probes for all the known cystic fibrosis mutations can be located on a single chip and a result can be obtained in a few hours.

A DNA microarray is an ordered arrangement of DNA capture probes fixed to a solid surface such as a glass slide or a gold contact on a silicon chip. The capture probes can vary in size from small oligonucleotides for sequence identification to cDNA for gene expression investigation. An array can be made up of thousands of elements, each containing thousands of probe molecules. For example, a commercial microarray slide contains 2,400 elements, each a probe for a different human gene.

Oligonucleotides can be synthesised on the matrix using photolithography techniques or pre-synthesised and then attached. An example of attachment to a glass slide is to pre-coat the slide with silanised polylysine. This imparts a positive charge to the surface, which will attract the negatively charged DNA. About 1 nl samples of each probe are added to give spots about 100 μm in diameter. The DNA is then fixed to the slide coating with UV light. Attached cDNA probes are partly converted to single strands by heat or alkali.

Attachment to a spot on a chip can be carried out by flooding the chip with a probe solution and making the spot on the chip electrically positive. The DNA migrates to the spot where it is chemically attached. After washing, the process is repeated with the next spot.

When a sample is added to an array it will bind to a probe if it contains a DNA sequence that will hybridise with the probe. After washing, the hybridised probe is identified. With most clinical assays, a labelled signal probe is added to bind to part of the target DNA not hybridised with the capture probe. The label can either generate a signal itself (eg. a fluorophore) or be an enzyme that can generate amplified signal. The location of the signal is then carried out. In the case of a fluorescent label, or product, the array elements are scanned with a laser beam and results processed by a computer.

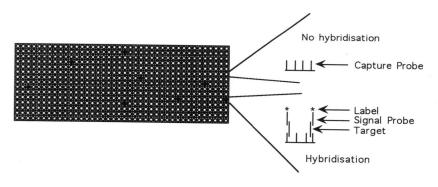

In some types of assay, labelled PCR products of DNA segments of interest are added to the array. In this case, different coloured fluorescent labels can be used in the PCR to further enhance the identification of a mutation etc. With gene expression, reverse transcriptase and fluorescent nucleotides can be used in the identification of mRNAs in a cell extract.

One microchip method uses the capture probe attached by a molecular "wire", in a chemical insulator, to a gold electrode. After the target binds, a signal probe is added and this binds to a free part of the target. The signal probe is labelled with ferrocene. This increases the electrical conductivity through the target and probe. When a small voltage is applied, a current flows through the gold electrode but not through the electrodes connected to unbound probes.

Autoradiography can be carried out on the microarray if the target or signal probe is labelled with a radioactive isotope. [33]P is commonly used as this has a low energy radioactive emission. Hence, the activity is confined to the array element.

Another interesting technique is the use of the equivalent of miniature interference filters. The capture probes are attached to the surface of the filters. Light of a specific colour is reflected from the surface of the filters. After the target has bound, an enzyme labelled signal probe is added. This binds to part of the target. After washing, the substrate is added and the enzyme converts this to a solid, which deposits as a thin film on the surface of the filter. This alters the optical characteristics of the filter, which now reflects light of a different colour.

EXERCISES

RFLP analysis, involving DNA extraction, digestion, electrophoretic separation of the fragments, membrane transfer and probing takes several days to complete. Hence, the methods covered here involve PCR techniques that can be completed in a day. One exercise will look at a VNTR sequence that can be used in forensic analysis. The other exercise uses the Johnson and Johnson kit to look for cystic fibrosis mutations.

EXERCISE 12-1

AMPLIFICATION OF THE D1S80 LOCUS USING P.C.R.

The D1S80 locus is a site on chromosome 1 that shows marked polymorphism. This polymorphism is due to a variable number of 16 base pair repeats (VNTRs). Polymorphism at this site can be used in forensic work for the identification of evidence or for paternity testing.

DNA will be obtained from a haemolysed blood sample and metal ions removed by an ion exchange resin (Chelex). The D1S80 site is then amplified by 25 PCR cycles in a Tris-HCl buffer containing KCl, $MgCl_2$ and gelatine. Primers of 30 and 31 nucleotides are used to select the D1S80 site. Precautions will be taken to prevent contamination and a negative control will be run to monitor possible contamination.

The PCR product will be subjected to electrophoresis in a 2.5% modified agarose gel. Ethidium bromide is included in the gel in order to locate the PCR products after electrophoresis. The dye intercalates between the bases of the DNA. This binding does retard the migration of the PCR products by about 15% when a 0.5 mg/l solution is used, as is the case in this exercise.

After electrophoresis, the gel is exposed to UV light and the DNA - ethidium bromide complex fluoresces. If a mercury lamp is used, the 254 nm emission line is absorbed by the DNA and the energy passed to the dye; whereas, the 366 nm line is absorbed by the dye itself. In both cases, orange light is emitted, with a maximum emission at 590 nm. A minimum of about 10 ng of DNA is required for detection of the ethidium bromide complex in any band.

The gel can be photographed while being exposed to UV light. Product migration is obtained by measuring the distance of the fluorescent band from the origin.

The original method used a 6% (T6, C3) polyacrylamide gel, and silver staining of the run gel. This method does produce better resolution, and is a little more sensitive, than using agarose with ethidium bromide staining. However, agarose gel electrophoresis is simpler and avoids the use of the neurotoxin acrylamide. One of the new agaroses, NuSieve 3:1, will be used in this exercise. The 2.5% gel, in Tris-borate-EDTA buffer, gives reasonably good resolution of the PCR products.

The PCR products vary in size between 375 and 850 base pairs. One or two bands should be seen in each sample; a band for each chromosome one. Xylene cyanol and bromphenol blue markers are included with the PCR products, together with a strong sucrose solution, for electrophoresis. Bromphenol blue migrates at about the same rate as 70 base pairs of DNA and the xylene cyanol migrates at a rate equivalent to about 700 bp of DNA. The sucrose increases the density of the sample so that it easily sinks to the bottom of the well in the agarose gel. DNA size markers are also included in separate lanes.

REAGENTS AND EQUIPMENT:

1.   Sterile Glass-Distilled Water.
2.   0.5% Chlorhexidine in 70% Ethanol.
3.   5% Chelex Suspension.
         Bio-Rad Cat No. 143-2832.
             5 g suspended in 100 ml of reagent 1.
4.   2 x PCR Master Mix.
     (a)   10 x Buffer (100 mM Tris-HCl [pH 8.4], 500 mM KCl, 15 mM $MgCl_2$, 0.1% gelatine).
           Dissolve 1.211 g of Tris (Tris(hydroxymethyl)amino-methane) in 50 ml of sterile glass-distilled water. Add 34.4 ml of 100 mM HCl and mix. Add 3.73 g of KCl, 143 mg of $MgCl_2$ and 100 mg of gelatine, and mix to dissolve the solids. Make to 100 ml with sterile glass-distilled water.
           A commercial PCR 10 x buffer can be used (Eg. Sigma Cat No. P 2192 or the buffer supplied with DNA polymerase).
     (b)   20 mM dNTP mixture.
           Pharmacia Biotech Cat No. 27-2094-01.
     (c)   Primers:
           MCT 118-1   5'GGGAAACTGGCCTCCAAACACTGCCCGCCG3'.
           MCT 118-2   5'GGGTCTTGTTGGAGATGCACGTGCCCCTTGC3'.
     Mix 1.0 ml of (a) and 400 µl of (b), in a sterile tube. Add 5.0 nmole of each primer and make to 5.0 ml with sterile glass-distilled water. Divide into 5 x 1.0 ml aliquots, in sterile tubes, and freeze at -20°C.
5.   Taq DNA Polymerase.
         Eg.   Pharmacia Biotech Cat No. 27-0799-01.
               (250 units, at 5,000 units/ml.)
6.   Light Mineral Oil (Liquid Paraffin).
         Eg.   Sigma Cat No. M 3516.
7.   Tris-Borate, EDTA Buffer, pH 8.3.
         Prepare a 5 x TBE (445 mM Tris-borate, 12.5 mM EDTA) buffer as follows:
             Dissolve 54.5 g of Tris, 27.8 g of boric acid and 4.66 g of EDTA in about 900 ml of distilled water. Make to 1 l with distilled water.
8.   Ethidium Bromide Solution (1.0 g/l).
         Dissolve 10 mg in 10 ml of distilled water.
             Ethidium bromide is a known carcinogen, so handle with care.
9.   Running buffer (0.5 X TBE).
         Add 1.0 ml of reagent 8 to 200 ml of reagent 7, add 1,800 ml of distilled water and mix well.
10. NuSieve 3:1 Agarose.
         FMC BioProducts Cat No. 50091.
11. DNA Size Standard.
         Eg.   Bio-Rad AmpliSize (50 - 2,000 bp), Cat No. 170-8200.

12.   Sucrose-Dye solution.
          Dissolve 60 g of sucrose in about 40 ml of distilled water and make to 100 ml. Add 250
          mg of xylene cyanol and 250 mg of bromphenol blue and mix until fully dissolved.

Plastic Disposable Gloves.
Cotton Wool.
Blood Lancets (or Glucolet etc).
Pasteur Pipettes.
5 μl Capillary Pipettors with Capillaries.
20, 30, 50, 75, 200 and 1,000 μl Pipettors with Tips.
Sterile 20 μl Aerosol Barrier Pipette Tips.
          Eg.   Xcluda, from Bio-Rad.
Sterile 1.5 ml Microcentrifuge Tubes.
          Eg.   Eppendorf Cat No. 0030 102.002.
Microcentrifuge.
          Eg.   Eppendorf 5410.
Thermal Cycler.
          To hold 500 μl tubes.
                    Eg.   Perkin-Elmer's Thermal Cycler 480 or Eppendorf's Mastercycler 5330 plus.
Sterile Amplification Tubes.
          Eg.   Perkin-Elmer's GeneAmp Thin-Walled Reaction Tubes (supplied sterilised) or
                    Eppendorf's Safe-Lock Microcentrifuge Tubes.
500 ml Beaker.
Heater/Magnetic Stirrer with Flea.
20 ml Pipette.
200 ml Measuring Cylinder.
Balance and Weighing Cups.
Plastic Food Wrap.
Vortex Mixer.
Autoclave Tape.
Electrophoresis Apparatus.
          Eg.   Bio-Rad Sub-Cell Electrophoresis Cell with Gel Tray [15 x 20 cm] and 15 well, 1.5 mm
                    thick comb (Cat No. 170-4304) or Bio-Rad Sub-Cell GT System with Tray and Gel Caster
                    [15 x 20 cm] (Cat No. 170-4483).
Power Pack.
          Eg.   Bio-Rad PowerPac 300, Cat No. 165-5051.
UV Transilluminator and Camera or a Mercury Lamp (with or without a 254 nm or 366 nm Filter).

PROCEDURE:

Wear plastic gloves when handling the sample and separate gloves when handling the PCR
product.

DNA Isolation.

1.   Pipette 1.0 ml of sterile glass-distilled water into a sterile 1.5 ml micro-centrifuge tube.

2.   Thoroughly wash your hands.

3.   Swab a finger with 0.5% chlorhexidine in 70% ethanol and allow to dry.

4.   Jab the finger with a sterile blood lancet or a finger puncture device such as "Glucolet".

5.   Allow a large drop of blood to form.

6.   Collect 5 μl of blood with a capillary sampler and add this to the tube.

7.   Close the cap and incubate the tube at room temperature, with occasional gentle mixing.

8. Centrifuge at 10,000 g for 3 minutes in a microcentrifuge.

9. Remove all but 20 - 30 µl of the supernatant with a Pasteur pipette.

10. Swirl the Chelex reagent to produce an even suspension and then add 200 µl to the tube.

11. Cap the tube, mix and place in an expanded polystyrene sheet into which appropriate size holes have been drilled.

12. Incubate the tubes by floating the sheet on the surface of a 56ºC waterbath, for 30 minutes, to chelate the metal ions.

13. Vortex the tubes rapidly for 10 seconds and replace the tubes in the polystyrene sheet.

14. Float the sheet on a simmering (100ºC) waterbath for 8 minutes, to denature the DNA.

15. Vortex for 10 seconds and then centrifuge the tubes at 10,000 g for 3 minutes.

PCR.

16. 30 minutes before use, turn on the thermal cycler.

17. 10 minutes before use, pre-warm the thermal cycler to 95ºC.

18. Add 30 µl of sterile glass-distilled water to each sterile amplification tube (500 µl micro tube with lid) and 50 µl to an extra tube (negative control). Label the lids of the tubes, not the sides, so that heat transfer is not effected during the PCR cycles.

19. Using a pipettor with aerosol barrier tips, add 20 µl of the isolated DNA, from step 15, to appropriately labelled tubes.

20. Thaw the 2 x master mix and add 100 units (20 µl of 5,000 units/ml) of *Taq* polymerase, per ml, and mix gently by inversion.

21. Add 50 µl of this mixture to each tube, cap and mix by inversion.

22. Add 50 µl of light mineral oil (liquid paraffin) to prevent evaporation during the heating cycles.

23. Centrifuge for 5 seconds at 12,000 g in a microcentrifuge (Use 1.5 ml tubes, with their caps removed, as insert liners of the centrifuge wells) to move all the liquid to the bottom of the tubes and separate the 2 liquid layers.

24. After centrifugation, place the tubes in the thermal cycler, and run the cycler as follows:

Initial   :   95ºC for 5 minutes.

25 cycles :   { 65ºC for 1 minute.
                 72ºC for 1 minute.
                 95ºC for 1 minute.

Cooling :   2°C per minute to 50°C.

Then rapidly to 4°C, if electrophoresis is not to be carried out immediately.

25. Centrifuge for 5 seconds at 12,000 g in a microcentrifuge, to drive all the aqueous solution below the mineral oil. The samples can be stored at -20°C prior to electrophoresis.

Electrophoresis Gel Preparation.

26. Dilute 20 ml of 5 x TBE buffer with 180 ml of distilled water to give a 0.5 x TBE buffer.

27. Add 150 ml of this 0.5 x TBE buffer to a 500 ml beaker, containing a stirrer flea, on a magnetic stirrer-heater.

28. Weigh out 3.75 g of NuSieve 3:1 agarose.

29. With rapid stirring, sprinkle the agarose onto the surface of the buffer, to produce an even suspension, at room temperature.

30. Cover the beaker with plastic wrap and pierce a small hole in the wrap for ventilation.

31. Weigh the beaker.

32. Bring the mixture to the boil, with moderate stirring, and simmer until the agarose is dissolved (about 10 minutes).

33. Reweigh the beaker, and add hot distilled water to bring the weight back to its original value.

34. When the solution has cooled to about 60°C, add 75 μl of 1 g/l ethidium bromide, with stirring.

35. Either tape the ends of the gel tray with autoclave tape or place the tray in a commercial gel caster, on a level surface.

36. Pour the agarose solution to a depth of 3 - 4 mm.

37. Place the 1.5 mm thick, 15 well comb near one end of the gel (there should be 0.5 - 1.0 mm of gel below the bottom of the comb) and allow the agarose to gel at room temperature.

38. After the gel has set, remove the tape (if used) and transfer the gel to the electrophoresis apparatus (comb at the cathode end) and add the running buffer until the buffer surface is about 2-3 mm above the gel. (Make sure that the electrophoresis apparatus is on a level surface.) Gently remove the comb.

Electrophoresis.

39. Add 5 μl of the sucrose-dye solution to 500 μl microcentrifuge tubes (1 for each sample + 2 for the size standards).

40. Carefully open the cap of the tubes from step 25, and using a different (from step 19) 20 μl pipettor with a separate aerosol barrier tip for each PCR product, add 20 μl of the PCR product to a tube from step 39. Cap and mix well.

41. Add 20 μl of the DNA size standard to 2 tubes from step 39, cap and mix well.

42. Centrifuge for 5 seconds at 12,000 g in a microcentrifuge, to drive all the liquid to the bottom of the tubes.

43. Using a 20 μl pipettor with separate aerosol barrier tips, load the samples into the gel wells. The standards should be loaded into the outer wells.

44. Measure the distance between the tank electrodes, in centimetres, and place the lid on the tank.

45. Connect the power supply and apply 5 volts per cm measured in step 44.

46. When the bromphenol blue dye has migrated most of the length of the gel (2 to 3 hours), turn off the power supply and remove the tank lid.

47. Remove the gel, on its tray, and submerge the gel in distilled water for 20 minutes.

48. Repeat step 47 with fresh distilled water.

49. Place the gel on a UV transilluminator and photograph. Alternatively the gel can be viewed with a mercury lamp (or a Hg lamp with a 254 (or 366) nm filter).

Steps 47 and 48 are optional, but are recommended by FMC BioProducts.

The migration distance of the bands can be measured on the gel or on the photograph.

The most common alleles found in Caucasians are:

Allele		Population Frequency
1	:	1%
3	:	21%
5	:	2%
6	:	3%
7	:	2%
8	:	1%
9	:	31%
10	:	4%
11	:	2%
12	:	1%
13	:	4%
14	:	6%
15	:	2%
16	:	8%
17	:	1%
18	:	1%

EXERCISE 12-2

CYSTIC FIBROSIS MUTATION SCREENING USING A.R.M.S.

As mentioned on page 12-12, cystic fibrosis is a defect in ion transport across epithelial membranes as a result of mutations in the Cystic Fibrosis Transmembrane Conductance Regulator (CFTR) gene.

CFTR protein is a chloride channel on the secretory side of exocrine epithelial cells. The channel is controlled by a regulatory domain and two ATP-binding domains. The regulatory domain has to be phosphorylated for the channel to open. Cyclic AMP induces this phosphorylation. ATP must also be bound to the ATP-binding domains for chloride to pass through the channel.

Normally, when too much liquid is on the secretory side of the cell, sodium channels in the membrane open to allow sodium into the cell. Chloride and water then follow. If too much water is absorbed, chloride is then removed, via the CFTR channel. In cystic fibrosis, this channel is missing or inoperative, and so chloride is not secreted. Excessive amounts of sodium are also absorbed into the cell, by an unknown mechanism, and pumped out on the inner side. These processes, in the lungs, are illustrated below:

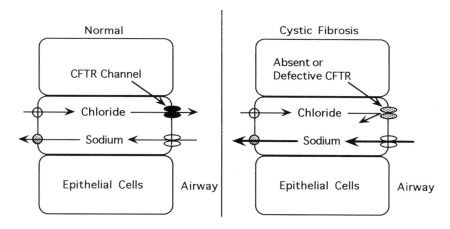

The processes, seen in cystic fibrosis, lead to dehydration of the secreted fluid in the lungs and pancreas (and other secretory tissues) and to the formation of thick mucus plugs in these tissues. The mucus in the lungs is a site for bacterial infection, which progressively gets worse. The plugs in the pancreas block the secretion of pancreatic enzymes, which results in malabsorption. As a result, mainly of respiratory problems, only about 70% of patients survive into adult life.

Cystic fibrosis is one of the most common genetic diseases in Caucasians, affecting about 1 in 2,000 live births. It is uncommon in Africans and Asians.

As previously mentioned, it is an autosomal recessive disease and so defective genes have to be inherited from both parents. This means that one Caucasian in 23 carries a defective gene. Carriers show no symptoms. When both parents are carriers, each child has a 25% chance of having cystic fibrosis, a 50% chance of being a carrier and a 25% chance of not being affected (ie. having two normal CFTR genes).

The CFTR gene is located on chromosome 7 and is 250 kb in size with 27 exons. It codes for a protein with a molecular weight of 1,480 Daltons. Over 500 mutations in the gene have been observed. Hence, it is not possible to screen for all mutations. Where the mutation is not known, indirect gene tracking can be used, as discussed on page 12-10.

The ΔF508 mutation accounts for about 70% of all the CFTR mutations in Caucasians. 10% comprise one of 6 fairly common mutations and the remaining 20% comprise uncommon mutations. However, there is considerable variation between countries. For example, 82% of the mutations are ΔF508 in Denmark.

Of the cystic fibrosis children born in Australia, 95% are detected 3-4 days after birth, using an ELISA screening method for excessive levels of trypsin and/or trypsinogen. 49% of these patients have 2 ΔF508 mutations. 42% have one ΔF508 and one other mutation. Only 9% have two other mutations.

Parents, who already have a child with cystic fibrosis, can be tested for the common mutations and then the foetus can be screened using amniocentesis or chorionic villi sampling. As mentioned above, indirect gene tracking, using genetic markers, can be used where parents have uncommon mutations.

As about one in 23 Caucasians has a CFTR mutation, population screening for carriers of the common mutations has been suggested. This has been instituted in some countries.

Therefore, in this exercise, you will screen your DNA for the ΔF508 mutation and three fairly common mutations found in northern Europe, using the Zeneca CF(4)m-PCR kit.

The DNA will be isolated from buccal epithelium cells and the 4 mutation sites amplified using ARMS (Amplification Refractory Mutation System, Allele Specific Amplification) described on pages 12-21 and 12-22. A normal DNA control (provided with the kit) and a negative control (containing no DNA) will also be amplified.

Specific mutation sites are amplified using paired vials. Vial A will amplify the normal ΔF508 + 621+1G→T and mutant G551D + G542X sites. Hence, normal DNA will yield 508 (160 bp) and 621 (367 bp) products.

Vial B will amplify mutant DF508 + 621+1G→T and normal G551D + G542X sites. Hence, normal DNA will yield 542 (255 bp) and 551 (283 bp) products.

After 35 PCR cycles, the product is subjected to electrophoresis on 3% NuSieve 3:1, in Tris-borate-EDTA buffer, containing ethidium bromide.

As mentioned above, the PCR products vary in size between 160 and 380 base pairs. A bromphenol blue marker is included with the PCR products, together with a strong sucrose solution, for electrophoresis. Bromphenol blue migrates at about the same rate as 40 base pairs of DNA in this gel. The sucrose increases the density of the sample so that it easily sinks to the bottom of the well in the agarose gel. DNA size markers are also included in separate lanes.

The run gel can be photographed while being exposed to UV light. Product migration is compared with that from the normal control and with the diagrams and photograph in the instruction booklet provided with the kit.

REAGENTS AND EQUIPMENT:

1. Zeneca Diagnostics CF(4)m-PCR Kit.
    Cat No. 834 9094 (25 tests).
    This includes:
    (1) Vial A (Yellow) x 25.
        A primers (normal 508 and 621+1, mutant 551 and 542) and dNTPs.
    (2) Vial B (Blue) x 25.
        B primers (normal 551 and 542, mutant 508 and 621+1) and dNTPs.
    (3) Buffer.
    (4) *Taq* Polymerase.
    (5) Normal Control.
    (6) Instruction Booklet.
2. Sterile Glass Distilled Water.
3. 4% Sucrose Solution.
    Dissolve 8 g of sucrose in about 150 ml of reagent 2 and make to 200 ml with reagent 2.
4. NaCl-EDTA Solution.
    Dissolve 58 mg of NaCl and 372 mg of EDTA in 100 ml of reagent 2.
5. 50 mM NaOH.
    Dissolve 200 mg of NaOH in 100 ml of reagent 2.
6. 1.0 M Tris-HCl Buffer (pH 7.5).
    Dissolve 12.11 g of Tris (Tris(hydroxymethyl)amino-methane) in about 50 ml of reagent 2. Add 7 ml of 10 M HCl and stir on a magnetic stirrer, with a pH electrode. Add the 10 M HCl dropwise until the pH reaches 7.5. Make to 100 ml with reagent 2.
6. Light Mineral Oil (Liquid Paraffin).
    Eg.   Sigma Cat No. M 3516.
7. Tris-Borate, EDTA Buffer, pH 8.3.
    Prepare a 5 x TBE (445 mM Tris-borate, 12.5 mM EDTA) buffer as follows:
        Dissolve 54.5 g of Tris, 27.8 g of boric acid and 4.66 g of EDTA in about 900 ml of distilled water. Make to 1 l with distilled water.
8. Ethidium Bromide Solution (1.0 g/l).
    Dissolve 10 mg in 10 ml of distilled water.
        Ethidium bromide is a known carcinogen, so handle with care.
9. Running buffer (0.5 x TBE).
    Add 1.0 ml of reagent 8 to 200 ml of reagent 7, add 1,800 ml of distilled water and mix well.
10. NuSieve 3:1 Agarose.
    FMC BioProducts Cat No. 50091.
11. DNA Size Standard.
    Eg.   Bio-Rad AmpliSize (50 - 2,000 bp), Cat No. 170-8200.
12. Sucrose-Dye solution.
    Dissolve 60 g of sucrose in about 40 ml of distilled water and make to 100 ml. Add 250 mg of bromphenol blue and mix until fully dissolved.

Plastic Disposable Gloves.
Sterile 25 ml Beakers.
    With graduations at 10 ml, for the 4% sucrose.
Sterile 12 or 15 ml Transparent Plastic Centrifuge Tubes, with Caps.
Bench Centrifuges.
Pasteur Pipettes.
5, 20, 100 and 500 µl Pipettors with Tips.
5 µl Capillary Pipettors and Tips.
Adjustable Pipettors, with Tips.
    Set to 108 and 12 µl.
Sterile 20 µl Aerosol Barrier Pipette Tips.
    Eg.   Xcluda, from Bio-Rad.

Sterile 1.5 ml Microcentrifuge Tubes.
        Eg.   Eppendorf Cat No. 0030 102.002.
400 or 500 ml Microcentrifuge Tubes.
        Eg. Eppendorf Cat No. 0013 036.004 (400 ml).
Microcentrifuge.
        Eg.   Eppendorf 5410.
Thermal Cycler.
        To hold 500 μl tubes.
                Eg.   Perkin-Elmer's Thermal Cycler 480 or Eppendorf's Mastercycler 5330 plus.
500 ml Beaker.
Heater/Magnetic Stirrer with Flea.
20 ml Pipette.
200 ml Measuring Cylinder.
Balance and Weighing Cups.
Plastic Food Wrap.
Vortex Mixer.
Autoclave Tape.
Electrophoresis Apparatus.
        Eg.   Bio-Rad Sub-Cell Electrophoresis Cell with Gel Tray [15 x 20 cm] and two 15 well, 1.5 mm
               thick combs (Cat No. 170-4304) or Bio-Rad Sub-Cell GT System with Tray and Gel Caster
               [15 x 20 cm] (Cat No. 170-4483).
Power Pack.
        Eg.   Bio-Rad PowerPac 300, Cat No. 165-5051.
UV Transilluminator and Camera or a Mercury Lamp (with or without a 254 nm or 366 nm Filter).

PROCEDURE:

Wear plastic gloves when handling the sample and separate gloves when handling the PCR
product.

DNA Isolation.

1.   Take 10 ml of 4% sucrose into the mouth and move around the mouth for 20 seconds.

2.   Collect the fluid in a sterile clear plastic centrifuge tube.

3.   Cap the tube and centrifuge at top speed in a bench centrifuge for 10 minutes.

4.   Remove the supernatant with a Pasteur pipette and discard.

5.   Add 500 μl of the NaCl-EDTA solution and resuspend the cells by vortexing the tube.

6.   Transfer the sample to a sterile 1.5 ml microcentrifuge tube.

7.   Centrifuge at 10,000 g for 3 minutes in a microcentrifuge.

8.   Remove the supernatant with a Pasteur pipette and discard.

9.   Add 500 μl of 50 mM NaOH. Cap the tube and resuspend the cells by vortexing.

10.   Place the tubes in an expanded polystyrene sheet into which appropriate size holes have
        been drilled.

11.   Incubate the tubes by floating the sheet on a simmering (100°C) waterbath, for 5
        minutes.

12. Vortex the tubes rapidly for 10 seconds.

13. Add 100 μl of 1.0 M Tris-HCl buffer (pH 7.5) and cap the tube.

14. Vortex for 10 seconds and then centrifuge the tubes at 10,000 g for 3 minutes.

PCR.

15. 30 minutes before use, turn on the thermal cycler.

16. 10 minutes before use, pre-warm the thermal cycler to 94°C.

17. For each sample, select A and B amplification vials from the kit. Label the lids; not the sides, so that heat transfer is not effected during the PCR cycles.

18. Centrifuge in a microcentrifuge for 5 seconds at about 12,000 g, to bring the liquid to the bottom of the tubes (Use 1.5 ml tubes, with their caps removed, as insert liners of the centrifuge wells).

19. Using a pipettor with separate aerosol barrier tips, add 5 μl of the prepared DNA, from step 14, to the 2 tubes.

20. Add one drop of light mineral oil (liquid paraffin), to prevent evaporation during the heating cycles, and re-cap the tubes.

21. Set up a pair of tubes for the normal control (provided with the kit) and a negative control pair with just the mineral oil.

22. Place the tubes in the thermal cycler, held at 94°C.

23. Prepare the polymerase reagent as follows:

> For each 10 samples (20 tubes), add the following to a sterile 1.5 ml microcentrifuge tube:
> 108 ml of sterile glass distilled water,
> 12 ml of ARMS buffer and
> 5 μl of *Taq* polymerase.
> Mix well.

24. After the vials have been at 94°C for 5 minutes, remove them (one at a time) and add 5 μl of the polymerase reagent to the lower (aqueous) layer. Replace the vial in the thermal cycler.

25. When the polymerase has been added to all vials, run the following PCR programme:

$$
34 \text{ cycles} \quad : \quad \left\{ \begin{array}{l} 60°C \text{ for 1 minute.} \\ 72°C \text{ for 1 minute.} \\ 94°C \text{ for 1 minute.} \end{array} \right.
$$

Then

$$
1 \text{ cycle} \quad : \quad \left\{ \begin{array}{l} 60°C \text{ for 1 minute.} \\ 72°C \text{ for 10 minute.} \end{array} \right.
$$

Then cool to room temperature; or 4°C, if electrophoresis is not to be carried out immediately.

Electrophoresis Gel Preparation.

26.   Dilute 20 ml of 5 x TBE buffer with 180 ml of distilled water to give a 0.5 x TBE buffer.

27.   Add 150 ml of this 0.5 x TBE buffer to a 500 ml beaker, containing a stirrer flea, on a magnetic stirrer-heater.

28.   Weigh out 4.5 g of NuSieve 3:1 agarose.

29.   With rapid stirring, sprinkle the agarose onto the surface of the buffer, to produce an even suspension, at room temperature.

30.   Cover the beaker with plastic wrap and pierce a small hole in the wrap for ventilation.

31.   Weigh the beaker.

32.   Bring the mixture to the boil, with moderate stirring, and simmer until the agarose is dissolved (about 10 minutes).

33.   Reweigh the beaker, and add hot distilled water to bring the weight back to its original value.

34.   When the solution has cooled to about 60°C, add 75 µl of 1 g/l ethidium bromide, with stirring.

35.   Either tape the ends of the gel tray with autoclave tape or place the tray in a commercial gel caster, on a level surface.

36.   Pour the agarose solution to a depth of 3 - 4 mm.

37.   Place one 1.5 mm thick, 15 well comb near one end of the gel (there should be 0.5 - 1.0 mm of gel below the bottom of the comb) and a second comb just over halfway down the gel. Allow the agarose to gel at room temperature.

38.   After the gel has set, remove the tape (if used) and transfer the gel to the electrophoresis apparatus (first comb at the cathode end) and add the running buffer until the buffer surface is about 2-3 mm above the gel. (Make sure that the electrophoresis apparatus is on a level surface.) Gently remove the combs.

Electrophoresis.

39.   Add 5 µl of the sucrose-dye solution to a 400 or 500 µl microcentrifuge tubes (1 for each PCR vial + 4 for the size standards).

40.   Centrifuge the vials from step 25, in a microcentrifuge for 5 seconds at about 12,000 g, to bring the liquid to the bottom of the vials, with the aqueous phase below the mineral oil.

41. Carefully open the cap and using a 20 µl pipettor with an aerosol barrier tip, transfer 20 µl of the PCR product to the tubes in step 39. Cap and mix well.

42. Repeat step 41 with the other vials, but using fresh aerosol barrier tips.

43. Add 20 µl of the DNA size standard to 4 tubes from step 39, cap and mix well.

44. Centrifuge for 5 seconds at about 12,000 g in a microcentrifuge, to drive all the liquid to the bottom of the tubes.

45. Carefully open the cap and using a 20 µl pipettor with separate aerosol barrier tips, load the samples into the gel wells. The standards should be loaded into the outer wells.

46. Measure the distance between the tank electrodes, in centimetres, and place the lid on the tank.

47. Connect the power supply and apply 5 volts per cm measured in step 46.

48. When the bromphenol blue dye has migrated about 80 mm (about 90 minutes), turn of the power supply and remove the tank lid.

49. Remove the gel, on its tray, and submerge the gel in distilled water for 20 minutes.

50. Repeat step 46 with fresh distilled water.

51. Place the gel on a UV transilluminator and photograph. Alternatively the gel can be viewed with a mercury lamp (with or without a 254 or 366 nm filter).

Steps 49 and 50 are optional, but are recommended by FMC BioProducts.

Compare your results with the normal control results.

Compare the results with the diagrams and photo on pages 12 to 14 of the Zeneca Diagnostics instruction booklet provided with the kit.

2 or 3 distinct bands should be seen in each track; except with the negative control, which should have no bands.

The normal control should show the normal genotype:

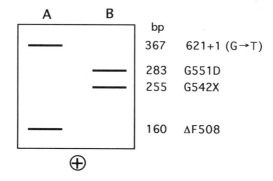

12-43

The position of the product bands should correspond with the correct base pair size, as indicated by the size markers.

Should your sample contain a mutation, you should consult your lecturer.

GENERAL REFERENCES:

BOOKS

Kirby, L. T., (1992), DNA Fingerprinting - An Introduction, W. H. Freeman and Co.

Brock, D. J. H., (1993), Molecular Genetics for the Clinician, Cambridge University Press.

Heim, R. A. and Silverman, L. M., (1994), Molecular Pathology, Carolina Academic Press.

Leder, P., Clayton, D. A. and Rubenstein, E., (1994), Scientific American Introduction to Molecular Medicine, Scientific American Inc.

Strachan, T, and Read, A. P., (1996), Human Molecular Genetics, Bios Scientific Publishers.

ARTICLES

Wenham, P. R., (1992), DNA-Based Techniques in Clinical Biochemistry: A Beginner's Guide to Theory and Practice, Ann. Clin. Biochem., **29**, 598-624.

Trent, R. J., (1993), Clinical and Laboratory Applications of Molecular Biology, Clin. Biochem. Rev., **14**, 56-61.

Dowton, S. B. and Slaugh, R. A., (1995), Diagnosis of Human Heritable Diseases - Laboratory Approaches and Outcomes, Clin. Chem., **41**, 785-794.

Sinclair, B., (1999), Everything's Great When It Sits on a Chip - A Bright Future for DNA Arrays, The Scientist, **13**, 18-20.

Kricka, L. J., (1999), Nucleic Acid Detection Technologies - Labels, Strategies and Formats, Clin. Chem., **45**, 453-458.

Specific References:

Florijn, R. J., Bonden, L. A. J., Vrolijk, H., Wiegant, J., Vaandrager, J.-W., Baas, F., den Dunnen, J. T., Tanke, H. J., van Ommen, G.-J. B. and Raap, A. K., (1995), High-resolution DNA Fibre-FISH for genomic DNA mapping and colour bar-coding of large genes, Human Molecular Genetics, **5**, 831-836.

Tazelaar, J. P., Friedman, K. J., Kline, R. S., Guthrie, M. L. and Farber, R. A., (1992), Detection of α_1-Antitrypsin Z and S Mutations by Polymerase Chain Reaction-Mediated Site-Directed Mutagenesis, Clin. Chem., **38**, 1486-1488.

Wood, N., Tyfield, L. and Bidwell, J., (1993), Rapid Classification of Phenylketonuria Genotypes by Analysis of Heteroduplexes Generated by PCR-Amplifiable Synthetic DNA, Human Mutation, 2, 131-137.

Reyes, A. A., Carrera, P., Cardillo, E., Ugozzoli, L., Lowery, J. D., Lin, C.-I. P., Go, M., Ferrari, M. and Wallace, R. B., (1997), Ligase Chain Reaction Assay for Human Mutations: The Sickle Cell by LCR Assay, Clin. Chem., 43, 40-44.

Cheung, V. G., Morley, M., Aguilar, F., Massimi, A., Kucherlapati, R. and Childs, G., (1999), Making and Reading Microarrays, Nature Genetics, 21, 15-19.

Amplification of the D1S80 locus using PCR.

Eisenberg, M. and Chimera, J. A., (1992), Human Identification by DNA Analysis - The Amplification Fragment Length Polymorphism Analysis Technique, Chapter 9 in Diagnostic Molecular Pathology - A Practical Approach, Vol. 2, Eds. Herrington, C. S. and McGee, J. O'D., p 183-192, IRL Press.

Cystic fibrosis mutation screening using ARMS.

Anon., (1995), CF(4)m-PCR Kit Instruction Booklet, Zeneca Diagnostics.

CHAPTER 13

IMMUNOLOGICAL AND RADIOISOTOPE TECHNIQUES

IMMUNOLOGICAL TECHNIQUES

As the name implies, immunological techniques use antibodies. The techniques examined in this chapter involve the use of antibodies as reagents. Antibodies were discovered in 1898 and first used as reagents in 1901, when agglutination reactions were introduced for blood group testing. However, it was not until after the Second World War that the full potential of antibodies as reagents was realised. Plate diffusion techniques were introduced in the late 1940s, immunoelectrophoresis and radioimmunoassay in the 1950s, and immunoradiometric assays and alternative labels (to radioisotopes) in the 1960s. The 1970s saw the development of monoclonal antibodies and more alternative labels. Immunosensors and Western blotting were introduced in the 1980s.

ANTIBODY PRODUCTION

Polyclonal Antibodies.

The antigen, to which the antibody is to be produced, must be as pure as possible. This is because antibodies will also be developed against any impurities. These undesired antibodies can decrease the specificity of the assay. A variety of animals can be used for antibody production. These include guinea pigs, rabbits, sheep, goats, horses, chickens and monkeys.

The animal is injected with antigen mixed with "Freund's Adjuvant" (A mixture of mineral oil, waxes and killed bacteria), which prolongs dispersion in the body and enhances antibody response. 0.2-2.0 mg of antigen are usually given in the first injection.

A few weeks later a second injection is given. Less pure antigen can be used, to boost the monoclonal cell lines. This, and subsequent injections, are given with adjuvant containing no killed bacteria.

After several months, blood is collected and allowed to clot. The serum is separated and diluted for use. Serum proteases and complement can be inactivated by heating to 56°C.

This antiserum contains antibodies to a variety of sites (epitopes - antigenic determinants) on the antigen. Different antibodies to the epitope are also produced by different cell lines. These will have slightly different binding characteristics for the antigen. Hence, polyclonal antibodies contain a number of antibodies, with different characteristics, to the antigen.

Proteins and complex carbohydrates are antigenic and, hence, can be used directly as antigens. However, small molecules (such as drugs) will not be recognised by the immune system as foreign. Hence, they can not be used directly to produce antibodies.

However, these small molecules can be attached to large molecules, such as albumin from a different animal. The attached small molecules are known as haptens. This conjugate can then be injected into rabbits etc. Antibodies are produced to a number of sites on the large molecule, including the hapten. These latter ones are active in the assay.

The sensitivity of an assay is governed by the affinity constant (K) of the antibody. The affinity constant is how far the equilibrium reaction, below, is driven to the right (ie. the strength of attraction between antibody (Ab) and antigen (Ag)):

$$Ag + Ab \underset{k_2}{\overset{k_1}{\rightleftharpoons}} Ag\text{-}Ab$$

k_1 and k_2 are rate constants.

At equilibrium:

$$k_1[Ag][Ab] = k_2[Ag\text{-}Ab]$$

$$K = \frac{k_1}{k_2} = \frac{[Ag\text{-}Ab]}{[Ag][Ab]}$$

Immunoassay sensitivity is inversely proportional to K. For example, the maximum radioimmunoassay sensitivity is approximately:

$$0.1 \times \frac{1}{K} \quad mol/l.$$

The maximum K for polyclonal antibodies is about 10^{12} litres/mole. Hence, the sensitivity limit of radioimmunoassays is:

$$\approx 0.1 \times 10^{-12} \ mol/l$$

$$\approx 10^{-13} \ mol/l$$
$$(\approx 10^8 \ molecules/ml)$$

Monoclonal Antibodies.

B lymphocytes and plasma cells (which produce antibodies) will only undergo a limited number of cell divisions. Hence, they will produce only small amounts of antibody, if grown in culture, before they die out. However, myeloma cells will undergo unlimited cell division. Myeloma cells are cancerous plasma cells. Most of these produce antibodies of their own; however, some do not. These can be used to form hybridomas with the B cell of interest.

Hybridomas, formed between these two cell types, provide the specificity of the B cell and the "Eternal life" of the myeloma cell.

The animal (usually a mouse or rat) is immunised as for polyclonal antibody production. However, less pure antigen can be used. The spleen is removed and its B lymphocytes are fused with (mouse or rat) myeloma cells. After fusion, the extra chromosomes in the hybridoma cells are rapidly shed.

The myeloma cells lack hypoxanthine-guanine phosphoribosyl transferase (HGPRT). HGPRT is one of the purine salvage pathway enzymes. About 90% of purines are recycled back to nucleotides, via the salvage pathways, and only about 10% of nucleotides are directly synthesised.

The cells are grown on a medium containing hypoxanthine, aminopterin and thymidine (HAT). Aminopterin is a folic acid antagonist and inhibits the formation of the purine and thymine nucleotides. Hypoxanthine is the substrate for purine salvage. Hence, only cells containing HGPRT will grow in the HAT medium. Unfused myeloma cells will not grow. Unfused B cells soon die or are outgrown by the hybridomas.

Hybridomas that generate the antibody of interest are then selected. These are then screened for cross-reactivity and binding affinity. The one with least cross-reactivity and greatest affinity constant is then grown in culture. These cells secrete the antibody into the culture medium, from where it can be harvested.

These selected hybridomas will grow in "Limitless supply". They produce a single antibody species with a constant affinity and avidity (the strength of antibody-antigen bond, after the immune complex is formed) for the antigen. However, the affinity constants tend to be lower than those of polyclonal antibodies. There is less cross reactivity and non-specific binding with monoclonal antibodies. Equilibrium in monoclonal antibody-antigen reactions is reached more rapidly, than with polyclonal antibodies. However, individual monoclonal antibodies do not form precipitation lattices used in immunoprecipitation techniques.

Antigen-Antibody Binding.

Antigen-antibody binding involve three types of interaction:

1.   Van der Waals-London Dipole-Dipole Interactions.

Molecules containing polarisable groups can form transient dipoles. This induces dipole formation in adjacent molecules with a net attractive force between the molecules. The force is inversely proportional to temperature. However, if the temperature is kept in the range 25 - 35°C, antigen-antibody binding is fairly constant. These forces operate over short distances (4-6 nm).

2.   Hydrophobic Interactions.

Hydrophobic parts of molecules interact, in an aqueous environment, to attain the lowest free energy level (2nd law of thermodynamics).

3.   Ionic Interactions.

Ionic bonding occurs between charged groups, mainly $-COO^-$ and $-NH_3^+$. If the pH is kept in the range 6-8, antigen-antibody binding is fairly constant. Ionic interaction force is considerably reduced in solutions with a high dielectric constant (ie. a salt solution). Hence, variations in ionic strength have a marked effect on antibody-antigen interactions.

Cation salts inhibit antibody binding to a cationic group on an antigen, by binding to the anionic group on the antibody. This inhibition follows the order: $NH_4^+ > K^+ > Na^+ > Li^+$. Lithium has the lowest ionic radius, but has the largest radius of hydration. Hence, it has least interaction with the anionic group on the antibody. A similar effect is seen with anions, where inhibition follows the order: $I^- > Br^- > Cl^- > F^-$.

The Effect of Polymers.

Polymers tend to increase the rate of antibody-antigen complex formation, and actual complex formation with low-avidity antibodies. The polymer does not form part of the complex. However, it effectively increases the antibody and antigen concentrations. Hence, antibody-antigen collisions are more likely. Polyethylene glycol (PEG) is the most commonly used polymer.

IMMUNOPRECIPITATION ASSAYS

Immunoprecipitation assays require polyclonal antibodies and large, multi-epitope antigens so that precipitation lattices are formed:

$$Ag_n + Ab \rightleftharpoons Ag_nAb \rightleftharpoons Ag_aAb_b$$

Ag_n :   Multi-epitope antigen.
Ab   :   Bivalent antibodies.
a and b   :   The number of antigen and antibody molecules in the immune complex.

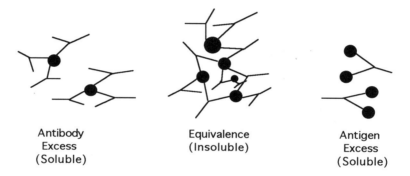

Antibody
Excess
(Soluble)

Equivalence
(Insoluble)

Antigen
Excess
(Soluble)

Equivalence is obtained when there are 2 to 3 antibody molecules for each antigen. If one plots the amount of precipitate against the concentration of antigen, for a fixed amount of antibody, one obtains the type of curve shown below:

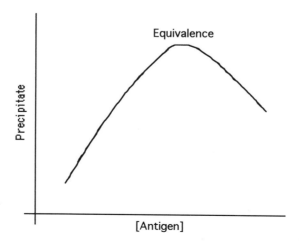

Passive Gel Diffusion.

Agar, or agarose, is used as a support medium to stabilise diffusion of antigen and/or antibody from wells in the gel. Antigen-antibody complexes precipitate when there is a moderate antibody excess (ie. equivalence). The precipitin bands may form and redissolve many times before the band becomes stable, during the diffusion process.

Double Immunodiffusion.

Antibody and antigen are in separate wells. Where they meet, a precipitin line is seen where antibody and antigen are in equivalence. Eg. The Ouchterlony technique for identification of a protein antigen:

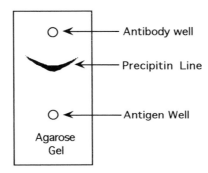

The precipitin line is closer to the well of lower concentration and concave towards the higher molecular weight component.

Single Immunodiffusion.

Single immunodiffusion is used to quantitate the amount of a known antigen. Eg. Radial immunodiffusion, where antigen diffuses into an antibody impregnated gel:

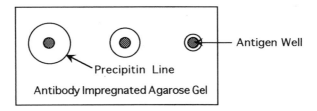

The diameter^2 of the precipitin ring is proportional to the antigen concentration:

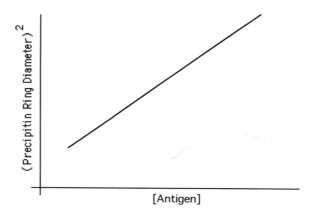

Immunoelectrophoretic Techniques.

These techniques involve the initial separation of the sample proteins by electrophoresis. These separated proteins are then allowed to diffuse towards antibodies that are diffusing from a trough (Immunoelectrophoresis) or they are driven into a gel containing an antibody (Electroimmunoassay) or antibodies (Crossed Immunoelectrophoresis) by an electric potential.

Immunoelectrophoresis.

After normal electrophoresis, on an agarose gel, a trough is cut in the gel, parallel to the electrophoretic run. This trough is then filled with antiserum and the separated proteins and the antiserum antibodies are allowed to diffuse towards each other. Where they meet, and equivalence is reached, precipitin lines are formed. This is illustrated, below, for serum proteins separated on an agarose gel and then allowed to diffuse towards a whole serum antiserum:

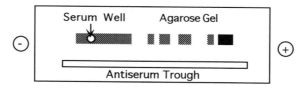

The precipitin lines can then be stained with a protein stain. The type of result obtained is illustrated below:

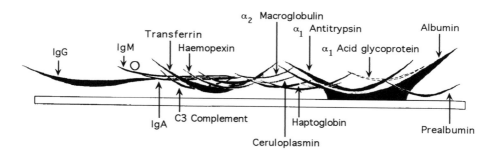

This technique can be used to detect the presence or absence of specific proteins.

Crossed Immunoelectrophoresis.   (2-Dimensional Immunoelectrophoresis).

Sample (Eg. serum) proteins are separated by normal electrophoresis in an agarose gel. A strip of the agarose gel (containing the sample proteins) is then removed and placed on the side of a glass sheet. Fresh agarose, containing anti-sample antibodies, is then poured to fill the plate.

The plate is then electrophoresed at right angles to the original run. The proteins migrate into the agarose until they are fully precipitated by the antibodies. Each protein appears as a separate peak.

The area under the peak is proportional to the concentrations of the protein in the sample and inversely proportional to the concentration of its specific antibody in the gel.

The technique is mainly used for detecting quantitative or qualitative changes in specific plasma proteins (eg. α_1-antitrypsin).

Proteins with concentrations down to 100 µg/l can be detected by this method.

After the second run, the excess antibody is eluted from the gel. The gels are then dried and the protein precipitin lines stained with a protein stain, such as Coomassie Brilliant Blue. Illustrated below is an example of crossed immunoelectrophoresis of human serum with an anti-human serum antiserum:

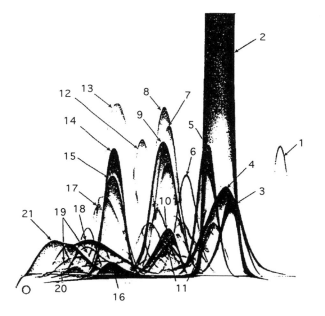

1 : Prealbumin
2 : Albumin
3 : Orosomucoid (α_1-Acid-
      Glycoprotein)
4 : α_1-Lipoprotein (HDL)
5 : α_1-Antitrypsin
6 : α_1-Antichymotrypsin
7 : Gc-Globulins
8 : α_2-HS-Glycoprotein
9 : Haptoglobin
10 : α_2-Macroglobulin
11 : Ceruloplasmin
12 : Antithrombin III
13 : Haemopexin
14 : Transferrin
15 : C3 Complement
16 : β-Lipoprotein (LDL)
17 : Properdin Factor B
18 : IgA
19 : C4 Complement
20 : IgM
21 : IgG

Electroimmunoasssay (Laurell's Rockets).

The antigenic protein is added to wells in an agarose gel containing antibodies to the protein. The protein is then driven into the gel by an electric force. Precipitin lines are formed at the equivalence points, but these redissolve as more antigen enters the area. Hence, the precipitin line moves up the gel as new areas of equivalence are formed until a final steady state is reached. These precipitin lines are rocket shaped. Illustrated below are the results of a set of 5 standards and duplicate samples of the unknown:

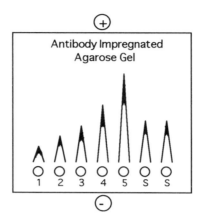

This technique is used for the quantitation of a specific antigen (Eg. a plasma protein). The height of the "rocket" is directly proportional to the log of the concentration of the antigen:

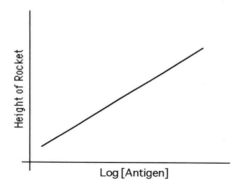

Particle Suspension Assays.

In these assays, the antibody-antigen complex is held in suspension and the amount of light absorbed or scattered is measured. This is proportional to the concentration of the analyte (antigen) in the sample. Where light absorption is used, the technique is known as "Turbidimetry":

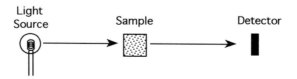

Accuracy and precision, using turbidimetry, are not very good. Hence, light scatter is commonly used to measure the amount of precipitate generated in these types of assay. This technique is known as "Nephelometry":

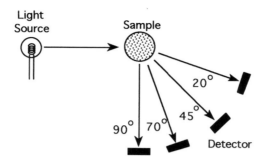

As can be seen, the scattered light can be measured at a variety of angles. Antibody-antigen particles are large and exert Rayleigh-Debye scattering, where the diameter of the particle is equal to, or a bit less than, the wavelength of light used in the instrument. Rayleigh-Debye scattering is illustrated below:

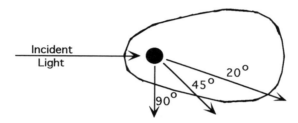

As can be seen, most of the scatter is in the forward direction, with least scatter at 90°.

An example of an instrument using this technique is the Beckman "Array" nephelometer. This uses a tungsten lamp with a 420 -560 nm filter and the amount of scattered light is measured at 70°.

The rate of change in light scatter, with time, is determined as this minimises the effects of physical and chemical interferences and leads to improved precision.

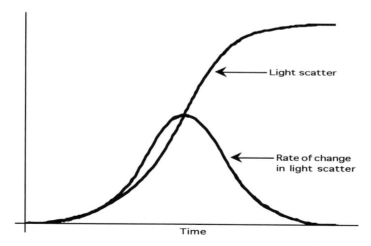

The response of peak rate of change in light scatter, with increasing antigen concentration, is illustrated below:

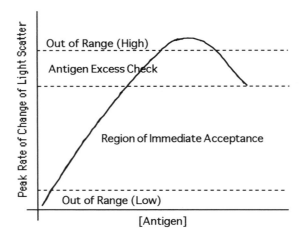

As can be seen, the standard curve becomes horizontal and then decreases. This is due to an excess of antigen and equivalence is not reached, as all the antibody is used up. With high values (in the antigen excess region), the instrument add extra antigen. The response is illustrated below:

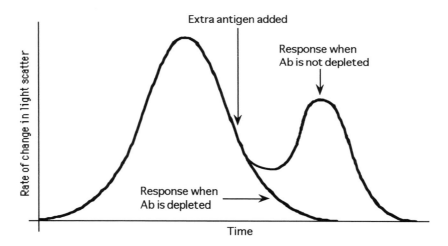

As can be seen; if the second peak is generated, the original result is accepted. If not, it is rejected.

Competitive Immunoprecipitation Techniques.

The antigen of interest is bound to a solid support that forms a fine suspension. Antibodies to the antigen cause the particles to aggregate. Free antigen, in the sample or standards, will compete with solid support antigen for a limited amount of antibody. Hence, as the amount of free antigen increases, the amount of aggregate decreases.

Eg.   The Beckman Immunochemistry System.

The analyte antigen (a hapten, such as a drug) is attached to a protein molecule, which then forms a conjugate "Polyantigen". The antibody binds the polyantigen to form large particles that scatter light. Antigen, in the sample or standard, competes with the conjugate for the antibody:

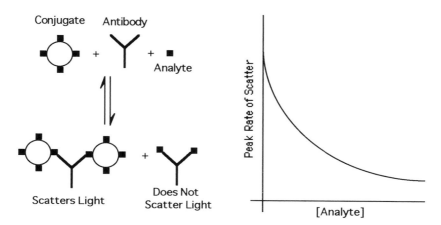

Latex Particle Agglutination Immunological Assays.

The term "Latex" is normally used to describe a rubber suspension. However, it can also be used to describe a suspension of small polystyrene beads. Antigen (or antibody) can be attached to the particles. The appropriate antibody (or antigen) causes the particles to agglutinate.

Eg.   The Roche Abuscreen "OnTrack" System.

This is latex agglutination inhibition immunoassay for the detection of drugs of abuse in urine. The antigen (drug) is attached to the latex particles. Sample, latex particles and antibody are mixed together. If the sample does not contain the drug, the latex particles will aggregate. If enough drug is present, the antibody binds this drug and not enough antibody is left to agglutinate the latex. Hence, the latex remains as a fine suspension:

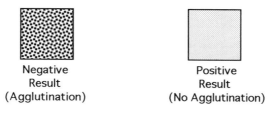

LABELLED IMMUNOLOGICAL ASSAYS

A label, which is used in these types of assays, is attached to the antigen or antibody to follow the antibody-antigen reaction.

The specificity of these assays is imparted by the use of antibodies. Even though antibodies do not have absolute specificity, they can be selected to have a very high degree of specificity for the analyte of interest. Alternatively, less specific antibodies can be used to detect a class of antigens, such as drugs (Eg. benzopdiazepines).

The sensitivity of these assays is imparted by the label. Labels can be radioisotopes, enzyme, fluorophores, luminescent chemicals and others (such as the nitroxide free radical).

The relative specific activity of the major labels are summarised below:

^{125}I.

One event per minute per 125,000 labelled molecules.

^{3}H.

One event per minute per 20,000,000 labelled molecules.

Enzymes.

Determined by the enzyme "Amplification Factor" and the detectability of the reaction product.

Eg.   The measurement of thyroid stimulating hormone (TSH) using the rate of generation of a violet formazan:

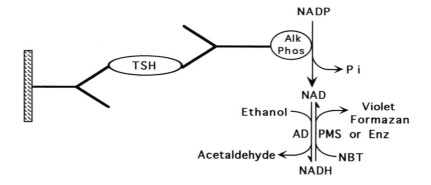

Luminescent Chemicals.

One event per labelled molecule.

Fluorescent Chemicals.

Many events per labelled molecule.

The sensitivity of the assay includes the ability to detect the events as well as the specific activity (amount of label per unit amount of labelled antigen or antibody). For example, most radioisotope decay events can be detected, but many fluorescent events are needed for detection.

Labelled immunological assays can be subdivided into immunoassays and immunometric assays.

Immunoassays.

In these assays, the antigen is labelled. A competitive system is used and either the bound or free label can be measured after equilibrium is reached.

$$\text{Antigen}^O$$
$$+ \quad \text{Antibody} \quad \rightleftharpoons \quad \text{Antigen}^O\text{-Antibody}$$
$$\text{Antigen}^* \qquad\qquad\qquad\qquad \text{Antigen}^*\text{-Antibody}$$
$$\text{(Free)} \qquad\qquad\qquad\qquad\qquad \text{(Bound)}$$

Antigen^O is the unlabelled antigen (standards or unknowns) and Antigen^* is the labelled antigen.

The sensitivity of this assay system increases as the antibody concentration decreases. However, precision deteriorates with very low antibody concentrations.

Immunometric Assays.

In these assays, the antibody is labelled. A non-competitive system is normally used and the bound label is measured. With this type of assay, the sensitivity increases as the antibody concentration increases. However, this plateaus off with excess antibody.

$$\text{Antigen} \quad + \quad \text{Antibody*} \quad \rightleftharpoons \quad \text{Antigen-Antibody*}$$
$$\text{(Excess)}$$

These labelled immunological techniques can also be subdivided into heterogeneous and homogeneous assays.

Heterogeneous.

The activity of the label is not changed by the binding of the antigen by the antibody. Therefore, the bound and free label must be separated before quantitation.

Homogeneous.

The activity of the label is changed by the binding of the antigen by the antibody. Therefore, there is no need to separate the free from the bound label.

This has the advantage that the technique is usually quicker, as there is one less analytical step. However, the technique does have the disadvantages that the background activity is usually higher and there is usually more endogenous interference from within the sample. These disadvantages can be eliminated by including a separation step in the assay. However, this nullifies the advantage of the system.

A number of methods can be used to separate the bound and free label. These include:

1.   Precipitation and Centrifugation.

The antibody-antigen complex can be salted out, using ammonium sulphate or acetone. In this procedure, the proteins are dehydrated and precipitate. The complex tends to precipitate before the free antibody. However, the process is non-specific and often incomplete.

2.   Adsorption and Centrifugation.

The free antigen is adsorbed onto dextran coated charcoal. It passes through the pores in the dextran and binds to the charcoal. The antibody-antigen complex is too large to pass through the pores and remains in solution.

The process works almost as well with uncoated activated charcoal. Here the pores in the charcoal only allow the free antigen to enter and be bound.

3.   Solid Phase Antibody or Antigen*.

With immunoassays, the antibody or labelled antigen is covalently linked to a solid particle. Commonly used particles include magnetic particles, dextran, cellulose and agarose. After the attainment of equilibrium, the bound label can be separated from the free by a magnet or by centrifugation.

Antibodies can also be adsorbed onto plastic tubes, plates or discs. After equilibrium is reached, the free antigen* can be separated from the bound by decantation.

4.   Double Antibody.

A second Antibody is raised to the immunoglobulins of the first animal.

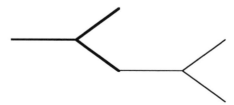

This double antibody can be used as the binding reagent in an immunoassay. Alternatively, the second antibody can be added after the initial incubation of the first antibody and antigen.

The double antibody-antigen precipitate still requires a reasonably large g force to sediment, during centrifugation.

5.   Double Antibody Solid Phase.

The double Antibody is linked to a solid phase such as cellulose or dextran. This provides a large particle that can easily be sedimented by centrifugation, after the attainment of equilibrium.

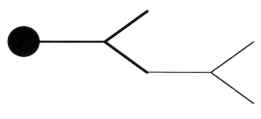

RADIOIMMUNOASSAY (RIA)

The first RIA technique was published, in 1959, by Berson and Yalow. Their technique was an assay for plasma insulin.

The technique involves the use of an amount of radiolabelled analyte equivalent to a value low in the analytical range. The labels are usually ^{125}I or ^{3}H. These are discussed in more detail in the section of radioisotopes, later in this chapter.

The labelled antigen (Antigen*) is incubated with a limited amount of antibody and the standards or unknowns (Antigen^O):

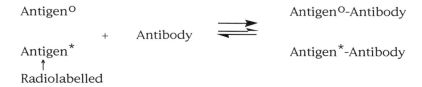

The antibody concentration is usually adjusted so that about 50% of Antigen* is bound when no Antigen^O is present (ie. for the zero standard). This does not give maximum sensitivity, but yields an assay with good precision. Greater sensitivity is achieved with less antibody, but the precision of the assay is decreased.

Tabulated below is the percentage of the added Antigen* in the antibody-antigen complex, as the unlabelled (standard) Antigen^O is increased:

[Antigen*]	[Antigen^O]	% Antigen* bound
1	0	50
1	1	25
1	2	17
1	4	10
1	9	5
1	49	1

A plot of these results yields an exponential curve:

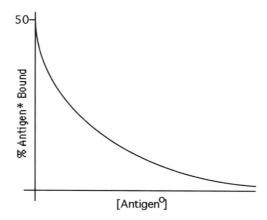

RIA results can be plotted in a number of ways:

Linear/Linear Plot.

With a linear/linear plot, the actual results obtained from the counter are plotted against the analyte concentration:

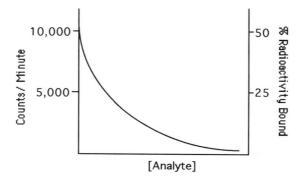

Due to the decay of the isotope, the counts per minute will decrease with time. Hence it is better to plot the percentage of the added radioactivity that is bound by the antibody, against the analyte concentration.

Linear/Log Plot.

In an attempt to linearise the results, the [Analyte] axis can be converted to a log scale. However, this generates a sigmoidal curve, with only the central section being a straight line:

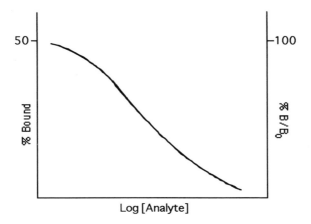

Log [Analyte]

% Bound (Bound counts per minute /Total counts per minute, x 100) can change with time, due to damage to the antigen* or to the antibody. Antibodies are usually fairly stable. However, radiolabelled antigens do become damaged, due to the effect of ionising particles generated during radioactive decay.

The ratio B/B_O is more stable with time; where B is the counts in the bound fraction (Standards or Unknown) and B_O is the counts in the bound fraction from the zero standard. Therefore, B/B_O for the zero standard is 100%.

Logit-Log Plot:

Formulae can be used to convert a sigmoidal curve into a straight line. The most commonly used is that suggested by Rodbard in 1968. This is a logit transformation:

$$\text{Logit } x = \text{Log}_e \frac{x}{1-x}$$

$$\therefore \text{Logit } B/B_O = \text{Log}_e \frac{B/B_O}{1-B/B_O}$$

Hence, a plot of % B/Bo against log [Analyte] will give a straight line:

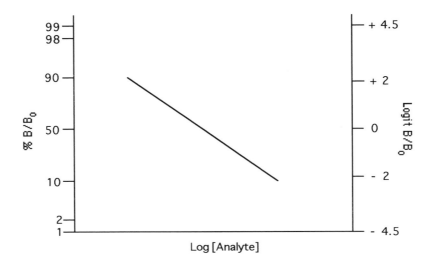

This plot is linear between 10 and 90% of B/B_0. Hence, the results are usually plotted on logit/log paper, or a computer linked to the counter will do this plot and print out the unknown results in the desired units.

IMMUNORADIOMETRIC ASSAYS (IRMAs)

In this type of immunological assay, the antibody is labelled with a radioisotope. The first assay of this type was published in 1968 by Miles and Hales. This also was an assay for insulin. An excess of a labelled, purified anti-insulin antibody was incubated with the sample. After the reaction had gone to completion, the excess antibody was removed by adding a solid phase insulin and centrifugation:

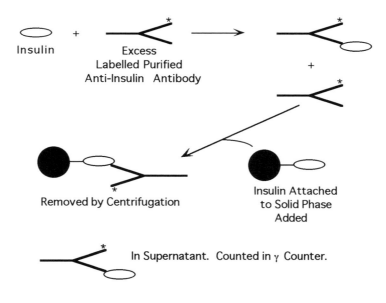

The results obtained are illustrated below:

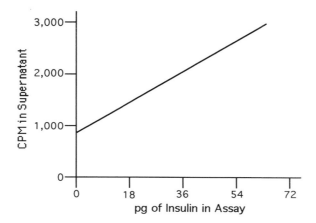

The sensitivity of the assay was limited by the non-specific binding (the zero standard value). With the introduction of monoclonal antibodies, in 1975, pure antibodies became available and the non-specific binding was greatly reduced. Hence, immunometric assays use monoclonal antibodies.

Most current IRMAs are "Two-Site" IRMAs. Two monoclonal antibodies are used; one is attached to a solid phase and the other labelled with ^{125}I. The system usually involves two incubations; the first with the labelled antibody and the second with solid phase antibody:

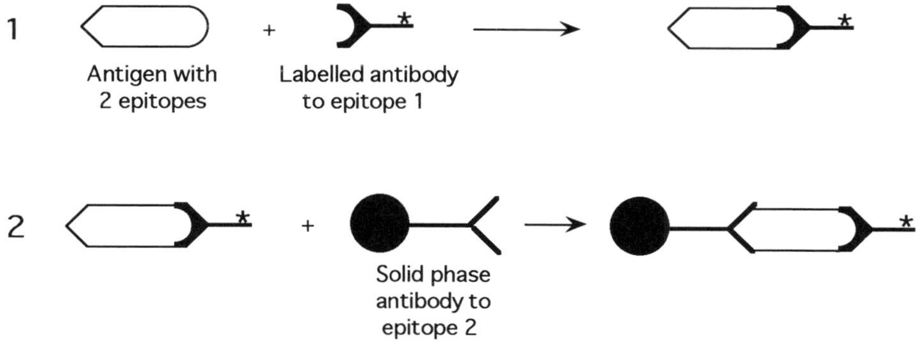

After the second incubation, the solid phase antibody is sedimented by centrifugation. In some assay systems, both antibodies are combined for a single incubation. Centrifugation sediments the antigen with its attached labelled antibody. This can then be washed and counted.

The count rate is directly proportional to the antigen concentration. A linear/linear plot is usually used for low valves and a log/log plot for other values, so that full assay range can be covered in two plots.

13-19

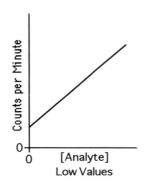 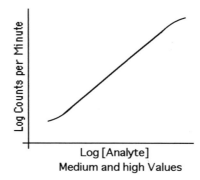

[Analyte]	Log [Analyte]
Low Values	Medium and high Values

As can be seen, the ends of the log/log plot do tail off slightly to give a slightly sigmoidal curve.

IRMAs have a number of advantages and disadvantages, when compared with RIAs. These are:

Advantages.

1.   Ease of Radiolabelling.

Antibody molecules contain a number of tyrosine residues. These can easily be iodinated (See the section of radioisotopes, later in this chapter). Most small antigens are difficult to label. They may require the attachment of a labelled molecule.

2.   Faster Reaction Rates.

Excess antibody drives the reaction to completion faster than equilibrium is reached in RIA. Monoclonal kinetics is simpler than polyclonal kinetics. Hence, the reaction rapidly progresses to completion.

3.   Expanded Working Range.

3-4 orders of concentration are obtained with IRMAs; whereas, only 2-3 orders of concentration are obtained with RIAs. This is illustrated below, by comparing the precision profiles of an IRMA and a RIA for the same analyte:

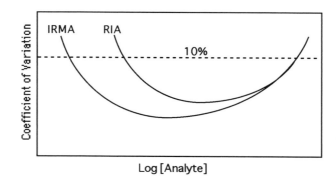

As can be seen, the precision rapidly deteriorates at both ends of the analytical range. However, the IRMA precision is maintained to about one order of magnitude lower. Where the coefficient of variation is 10%, or higher, the assay results will be unreliable.

4.   Increased Sensitivity.

With IRMA, all the antigen is in a labelled complex; whereas, only a proportion of the complex is labelled in RIA, and limiting amounts of antibody are used.

IRMA sensitivity depends upon the level of non-specifically bound antibody. The sensitivity increases as the non-specific binding is reduced.

5.   Higher Specificity for Analyte.

Recognition of the analyte is by two antibodies in IRMA, instead of one in RIA. Hence, the assay system is more specific.

However, interferents may bind to one of the antibodies and thus reduce the amount available to bind the analyte. This will lead to low results.

6.   Improved Assay Robustness.

IRMAs are less sensitive to temperature variations than RIAs.

Disadvantages.

1.   Analyte Molecular Size.

The antigen (analyte) must have two antigenic determinants (epitopes).

2.   "High-Dose Hook" Effect.

Standard curves invert at very high analyte concentrations:

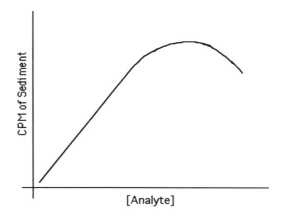

The solid phase-antibody becomes limiting at high analyte concentrations. This leaves free analyte in solution. This free analyte then competes for the labelled antibody:

3.   Robustness.

Monoclonal antibodies exhibit narrow optimum pH and ionic strength ranges, compared with polyclonal ones.

4.   Anti-immunoglobulin Antibodies in the Sample.

Some individuals have antibodies, in their plasma, that recognise mouse immunoglobulins. These can bind both assay antibodies, if mouse monoclonal antibodies are used in the assay. This leads to high results.

5.   Relatively High Consumption of Pure Antibody.

Only small amounts of antibody are needed for RIA, and this antibody does not need to be purified. However, relatively large amounts of pure antibody are required for IRMAs.

6.   An Extra Incubation Step.

Most IRMAs use two incubation steps; whereas, only one is needed for RIA.

ALTERNATIVE IMMUNOASSAYS

"Alternative immunoassays" is a term that has been used to cover immunoassays and immunometric assays that do not use radioisotopes as labels. The main reason for their use is to avoid the hazards associated with radioisotopes.

Using ^{125}I, as the label, the limit of analyte concentration in the assay mixture is approximately 100 fmol/l for RIA and 10 fmol/l for IRMA. Assuming that plasma is diluted 1 in 10 in the assay mixture, this limit covers all the commonly assayed analytes in clinical biochemistry. The detection of some tumour, viral or bacterial antigens has been suggested as a requirement for more sensitive assays.

Theoretically, non-isotopic labels can give greater sensitivity. However, most alternative immunoassays have less sensitivity than RIA or IRMA.

ENZYME LABELS.

A whole variety of enzyme labelled immunoassay and immunometric assay techniques have been introduced. Most of these use enzymes that generate coloured or UV absorbing products. Two of these will be examined.

Fluorescent or chemiluminescent products are generated in some assay systems. Most fluorescent systems use alkaline phosphatase as the label. After separating the bound and free label, the substrate (4-methyl umbelliferyl phosphate) is added. The enzyme hydrolyses off the phosphate to leave 4-methyl umbelliferone, which fluoresces. Hence, the rate of increase in fluorescence can be measured.

Two examples of chemiluminescent products are discussed under the section on chemiluminescent labels.

Enzyme-Linked Immunoadsorbent Assay (ELISA).

ELISA is a commonly used series of techniques for the identification and semi-quantitation of a variety of analytes. The two major techniques are outlined below:

(a)   The Antigen is labelled.

This is an immunoassay technique for the semi-quantitation of an analyte (antigen). The enzyme label is usually alkaline phosphatase, peroxidase, or urease.

The antibody to the antigen is bound to a support, such the walls of a tube or microtitration plate wells. The binding is usually by adsorption, but can be by chemical attachment.

The sample is added, together with a fixed amount of labelled antigen. Hence, there is competition between the labelled antigen and the antigen in the sample for the limited amount of bound antibody.

After equilibrium is reached, the tubes or wells are washed and the substrate added. The generation of a coloured product is observed. Enzyme activity is inversely proportional to the amount of antigen in the sample and hence the rate of colour development can be used to semi-quantitate the amount of analyte present. The steps in the assay are illustrated below:

1. Adsorb antibody to the surface of the wells.

2a. Add enzyme labelled antigen   2b. Add enzyme labelled
    + Standard or "Unknown".   antigen.

     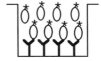

3. Wash, add substrate and measure product (ο).

(b)   The Antibody is labelled.

This is an immunometric assay technique and is usually used for the detection of the antigen in a sample.

An excess of the first antibody is chemically bound or adsorbed to a support, such as a microtitration well. The sample is added and the antigen, if present, will bind to this antibody. The second (enzyme labelled) antibody is added, or is already present. This then binds to the attached antigen. After washing, the substrate is added and the wells observed for the presence of coloured product (o):

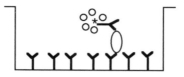

The technique can be used semi-quantitatively, as the enzyme activity is directly proportional to the amount of antigen in the original sample.

An example of this type of assay is the CSL Snake Venom Detection Kit for the identification of the type of snake that has bitten a patient. The kit consists of an 8 well microtitration strip. One specific antibody is attached to each well wall. The second (peroxidase labelled) antibody is freeze dried in the well. Two drops of the diluted sample (preferably a bite site swab or urine) are added, mixed and allowed to stand for 10 minutes. The wells are then washed and the substrates (o-tolidine) and peroxide added. The wells are observed for 10 minutes. The first well that turns blue indicated the type of snake that has bitten the patient. This information can then be used in the selection of which anti-venom should be used to treat the patient. A positive reaction is illustrated below:

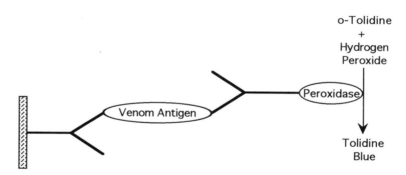

Enzyme Multiplied Immunoassay Technique (EMIT).

This is a homogeneous immunoassay system for haptens, such as drugs. The Hapten (Drug) is attached to an enzyme, without reducing the enzyme activity. The antibody recognises both the labelled and unlabelled hapten in the standards or samples.

Binding of the labelled hapten, by the antibody, reduces the enzyme's activity. This is due to a blocked active site and/or a conformational change in the enzyme.

This is technique used in the Syva EMIT®-Tox system. This is a qualitative, or semi-quantitative, assay system for the presence of drug groups in serum or plasma. Different drugs, or their metabolites, in a drug group give different levels of response in the assay. The enzyme used is a bacterial glucose-6-phosphate dehydrogenase, which catalyses the following reaction:

$$\text{G-6-P} + \text{NAD}^+ \xrightarrow{\text{G-6-P D}} \text{Gluconolactone-6-P} + \text{NADH} + \text{H}^+$$

Hence, the reaction can be followed by observing the increase in absorbance at 340 nm. The rate of increase in absorbance indicates the level free drug-enzyme conjugate in the reaction mixture.

If a member of the drug group is present in the sample, it will compete with the conjugated drug for the limited amount antibody. Therefore, as the concentration of the drug in the sample increases, more enzyme-conjugated drug remains free. Hence, the enzyme activity increases:

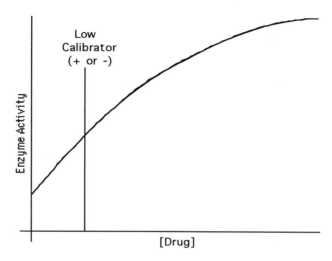

As can be seen, low (+ or -) cut-off calibrators are usually used to decide if the drug is present, or not, in significant quantities.

FLUOROPHORE LABELS.

A variety of fluorophore labelled immunoassay and immunometric assay techniques have been introduced. Two of these will be examined.

<u>D</u>elayed <u>E</u>nhanced <u>L</u>anthanide <u>F</u>luorescence <u>I</u>mmuno<u>a</u>ssay (DELFIA).

This system is marketed by Pharmacia, which uses the long fluorescent life of a europium complex. This complex consists of a β-diketone and tri-octylphosphine oxide surrounding a europium ion. In the Pharmacia instrument, the complex is excited with a 340 nm light pulse of less than one microsecond. The fluorescence emission is measured at 613 nm during the period of 400 to 800 microseconds after the flash. Fluorescence from any other compound present will have decayed before measurement is made. 1,000 μs (1 ms) after the flash, the cycle is repeated:

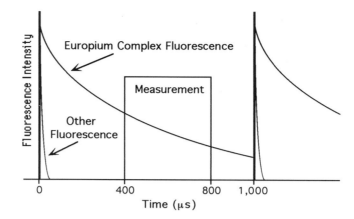

The cycles are repeated 1,000 times and the fluorescence values added. Hence, a result is obtained after one second.

The europium can not be directly attached to the compound that it is being labelled. Hence, EDTA is attached and the europium is held to the EDTA by ionic links. After the assay reaction has gone to completion, An "Enhancement" solution is added. This dissociates the europium from the EDTA and forms the fluorescent complex.

Fluorescence Polarisation Immunoassay.

Abbott's TD_X is an example of this type of assay. This is a homogeneous immunoassay system for drugs. The drug is labelled with fluorescein, which absorbs blue light (485 nm) and emits green light ($\cong$ 540 nm). The layout of the instrument is illustrated below:

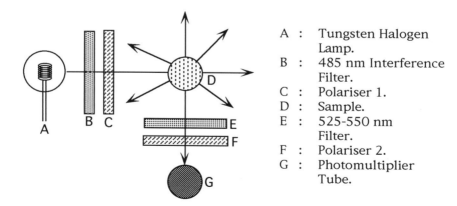

A : Tungsten Halogen Lamp.
B : 485 nm Interference Filter.
C : Polariser 1.
D : Sample.
E : 525-550 nm Filter.
F : Polariser 2.
G : Photomultiplier Tube.

The labelled drug (Drug*) competes with the unlabelled drug (Drug°) in the standards, or unknowns, for a limited amount of antibody:

Drug*
          +   Ab   ⟶   Ab-Drug*

Drug°   Ab-Drug°

The Ag-Drug* is large and, therefore, rotates slowly in solution. Hence, the orientation of the fluorescein label does not change in the time between excitation and emission. As a result, the emitted light is still polarised and will pass through polariser 2. The Drug* is relatively small and will rotate more rapidly. Hence, the orientation of the label will change in the time between excitation and emission. As a result, the emitted light will no longer be polarised. Therefore, less light will pass through polariser 2. Hence, the detected light emission decreases as the analyte concentration increases:

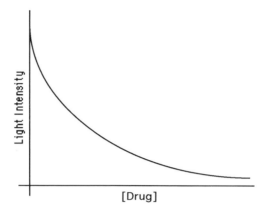

CHEMILUMINESCENT LABELS.

Chemiluminescence is used in two ways for monitoring immunoassays and immunometric assays. The luminescent chemical can be used directly as a label or the chemical is used as a substrate, where the actual label is an enzyme. Examples of both these uses will be examined.

Acridinium esters are used as labels by Ciba-Corning. These can be attached to the antibody or antigen. After the completion of the reaction, or attainment of equilibrium, the bound and free forms of the label are separated. Light is generated by the addition of hydrogen peroxide and sodium hydroxide. The reactions are illustrated below:

R is the antigen or antibody.

* Reaction Initiation : 1. Hydrogen peroxide in nitric acid. 2. Sodium hydroxide.

The hydrogen peroxide is in nitric acid to maintain its stability. An excess of sodium hydroxide is then added to start the reactions.

As can be seen, the product (10-methylacridone) is released from the antibody or antigen and emits a photon with a wavelength of about 430 nm.

The light pulse that is generated is rather short, as illustrated below:

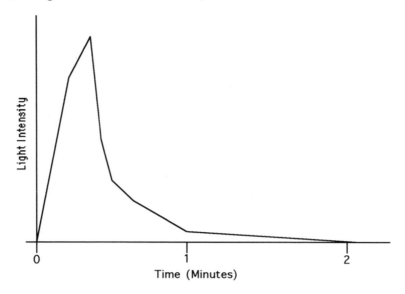

The amount of light emitted during this pulse is determined, using a photomultiplier system with a photocathode.

Where the luminescent chemical is used as a substrate, the light emission lasts for much longer and light intensity is measured.

DPC Corporation uses alkaline phosphatase as the enzyme label and adamantyl dioxethane phosphate as the substrate. Hydrolysis liberates the phosphate and an unstable adamantyl dioxethane anion. This then releases light.

Amersham International uses peroxidase as the enzyme label and luminol as the chemiluminescent substrate. The peroxide catalyses the oxidation of luminol, in the presence of hydrogen peroxide, alkali and an enhancer (Eg. 6-hydroxybenzothiazole). This is illustrated, below, for an immunometric assay:

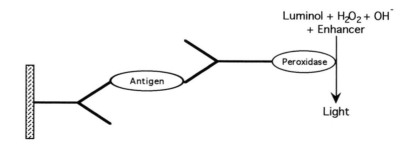

IMMUNOPROBES AND IMMUNOSENSORS

Immunoprobes.

Immunoprobes use the technique of ambient analyte immunoassay. In this technique, antibodies to the analyte (Eg. free hormone in plasma) are attached to a probe. The proportion of antibody sites that are occupied is proportional to the free hormone concentration and the antibody's affinity constant. This is independent of the sample volume.

The probe is added to neat sample and the hormone is allowed to bind to the antibody. The bound and free antibody sites are then quantitated. Where there is uniform distribution of antibody, and the same total amount of antibody, per probe, a radioisotope technique can be used. One probe is incubated with the sample. After equilibrium has been reached, the probe is washed. This probe, and a second probe, are then incubated with an excess of radiolabelled analyte, so that all the binding sites are filled. After washing, the radioactivity of the probes is determined and the proportion of sites occupied by the sample analyte is calculated. This is directly proportional to the analyte concentration.

The volume of sample must be large enough for the binding of analyte to the probe not to produce a significant drop in analyte concentration in the sample. Where this technique is used for the determination of free hormone levels, the probes should also not bind enough free hormone in the sample to produce a significant effect on the equilibrium between free and plasma protein bound hormone.

Immunosensors.

Immunosensors use antibodies bound to transducers, to detect antigen binding. The transducer detects changes in the physical properties of the antibody when the antigen is bound. These changes can be electrical, optical or a weight increase.

Transducers that monitor current changes are known as field effect transistors or ImmunoFETs. The most successful electrical transducers observe piezo-electric effects. One of this type of transducer uses antibody-coated crystals. The binding of antigen to the antibody produces a resonant frequency change in the crystal and a surface acoustic wave change.

Optical transducers use plasmon resonance effects or Evanescent wave systems.

Plasmon resonance involves total internal reflection of light at the interface of two media having different refractive indices. This reflection occurs when the angle of the incident light is greater than a certain value and the light is coming from the medium with the higher refractive index. If the interface is coated with a thin layer of a conducting metal, such as gold, plasmon resonance can occur. At a specific angle, for light of a specific wavelength, there is a marked reduction in reflected light:

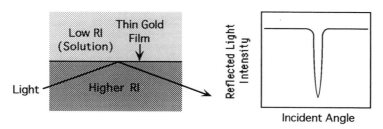

If the gold film is coated with antibodies, the incident light at a specific angle will be absorbed. If these antibodies then bind their antigens, this wavelength is changed (due to an effective change in refractive index). Hence, if the first wavelength is used in the instrument, the reflected light will increase as the antigen binds to the antibody.

Evanescent wave transducers involve an optical fibre coated in antibody. Light is fully transmitted along the fibre, when no antigen is bound, due to total internal reflection. Where the antigen is bound to the antibody there is an effective increase in the refractive index of the liquid surrounding the fibre. Hence, some light is no longer reflected and leaves the fibre. Therefore, less light reaches the end of the fibre as the amount of bound antigen increases:

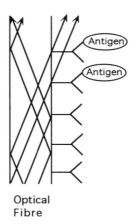

Optical
Fibre

Current immunosensors lack sensitivity.

RADIOISOTOPE TECHNIQUES

Radioisotopes are naturally occurring, or artificially produced, isotopes that exhibit radioactivity. When these isotopes decay, they emit α or β particles and/or γ rays. An α particle is a helium nucleus (2 protons and 2 neutrons), a β particle is an electron and a γ emission is a high energy photon.

The major use of radioisotopes in clinical biochemistry is as labels of DNA and of antigens or antibodies in immunoassays. These isotopes are β and/or γ emitters.

^{32}P, ^{33}P, ^{35}S and ^{3}H are used to label DNA probes and can easily be incorporated into the probe during synthesis. These isotopes are β emitters, with ^{32}P producing the highest energy and ^{3}H the least energy. The high energy emitters are localised by placing a photographic film in contact with the source. ^{3}H labelled probes are used to localise insitu DNA sequences. In this case, a photographic emulsion is poured directly onto the specimen.

The most commonly used isotope in immunoassays is ^{125}I. This is a low energy γ emitter. ^{3}H is sometimes used as the label, especially with steroids. Other isotopes, such as ^{57}Co labelled vitamin B_{12}, can be used where the element is part of the structure of the analyte.

γ and β emitting isotopes will be illustrated by discussing ^{125}I and ^3H.

$$^{125}I$$

Decay Process.

The ^{125}I nucleus is proton rich and captures an orbiting electron (electron capture), from the inner (K) electron shell, to produce an energy rich nucleus. Part of this energy is released as a neutrino (v). Due to the depletion of an electron on the inner electron shell, the atom is still in an unstable state. Therefore, electrons move from the outer shells to the depleted inner shell. During this process, γ and/or X-ray photons are emitted.

$$^{125}_{53} I + ^{0}_{-1}e \longrightarrow ^{125}_{52} Te + \gamma \text{ photon(s)} + ^{0}_{0}v$$

As can be seen the atomic weight is unchanged, but the atomic number has been reduced by one, as one proton has been converted to a neutron. Hence, tellurium has been formed.

The energy associated with the γ emission is 35.5 keV; whereas, the total decay energy is 149 keV.

^{125}I has a half life of 60 days. This means that the isotope has a relatively short shelf life. However, compounds labelled with ^{125}I have relatively high specific activity (radioactivity per unit mass).

Labelling.

Due to the fact that the amino acid tyrosine readily reacts with free iodine, ^{125}I can be used to iodinate tyrosine in proteins (antibodies) or tyrosine that can then be attached to the analyte. Some drugs will react directly with iodine and can, therefore, be labelled directly.

The mechanism for labelling is to generate free $^{125}I_2$ by oxidation of the 125 iodide. After iodination, the excess $^{125}I_2$ is converted back to $^{125}I^-$ by a mild reducing reagent. This also stops any oxidation of -SH groups in the protein.

Iodine can be incorporated into either/or both of the 2 positions shown below. However, only one iodine is usually incorporated:

$$HO-\underset{\underset{^{125}I}{}}{\overset{\overset{^{125}I}{}}{\bigcirc}}-CH_2-\underset{\underset{H}{}}{\overset{\overset{COOH}{}}{C}}-NH_2$$

Attachment of iodinated tyrosine to analytes is usually by using the methyl ester and attachment via the amino nitrogen.

After iodination, the labelled analyte, or antibody, is purified to separate the product from damaged product and unreacted isotope.

Quantitation.

As already mentioned, ^{125}I is a γ emitter. The radioactivity is quantitated by counting the number of γ photons emitted in a given time. This is achieved in a crystal scintillation counter.

The sample is placed in a well that is in the centre of a thallium activated sodium iodide crystal or in a well between two thallium activated sodium iodide crystals. The crystals are surrounded by an aluminium housing that prevents water vapour reaching the hygroscopic crystal, as well as shielding it from external light. A photomultiplier is linked to each crystal by a quartz window. The process of crystal scintillation is explained below:

1.   A proportion of the γ radiation is absorbed by the crystal. High atomic number elements (Eg. Thallium [81] and Iodine [53]) have greater absorption than lower ones, due to the large number of orbiting electrons. A direct hit by the γ photon on an orbiting electron knocks it out of orbit, producing a secondary electron with energy just below that of the γ-ray. This is known as photoelectric absorption.

     Where all the energy of the γ photon is not absorbed by the electron, the photon will be deflected and continue on with reduced energy. The secondary electron, in this case, will have less energy than that produced by photoelectric absorption. This type of interaction is known as Compton scattering. The former process is the most important where the absorbing material has a high atomic number and the γ radiation is below 0.5 MeV, as is the case with ^{125}I.

2.   These secondary electrons travel a short distance within the crystal, imparting some of their energy to other electrons in their path.

3.   Some of these excited electrons, in returning to their ground state, lose part of their excess energy as photons with energy in the UV or visible range. The number of photons produced is proportional to the energy of the γ-ray.

4.   A proportion of these photons strike the cathode of a photomultiplier (PM) that is in close contact with the crystal, causing the emission of electrons.

5.   These electrons are amplified through a number of dynode stages to produce a short duration voltage drop at the anode.

6.   Where 2 crystals are used, these pulses are detected by both photomultipliers at approximately the same time. The pulse height, seen by each PM, may be different, due to a different number of photons striking the cathodes of the PM tubes.

     Therefore, the two pulse heights are added, after amplification, by the summation circuit. This produces one pulse, representing a single radioactive disintegration.

     The height of these summed pulses is proportional to the energy of the original γ photon.

7.   These pulses are then simultaneously analysed by pulse height analysers and the desired pulses are counted on the scalers. The system is illustrated below:

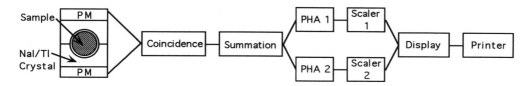

Within each channel, the pulse is amplified by a linear amplifier (modified by a "Gain" control) and the desired pulses are selected by an energy window, using lower and upper pulse-height discriminators. These are then passed to the scalers.

8.   The scaler provides the data that is displayed by the instrument and is fed to a computer and/or the printer.

The volume of the solution counted should be kept constant, as the efficiency of the counting process is reduced with large volumes:

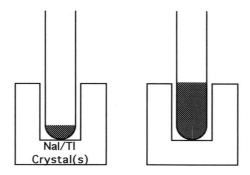

As can be seen, a greater proportion of the emissions from the larger volume sample will be lost, and not interact with the crystal(s).

Tritium (^{3}H)

Decay Process.

Tritium has a neutron rich nucleus (one proton and two neutrons). One of these neutrons decays to produce a proton and an electron (the β particle). However, an antineutrino is also produced and the decay energy is divided between the β particle and the antineutrino ($\overline{\nu}$). Hence, the emitted β particle can have a number of energies.

$$^3_1\text{H} \longrightarrow \ ^3_2\text{He} + \ ^0_{-1}\text{e} + \ ^{-0}_0\overline{\nu}$$

As can be seen, the atomic weight has remained constant, but the atomic number has increased by one.

The maximum energy of the β emission is 18.6 keV. The average emission is only 6 keV.

The energy of the β particle determines its penetrating power. For example, a β particle from ^3H will not penetrate the skin; whereas, one from ^{32}P will penetrate clothes and skin and into the underlying tissue.

^3H has a half-life of 12.3 years. This means that the isotope has a relatively long shelf life. However, compounds labelled with ^3H have relatively low specific activity.

Labelling.

Tritiated compounds are usually obtained commercially where one, or more, hydrogen atoms in the molecule has been replaced by tritium.

Quantitation.

Due to the short penetrating power of most β particles (especially that of tritium) crystal scintillation counters can not be used. However, the β particles can be made to excite fluorophores in solution. Hence, liquid scintillation counters are used for low and medium energy β emitters.

In the liquid scintillation system, the sample is added to an organic liquid scintillator solution. The β particles, emitted from the sample, cause the scintillator to emit short pulses of light, which are detected by a photomultiplier. The simplest system consists of an aromatic fluorescent solute dissolved in an aromatic solvent. The process of scintillation is illustrated and explained below:

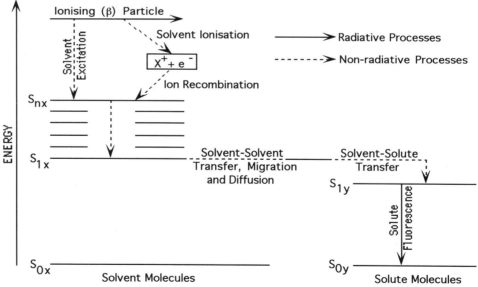

The ionising particle causes excitation of solvent (X) molecules and ionisation of solvent molecules ($X \rightarrow X^+ + e^-$). The ionised solvent molecules rapidly recombine with slow electrons to yield excited solvent molecules with the energy levels S_{nx}. This is followed by rapid internal energy conversion to the lowest excited singlet state of the solvent molecules, S_{1x}, and the loss of excess energy as heat. The energy of the excited solvent molecules is transferred to other solvent molecules in close proximity and thence to the solute (Y) molecules. The lowest excited singlet state of the solute molecules must be below that of the solvent to ensure a high efficiency of energy transfer. On transfer, the energy of the solvent returns to the ground state (S_{0x}) and the energy of the solute is elevated to its lowest excited singlet state (S_{1y}). The solute molecule on returning to its ground state (S_{0y}) liberates its excess energy as fluorescence photons.

1. The energy of the β particle, emitted from the isotope, is lost by ionisation and excitation of solvent molecules. About 10% of these excited solvent molecules rapidly reach the lowest excited π electronic singlet state. The number of molecules in this state is proportioned to the energy of the emitted β particle.

2. Some of these solvent molecules collide with solute molecules raising their energy to the lowest excited π electronic singlet state.

3. A portion of these excited solute molecules return to their ground state by emitting fluorescence photons.

4. A fraction of these photons strike the cathode of the photomultiplier tube causing the emission of electrons.

5. The emission of electrons is amplified by a series of dynodes until a final, short duration, voltage drop is produced at the anode. These are counted on the scaler of the instrument, after being processed by the pulse height analyser.

The strength of the voltage drop at the anode (pulse height) is proportional to the energy of the emitted β particle. The instrument can be adjusted to select scintillations of desired energy levels. This is done with the gain and window settings of the pulse height analysers. The gain amplifies the anodic pulses and the window is set by lower and upper discriminators.

Most instruments are fitted with 2, or more, pulse height analysers (Channels 1,2,3 etc; or red, green and blue channels). Pulses of the desired energy are then passed onto the scalers for counting.

As photomultipliers tend to produce "dark noise", two photomultipliers are used in most instruments. These are linked so that only light flashes are counted; ie. only impulses occurring at the same instant of time in both photomultipliers are fed to the pulse height analysers after being added together.

Illustrated below, is a functional diagram of a 2 channel instrument:

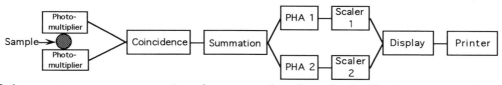

One of the most common aromatic solvents used in liquid scintillation systems is toluene. Triton X - 100, which renders the toluene able to mix with a limited amount of aqueous solution, is usually included in the scintillation mixture.

A commonly used fluorescent solute is 2,5-diphenyloxazole (PPO). It has a fluorescence emission maximum at 365 nm.

A typical scintillation solution has the following composition:

PPO	:	6 g/l toluene.
Toluene	:	2 parts.
Triton X-100	:	1 part.

10 ml of this scintillation solution will mix with up to 1.5 ml of aqueous sample.

A secondary solute can be used in addition to the primary fluorescent solute. This has an absorption maximum at the emission maximum of the primary solute and, therefore, has maximum fluorescence emission at a longer wavelength.

Light from the fluorescence of the primary solute is absorbed by the secondary solute and some energy is also transferred from the primary to the secondary by dipole-dipole resonance transfer. The secondary solute then emits light detected by the photomultipliers.

A common secondary solute is 1,4-di-(2-(5-phenyl-oxazolyl))-benzene (POPOP). This has a maximum absorption at 365 nm and a fluorescence emission maximum at 415 nm.

The use of a secondary solute depends on the response characteristics of the photomultiplier. With a bialkali cathode, as in modern instruments, the use of POPOP is not recommended unless coloured solutions are being used, as it lowers the efficiency of the system. Coloured solutions tend to absorb lower wavelength light more than longer wavelength light.

Quenching.

Fluorescence quenching is the major problem with liquid scintillation counting. It can be divided into two groups:

1.   Impurity quenching.

This is due to collisions between quencher molecules and excited solvent or excited fluorescence solute molecules, thus removing energy that would be lost as fluorescence.

2.   Colour quenching.

This is due to the absorption of fluorescent light by the coloured molecules, thus reducing the light reaching the photodetectors. A secondary solute tends to reduce this, as the longer wavelength of the emitted light is usually less absorbed by coloured solutions.

Quench Correction.

There are three main methods for quench correction. These are:

1.   Internal Standardisation.

This is the best method, but is also the most time consuming and costly. A known quantity of radiation, of the isotope being used, is added to the sample after it has been counted. The sample is counted again, the difference in counts being due to the added activity. The counting efficiency can thus be calculated and this figure used to correct the original count to disintegrations per second.

2.   Channels Ratio.

A plot of count rate against pulse height, for a given isotope, gives a rather flat curve that falls off to zero as the pulse height increases. The effect of quenching on this curve is to compress it so that the majority of the spectrum is in the lower pulse height region. As a result, the ratio of two different counting channels will change as quenching increases. This is illustrated below:

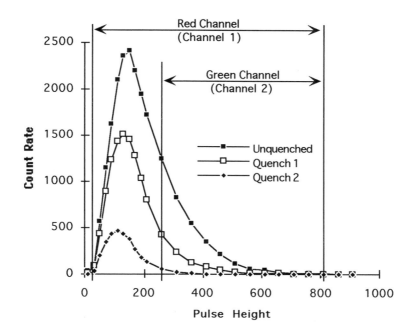

Frequency of tritium pulse heights with increased quenching.
Quench 2 is greater quenching than quench 1.

One channel, the red (Channel 1), is set for optimum counting of the isotope and the other channel, the green (Channel 2), is set with a narrower window so that a plot of the percentage efficiency against the ratio of the green/red (2/1) counts gives as close to a straight line as possible.

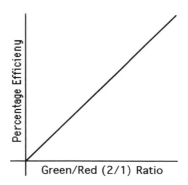

3.   External Standardisation.

In this method a γ emitter, ^{226}Radium, is brought into close proximity to the vial being counted. The sample is counted first with the γ source shielded behind a thick lead shield. The source is then brought into position and the sample is counted again.

Channel 1 (red channel) is set for optimum counting efficiency of the isotope. The second channel (green channel) is often set with its lower discriminator set above the maximum β energy of the isotope being used. The upper discriminator is set to infinity. This channel then counts the γ emission.

A plot of the % efficiency against the γ counts gives a straight line.

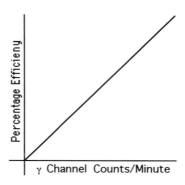

Counting Statistics.

Radioactive disintegrations follow a Poisson distribution. The standard error of the mean ($\bar{x}$) of n counts on the same sample is:

$$\sqrt{\frac{\bar{x}}{n}}$$

The confidence limits of this mean:

$$\bar{x} \pm d\sqrt{\frac{\bar{x}}{n}}$$

Where d (standard deviation) covers the desired percentage. For example, 95% confidence limits are obtained when d = 1.96.

For a single count (x), the standard error becomes:

$$\sqrt{x}$$

Hence, 95% confidence limits are given by:

$$x \pm 1.96\sqrt{x}$$

Eg. 1. For 10,000 counts, 95% confidence limits = 10,000 ± 196. ie. one can be 95% sure that the true count lies between 9,804 and 10,196 (approximately a possible 2% error).

2. For 100 counts, 95% confidence limits = 100 ± 19.6. ie. one can be 95% sure that the true count lies between 80 and 120 (a possible 20% error).

Therefore, it is desirable to obtain about 10,000 counts to keep counting error to a reasonable level.

Background and Precision.

If an empty tube, or liquid scintillation vial containing scintillant only, is counted, a finite count is obtained. This is the background count and is mainly due to cosmic radiation. This background limits the ability of the instrument to obtain reliable values for low activity samples, due to variations in the background count.

When the source strength is 5-10 times the background, the variation in background has very little effect on precision. When the background is equal to the source strength, the standard error is doubled. To obtain the same precision, the counting time should be increased by a factor of 2^2; ie. 4.

The standard error increases exponentially as the source strength falls below background.

Hence, in order to avoid very long counting times, the samples counted should have a reasonable amount of radioactivity.

Radioisotope Safety.

There are two hazards associated with the use of radioisotopes. These are:

1.   Radiation.

This is the only hazard from a sealed (closed) source. All γ and high energy β sources are a potential hazard and, therefore, care must be taken with the use of these materials.

2.   Contamination.

This is an extra problem with unsealed (open) sources, such as powders, liquids and gases. These can enter the body via the lungs, skin and/or mouth. This is a problem with all radioisotopes. Hence, a tritiated compound is not a hazard when it is in a container. However, it is a problem if absorbed through the skin and/or lungs or is ingested. It is then that the ionising action of the emitted β particles causes tissue damage.

The effects of radiation on tissues.

Water is the main component of tissues and its ionisation produces ion pairs:

$$H_2O \rightarrow H_2O^+ + e^-$$
$$e^- + H_2O \rightarrow H_2O^-$$

Free radicals (OH^o and H^o) are also produced and these do most of the tissue damage. The major problem is the result of DNA damage that leads to cell death or to malignant mutations.

Different cell types have different sensitivity to radiation. For example, mature lymphocytes, erythroblasts and spermatogonia are the most sensitive. Muscle and nerve are the most resistant.

Shielding.

Radiation is greatly reduced by shielding. Hence, shielding should always be used where possible.

β shielding.

1 cm thick glass or 2 cm thick Perspex will absorb all β particles with energies below 4 MeV.

γ shielding.

γ shielding should be as close as possible to the source. Lead is the most common shield material. The thickness of lead required to absorb 90% of the radiation is tabulated below:

γ energy (MeV)	Thickness (cm)
0.5	1.25
1.0	3.5
1.5	5.0
2.0	6.0

Body Contamination.

Factors affecting the hazard of isotopes in the body include:

1. Amount of isotope.
2. Type of decay.
3. Decay energy.
4. Isotope half-life.
5. Biological half-life.
6. Organ concentration.
   Eg. Iodine in the thyroid.

Monitoring Contamination.

Bench surfaces should be monitored on a regular basis. Geiger tubes will detect all radiation except those from low level β emitters (Eg ^3H, ^{14}C and ^{35}S). Contamination with low energy β emitters can be detected by counting a swab, from the suspect area, in a liquid scintillation counter. Contaminated clothing and hands etc. can be monitored as above.

Film Badge.

Film badges estimate the type of radiation and energy level in exposed individuals. These should be worn by all users of X-ray equipment and people handling sources of 50 μCurie (2 MBq), or more, on a regular basis. These badges will not pick up low energy β radiation, such as that from ^3H.

Radioisotope Laboratory Rules.

The use of radioisotopes should be confined to a dedicated laboratory. If this is not possible, a section of a laboratory should be set aside for radioisotope work. A set of laboratory rules should be drawn up for use in the laboratory. A suggested set of rules can be found in Appendix 2.

EXERCISES

Immunoelectrophoresis and electroimmunoassay exercises will demonstrate immunoprecipitation techniques. Commercial kits will be used to demonstrate radioisotope and enzyme labelled immunoassay techniques. ^{125}I energy spectrum and ^3H quench correction exercises are also included.

EXERCISE 13-1

IMMUNOELECTROPHORESIS OF SERUM PROTEINS

Immunoelectrophoresis has been discussed earlier in this chapter (page 13-6). In this exercise, the technique will be carried out, using an agarose gel, on a glass microscope slide. Two sample wells and an antiserum trough will be cut in the gel. A serum (or plasma) sample and a marker will be run, using standard electrophoresis. The trough is then filled with antiserum and the slides incubated overnight, at room temperature, in a moist chamber.

The excess antibody is then removed by incubating the slide in 0.9% saline. This is followed by the drying and staining of the gel precipitin lines. After the removal of the excess stain, the slides can be dried and photographed.

The technique can be used to identify the type of immunoglobulin produced in monoclonal gammopathies and to detect specific protein deficiencies (Eg. α_1-Antitrypsin). In these techniques, normal serum is added to one well and the patient's serum to the other well. An alternative method is to have one well in the centre of the slide and troughs on either side. One trough is filled with an anti-human serum antiserum and the other with an anti-protein of interest antiserum.

REAGENTS AND EQUIPMENT:

1. Barbital Buffer, pH 8.6, 100 mmol/l.
    Dissolve 5.52 g of barbital (diethylbarbituric acid) in about 1 l of boiling distilled water. Add 35.0 g of sodium barbital. When dissolved, cool and make to 2 l with distilled water.
    Store at 4°C.
2. Normal Serum or Plasma.
    Keep on ice.
3. Electrophoretic Marker.
    Dissolve 10 mg of bromphenol blue in 10 ml of 0.9% NaCl (Reagent 5). Dissolve 1.0 g of bovine serum albumin in 1.0 ml of the bromphenol blue solution.
    Store at 4°C.
4. Agarose Solution.
    Suspend 1.0 g of agarose in 50 ml of reagent 1 + 50 ml of distilled water, in a 500 ml conical flask. Heat, with stirring, to gentle boiling so that the agarose dissolves. Allow to cool in a 60°C waterbath. Place a 5 ml graduated pipette in the flask.
    Prepare just before use.
5. Anti-Human Serum Antiserum.
    Keep on ice.

6.	0.9% Saline.
	Dissolve 9.0 g of Analar NaCl in distilled water and make to 1 l.
7.	Amido Schwarz Stain.
	Dissolve 4.0 g of anhydrous sodium acetate in 500 ml of distilled water, add 1.4 ml of glacial acetic acid and mix well. Add 250 mg of Amido Schwarz (Naphthol Blue Black) to this solution and mix until dissolved.
8.	2% Acetic Acid.
	Mix 10 ml of glacial acetic acid with 490 ml of distilled water.

Microscope Slides, 25 x 75 x 1.2 mm (1 x 3").
	Thoroughly clean and dry.
Levelling Plate.
	Or level surface.
Pasteur Pipettes and Teats.
	+ One Pasteur pipette attached to a suction pump.
Trough Cutter.
	Tape a razor blade to each side of a 1.2 mm thick microscope slide so that part of the blades protrudes below the edge of the slide. This will cut a trough about 1.5 mm wide.
1 and 2 µl Capillary Pipettors with Capillaries.
Electrophoresis Apparatus with Water Cooled Plate and Constant Current Power Pack.
	Eg.	LKB system (2117 Multiphor tank and power pack).
Wicks.
	Whatman No. 1 chromatography paper, wide enough to cover most of the electrophoresis plate and long enough, when folded double, to reach from the buffer to about the centre of the plate. Interleaf 3 of these doubled sheets.
Dish, partly filled with the buffer, for wick soaking.
Filter Paper (18.5 cm diameter).
37°C Incubator Oven.
Water Bath at 60°C.
100 µl GC Syringe.
Beaker Containing Distilled Water for Washing.
Humid Containers.
	Food storage boxes, with lids. Place distilled water dampened filter paper in the bottom of the dishes and attach the lids.
Dishes for Washing.
	Eg. Food storage boxes or Petri dishes.
Dish for Staining and Destaining.
	Eg. Petri dishes.
Bench Covered with Clean Brown Paper.
(Camera).

PROCEDURE:

1.	Place a clean, dry 25 x 75 mm (1 x 3") microscope slide on a level surface (Eg. a levelling table). Using the 5 ml graduated pipette (in the agarose solution), transfer 2.5 ml of the hot agarose solution onto the slide; starting at the centre and then moving the pipette around the edges, with the tip just above the surface of the slide. Make sure that the agarose is evenly distributed on the surface of the slide. Allow to set for about 15 minutes.

2.	Draw a rectangle template, 25 x 75 mm, on a piece of paper, add the application dots and centre line as illustrated below:

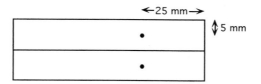

3. Place the slide on the template. Using a Pasteur pipette, suck out the 2 application wells (1-2 mm diameter) from the agarose.

4. Using 2 razor blades, taped to a microscope slide, cut a trough ($\cong$ 1.5 mm wide) in the agarose, along the midline of the slide, but leaving a few mm at each end. Do not remove the trough gel at this stage.

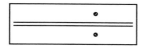

5. Using a capillary pipettor, add 2 µl of serum (or plasma) to one well and 1 µl of the protein marker to the other well.

6. Make sure that the electrode chambers are filled with buffer (ie. the buffer must be above the well dividers for both electrodes) and that the cooling water is flowing.

7. Soak the wicks (6 thicknesses of Whatman No. 1 chromatography paper) in barbital buffer and lay them on the electrophoresis plate so that they contact the buffer and leave a 65 mm gap between the wicks.

8. Invert the slide and place it in the electrophoresis apparatus so that both ends contact the wicks and the samples are towards the cathode (negative) end. Gently push down the slide to make good electrical contact with the wicks.

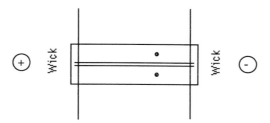

9. When the apparatus is full, place the lid on the tank and run the electrophoresis at a constant current of 6 mA per slide.

10. When the marker has migrated about 30 mm (about one hour), turn off the power and remove the slides.

11. Carefully suck out the agarose in the antiserum trough, with a Pasteur pipette attached to a suction pump.

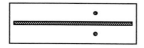

12. Add 80 μl of anti-human serum anti-serum to the trough with a 100 μl GC syringe (thoroughly wash after use).

13. Place the slides in a humid container (food storage boxes containing dampened filter paper) and incubate at room temperature, overnight. White precipitin lines, in the gel, should be seen after incubation.

14. Soak the slides in 0.9% saline (~ 100 ml each) for 24 hours and then distilled water for 2 hours.

15. Cover the slides with filter paper and allow them to dry overnight in a 37°C incubator.

16. Place the slides in the Amido Schwarz stain (0.05% in acetate buffer) for 30 minutes.

17. Pour off the stain and remove the excess stain from the gel with washes in 2% acetic acid. Allow the slides to dry. These can then be photographed and/or stored.

Identify the protein precipitin lines.

EXERCISE 13-2

ELECTROIMMUNOASSAY OF PLASMA ALBUMIN

Electroimmunoassay (Laurell's Rockets) is a technique that can be used to determine the concentration of a specific protein in a complex protein mixture, such as plasma. This technique was discussed on pages 13-7 and 13-8. This exercise will demonstrate the technique for the determination of the concentration of albumin in serum or plasma.

As discussed, the samples are loaded into wells that have been cut into an agarose gel containing a specific antiserum directed against the protein of interest. A rabbit anti-human albumin antiserum is used in this exercise.

The agarose gel is prepared in a buffer of pH 8.4, on a 50 x 50 mm GelBond base. The albumin is negatively charged at this pH and will migrate towards the anode on electrophoresis.

As the albumin (antigen) migrates within the gel, to reacts with the antibody to form an antibody-antigen complex. This complex precipitates within the gel.

As more antigen moves into the zone of the antibody-antigen complex, the antibody-antigen equivalence is displaced and the original complex dissolves. This antigen now moves into areas of fresh antibody and is again precipitated.

The amount of antigen in the leading edge is successively diminished until a final equivalence of the antigen-antibody complex is reached. At this point, a stable precipitate is formed with no further migration. Cones or rocket shaped bands of precipitate can be seen in the gel, but these are usually fairly faint.

The excess antibody in the gel is then eluted by incubating the gel with 0.9% NaCl. After a water wash and the drying of the gel, the precipitin "Rockets" will be stained with Coomassie Blue, which makes them easy to see.

The height of these rockets is directly proportional to the logarithm of the concentration of the antigen. Hence, their heights will be measured, a standard curve drawn and the original serum (or plasma) albumin concentration calculated.

Proteins with concentrations on low as 1 mg/l can be measured by this technique.

REAGENTS AND EQUIPMENT:

1. 30 mM Barbital Buffer, pH 8.4.
   Dissolve 1.84 g of barbital (diethylbarbituric acid) in about 1 l of boiling distilled water. Add 10.3 g of sodium barbital. When dissolved, cool and make to 2 l with distilled water.
   Store at 4°C.
2. 1% Agarose Solution.
   Suspend 1.0 g of agarose in 100 ml of reagent 1, in a 500 ml conical flask. Heat, with stirring, to gentle boiling so that the agarose dissolves. Allow to cool in a 52°C waterbath. Place a 5 ml graduated pipette in the flask.
   Prepare just before use.
3. 40 g/l Human Albumin Standard.
   Dissolve 4.0 g of human albumin in about 50 ml of distilled water and make to 100 ml.
   Keep on ice.
4. Rabbit Anti-human Albumin Antiserum.
   The actual volume of antiserum used in the gel may have to be altered, depending on the strength of the antiserum. If the strongest standard "rocket" runs as far as the anode wick, more antiserum should be included in the gel. If the "rockets" are short, less antiserum should be used.
   Keep on ice.
5. Human Serum or Plasma.
   Keep on ice.
6. Stain.
   Dissolve 1.5 g of Coomassie Brilliant Blue R-250 in 220 ml of methanol. Add 220 ml of distilled water and 55 ml of glacial acetic acid. Filter, using Whatman No. 1 paper.
7. Destain.
   Mix 400 ml of methanol, 400 ml of distilled water and 100 ml of glacial acetic acid.
8. 0.9% NaCl.
   Dissolve 9.0 g of Analar NaCl in distilled water and make to 1 l.

GelBond.
   (Pharmacia Biotech Cat No. 80-1129-32).
   Cut into 50 x 50 mm squares.
Test Tubes.
Wassermann Tubes.
5 and 10 ml Graduated Pipettes.
1 μl Capillary Pipettors with Capillaries.
20, 100 and 1,000 μl Pipettors and Tips.
Electrophoresis Apparatus with Water Cooled Plate and Power Pack.
   Eg. LKB system (2117 Multiphor tank and power pack).
Wicks.
   Whatman No. 1 chromatography paper, wide enough to cover most of the electrophoresis plate and long enough, when folded double, to reach from the buffer to about the centre of the plate. Interleaf 3 of these doubled sheets.
Water Bath at 52°C.
Beaker Containing Distilled Water for Washing.
Filter Paper (~ 9 cm diameter).
Paper Towels.

Glass TLC Plates.
500 ml Plastic Bottle filled with Water, (As a 500 g weight).
Hair Dryer.
Dishes for Washing.
     Eg. Food storage boxes or Petri dishes.
Dish for Staining and Destaining.
     Eg. Petri dishes.
Paper Tissues.
Bench Covered with Clean Brown Paper.

PROCEDURE:

1.   Using the 30 mM barbital buffer, dilute the 40 g/l human albumin standard as follows:

Dilution	Albumin Solution	Buffer
1 in 75	100 µl of 40 g/l	7.4 ml
1 in 100	100 µl of 40 g/l	9.9 ml
1 in 200	1.0 ml of 1/100	1.0 ml
1 in 300	1.0 ml of 1/100	2.0 ml
1 in 500	1.0 ml of 1/100	4.0 ml
1 in 1,000	1.0 ml of 1/500	1.0 ml

Mix the contents of each tube, after adding the 2 reagents.

2.   Using the 30 mM barbital buffer, dilute the serum or plasma as follows:

Dilution	Serum or Plasma	Buffer
1 in 50	100 µl of neat	4.9 ml
1 in 100	1.0 ml of 1/50	1.0 ml
1 in 500	1.0 ml of 1/50	9.0 ml

Mix the contents of each tube, after adding the 2 solutions.

3.   Place a test tube in the 52°C water bath.

4.   Leave for three minutes and then add 5.0 ml of the warm 1% agarose solution. This transfer should be made as quickly as possible to avoid the agarose setting in the pipette.

5.   Place a 50 x 50 mm piece of GelBond (hydrophilic surface upwards) on a level surface (Eg. a levelling plate).

6.   Add 30 µl of rabbit anti-human albumin antiserum to the test tube in the water bath and briefly vortex to ensure even dispersion of the antibody. Return the tube briefly to the water bath to allow any bubbles to float to the surface.

7. Remove the test tube from the water bath and pour the contents of the tube onto the GelBond. Keep the neck of the tube close to the plate and pour slowly. Surface tension will keep the liquid on the surface of the GelBond.

8. Allow the gel to set for about 15 minutes.

9. Draw a 50 x 50 mm template square, on a piece of paper, and mark the application spots, 5 mm apart and 10 mm from the bottom:

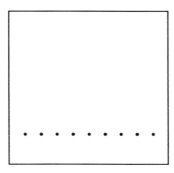

10. Place the gel on the template and suck out the application wells with a Pasteur pipette.

11. Using a capillary pipettor, add 1 µl of each of the albumin and plasma dilutions to separate wells. Wash the capillary, 2 or 3 times with distilled water between each application and dry with a paper tissue.

12. Place the gel on the cooling plate of the electrophoresis apparatus, so that the application holes are on the cathode (negative) side.

13. Align 3 gels (or more, if using a larger apparatus) on the cooling plate.

14. Make sure that the electrode chambers are filled with the 30 mM barbital buffer (ie. the buffer must be above the well dividers for both electrodes) and that the cooling water is flowing.

15. Soak the wicks (6 thicknesses of Whatman No. 1 chromatography paper) in 30 mM barbital buffer.

16. Lay them on the electrophoresis gels to cover about 5 mm of the gel and make contact with the buffer.

17. Hold the wicks in place with glass TLC plate.

18. Apply 200 volts and run for $2^{1}/_{2}$ hours.

19. At the end of the electrophoretic run, observe the protein precipitation "rockets". These can be very faint.

20. Soak the gels overnight in 0.9% NaCl (~100 ml each) and then for 2 hours in distilled water.

21. Place the gel on a glass plate. Place a sheet of distilled water dampened filter paper on the gel and then a wad of paper towels on top of this. Carefully add another glass plate and a 500 g weight.

22. Leave for 15 minutes and then dry the gel with a hair dryer.

23. Place the dried gel in a dish and cover with Coomassie Brilliant Blue R-250 (0.2 % in destaining solution) for 5 minutes.

24. Pour the stain back into its bottle and add destain (methanol : acetic acid : water [5 : 1 : 5]) to the dish. Swirl to elute the unbound stain and then discard the wash.

25. Repeat the destaining until the background is clear.

26. Stand the gel upright on a wad of paper towels and allow to dry.

Measure the heights of the rockets and plot these values against the concentration of albumin on semi-log paper. Read off the diluted serum plasma albumin concentrations. Calculate the serum or plasma albumin concentration in the original (undiluted) sample.

EXERCISE 13-3

ENERGY SPECTRUM OF ^{125}I AND COUNTER EFFICIENCY

^{125}Iodine is a proton-rich (neutron-deficient) isotope. As a result, it decays by electron capture from the K shell. Electrons then move in from the outer shells to replace this electron. These electron movements result in the emission of γ-ray and/or X-ray photons. The product of this decay is tellurium.

$$^{125}_{53}I \rightarrow \ ^{125}_{52}Te$$

The energy and abundance of the major photon emissions are listed below:

Photon	Energy (keV)	Emissions/100 disintegrations
γ-ray	35.5	6.8
X-rays:		
K_α^1	27.5	73.8
K_α^2	27.2	37.8
K_β^1	31.0	19.9
K_β^2	31.8	4.1

The NaI (Tl) detector can not separate these closely spaced energies. As a result, the pulse-height spectrum shows a single asymmetric peak, centred between 28 and 29 keV. This peak is known as the "Photopeak".

^{125}I also has a second peak that lies between 50 and 60 keV. This peak is called the "Sum peak".

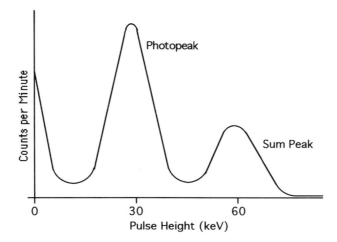

A small proportion of the sum peak is due to the emission of a photon from two separate atoms, within the resolving time of the detector. These are then processed by the instrument as a single pulse with the combined energy of both photons. This effect is negligible with count rates below 100,000 counts per minute, but increases with the increase in count rate above this level.

The major contribution to the sum peak is that a large number of ^{125}I decays are associated with the emission of two photons from the same atom. Both are emitted within the resolving time of the detector and, hence, are seen as a single pulse with the combined energy of both photons. This phenomenon is independent of the count rate.

The photopeak and sum peak will be examined, in this exercise, by scanning the energy of the photon emissions. This is carried out by counting the number of pulses within a narrow energy range (window) and increasing the energy level that is examined by the window.

Examination of the energy spectrum of a radioisotope is essential for optimum setting of the instrument.

With poor counting efficiency, one of the pair of photons, emitted by a single ^{125}I atom, may not be detected. As a result, the decay will be seen in the main photopeak region. As the efficiency of the system increases, a greater proportion of the pairs are detected and, hence, these appear in the sum peak region. Therefore, as the efficiency increases, the sum peak increases and the photopeak decreases.

Eldridge devised the following equation for the calculation of the actual number of disintegrations per second (Bq):

$$Bq = \frac{(P + 2\ S)^2}{4\ S}$$

Where   Bq   =   The activity of the sample, in Becquerel (dps).
             P   =   Counts per second of the photopeak.
             S   =   Counts per second of the sum peak.

The percentage efficiency can, therefore, be calculated from the equation:

$$\% \text{ efficiency} = \frac{100 \times T}{Bq}$$

Where   T   =   The total counts per second (ie. P + S).

The Instrument will, therefore, be set to count the photopeak and the sum peak. From these values, the percentage efficiency will be calculated.

REAGENTS AND EQUIPMENT:

1. ^{125}I Source.
           The most convenient source is the total counts tube from an assay using ^{125}I (Eg. Exercise 13-4).

γ-Counter.
           This exercise is written for a Packard Instrument Company Auto-Gamma system (5220), with 2 channels (red and green). This instrument was calibrated, during manufacture, so that a full-scale setting on the discriminator corresponds to 200 keV, on a Gain setting of 100%. On 10% Gain, the discriminator corresponds to 2 MeV on its full-scale setting, etc.
           Different instruments will probably need different settings for this exercise.

PROCEDURE:

The Radioisotope Laboratory Rules (Appendix 2) should be read and used in conjunction with this exercise.

^{125}I Spectrum.

1.   Set the following on the Spectrometer (3002):

Preset Time	:	0.5.
Minutes - Seconds	:	Minutes.
Mode Selector	:	Auto Preset.
Preset count	:	900,000.
Reject	:	Off.
Mains Power	:	HV.
Display	:	Time.

Discriminators:

Background Subtraction	:	0.
Gain	:	100%.
Window	:	2%.
A	:	0.
C	:	10.

2. Set the following on the Automatic Control Unit (578):

Sample Changer : Stop.
Display : On.

3. Place the ^{125}I source (Eg. a "TC" tube, from exercise 13-4, in the instrument. Place a green "start" insert in front and a red "stop" insert behind the sample.

4. Switch the sample changer to "Operate". If the instrument does not select the sample, press "Group Reject".

5. Once the sample is in the counting well, turn the sample changer to "Repeat" and the Mode Selector to "Manual Preset".

6. Press the "Reset-Start" button. Record the red and green channel counts.

7. After the count, reset the lower discriminators:

A : 20.
C : 30.

8. Press the "Reset - Start" button.

9. Repeat steps 7 and 8 increasing the discriminator settings in steps of 10, until a setting of 400 is reached. (ie. the next settings are A : 40 and C : 50).

The full scale on the discriminator corresponds to 200 keV. Therefore, 10 units correspond to 2 keV. The window is 2% (ie. 2% of 1,000) and corresponds to 20 units (4 keV). The centre of the window is, therefore, the lower discriminator (A or C) value plus 2 keV. (The instrument is calibrated on centre-of-the-window values).

Plot the number of counts (in 30 seconds) against the energy level of the centre of the window.

Counting Efficiency.

1. Set the red channel to count the photopeak:

Window : A - B.

A : Set this to the value of the lower discriminator that gave the lowest count between 0 and the photopeak. (As determined above).

B : Set this to the value of the lower discriminator (plus 10) to give the lowest count between the photopeak and the sum peak.

2. Set the green channel to count the sum peak:

Window : C - D.

C : Set as for B, above.

13-51

D   :   Set this to the value of the lower discriminator (plus 20) where the spectrum has returned to the background, to the right of the sum peak (As determined above).

3.   Count the source tube for 1 minute.

4.   Count an empty tube for 1 minute (Background).

From the results, calculate:

i.   The total counts per second ([the photopeak - its background] + [the sum peak - its background])

ii.   The activity in Bq, from the formula above.

iii.   The percentage efficiency of the counter, from the formula above.

EXERCISE 13-4

RADIOIMMUNOASSAY OF SERUM DIGOXIN

Digoxin is the most widely prescribed cardiac glycoside used for the control of congestive heart failure and certain abnormalities in cardiac rhythm. It is purified from digitalis, an extract from the leaves of foxgloves. Cardiac glycosides are given to increase the contraction force of heart muscle and to decrease heart rate.

The action of digoxin is to increase the intracellular calcium concentration in heart muscle, resulting in the increased strength of the heartbeat. The drug does this by inhibiting the sodium-potassium pump, giving rise to an increase in intracellular sodium concentration. This decreases the sodium gradient across the plasma membrane and thus reduces the driving force of the sodium-calcium exchanger. Hence, less calcium is pumped out of the cell. The higher intracellular sodium concentration increases the sodium gradient across the inner mitochondrial membrane. (The mitochondrial matrix is low in sodium). This increases the driving force for the mitochondrial sodium-calcium exchanger and more calcium passes from the mitochondrial matrix to the cytosol.

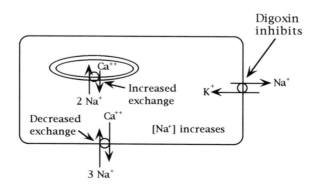

Digoxin is given orally in an initial dose to bring the plasma level into the therapeutic range. About 85% of this dose is absorbed rapidly, if gastrointestinal function is normal. Digoxin is not metabolised to any significant extent. It is mainly removed from the body by excretion in the urine. The blood level is maintained by subsequent daily doses.

Without direct determination, it is difficult to assess the blood level and glycoside toxicity can easily result due to digoxin's narrow therapeutic range. This is especially true if the patient shows signs of malabsorption or has renal failure.

It is estimated that, without the determination of the plasma level of the cardiac glycoside being used, more than 30% of patients are over or under digitalised. The direct assay of the serum digoxin is, therefore, an asset in maintaining the correct blood level.

Digoxin consists of a steroid structure, known as digoxigenin, onto which a chain of three sugar residues (digitoxose) is attached by an ether link onto position 3 of the steroid structure.

Digoxigenin   Digitoxose

Steroid structures are too small to produce an antibody response in animals and so digoxin is conjugated with albumin to produce the antigen. This is usually via portion 6 on the steroid structure, so that most of the characteristic sites on the molecule remain exposed. The antibody selected is one that recognises the digoxigenin part of the conjugate.

The therapeutic range (1-2 µg/l) of digoxin is below the detection level of most non-radioisotope labelled immunoassays. Hence, RIA is commonly used in the therapeutic monitoring of plasma digoxin levels in patients. There are a number of RIA digoxin kits on the market. The one selected here is the Amerlex kit; originally produced by Amersham International, but now produced by Ortho-Clinical Diagnostics (a Johnson and Johnson Company).

The antibody, used in this kit, is produced in rabbits and has been covalently attached to a polymer of uniform diameter (Amerlex). This facilitates the easy separation of the bound and free digoxin after the attainment of equilibrium.

The labelled digoxin is tagged with [125]I. As iodine cannot be directly incorporated into the digoxin structure, a tyrosine derivative has to be used. The tyrosine is usually attached to the terminal sugar. The tyrosine is then iodinated with [125]I. This leaves the digoxigenin structure free to act as the antigenic material.

The assay is based on the competitive binding of a fixed amount of labelled digoxin and a variable amount of unlabelled digoxin (the standards and the unknown) with a limited amount of antibody.

$$\text{Digoxin} + \text{Antidigoxin Antibody} \rightleftharpoons \text{Digoxin-Antidigoxin Antibody Complex}$$

The actual amount of antibody is usually adjusted so that there is enough to bind about half the labelled antigen (digoxin) when no unlabelled antigen is present (the zero standard). As unlabelled antigen is added to the system, this competes with the labelled antigen for the limited antibody sites; both having the same affinity for the antibody. Therefore, as the amount of unlabelled antigen is increased, the amount of labelled antigen bound to the antibody falls off exponentially.

A fixed, limiting, quantity of the Amerlex-antibody suspension is, therefore, incubated at room temperature with a fixed quantity of ^{125}I labelled digoxin and a variable quantity of unlabelled digoxin (the standards and unknown). Incubation is carried out in plastic tubes, as glass tubes tend to adsorb some of the labelled digoxin onto the walls of the tubes.

Equilibrium is reached after 30 minutes. The Amerlex-antibody suspension, with its bound digoxin, is then separated by centrifugation. After centrifugation, the supernatant, containing unbound digoxin, is discarded. The radioactivity of the precipitate is then determined in a γ counter.

A series of standard digoxin solutions (provided with the kit) are treated in the same way as the unknown and control serum samples. The digoxin level of the serum can be determined from a standard curve.

A background (Bg) count is also carried out on an empty tube. The total radioactivity added to each tube is also determined (TC). The zero standard is usually 40-60% of this figure. However, for this kit, it is higher.

Serum containing no digoxin is included in all the standards so that the equilibrium conditions are comparable with those for the unknown serum samples. The standards are, therefore, made up in human serum. These standards have the nominal values of 0, 0.3, 0.5, 0.7, 1.5, 2.5 and 5.0 µg/l of serum. However, the exact values are stated on the vial labels and on the plotting sheets.

Case History.

The serum sample is from a 75 year-old man being treated with digoxin for congestive heart failure. Due to the very narrow therapeutic range, and that some of the symptoms of overdose are similar to underdose, therapeutic drug monitoring is very important. Both overdose and underdose can be life threatening. Hence, the sample has been provided for therapeutic drug monitoring.

REAGENTS AND EQUIPMENT:

1.   Digoxin Kit.
        This exercise is written for the Ortho-Clinical Diagnostics' Amerlex kit (Code IM 2031, Cat. No. 141 8755). However, any ^{125}I digoxin kit can be used. However, the procedure may be different; in which case, the procedure that comes with the kit should be followed.

The Amerlex kit contains:
      Labelled digoxin solution (red).
      Antibody suspension (blue).
      6 standards.
      Result plotting sheets (linear/linear, with the actual standard values marked).

2.   Patient's Serum.
      Use a control serum with a stated value in the therapeutic range. Place this in a bottle marked "Patient's Serum".
      Keep on ice.

3.   Control Serum.
      Use one with a stated value. Value to be given to the students.
      Keep on ice.

Plastic Wassermann Tubes + Racks.
Bench Centrifuges.
Forceps.
Plastic Gloves.
50 and 200 ml Pipettors and Tips.
Radioisotope Waste Beaker.
Waste Bag.
      For contaminated tips, paper tissues and gloves. Attach radioactive label tape to the bag.
Paper Tissues
Vortex Mixer.
γ Counter.
      This exercise is written for a Packard Auto-Gamma counter. However, any γ counter can be used. This should be set for optimum ^{125}I counting.
Linear-Linear and Logit-Log Graph Paper.
      Photocopies of the plotting sheets can be used as the linear-linear paper.

PROCEDURE:

This exercise can be carried out in conjunction with exercise 13-3.

The Radioisotope Laboratory Rules (Appendix 2) should be read and used in conjunction with this exercise.

1.   Set up the following, in duplicate, in plastic Wassermann tubes, in the order shown (A Total Counts [TC] tube is then ready for counting in exercise 13-3):

	Standards							Unknown Serum	Control Serum
	0	0.3	0.7	1.5	2.5	5.0	TC	US	CS
μl of appropriate standard	50	50	50	50	50	50	-	-	-
μl of appropriate serum or plasma	-	-	-	-	-	-	-	50	50
μl of ^{125}I-digoxin (Red)	200	200	200	200	200	200	200	200	200
μl of Amerlex-antibody suspension (Blue) (Gently shake the bottle before pipetting)	200	200	200	200	200	200	-	200	200

13-55

2. Thoroughly mix the tubes on the vortex mixer.

3. Incubate the tubes at room temperature for 30 minutes. (The tube colour should be a deep purple).

4. Centrifuge these tubes, at top speed, in the bench centrifuge for 15 minutes.

5. Decant the supernatant from these tubes into a beaker and keep the tubes inverted for step 6.

6. Drain the tubes for 5 minutes by inverting them on a wad of paper tissues and then wipe any residual liquid from the mouth of the tubes.

7. Count all the tubes, tubes "TC" and 2 empty tubes (Bg) for 1 minute each, in a g counter, using the ^{125}I setting. If using a Packard Auto-Gamma, and having completed exercise 13-3, the procedure is:

    (a) Set the window A - B:

        A :   As set for exercise 13-3.
        B :   Set to the D setting from exercise 13-3.

    Leave the Gain on 100%.

    Alternatively the following settings can be used:

Spectrometer:

Preset time	:	1 minute.
Mode	:	Auto Preset.
Red and Green Channels:		
Background subtraction:		0.
Gain	:	100%.
Window	:	A - B.
A	:	50.
B	:	400.
Display	:	Red Channel.
Preset count	:	900,000.
Reject	:	Off.
Mains Power	:	HV.

Automatic Control Unit:

Sample changer	:	Stop.
Display	:	On.

    (b) Place a green "start" insert before the samples and a red "Stop" insert after the samples.

    (c) Switch the sample changer to "Operate". If the instrument does not select the first sample, press "Group Reject".

8. Record the actual standard values and the quoted control serum value.

Calculations:

1. Subtract the average background (Bg) count from all other counts.

2. Calculate the activity of standard 0 as a % of "TC".

3. Plot the count for each standard tube against its digoxin concentration on linear-linear paper provided (ie. 2 values for each standard) and draw the curve of best fit. This illustrates the competition between labelled and unlabelled digoxin for the limited number of antibody binding sites.

4. Calculate each standard count as a percentage of the average zero standard count (ie. 2 values for each standard, except the zero standard).

5. Plot these values against the digoxin concentration on the log-logit graph paper provided. Draw the straight line of best fit.

6. Read off the average assay values of the unknown sample and the control.

7. The molecular weight of digoxin is 781. Calculate the concentration of digoxin in the unknown and control sera in nmol/l.

Therapeutic Range:

In a study in the UK, serum samples from 219 patients, of known clinical status, were assayed for digoxin, using the Amerlex Digoxin RIA kit. The study suggests that the borderline concentration between toxic and therapeutic levels is 2.0 µg/l. It also suggests that the borderline concentration between underdigitalised and therapeutic levels is 1.1 µg/l.

Amersham International, however, suggest that each laboratory should establish their own discrimination levels, as these may vary for different patient populations.

Abdominal pain, vomiting, diarrhoea, impaired colour vision, slow heart rate and abnormal heart rhythms are symptoms of overdose. Anti-digoxin antibody Fab fragments can be injected into the patient to treat severe digoxin poisoning. These complex with digoxin and the resulting complexes are small enough to pass through the glomeruli and be excreted in the urine.

Fatigue and breathlessness are common symptoms of congestive heart failure. Oedema is usually present. Tachycardia and hypotension may be seen.

Note:

Samples should not be taken within 6 hours of the last dose of digoxin. Serum samples can be stored at 2-4°C for 48 hours; but should be frozen, if stored for a longer period.

Additional Calculation:

Calculate the mass of ^{125}I-digoxin used in the assay tubes; ie. the mass of unlabelled digoxin required to halve the zero standard activity.

EXERCISE 13-5

3H QUENCH CORRECTION

As discussed on pages 13-36 to 13-38, quenching can be a problem with liquid scintillation counting. As mentioned, there are three main methods for quench correction: internal standardisation, channels ratio and external standardisation. The latter two methods will be examined in this exercise.

A series of quenched standards will be prepared by adding chloroform to a known amount of ^3H_2O. For the channels ratio method, these vials are counted in two channels. One channel is set for optimum counting of the isotope and the other channel is set with a narrower window so that a plot of the percentage efficiency against the ratio of the two channel counts gives as close to a straight line as possible.

With external standardisation, the main channel is set for optimum counting efficiency of the isotope and the second channel is used to count the γ emissions from the external radium source. The quenched standards are counted and percentage efficiency is plotted against the γ counts.

The procedures for a Packard 2211 liquid scintillation counter and for a LKB RackBeta are described in the procedure. The procedure will have to be modified for a different counter.

In the Packard 2211, the sample vials are loaded into a sample belt. The sample to be counted is lowered into the counting chamber and viewed by two bi-alkali cathode photomultipliers. The light flashes, resulting from each disintegration, are processed as described on pages 13-34 to 13-36, by two channels (red and green). A ^{226}Ra external standard is located behind a lead shield and can be moved into position alongside the sample (the "In" position) for quench correction. Time, or counts, in either channel can be displayed and this information is sent to a printer at the end of the counting period.

With the LKB RackBeta, sample vials are loaded into racks that hold 10 vials. These are then placed into the instrument. A conveyor system then moves the samples into position under the detectors. The sample to be counted is lifted into the counting chamber by an elevator.

The sample is viewed by a matched pair of bi-alkali photomultiplier tubes. A ^{226}Ra external standard is located behind a lead shield and can be moved into position alongside the sample by pneumatic pressure applied to a stainless steel tube.

The photomultiplier high voltage is automatically adjusted to maintain the correct energy output for events detected by the instrument. Two LEDs, adjacent to the photomultipliers, are pulsed at 100 flashes per second. The detected energy levels are then compared with preset values and the photomultiplier HV adjusted until these values are reached. As the HV is automatically adjusted, there is no need for it to be on all the time. Hence, the HV is only turned on when the sample is in the measuring chamber. This increases the life of the photomultiplier tubes.

The energy spectrum of the sample being counted is displayed by the instrument. The horizontal (energy) scale is logarithmic and the spectra of unquenched ^3H, ^{14}C and ^{32}P are illustrated. This gives easy identification of the isotope being counted.

The numeric display illustrates sample number, time, counts, cpm etc.

The instrument is controlled from a computer, and results are printed out on a printer.

The computer contains 2 main programmes:

1	:	Standard counting.
2	:	Spectrum plot.

There are also 5 default counting programmes in the system:

1	:	^3H counting with channels ratio.
2	:	^{14}C counting with channels ratio.
3	:	^3H counting with ESR.
4	:	^{14}C counting with ESR.
5	:	Mixed ^3H and ^{14}C counting with ESR.

Any of the parameters (eg. counting time, window settings etc), within a programme, can easily be changed.

REAGENTS AND EQUIPMENT:

1. Scintillation Solution.
    Dissolve 6.0 g of 2,5-diphenyloxazole (PPO) in 1 l of toluene. Add 500 ml of Triton X-100 and mix well.
    Alternatively, a commercial scintillation solution, such as Packard's InstaGel, Ultima Gold and Ultima Gold LLT (Low Level Tritium) can be used. These are more efficient than the above solution, but are more expensive.
    Place in a dark bottle with a dispenser set to 10.0 ml.
2. Standardised Tritiated Water.
    Eg. Amersham International's Liquid Scintillation Standard Solution, Cat. No. TRR9.
3. Chloroform.

Glass Liquid Scintillation Vials (20 ml, 28 x 61 mm) with Caps.
20, 50, 100, 200 and 1,000 μl Pipettors and Tips.
Paper Tissues.
Disposable Plastic Gloves.
Radioisotope Waste Bottle.
    For used scintillation solution. Attach a radioactive label.
Waste Bag.
    For contaminated tips, paper tissues and gloves. Attach radioactive label tape to the bag.
Liquid Scintillation Counter.
    Any liquid scintillation counter can be used. However, these notes are written for a Packard 2211 and a LKB RackBeta. The exercise should be able to be adapted for use with other counters.

PROCEDURE:

The Radioisotope Laboratory Rules (Appendix 2) should be read and used in conjunction with this exercise.

1. Prepare the standardised tritiated water (2,000 disintegrations per second per ml) as follows:

(a) Read the specific activity and date from the bottle of standardised tritiated water.

(b) Calculate the percent remaining from the formula:

$$\% \text{ remaining} = \frac{100}{\text{Antilog} \dfrac{(0.3010 \ t)}{h}}$$

Where   t   =   The time elapsed, in months.
            h   =   The half-life, in months (147).

(c) Calculate the current specific activity.

(d) Dilute enough of this tritiated water, with distilled water, to give about 10 ml of 2,000 d/s/ml (2,000 Bq/ml).

2. Add 10.0 ml of scintillation solution from the dispenser to seven vials. Label the caps S1 to S7.

3. Add 1.0 ml of standardised tritiated water (2,000 Bq/ml).

4. Add chloroform (the quenching agent) as shown below:

Vial	S1	S2	S3	S4	S5	S6
µl of $CHCl_3$	20	50	100	200	400	600

5. Cap the vials and shake until clear.

Packard 2211.

Channels Ratio.

Set the following:

Automatic Control Unit, No. 2008.
     Sample Changer   :   Stop.
     External Standardisation   :   Out.

Spectrometer Control Unit, No. 2002.
     Mode selector   :   Auto Preset.
     Preset time   :   2 minutes.
     Preset Count   :   900,000.
     Reject   :   Off.
     Main Power   :   HV.
     Display   :   Red Channel
     Red Channel
         Gain   :   50%.
         Window   :   A - B.
         A   :   40.

```
          B            :      800.
    Green Channel
          Gain         :      50%.
          Window       :      C - D.
          C            :      255 (270 when the instrument was new).
          D            :      800.
```

6.   Load the sample vials into the sample changer. Place the green group "start" insert ahead of the samples and the red group "end" insert at the end of the samples.

7.   Switch "Sample Changer" to "Operate".

> The printer will print out:
> Sample number.
> Time.
> Red channel counts.
> Green channel counts.

Calculations:

(i).   Calculate the percentage efficiency, for each vial, using the count in the red channel (240,000 counts = 100%).

(ii).   Divide the green channel counts by the red channel counts.

Plot the percentage efficiency against the channels ratio.

External Standardisation.

8.   Reset the instrument as follows:

```
          External Standardisation         :      In.
          Preset time                      :      1 minute.
          Green channel
                Gain (2 concentric switches)   :    1%.
                Window                         :    C-∞.
                Lower discriminator (C)        :    110.
```

9.   Count all the quenched standards.

10.   After counting, return "External Standardisation" to "Out".

Plot the % efficiency of the quenched standards (as determined in the channels ratio part of this exercise) against the green channel counts determined above.

Shutdown Procedure:

(i).   Remove all samples from the sample chamber.

(ii).   Place in the sample changer conveyor belt:

    1.   A green group "Start" insert.
    2.   An empty vial.

3.   A red group "End" insert.

(iii).   With the "Sample Changer" on "Operate" and the "Mode Selector" on "Auto Preset", press the "Group Reject" button.

(iv).   When the vial has gone down into the counting area and the count started, select the following:

        (a)   Sample Charger   :   Stop.
        (b)   Mode Selector   :   Manual preset.
        (c)   Press the "Stop" button.

(v).   Turn the "Display" to "Off" and close the lid to the sample chamber.

LKB RackBeta.

6.   Count the vials (from step 5) for 2 minutes, as follows:

(a)   Turn on the computer (the programme is automatically booted).

(b)   Select the printer (F7 on the computer), turn on the printer and make sure that it is on line.

(c)   Programme the run as follows:

Ready ->		P	"Enter"
Parameter Group ->		C3	"Enter"
ID: ->		Your initials	"Enter"
Line->		1	"Enter"
Program Mode 1->			"Enter"
Count Mode	3->		"Enter"
Listing	Y->		"Enter"
Time 60->		120	"Enter"
Counts 1 900000->			"Enter"
Auto Window	N->		"Enter"
Ratio Monitor	N->		"Enter"
Channel 1	8-110->		"Enter"
Channel 2	0-0->	60-110	"Enter"
Channel 3	100-135->		"Enter"
Channel 4	135-184->		"Enter"
External Std Time 30->			"Enter"
External Std Counts 900000->			"Enter"
Print	1,2,5,7,21,22,8,10->	1,2,4,12,21	"Enter"

C clears previous parameters, and parameter group 3 sets the instrument for ^{3}H counting, with external standardisation.

The printer numbers correspond to the following:

      1   :   Sample Position.
      2   :   Time (Seconds).

4   :   Counts in Channel 1.
12   :   Counts in Channel 2.
21   :   External Standard Ratio.

(d)   Wipe the outside of the vials with paper tissues and load them into the counting racks orientated so that the 2 rectangular slots on the rack face you.

(e)   Load the racks into the right hand side of the instrument.

(f)   Load an empty rack with a "Stop" (solid) plug on its left hand side into the instrument, after the sample racks.

(g)   Type A3 and press "Enter" to start the counting.

Should you need to stop the counting, at any stage, type O and press "Enter".

7.   After the counting has finished, remove your samples.

Calculations:

Channels Ratio.

1.   Calculate the percentage efficiency, for each vial, using the count in the channel 1 (240,000 = 100%).

2.   Calculate the channels ratio (Channel 2 counts ÷ Channel 1 counts).

3.   Plot the percentage efficiency against the channels ratio.

External Standardisation.

The actual γ counts are not available in the LKB RackBeta. However, an External Standard Ratio (ESR) is provided. This is calculated as follows:

$$ESR = \frac{S_4 - I_4}{S_3 - I_3}$$

Where S_4 = Counts in channel 4 with the radium pellet in place.
              I_4 = Counts in channel 4 with the radium pellet shielded.
              S_3 = Counts in channel 3 with the radium pellet in place.
              I_3 = Counts in channel 3 with the radium pellet shielded.

            Channel 3 is set to count medium energy events and channel 4 is set to count high energy events.

Plot the % efficiency of the quenched standards against the ESR values.

Notes:

i.   Background subtraction is not necessary, as the activity is high enough so that background radiation does not have a significant effect on the counts.

13-63

ii.   The scintillation solution should be kept in a dark bottle and the vials not exposed to too much light, so that photoluminescence is avoided.

iii.   The glass of the scintillation vials must be clean, so that light is not absorbed. Wipe each vial with paper tissues before you load them into the sample changer or racks.

EXERCISE 13-6

COMPETITIVE PROTEIN BINDING ASSAY OF URINE CYCLIC AMP

This assay is not strictly an immunoassay, as antibodies are not used. However, a specific cAMP binding protein is used instead of an antibody. This is the regulator subunit of a cAMP-dependent protein kinase. This type of assay is known as a competitive protein binding assay.

Adenosine 3'5'-cyclic monophosphate (cAMP) has been known, since the early 1960's, to act as a second messenger in producing an intracellular response to the primary message of several hormones.

The hormone - cAMP - physiological effects interactions are shown below:

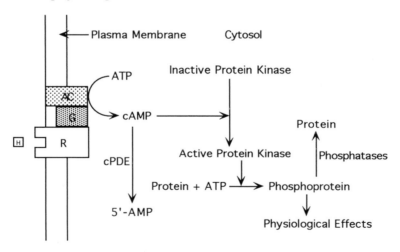

The GTP regulatory site (G) transfers the information, that the hormone (H) has bound to its receptor (R), on to the adenylate cyclase (AC). This then converts ATP to cAMP. cAMP then binds to the regulator subunits of inactive protein kinases. This binding activates the kinase, which then phosphorylates specific proteins that have physiological effects.

To reverse the effect, cAMP is inactivated by the action of cyclic nucleotide phosphodiesterase enzymes (cPDE), and the phosphoproteins are dephosphorylated back to the original form by the action of phosphatases.

cAMP is present in blood and urine as well as in other body fluids. The level in the blood and urine is increased by the action of parathyroid hormone (PTH) on the renal cortex, glucagon on the liver and as a result β-adrenergic stimulations.

Cyclic AMP in the urine is observed from two sources, filtered and nephrogenous. The nephrogenous source is due to active secretion. This is dependent upon the action of PTH and accounts for 20-60% of the total urinary cAMP in normal subjects.

The intravenous injection of 200 MRC units of bovine PTH results in a massive increase in plasma and urinary cAMP levels (up to 200 fold in urine and 50 fold in plasma) in normal subjects. Patients suffering from pseudohypoparathyroidism, where the action of PTH on the kidneys is blocked, show less than a four-fold rise in urine cAMP and a two-fold rise in plasma cAMP concentrations.

Nephrogenous cAMP levels in the urine have been used in the investigation of hyper- and hypoparathyroidism. Hyperparathyroidism showing an increase and hypoparathyroidism showing a decrease.

Plasma levels of cAMP rise after the intravenous injection of glucagon. The extent of the rise is greater in patients with biliary obstruction than in normal subjects or patients with hepatitis. This is partially because the liver is the major organ for the removal of cAMP from the plasma into the bile.

For urine values of cAMP to be meaningful, accurately timed urine collections and adequate renal function is essential. Efforts to overcome these problems have suggested the use of [cAMP]/[creatinine] ratios, or [cAMP] per litre of glomerular filtrate, as a means of expressing urinary cAMP output.

The Amersham International cAMP Assay Kit is used in this exercise. This uses the regulator subunits of the protein kinase, isolated from bovine muscle, as the binding agent.

A limited amount of binding agent is incubated with a constant amount of tritiated cAMP (the ^{3}H is in position 8 of the purine) and variable amounts of unlabelled cAMP (the standards and unknowns).

The amount of binding protein used in the assay is adjusted to bind about half of the ^{3}H-cAMP, when no ^{1}H-cAMP is present.

Both the labelled and unlabelled cAMP compete for the limited number of binding sites on an equal basis, as the ^{3}H-cAMP has the same binding affinity as ^{1}H-cAMP:

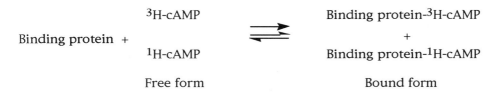

Binding protein + ^{3}H-cAMP   ⇌   Binding protein-^{3}H-cAMP

                     ^{1}H-cAMP   +

                                         Binding protein-^{1}H-cAMP

          Free form   Bound form

As the concentration of unlabelled cAMP increases, it displaces the ^{3}H-cAMP from the binding protein. Hence, the radioactivity of the free form increases, and the bound form decreases, as the concentration of ^{1}H-cAMP is increased.

After equilibrium is reached, the bound and free forms are separated by binding the free form onto dextran coated charcoal. The dextran acts as a molecular sieve and only allows small molecules to reach the charcoal, where they are bound. The c-AMP-binding protein complex is too large to penetrate the dextran and, hence, remains in solution. However, the unbound cAMP penetrates the dextran and is bound by the charcoal.

After centrifugation, an aliquot of the supernatant (containing the bound form) is counted in the liquid scintillation counter.

Incubation and centrifugation are carried out at 2-4°C, in order to reduce dissociation of the binding protein-cAMP complex, once it is formed. This can happen during the separation of the bound and free forms, as the equilibrium is upset by adding the charcoal reagent.

You will be assaying the cAMP content of your own urine sample. It is important that an accurately timed sample is obtained. Diurnal variations in plasma and urine cAMP have been reported, but these appear to be relatively small (10 - 20% from the mean level).

A series of standard ^{1}H-cAMP are treated in the same way as the unknowns and hence the amount of cAMP in the urine can be calculated. The standards and unknowns are run in duplicate.

Three additional samples are also counted in duplicate. One of these (Bg) is a sample in which no radioactive material is present. The count of this vial is due to background radiation, mainly cosmic rays, and this value is subtracted from all other counts.

The second sample (TC) is to determine the amount of radioactivity of the ^{3}H-cAMP added to samples in the assay procedure. The value of Standard O should be about 40-60% of this figure. This means that 40-60% of the added ^{3}H-cAMP is bound to the binding protein when no unlabelled cAMP is present.

The third sample (NSB) is to determine the level of non-specific binding of the ^{3}H-cAMP to plasma proteins etc. In this assay procedure no binding protein is added. The value of this sample should be less than 5% of that obtained in sample "TC", indicating that at least 95% of the cAMP has been adsorbed onto the charcoal.

HCl is added to the scintillation vials to lower the pH to about 2. This increases the solubility of the constituents.

REAGENTS AND EQUIPMENT;

1. Cyclic AMP Kit.
    > Amersham International PLC kit number TRK.432. This kit contains freeze-dried reagents. These are reconstituted as follows:
    > Tris/EDTA Buffer.
    >> Dissolve in 25 ml of distilled water.
    > Binding Protein.
    >> Dissolve in 15 ml of distilled water.
    > ^{3}H-cAMP.
    >> Dissolve in 10 ml of distilled water.
    > cAMP Standard.
    >> Dissolve in 5 ml of distilled water.
    > Charcoal Reagent.
    >> Add 20 ml of cold distilled water and a stirrer flea. Stir, on a magnetic stirrer, in an ice bath.
2. Scintillation Solution.
    > Dissolve 6.0 g of 2,5-diphenyloxazole (PPO) in 1 l of toluene. Add 500 ml of Triton X-100 and mix well.
    > Alternatively, a commercial scintillation solution, such as Packard's InstaGel, Ultima Gold and Ultima Gold LLT (Low Level Tritium) can be used. These are more efficient than the above solution, but are more expensive.
    > Place in a dark bottle with a dispenser set to 10.0 ml.
3. 0.1 M HCl.

500 ml Beakers.
> For urine collection.

250 and 500 ml Measuring Cylinders.
> For urine volume measurement.

1 ml Bulb Pipettes.

50 ml volumetric Flasks.

Glass and Plastic Wassermann Tubes.

50, 100, 200, 500 and 800 μl Pipettors and Tips.

Bench Centrifuges.
> Refrigerated or in a cold room.

Cold Room and/or Refrigerator.

Glass Liquid Scintillation Vials (20 ml, 28 x 61 mm) with Caps.

Forceps.

Paper Tissues.

Vortex mixer.

Disposable Plastic Gloves.

Radioisotope Waste Bottle.
> For used scintillation solution. Attach a radioactive label.

Waste Bag.
> For contaminated tips, paper tissues and gloves. Attach radioactive label tape to the bag.

Liquid Scintillation Counter.
> Any liquid scintillation counter can be used. However, these notes are written for a Packard 2211 and a LKB RackBeta counter.

PROCEDURE:

The Radioisotope Laboratory Rules (Appendix 2) should be read and used in conjunction with this exercise.

If quench correction is to be used, exercise 13-5 should be carried out together with this exercise.

Urine Sample.

1. Note the exact time you pass the last urine before you come to the laboratory. (Empty the bladder as completely as possible).

2. Collect a urine sample in a 500 ml beaker and note the exact time. (Empty the bladder as completely as possible).

3. Measure the urine volume.

4. Dilute 1.0 ml of urine to 50 ml with distilled water, in a volumetric flask. (This diluted urine is used in the assay.)

Assay:

Work as a group to prepare the standards. These will be run in duplicate as well as duplicate samples of the diluted urines.

Preparations of Standards:

The standard solution contains 1,600 pmol of cAMP dissolved in 5.0 ml of H_2O.

1. Set up 4 Wassermann tubes labelled 1, 2, 4 and 8 and add 500 ml of Tris/EDTA buffer to each.

2. Add 500 ml of the standard cAMP to tube 8 and mix well.

3. Transfer 500 ml from tube 8 to tube 4 and mix well.

4. Repeat this serial dilution until 500 ml is transferred from tube 2 to tube 1. Mix tube 1.

The 50 µl used in the standard assays will contain:

Original standard	:	16 pmol/tube.
8	:	8 pmol/tube.
4	:	4 pmol/tube.
2	:	2 pmol/tube.
1	:	1 pmol/tube.

Assay Procedure:

1. Add 0.8 ml of 0.1M HCl and then 10.0 ml of scintillation solution, from the dispenser, to 18 (+ 2 x the number in the group) counting vials. Label 2 caps "TC" and 2 "Bg".

2. Set up the following plastic Wassermann tubes, in duplicate; kept cold in an ice-water bath:

	0	1	2	4	8	16	NSB	Diluted Urines
µl of buffer	50	-	-	-	-	-	150	-
µl of appropriate standard	-	50	50	50	50	50	-	-
µl of 1/50 urine	-	-	-	-	-	-	-	50
µl of ^{3}H-cAMP	50	50	50	50	50	50	50	50
µl of binding reagent	100	100	100	100	100	100	-	100

3. Vortex each tube for approximately 5 seconds.

4. Place the tubes in the refrigerator or cold room for $1^1/2$ hours.

5. After about $1^1/4$ hours, make sure that the charcoal reagent is being stirred in an ice-bath.

6. After the $1^1/2$ hours incubation, remove the tubes from the refrigerator, or cold room, and place in ice-water.

7. Add 100 µl of charcoal reagent to each tube and briefly vortex.

8. Within 1 to 6 minutes of adding the charcoal, centrifuge the tubes in a bench centrifuge, in the cold room or use a refrigerated centrifuge, at top speed for 10 minutes. Use forceps to load and unload the centrifuges.

9. Without disturbing the sediment, transfer 200 µl aliquots of the supernatant to counting vials (label the caps appropriately).

10. Add 50 µl aliquots of the ^{3}H-cAMP and 150 µl of water to two of the remaining counting vials (TC - The total activity used in each assay is calculated from these samples.)

11. Add 200 µl of H_2O to the last two vials (Bg - The background activity.)

12. Shake the vials until they are clear.

13. Wipe the outside of the vials with paper tissues and count them overnight in a liquid scintillation counter. The settings for the Packard 2211 and LKB RackBeta are shown below:

Packard 2211.

Use the settings from exercise 13-5, step 8, except:

> External standardisation : Auto.
> Preset time : 10 minutes.

The printer will print out:

> Sample number.
> Time (10).
> Red channel counts (external standardisation "out")*.
> Green channel counts (external standardisation "out").
> Red channel counts (external standardisation "in").
> Green channel counts (external standardisation "in")*.

> * Only these counts are used. See below as "red" and "green" counts.

After the count, turn the "External Standardisation" to "out".

LKB RackBeta.

Use the settings from exercise 13-5, except:

> 03 Time   ->   600.

Calculations:

i.   Packard: Subtract the "red" background (Bg) count for each "red" count from all samples.

      LKB: Subtract the background (Bg) count for Channel 1 from all other Channel 1 counts.

ii.   Quench correction is carried out as follows:

      Packard: Using the "green" counts, look up the percentage efficiency of each sample from the graph plotted in exercise 13-5. Correct the "red" counts to 100% efficiency (ie. disintegrations per 10 minutes). Calculate each sample activity in disintegrations per second (ie. divide by 600).

      LKB: Using the ESR, look up the percentage efficiency of each sample from the graph plotted in exercise 13-5. Correct the Channel 1 counts to 100% efficiency (ie. disintegrations per 10 minutes). Calculate each sample activity in disintegrations per second (ie. divide by 600).

iii. Calculate the total activity that was originally in the volume counted (200 µl out of a total of 300 µl); ie. multiply the value of "TC" by 2/3.

iv.   Check that the value of standard 0 is between 40 and 60% of that calculated in (iii).

v.　　Check that the value of "NSB" is less than 5% of the figure calculated in (iii).

vi.　　Subtract the "NSB" value from the standards and unknowns and use these figures for the C_O/C_X calculations below.

vii.　　Calculate the C_O/C_X value for each standard and for the unknowns, where:

$$C_O \quad = \quad \text{the d/s for standard 0.}$$
$$C_X \quad = \quad \text{the d/s for the standard or unknown.}$$

viii.　　Plot a graph of C_O/C_X for the standards against pmol cAMP per tube:

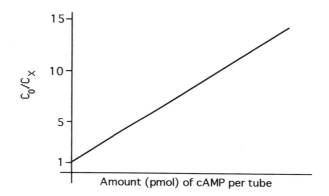

ix.　　Read off the unknown urine values.

x.　　Calculate the urinary output of cAMP in μmol/day for all members of the group.

Reference Range:

　　2-9 μmol/day.

Additional Calculations:

(a)　　Calculate the amount of ^{3}H-cAMP used in the assay tubes; ie. the amount of ^{1}H-cAMP required to halve the zero standard activity.

(b)　　Calculate the specific activity of the ^{3}H-cAMP in Bq (d/s) /pmol and in Bq/ng. (The molecular weight of cAMP is 329.)

Note.

If the quenching is about the same in all samples (similar "green" channel counts with the Packard or similar ESRs with the LKB), quench correction is not necessary. In this case, the actual counts can be used to calculate the C_O/C_X values. Calculation (b), above, can not be done without quench correction on the TC tube.

EXERCISE 13-7

ENZYME IMMUNOASSAY OF PLASMA BENZODIAZEPINES

Benzodiazepines are commonly used drugs for the treatment of anxiety and nervousness. Valium (diazepam) is one of the most commonly prescribed drugs. Other benzodiazepines commonly prescribed are:

> Oxazepam (demethylated diazepam).
> Chlordiazepoxide.
> Clonazepam.
> Flurazepam.
> Lorazepam.
> Nitrazepam.

The basic structure of these drugs are shown below:

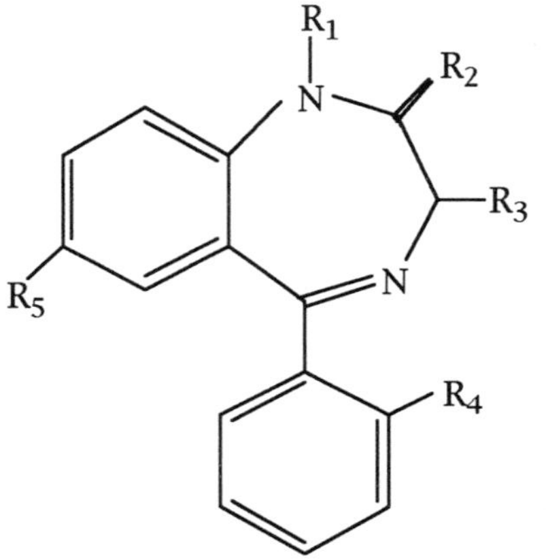

R_1 : H or CH_3
R_2 : O or $NHCH_3$
R_3 : H or OH
R_4 : H or Cl
R_5 : Cl or NO_2

These drugs all have similar pharmacological effects, but their pharmacokinetics are different; for example, chlordiazepoxide and oxazepam have shorter half lives than diazepam.

Fatalities, following oral ingestion, are usually rare, unless a combination of drugs is taken with the benzodiazepine.

The Syva EMIT®-TOX will be used, in this exercise, to semiquantitate the level of benzodiazepines in plasma. This assay is a homogenous enzyme immunoassay and has been described on pages 13-24 and 13-25.

In this system, the drug is covalently linked to an enzyme (bacterial glucose-6-phosphate dehydrogenase) so that the active site of enzyme is not adversely effected. The antiserum contains antibodies to the drug. The binding of the antibody to the drug-enzyme conjugate partially obscures the active site of the enzyme as well as producing a conformational change in the enzyme structure, thus reducing the enzyme activity.

Free drug will also be bound by the antibody. Hence, there is competition between the free drug and the drug-enzyme conjugate for the limited number of antibody sites. As the amount of free drug (standards or unknown) increases, less of the drug-enzyme conjugate is bound and, hence, the amount of available enzyme increases. Hence, at equilibrium, we have:

Drug		Antibody-Drug
+	+	+
Drug-Enzyme		Antibody-Drug-Enzyme
Active Enzyme		Inactive Enzyme

The glucose-6-phosphate dehydrogenase used in this assay is bacterial in origin. Unlike the mammalian enzyme, which uses NADP, the bacterial enzyme uses NAD. This avoids interference from G-6-PD in the patient's plasma. The enzyme catalyses the following reaction:

$$\text{G-6-P} + NAD^+ \rightleftharpoons \text{Gluconolactone-6-phosphate} + NADH + H^+$$

Hence, the reaction can be followed by observing the increase in absorbance at 340 nm.

The antiserum, used in the assay, was raised to diazepam, as the antigen, using sheep. However, a non-specific antiserum was selected, so that other benzodiazepines would be recognised, as well as diazepam. The binding affinity for other benzodiazepines is, however, less than that for diazepam. The approximate amounts of some benzodiazepines that give results equivalent to 1 mg/l diazepam are listed below:

Clonazepam	:	7 mg/l
Chlordiazepoxide	:	17 mg/l
Oxazepam	:	3 mg/l
Flurazepam	:	10 mg/l
Nitrazepam	:	3 mg/l
Lorazepam	:	10 mg/l

However, the actual level does vary from batch to batch. Hence, benzodiazepines, other than diazepam are expressed as diazepam equivalents in semiquantative assays.

Case History.

The plasma sample is from a 27 year old man. He had been involved in 3 hit-and-run accidents within a few hours. On admission to hospital, he claimed that he had taken about 25 five mg Valium tablets, that morning, with the intention of committing suicide. He had also consumed about half a bottle of whisky. The physician has requested a semi-quantitative determination of plasma benzodiazepines, to confirm the patient's story.

Diazepam is absorbed rapidly from the gastrointestinal tract to produce peak blood levels after about one hour. The half-life of the drug is about 10 hours. It is metabolised by N-demethylation and 3-hydroxylation to oxazepam. These metabolites also have pharmacological activity. Oxazepam is then excreted, mainly as its glucuronide in the urine, or as unconjugated metabolite in the bile, over the next few days.

A therapeutic dosage of 20 mg per day usually leads to a blood level of 0.2 to 0.4 mg/l. Toxicity is seen with levels above 5 mg/l. Ataxia, nystagmus and sedation are signs of overdose. Coma can be induced with very high levels of diazepam. Toxicity will be seen at lower blood levels, when alcohol is consumed with the diazepam, as in this case.

REAGENTS AND EQUIPMENT:

1. Syva EMIT® Tox Serum Benzodiazepine Assay.
    Cat. No. 7B119UL. The kit contains freeze dried and concentrated reagents. These are reconstituted as follows:

    Reagent A : Add 3.0 ml of distilled water and swirl to dissolve the contents of the vial.

    Reagent B : Add 3.0 ml of distilled water and swirl to dissolve the contents of the vial.

    Buffer   : Make the contents of the bottle up to 200 ml with distilled water and mix well.

    Store at 4°C.

2. Diazepam Solution (500 mg/l).
    Dissolve 10 mg of diazepam in 20 ml of Analar ethanol.

3. Patient's Plasma.
    Add 200 μl of reagent 2 to 50 ml of plasma or serum. Mix well.

4. Diazepam Standard (1 mg/l).
    Dissolve 200 mg of albumin (BSA or human) in about 40 ml of the buffer and make to 50 ml with the buffer. Add 100 μl of reagent 2 and mix well.

Wassermann Tubes.
20, 50, 100 and 500 μl Pipettors and Tips.
1 ml Graduated Pipettes.
Semi-micro Cuvettes.
Paper Tissues.
Plastic Cuvette Stirrers.
Spectrophotometers.
> Preferably a double beam instrument that has a kinetic mode. Single, or double, beam instruments can be used with recorders. Make sure that the recorder input is adjusted so that the same absorbance units are displayed on the recorder as on the spectrophotometer.

PROCEDURE:

Allow all reagents to reach room temperature before the assay.

1. Set up the following standards in Wassermann tubes, and mix well:

	1	2	3	4	5	6
μl of 1.0 mg/l diazepam standard in Tris-HCl buffer containing 4 g/l albumin	-	20	50	100	500	~1,000
ml of 55 mM Tris-HCl buffer, pH 8.0	1	0.98	0.95	0.9	0.5	-

2. Dilute 100 ml of the patient's plasma with 0.9 ml of Tris-HCl buffer and mix well (→ 1/10 plasma).

3. Dilute 100 ml of the diluted plasma from step 2 with 0.9 ml of Tris-HCl buffer and mix well (→ 1/100 plasma).

4.	To a glass semi-micro cuvette add 750 µl of buffer, 50 ml of standard 1 and 50 µl of Reagent A (antibody, glucose-6-phosphate and NAD in Tris-HCl buffer, pH 8.0). Mix well.

5.	Wipe the optical faces of the cuvette with paper tissues and place it in a spectrophotometer, set for kinetic measurements to record at 15 second intervals for one minute, at 340 nm.

6.	Add 50 µl of Reagent B (diazepam coupled glucose-6-phosphate dehydrogenase in Tris-HCl buffer, pH 8.0), mix with a plastic cuvette stirrer, close the spectrophotometer lid and start the run as quickly as possible.

7.	Repeat the assay with the other standards and then with the 1/10 and 1/100 plasma.

Record the 45 second minus the 15 second absorbance values for each standard and sample.

Plot a standard curve of absorbance difference against standard concentration, read off the unknowns and calculate the approximate level of diazepam in the patient's plasma.

Note.

Single or double beam spectrophotometers, connected to a recorder, can be used. Make sure that the recorder readings correspond to those of the spectrophotometer. Zero double beam spectrophotometers at step 5 or set single beam instruments to an absorbance of 0.1. Record for 2 to 3 minutes at a chart speed of 2 cm per minute, in step 6. Record the maximum rate of absorbance change for each sample, from the chart paper. Plot these values against the standard concentrations.

Diazepam Reference Values:

Therapeutic	:	0.1-1.0 mg/l.
Toxic	:	Above 5 mg/l.

GENERAL REFERENCES:

Milstein, C., (1980), Monoclonal Antibodies, Sci. Am., Oct. 56-64.

Varley, H., Gowenlock, A. H. and Bell, M., (1980), Immunological Methods, Chapter 5 in Practical Clinical Biochemistry, Vol. 1, 5th Ed., p 103-137, William Heinemann Medical Books.

Kricka, L. J., (1994), Principles of Immunochemical Techniques, Chapter 10 in Tietz Textbook of Clinical Chemistry, 2nd. Ed., Eds. Burtis, C. A. and Ashwood, E. R., p 256-282, W. B. Saunders Co.

Haaf, E., (1969), Laboratory Notes for Medical Diagnosis, No. 3 (Immunoprecipitation Methods), Behringwerke AG.

Chard, T., (1982), An Introduction to Radioimmunoassay and Related Techniques, (2nd Edition). Laboratory Techniques in Biochemistry and Molecular Biology, Vol. 6, Part 2. Eds. Work, T. S. and E., Elsevier/North Holland, Biomedical Press.

Collins, W. P., (1985), Alternative Immunoassays, Wiley Medical Publications.

Elkins, R. P., (1987), An Overview of Present and Future Ultrasensitive Non-Isotopic Immunoassay Development, Clin. Biochem. Rev., **8**, 12-23.

Elkins, R. and Chu, F., (1997), Immunoassays and Other Ligand Assays: Present Status and Future Trends, J. Int. Fed. Clin. Chem., **9**, 100-109.

Powsner, E. R., (1994), Basic Principles of Radioactivity and its Measurement, Chapter 9 in Tietz Textbook of Clinical Chemistry, 2nd. Ed., Eds. Burtis, C. A. and Ashwood, E. R., p 283-312, W. B. Saunders Co.

Immunoelectrophoresis.

Varley, H., Gowenlock, A. H. and Bell, M., (1980), Immunological Methods, Chapter 5 in Practical Clinical Biochemistry, Vol. 1, 5th Ed., p 562-563, William Heinemann Medical Books.

Prince Henry's Hospital, Melbourne (Australia), Method Sheets.

Electroimmunoassay (Laurell's Rockets).

Becker, W. and Sieber, (1975), Methods of Qualitative and Quantitative Immunoelectrophoresis, Hoechst (Behring Institute).

Varley, H., Gowenlock, A. H. and Bell, M., (1980), Immunological Methods, Chapter 5 in Practical Clinical Biochemistry, Vol. 1, 5th Ed., p 563-565, William Heinemann Medical Books.

^{125}I Energy Spectrum and Eldridge Calculation.

Packard Auto-Gamma Spectrometer Instruction Manual (p 4-3) and Appendix D (1972).

RIA of Serum Digoxin.

Ortho-Clinical Diagnostics Amerlex Digoxin Kit Booklet.

Ortho-Clinical Diagnostics is a Johnson and Johnson Company.

Quench Correction.

Herberg, R. I., (1965), Channels Ratio Method of Quench Correction in Liquid Scintillation Counting, Packard Technical Bulletin No. 15.

Packard and LKB Liquid Scintillation Counter Manuals.

CPBA of Urine cAMP.

Amersham International Cyclic AMP Assay Kit Booklet.

Wood, P. J., (1981), The Diagnostic Potential of Cyclic Nucleotide Measurements, Chapter 12 in Recent Advances in Clinical Biochemistry, **2**, 271-294, Eds. Alberti, K. G. M. M. and Price, C. P., Churchill Livingstone.

EMIT of Plasma Benzodiazepines.

Syva EMIT®-Tox Serum Benzodiazepine Assay Booklet.

EMIT® is a registered trademark of Syva. Syva is a subsidiary of Dade Behring.

CHAPTER 14

COULOMETRY, OSMOMETRY AND REFRACTOMETRY

COULOMETRY

Coulometry involves the measurement of the amount of electricity required to carry out an oxidation or reduction reaction. This amount is expressed in coulombs (Q). Coulombs are the product of current (in amperes) and time (in seconds). Q is related to the amount of substance oxidised or reduced in the system by the following equation:

$$Q = znF$$

Where:

z	=	The number of electrons involved in the oxidation or reduction of one molecule or ion.
n	=	The number of moles involved in the reaction.
F	=	Faraday's constant (96,487 coulombs/mole).

The Chloridometer.

The most common use of coulometry in clinical biochemistry is in the chloridometer. This is an instrument for measuring the chloride concentration in biological fluids. It uses the principle of coulometric titration.

When an electrical potential is applied across two silver electrodes, in a chloride solution, silver ions are generated at the anode:

$$Ag \rightarrow Ag^+ + e^-$$

The electrons are removed via the anode and silver ions pass into solution. The silver ions then react with chloride ions to form insoluble silver chloride. The rate of release of silver ions into solution is proportional to the electrical current flowing in the system. If the current flow is kept constant, as in this procedure, the amount of silver ions released is proportional to the time of current flow. Therefore, in this method of titration, silver ions are added to the solution at a constant rate from a silver wire anode. At the end point, free silver ions appear in solutions, as all the chloride has been used up.

The presence of this end point is detected by a biamperometric end point detection system. This consists of a pair of silver microelectrodes, in the assay solution, having a micro-ammeter in series with one of the electrodes. A voltage of 0.25 volts is applied across the electrodes. A small current flows in system due to the following:

Anode	:	$Ag \rightarrow Ag^+ + e^-$
Cathode	:	$Ag^+ + e^- \rightarrow Ag$

As can be seen, there is no net change in silver ion concentration. At the end point of the titration, the level of silver ions in solution rises with the net result of an increase in current flow between the two silver microelectrodes. The increase in current in the system activates a micro-switch that turns off the power to the main electrodes.

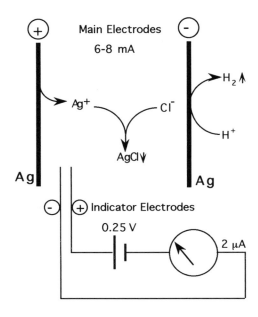

The titration is carried out in a solution containing nitric acid, acetic acid and gelatine, which is stirred by a small stirrer. The nitric acid provides the protons for cathodic reaction:

$$2 H^+ + 2 e^- \rightarrow H_2$$

The acetic acid tends to reduce the polarity of the solution and thus decrease the solubility of silver chloride. Gelatine tends to make the titration curve more linear by being preferentially adsorbed onto high spots on the anode, thus equalising the reaction rate on the anode surface. The gelatine reagent also contains thymol as an antibacterial agent and thymol blue as a pH indicator.

When the instrument is switched to titrate, the power is applied to the main electrodes so that a constant current, in the order of 6-8 mA, is flowing between them. The digital timer of the instrument is also switched on. About 2 µA flow in the microelectrode circuit until the end point is reached. Then the current rapidly rises until a preset value (usually 20 µA).

The instrument then switches itself, and the digital timer, off. The time of the titration is thus recorded and as this is proportional to the amount of silver ions released, it is also proportional to the chloride content of the assay solution. By comparing the titration time taken for a known chloride solution with that taken for the unknown solution, the chloride content of the unknown can be calculated. As with all titrations, a blank titration is also required.

OSMOMETRY

Osmometry is a term that is used for the measurement of the total molar concentration of a solution. Even though the name suggests that this measurement is achieved by the measurement of the osmotic strength of the solution, this is not usually so for body fluids.

When a substance is dissolved in water, the resultant solution will exert an osmotic pressure, if separated from pure water by a semipermeable membrane. In addition to this effect, the freedom of movement of the water molecules is reduced. This effect is seen as a decrease in the freezing point and vapour pressure of the solution and an increase in its boiling point.

If 6.023×10^{23} particles (Avogadro's number) are dissolved in one kilogram of water, a 1.00 molal solution is produced. This is a 1.00 osmolal solution if there are no solvent-solute or solute-solute interactions. As a result, the colligative properties of a 1.00 osmolal solution, compared with pure water, are changed as follows:

1. The freezing point is lowered by $1.858^\circ C$.
2. The boiling point is raised by $0.52^\circ C$.
3. The vapour pressure is lowered by 0.3 mm Hg.
4. The osmotic pressure is increased by 17×10^3 mm Hg.

Instruments that determine the total osmolality of body fluids use properties 1 and 3. Osmolality should not be confused with osmolarity. These are:

Osmolality : The number of osmoles per kilogram of water.
Osmolarity : The number of osmoles per litre of solution.

Osmolality is used in preference to osmolarity because it is independent of temperature. The volume of a solution increases with temperature and, hence, its osmolarity will fall. However, its osmolality will remain constant.

The osmolality of a solution is equal to ϕnC, where:

ϕ = The osmotic coefficient (activity coefficient).
n = The number of particles into which each molecule can dissociate, when in solution.
C = The concentration in mol/kg H_2O.

ϕ takes into account incomplete dissociation, association between solute and solvent as well as solute - solute interactions. For example, ϕ for glucose is 1.00, for urea is 0.94 and for NaCl is 0.93 (plasma concentration) or 0.91 (strong urine concentration).

The osmolality of plasma (Reference range: 275-295 milliosmoles (mOsm) per kg H_2O) is approximately made up of:

Cations:
Sodium	:	135	mOsm/kg H_2O
Potassium	:	4	"
Calcium	:	2	"
Magnesium	:	1	"

Anions:
- Chloride   :   100   mOsm/kg H_2O
- Bicarbonate   :   25   "
- Organic acids   :   5   "
- Proteins   :   1   "
- Phosphate   :   1   "
- Sulphate   :   0.5   "

Uncharged:
- Glucose   :   5   "
- Urea   :   5   "

Osmolal Gap (Osmotic Gap).

The osmolal gap (osmotic gap) is a useful concept. It is the difference between the measured osmolality and that calculated from the concentration of the major solutes in plasma. It is derived from:

$$\text{Measured osmolality} - (2[Na^+] + [\text{Glucose}] + [\text{Urea}])$$

The osmolal gap is normally less than 10 mOsm/kg H_2O. An elevated level is due to a decreased plasma water content (Eg. hyperlipidaemia or hyperproteinaemia) or to the presence of high concentrations of low molecular weight substances.

Freezing point instrument results should be used, rather than those from vapour pressure osmometers, for the calculation. This is especially important if volatile components are contributing to the osmolal gap.

CLINICAL USEFULNESS OF OSMOLALITY MEASUREMENTS.

PLASMA OR SERUM.

If plasma is used, heparin should be used as the anticoagulant. Low molecular weight anticoagulants, such as EDTA, citrate or fluoride-oxalate, will increase the osmolality of the sample.

Toxicology.

Where ingested xenobiotics are present in the plasma in millimolar amount, the osmolality of the plasma will be increased. This is seen as an increase in the osmolal gap. For example, an increase of 1.0 mOsm/kg H_2O is produced by:

Xenobiotic	Concentration (mg/l)
Methanol	26
Ethanol	43
Ethylene glycol	50
Acetone	55
Isopropyl alcohol	59

Aspirin (acetylsalicylic acid) is an example of a non-volatile xenobiotic that produces a moderate increase in the osmolal gap when taken in overdose. Aspirin is absorbed by the mucosal cells of the gastrointestinal tract and is then hydrolysed. Hence, it enters the circulation as salicylic acid. Mild toxicity is seen with a osmolality rise of 3 mOsm/kg H_2O, moderate toxicity with 6 mOsm/kg, severe toxicity with 8 mOsm/kg and coma and death (if not treated) with levels above 9 mOsm/kg H_2O.

Cerebral Oedema.

In the treatment of cerebral oedema, compounds that remain in the extracellular fluid can be given to increase the extracellular osmotic pressure. Hence, water is drawn out of the brain. Mannitol is commonly used for this purpose, and its level in the extracellular fluid can be monitored by observing the increase in the osmolal gap. This is usually increased by 10 mOsm/kg H_2O.

Hyponatraemia.

Hyponatraemia and an increased, or occasionally normal, plasma osmolality are seen in diabetes mellitus. For every mOsm/kg H_2O rise in osmolality due to increasing plasma glucose, the sodium falls by 0.6-0.7 mOsm/kg H_2O. Hence, there is a net increase in osmolality of 0.3-0.4 mOsm/kg H_2O. If the net increase is greater than this, there is increased water loss, compared with sodium; or other osmotically active molecules are present. If the increase is less, there is excess sodium loss.

Samples showing pseudohyponatraemia (Eg. hyperlipidaemia with sodium methods involving dilution) will show normal osmolality.

Trauma.

Patients with major injuries usually have elevated osmolalities (around 325 mOsm/kg H_2O). Values around 345 mOsm/kg H_2O are seen with patients in shock. If this value rises, the prognosis is poor.

Coma.

Very high plasma osmolality is seen in coma induced by renal failure, drug overdose or diabetes mellitus.

URINE.

Urine osmolality is a better method, than specific gravity, for evaluating the urine concentrating or diluting ability of the kidney. The reference range for a 24 hour specimen is 550-850 mOsm/kg H_2O. However, the most dilute urine can be only 50 mOsm/kg H_2O and the most concentrated 1,200 mOsm/kg H_2O.

The protein content of urine has very little effect on its osmotic strength, due to the high molecular weight of proteins. However, protein has a marked effect on the specific gravity of urine. Other than the presence of protein (or other high molecular weight material) in the urine, there is a linear relationship between the urine specific gravity and its osmolality.

If the patient has received no fluids for 12 hours, a urine sample will have an osmolality of greater than 850 mOsm/kg H_2O (SG greater than 1.025), if the renal concentrating ability is normal.

Increased Urine Output.

With polydipsia (increased water consumption), both plasma and urine osmolalities are low. The urine osmolality is usually below 100 mOsm/kg H_2O.

In diabetes insipidus, the plasma osmolality is elevated; whereas, the urine osmolality is decreased. The ratio of urine to plasma osmolality is between 0.2 and 0.7; whereas, the normal ratio is between 1 and 3. With diabetes mellitus, both plasma and urine osmolalities are elevated.

With chronic renal failure, the ability of the kidneys to concentrate urine is decreased and the urine output has a fairly constant osmolality. These patients usually have moderate polyuria.

Decreased Urine Output.

Decreased urine output can be due to dehydration or seen in the oliguric phase of acute renal failure. With dehydration, the urine osmolality is markedly elevated. If the renal failure is due to a pre-renal cause, such as a reduction in blood flow to the kidney (Eg. severe haemorrhage), the glomerular filtration rate will be reduced, but tubular function will be normal. Hence, the urine will have a high osmolality with a low sodium concentration. If the renal failure is due to tubular damage, the urine will have a low osmolality with a relatively high sodium concentration. These values are summed up below:

    Pre-renal:
        Urine [Na]   :   < 20 mmol/l.
        Urine osmolality   :   > 500 mOsm/kg H_2O.
        Urine/Plasma [Creatinine]   :   ≅ 40.

    Tubular Necrosis:
        Urine [Na]   :   > 50 mmol/l.
        Urine osmolality   :   < 500 mOsm/kg H_2O.
        Urine/Plasma [Creatinine]   :   < 40.

Urine Osmolal Gap.

The urine osmolal gap (osmotic gap) is calculated as follows:

$$\text{Measured osmolality} - (2[Na^+] + 2[K^+] + [\text{Glucose}] + [\text{Urea}])$$

This osmolal gap is useful in the assessment of renal tubular acidosis. In the normal production of an acidic urine, ammonia is generated in the renal tubules and passed into the lumen:

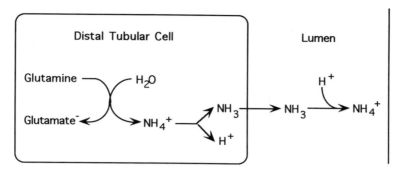

In renal tubular acidosis, there is a decreased bicarbonate (and an increased chloride) reabsorption in the proximal tubules. This results in an insufficient acidification in the distal tubule. As a result, ammonia production is reduced. Hence, less ammonium ions enter the urine. This is seen as a reduced urine osmolal gap.

OSMOMETERS.

FREEZING POINT OSMOMETERS.

The most commonly used type of osmometer in clinical biochemistry is the freezing point instrument. Freezing point has advantages over vapour pressure as it is not affected by ambient temperature (vapour pressure increases with temperature). Samples containing volatile components can not be accurately measured by vapour pressure depression; whereas, they can be accurately measured using a freezing point osmometer.

The sample is placed in a thermostatically controlled cooling bath or block at a temperature well below the freezing point of the sample. A thermistor probe is placed in the sample. The sample is allowed to cool below its freezing point (supercooled) to a specific temperature. The sample is then agitated by a fine wire or a mechanical pulse. This causes ice crystals to form. The latent heat of fusion warms the sample to the actual freezing temperature. This temperature is held until the sample is fully frozen and then the temperature falls further. This is illustrated below:

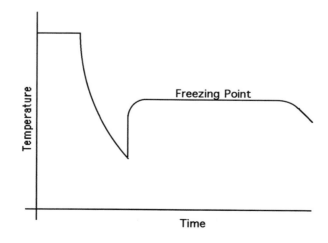

With some instruments, the freezing point temperature is recorded. This can then be converted to milliosmoles per kilogram water. However, most instruments are calibrated with solutions of known osmotic strength. These then give direct readings in mOsm/kg H_2O.

VAPOUR PRESSURE OSMOMETERS.

These instruments measure the decrease in the dew point temperature of a solution. This is directly proportional to the decrease in the vapour pressure of a solvent by the presence of solutes. The dew point is the temperature at which condensation of water, from an aqueous vapour, occurs.

The sample is applied to a filter paper disc, which is sealed into a vapourisation-condensation chamber. The temperature of a thermocouple within the chamber is then cooled to below the dew point by passing an electric current through the thermocouple. The current is then switched off and water vapour condenses on the thermocouple.

The heat of condensation raises the temperature of the thermocouple to that of the dew point. This is a stable temperature, because any further rise in temperature would cause water to evaporate from the thermocouple and thus cool it. These temperatures are illustrated below:

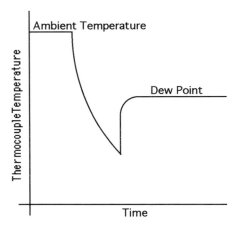

The voltage generated by the thermocouple is proportional to its temperature. This, in turn, is related to the vapour pressure and thus to the osmolality of the solution. Solutions of known osmolality are used to calibrate the osmometer.

Most instruments contain temperature-compensating circuits to reduce the effect of changes in ambient temperature.

REFRACTOMETRY

Refractometry is the measurement of the refractive index (index of refraction) of a transparent solid, liquid or gas. Its use in clinical biochemistry is applied to liquids such as plasma and urine. The refractive index of a medium is the ratio of the velocity of light in a vacuum to that in the medium. Using air, instead of a vacuum, only introduces a small error when refractive indices of liquids or solids are measured. Hence, air is commonly used as one of the media. Refractive index is usually measured as the change in angle as the light ray passes from one medium to the other:

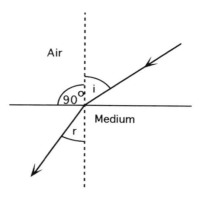

Sine i (the angle of incidence) and sine r (the angle of refraction) are proportional to the velocity of light in the two media. The refractive index is:

$$\frac{Sin\ i}{Sin\ r}$$

The refractive index is dependent upon the wavelength of light and the temperature. It decreases with increasing wavelength and increasing temperature. The change of refraction with wavelength is known as dispersion. Hence, the wavelength and temperature should be stated. For example, η_D^{20} is the refractive index at 20°C, using the D lines of sodium (589.0 and 589.6 nm).

If the light passes from the denser medium to air, an angle will be reached when the emerging ray will run along the surface. This is known as the critical angle. If the angle is further increased, there will be total internal reflection. These concepts are illustrated below:

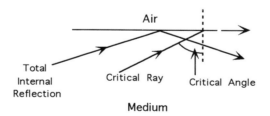

Refractive Index of Solutions.

Whereas the osmotic strength of a solution depends upon the number of particles present, the refractive index depends upon the mass of solute present. Hence, plasma proteins are the major contributors to the refractive index of serum or plasma; whereas, they are only minor contributors to the osmolality.

Refractive Index of Urine.

As discussed under the section on osmometry, the concentrating ability of the kidneys can be assessed using osmometry. Concentrating ability can also be assessed by measuring the specific gravity (relative density) of the urine. The most accurate way of doing this is to use a hydrometer (urinometer) that will measure specific gravities between 1.000 and 1.050.

There is a reasonable correlation between the specific gravity (SG) and the refractive index of urine. Hence, refractometers (calibrated in g/ml) can be used in the assessment of renal function, if osmometers or hydrometers are not available. The Atago Uricon-NE is an example of such an instrument. This is a hand held refractometer which has SG (1.000-1.050) and refractive index scales.

Some urine testing strips have SG segments. These actually measure ionic strength and are not very accurate. The specific gravity block on the Ames (Bayer) N-Multistix SG is impregnated with a pre-treated polyelectrolyte and bromthymol blue. Increasing the ionic strength of the urine displaces protons from the polyelectrolyte. These protons then produce a colour change with the indicator. This method is not reliable for urine with elevated levels of urea, glucose or protein. Alkaline urine also produces unreliable results.

Refractive Index of Plasma or Serum.

As mentioned above, the major contributors to plasma, or serum, refractive index are the plasma proteins. Hence, refractive index can be used to obtain a reasonably reliable value for the total plasma protein content.

The refractive index of plasma (or serum) increases with increasing protein content. This method of protein determination was first used in 1900 and offers a very quick method that can be carried out on a single drop of plasma or serum. Abnormally large amounts of urea or glucose, in the sample, will produce an elevated result. However, reasonably reliable results can be obtained from lipaemic samples.

Standard refractometers can be used, but it is more convenient to use a hand held refractometer, calibrated in g/l (or g/100 ml). These provide a rapid result that is suitable for monitoring changes in plasma protein concentration, or for obtaining a total protein value for plasma protein electrophoresis.

EXERCISES

One coulometry, two osmometry (one freezing point and one vapour pressure) and one refractometry exercises are included.

EXERCISE 14-1

THE DETERMINATION OF THE PLASMA CHLORIDE CONCENTRATION, USING COULOMETRY

Chloride is the major anion in extracellular fluid. Its plasma level is affected by the water and acid-base status of the patient. Low plasma levels are seen in salt-losing nephritis, Addisonian crisis, some types of metabolic acidosis and with prolonged vomiting. Raised plasma levels are seen in dehydration, renal tubular acidosis, prolonged diarrhoea, diabetes insipidus, adrenocortical hyperfunction and with salicylate poisoning.

Anion Gap.

The difference in the total value of the concentrations of the commonly measured cations and anions is called the anion gap. This is due to there being more unmeasured anions than cations. The major contributor being plasma proteins. The commonly used formulae for calculating the anion gap are:

(i) $[Na^+] - ([Cl^-] + [HCO_3^-])$.

(ii) $([Na^+] + [K^+]) - ([Cl^-] + [HCO_3^-])$.

The reference values for (i) are 8-16 mmol/l and for (ii) are 10-20 mmol/l.

As mentioned above, where a metabolic anion is produced in metabolic acidosis, the plasma level of chloride will fall and the anion gap will increase. Hence, an increasing anion gap is suggestive of metabolic acidosis. Increased plasma protein concentration, as in myelomatosis, will also give rise to an increased anion gap. A decreased anion gap is seen with low plasma albumin levels.

The elevated anion, and associated disease, resulting in a metabolic acidosis are tabulated below:

Disease	Elevated Anion
Diabetes (Ketoacidosis)	Acetoacetate β-hydroxybutyrate
Renal Failure	Sulphate Phosphate
Congestive Heart Failure (Tissue Hypoxia)	Lactate
Some poisonings. Eg. Aspirin Methanol	Salicylate Formate

In this exercise, the plasma (or serum) chloride concentration will be determined using a chloridometer. The exercise is written for a Buchler-Cotlove chloridometer. The procedure may have to be modified for a different instrument.

REAGENTS AND EQUIPMENT:

1.   100 mmol/l Chloride Standard.
      Dissolve 5.85 g of Analar NaCl in distilled water and make to 1 l.
2.   Plasma or Serum.
      Keep on ice.
3.   Acid Reagent.
      To 900 ml of distilled water, add 6.4 ml of concentrated nitric acid and 100 ml of glacial acetic acid. Mix well.
      Place in a bottle with an attached dispenser set to 4.0 ml.

4.   Gelatine Reagent.
        Dissolve 600 mg of gelatine, 10 mg of thymol and 10 mg of thymol blue in 100 ml of boiling water.
        Store at 4°C.

100 and 200 µl Pipettors with Tips.
Chloridometer.
        This exercise is written for the Buchler-Cotlove instrument (Buchler Instruments Inc), but similar instruments can be used. In which case, the procedure may need to be modified.
Polyethylene Cups (10 ml).
        Supplied with the chloridometer.

PROCEDURE:

Instrument Warm Up.

1.   Check that there is a fresh 1.5 volt, AA size battery for the indicator circuit, in the rear of the instrument, and that there is ample silver wire for the anode.

2.   Switch on, with the "Titration" switch in position 2.

3.   Move the red meter indicator to the left until it coincides with the micro-ammeter needle. This will switch off the stirrer and the timer.

4.   Set the red meter indicator to 20 µA and allow 10 minutes to warm up.

Titration.

5.   Carry out the titration (in duplicate) on distilled water, the standard and the plasma or serum, as follows:

        A.   Set up the titration mixture in a 10 ml plastic cup:

                i.   4.0 ml of acid reagent (0.1 M nitric acid in 1.75 M acetic acid).

                ii.   200 ml of gelatine reagent (0.6% gelatine containing 0.01% thymol and 0.01% thymol blue).

                iii.   100 µl of test solution.

        B.   Fully raise the cup so that the electrodes are in the titration mixture. Set the titration rate switch to "HI" and zero the time indicator.

        C.   Switch the "Titration" switch to position 1. Allow about 30 seconds for the current to stabilise. This should be below 5 µA.

        D.   Switch the "Titration" switch to position 2. The timer will start and switch off at the end of the titration, when the current rapidly increases to 20 µA. Record the titration time.

        E.   Thoroughly wash the electrodes and cup in distilled water, after the last titration.

Average the duplicate readings. Subtract the blank (water) reading from the other two values. Then calculate the plasma (or serum) value.

Reference Range:

Serum or plasma chloride   :   98-109 mmol/l.

Notes:

1. If the second blank titration time is significantly shorter than the first, the titration should be repeated until the blank values are similar.

2. The electrodes should be cleaned in 0.880 (SG) ammonia for 5 minutes if the instrument gives high blanks or if the current flow is unstable after the first 20 seconds of titration.

EXERCISE 14-2

THE EFFECT OF ETHANOL ON PLASMA OSMOLALITY

As already discussed, the presence of xenobiotics in the plasma (in millimolar amounts) will increase the osmolal gap. This increase can be used to assess the plasma level of the xenobiotic. In this exercise, the effects of increasing levels of ethanol on the osmolality of plasma will be examined. As vapour pressure osmometers can not be used for osmolality measurements where volatile components are present, a freezing point osmometer has to be used. Methods for the use of two osmometers are described.

REAGENTS AND EQUIPMENT:

1. 500 g/l Ethanol.
   Make 63.4 ml of ethanol to 100 ml with distilled water and mix well.
2. Plasma (or Serum).
   Any anticoagulant can be used for the plasma, as the initial osmolality is unimportant. Keep on ice.
3. 400 mOsm/kg H_2O Solution.
   Heat about 15 g of Analar NaCl, overnight, at 110°C. Allow to cool in a desiccator. Weigh out 12.68 g and add to a 1 l volumetric flask filled to the mark with distilled water + 2 ml (to allow for expansion to room temperature). Stir and invert until fully dissolved and evenly distributed.
   For Knauer Osmometer.
4. 50 mOsm/kg H_2O Solution.
   Heat about 5 g of Analar NaCl, overnight, at 110°C. Allow to cool in a desiccator. Weigh out 1.525 g and add to a 1 l volumetric flask filled to the mark with distilled water + 2 ml (to allow for expansion to room temperature). Stir and invert until fully dissolved and evenly distributed.
   For Advanced Instruments Osmometer.
5. 850 mOsm/kg H_2O Solution.
   Heat about 30 g of Analar NaCl, overnight, at 110°C. Allow to cool in a desiccator. Weigh out 27.22 g and add to a 1 l wide necked volumetric flask filled to the mark with distilled water + 2 ml (to allow for expansion to room temperature). Stir and invert until fully dissolved and evenly distributed.
   For Advanced Instruments Osmometer.

Wassermann Tubes.
10, 100, 200, 500 and 1,000 µl pipettors and Tips.

Freezing Point Osmometer.

Instructions are included for a Knauer and an Advanced Instruments Osmometer. The Knauer also requires measuring vessels. The advanced Instruments also requires a sampler with tips (sample cells) and chamber cleaners.

Other freezing point osmometers can be used. The current Advance Instruments equivalents are their 3D3 and 3300 osmometers. Specific instrument calibration and operating instructions should be followed.

Paper Tissues.

PROCEDURE:

1. Prepare the following ethanol solutions, in Wassermann tubes, and mix well:

g/l ethanol	50	100	150	200	250
µl of 500 g/l ethanol	100	200	300	400	500
µl of distilled water	900	800	700	600	500

2. Set up the following samples, in Wassermann tubes, and mix well:

g/l ethanol	0	0.5	1.0	1.5	2.0	2.5
ml of plasma	1.0	1.0	1.0	1.0	1.0	1.0
µl of distilled water	10	-	-	-	-	-
µl of 50 g/l ethanol	-	10	-	-	-	-
µl of 100 g/l ethanol	-	-	10	-	-	-
µl of 150 g/l ethanol	-	-	-	10	-	-
µl of 200 g/l ethanol	-	-	-	-	10	-
µl of 250 g/l ethanol	-	-	-	-	-	10

3. Calibrate a freezing point osmometer and measure the osmolality of the samples. Instructions for a Knauer Electronic Semi-micro Osmometer and an Advanced Instruments 3MOplus Micro Osmometer are given below. If a different instrument is used, its calibration and operating instructions should be followed.

Plot osmolality against plasma ethanol concentration and calculate, from the graph line, the concentration of ethanol that increases the osmolality by 1.0 mOsm/kg H_2O.

Knauer Electronic Semi-micro Osmometer.

This instrument uses a cooling block, the temperature of which can be controlled to adjust the rate of sample cooling. 50 or 150 μl samples can be accommodated in special tubes (measuring vessels). Freezing is initiated manually when a specific temperature is reached, by activating a vibrating rod. The instrument is calibrated with distilled water and a 400 mOsm/kg H_2O solution. Results are displayed on a meter.

Calibration.

1. Switch on the instrument and allow 3 minutes for the block to cool. Make sure that the cooling water is flowing (low flow rate).

2. Pipette 150 μl of distilled into the measuring vessel designed for 150 μl samples.

3. Attach the measuring vessel to the glass cone on the measuring head and lower this into the cooling aperture.

4. Switch the range selector to Δ.

5. When the sample temperature fall below 0°C, the meter needle starts to move. When it reaches the Δ position, briefly press the white vibrator button to induce freezing. When freezing starts, the meter needle moves to a lower value and holds this position.

6. Switch the range selector to 400.

7. Use a screwdriver to adjust the zero potentiometer (left front control) to read zero, before the needle starts to rise (decreasing temperature).

8. Raise the measuring head and allow the sample to thaw.

9. Rinse clean and dry the measuring vessel. Rinse and carefully dry the thermistor and vibrator rod with paper tissues.

10. Repeat steps 1-9, except use 150 μl of the 400 mOsm/kg H_2O solution. At step 7, adjust the 400 potentiometer (second from left front control) to read 400.

Sample Measurement.

11. Carry out steps 1-6, except use 150 μl of sample. Record the osmolality, from the meter reading, at step 6. If the needle goes off scale, during freezing, switch the range selector to 800.

12. Carry out steps 8 and 9.

Notes.

(i) If freezing takes place before the Δ position is reached, the temperature of the cooling block should be increased. ie. Decrease the cooling rate (control on the top of the instrument).

(ii)   If the sample does not freeze after brief stirring, the temperature of the cooling block should be reduced.   ie. Increase the cooling rate.

(iii)   The instrument should not be left on for long periods of time, or the cooling chamber will ice up.

Advance Instruments 3MOplus Micro Osmometer.

This instrument only requires 20 μl of sample and the assay is carried out on the sample in the sampler tip (sample cell). The sample is supercooled to a specific temperature and then given a mechanical pulse that starts the freezing process. Results are digitally displayed and can be printed by a printer, if attached. The instrument is calibrated with 50 and 850 mOsm/kg H_2O solutions.

Calibration.

The instrument is very stable and does not often require recalibration.

1.   Press the "Calibrate" button.

2.   Place a sample tip (sample cell) on the sampler and aspirate a sample of the 50 mOsm/kg H_2O solution. Discard this sample and aspirate another 20 μl.

3.   Blot the sides of the tip with a paper tissue. Also, blot any liquid protruding beyond the end of the tip.

4.   Remove the chamber cleaner from the sample port.

5.   Insert the sample tip into the sample port. Do not inject the sample.

6.   Press the "Start" button.

7.   After completion of the calibration, remove the sampler and discard the tip. Blot the sampler tip with a paper tissue.

8.   Insert a clean, dry chamber cleaner all the way into the sample port. Rotate it 2 or 3 times. Remove this cleaner and repeat the cleaning with another cleaner.

9.   Repeat steps 1-9.

10.   When the instrument displays "850 mOsm calibration", repeat the calibration (steps 1-9) using the 850 mOsm/kg H_2O solution.

11.   Repeat step 10 until the instrument displays "Calibration Complete".

Sample Measurement.

12.   Carry out steps 2-9 with each sample, instead of the standard solution. The result will be displayed in mOsm/kg H_2O.

Note.

This instrument can be left on for extended periods.

EXERCISE 14-3

THE MEASUREMENT OF URINE OSMOLALITY

As mentioned on page 14-5, urine osmolality can be used for evaluating the urine concentrating or diluting ability of the kidney. Hence, its measurement plays an important role in assessing renal function.

In this exercise, the urine concentrating ability of your kidneys will be assessed by measuring the osmolality of the first urine sample passed in the morning. A vapour pressure osmometer will be used. The principle of this type of instrument is discussed on page 14-8.

REAGENTS AND EQUIPMENT:

1.   290 mOsm/kg H_2O Solution.
    Heat about 15 g of Analar NaCl, overnight, at 110°C. Allow to cool in a desiccator. Weigh out 9.14 g and add to a 1 l volumetric flask filled to the mark with distilled water + 2 ml (to allow for expansion to room temperature). Stir and invert until fully dissolved and evenly distributed.
2.   1,000 mOsm/kg H_2O Solution.
    Heat about 40 g of Analar NaCl, overnight, at 110°C. Allow to cool in a desiccator. Weigh out 32.12 g and add to a 1 l wide necked volumetric flask filled to the mark with distilled water + 2 ml (to allow for expansion to room temperature). Stir and invert until fully dissolved and evenly distributed.

Vapour Pressure Osmometer.
    The procedure for the Wescor 5100C is described below. This instrument also requires a pipettor set to 8 µl (and tips) and 5-6 mm filter paper discs (Wescor Cat. No. SS-007).
    Other vapour pressure osmometers can be used. The assay can also be carried out using a freezing point osmometer. However, a vapour pressure osmometer provides the student with the experience of an additional technique.
Screw Top Containers.
    For urine collection.
Plastic Forceps.
Paper Tissues.

PROCEDURE:

Sample Collection.

1.   Have nothing to drink after your evening meal on the day prior to the laboratory session.

2.   Empty the bladder before going to bed and discard the urine.

3.   Collect a sample of urine passed just after waking. Normal eating and drinking can then be resumed.

Osmolality Measurement.

The calibration and operating instructions of a Wescor 5100C osmometer are described below. If a different instrument is used, its calibration and operating instructions should be followed.

Calibration.

The instrument should be left switched on at all times and should have been on for about 15 minutes before calibration.

4. Process the 290 mmol/kg H_2O solution as described in the operating instructions below. Set the display to 290 with the "Calibrate 290" control.

5. Process the 1,000 mmol/kg H_2O solution. Note the reading. For every 1.0 below 1,000, add 0.41 to 1,000. Set the "Calibrate 1000" to this value. It the reading is above 1,000: for every 1.0 above 1,000, subtract 0.41 from 1,000. Set the "Calibrate 1000" to this value. Alternatively, the nomogram (Figure 4-2) of the instrument manual can be used for this setting. This sets the correct "Slope".

6. Repeat step 4. This sets the correct "Offset".

Operating Instructions.

7. Open the sample chamber by rotating the sealing knob anticlockwise.

8. Withdraw the sample slide and make sure that the sample holder is clean and dry.

9. Using plastic forceps, place a sample disc (5-6 mm filter paper disc) in the sample holder.

10. Apply 8 µl of sample to the centre of the sample disc. (The disc should be fully saturated with a slight liquid meniscus on its surface). Make sure that no liquid touches the outer surface of the sample holder. (Start again, if this happens).

11. Push in the sample slide. (The "In Process" indicator will light).

12. Close the sample chamber by rotating the sealing knob 180° clockwise.

13. After about 90 seconds, the "In Process" indicator will go out and an audible tone will be sounded. Record the osmolality as displayed.

14. Open the chamber and withdraw the sample slide, as in steps 7 and 8. Use a paper tissue to remove the wet disc and to dry the sample holder.

Reference Value:

With normal renal function, the sample osmolality should be greater than 850 mOsm/kg H_2O.

EXERCISE 14-4

THE DETERMINATION OF THE PLASMA PROTEIN CONCENTRATION BY REFRACTOMETRY

In this exercise, a standard, and a hand held, refractometer will be used to determine the total protein concentration of a plasma or serum sample.

A number of formulae have been used to inter-relate protein content and refractive index. The formula used in this exercise is:

$$\text{Protein content in g/l} = 5,750 \, (RI_s - RI_w) - 18$$

Where:

RI_s = Refractive index of the sample.

RI_w = Refractive index of water at the same temperature (20°C).

18 is the correction factor for the salts and non-electrolytes in the sample.

This formula is derived from data supplied by Atago Optical Works and produced by Professor Matsumura of Tokyo Women's Medical College.

This exercise is written for an Atago Abbe refractometer and an Atago hand held (serum protein) refractometer. Other manufacturer's instruments operate on the same principles. Their operating instructions may be slightly different.

REAGENTS AND EQUIPMENT:

1.   Plasma or Serum.
       Keep on ice.
2.   Distilled Water.

Pasteur Pipettes with Teats.
Paper Tissues.
Abbe Refractometer.
       Atago Cat. No. 302 or similar.
       Circulate water, at 20°C, through the instrument by connecting it to a thermostatically controlled water bath.
Hand Held Serum Protein Refractometer.
       Atago Cat. No. 310 or similar.

PROCEDURE:

1.   Follow the procedures, for both types of instrument, as described below. Carry out 4 sets of readings when using the Abbe refractometer, and take the average, so that the 4th decimal place can be determined.

The Atago Abbe Refractometer.

In the Abbe refractometer, the sample is placed between 2 glass prisms. Light enters the top prism from an oblique angle to the base. Rays with angles less than the critical angle will pass through the liquid, the bottom prism and then pass on to the eyepiece. Whereas, rays with angles greater than the critical angle will undergo total internal reflection and be absorbed by the casing. This gives rise to light and dark sections seen in the eyepiece.

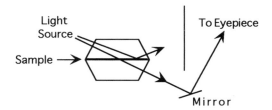

In the Atago Abbe refractometer, the surface of the top prism is rough ground so that light is refracted in all directions. This gives a better demarcation between the light and dark sections.

The critical angle increases as the refractive index of the sample increases. The instrument measures the critical angle as refractive index, rather than as an angle. This is done by moving the eyepiece, attached to a scale, until the interface between the light and dark sections coincides with the intersection of two cross-hairs, seen in the eyepiece:

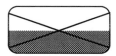

The refractive index is then read from the scale seen in the eyepiece below the light and dark sections.

2. Turn on the light source.

3. Check that water is circulating through the instrument from the thermostatically controlled water bath.

4. Check that the temperature is 20ºC on the refractometer thermometer.

5. Turn the "Collateral Prism Locking" knob (on the left-hand side) clockwise to unlock the prism and hinge open the prisms.

6. Place one drop of distilled water on the main prism. Close and lock the collateral prism.

7. Focus the refractometer scale by rotating the eyepiece.

8. Adjust the "Measurement" knob (bottom knob on the right-hand side) until the demarcation of light and shade appear in the field of vision.

9. Adjust the "Dispersion" knob (top knob on the right-hand side) until the demarcation line appears quite sharp.

10. Readjust the "Measurement" knob until the demarcation line is aligned with the cross-hairs (as shown above).

11. Read off the refractive index on the top scale. An average of 3-5 readings should be taken if the 4th decimal place is required. To do this, rotate the "Measurement" knob slightly and then realign the demarcation line with the cross-hairs and re-read the refractive index. (The bottom scale reads in grams of sugar per 100 ml.)

12. Open the prisms and wash both prisms with distilled water and dry with paper tissues.

13. Repeat the determination using the plasma, or serum, sample.

Calculate the plasma, or serum, protein content in g/l, from the above formula.

Note.

Reasonably reliable results are obtained with the refractometer maintained at a constant room temperature (18-25°C), if connection to a 20°C water bath is not possible.

The Atago Hand Refractometer.

In the hand refractometer, the sample is placed between a glass prism and a light absorbing plate. Rays with angles less than the critical angle will pass through the liquid and be absorbed; whereas, those with angles greater than the critical angle will undergo total internal reflection and pass on to the eyepiece. This gives rise to light and dark sections seen in the eyepiece.

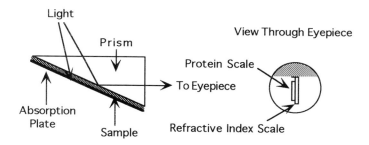

The light passes through a scale in the barrel of the instrument, so that the interface between light and dark sections can be read from the scale. The left hand scale is calibrated in g/100 ml and the right hand scale in refractive index.

14. Hold the refractometer horizontally with the cover plate upwards. Open the cover plate and place 2 drops of distilled water on the face of the prism. Close the cover plate and rotate the refractometer so that the cover plate is downwards.

15. Hold the refractometer towards the light and focus the scale by turning the eyepiece.

16. The interface boundary line should read a refractive index of 1.333 (right-hand scale). The scale can be adjusted with a small screwdriver, if the interphase does not read 1.333.

17. Wipe off the water with a soft tissue and fully dry.

18. Place two drops of plasma, or serum, of the prism and close the cover plate.

19. Read off the serum protein concentration from where the interphase line crosses the left-hand scale. Multiply this value by 10 to obtain g/l.

20. Wash the prism and the cover plate with distilled water to remove all traces of protein, then blot dry with paper tissues.

Reference Range:

Total serum protein : 59-82 g/l.

If plasma is used, the reference range will be the above plus the fibrinogen concentration which, on the average, is 3 g/l.

REFERENCES:

Coulometry.

Durst. R. A., (1994), Electrochemistry - Coulometry, part of Chapter 5 in Tietz Textbook of Clinical Chemistry, 2nd. Ed., Eds. Burtis, C. A. and Ashwood, E. R., p 177-178, W. B. Saunders Co.

Weissman, N. and Pileggi, V. J., (1974), Inorganic Ions - Chloride, part of Chapter 19 in Clinical Chemistry - Principles and Technics, 2nd. Ed., Eds. Henry, R. J., Cannon, D. C. and Winkelman, J. W., p 712-720, Harper and Row.

Buchler-Cotlove Chloridometer Instruction Manual.

Osmometry.

Garcia-Webb, P., (1988), Osmosis and Osmometry, Chapter 20 in Principles of Clinical Biochemistry - Scientific Foundations, 2nd. Ed., Eds. Williams, D. L. and Marks, V., p 289-296, Heineman Medical Books.

Freier, E. F., (1994), Osmometry, Chapter 6 in Tietz Textbook of Clinical Chemistry, 2nd. Ed., Eds. Burtis, C. A. and Ashwood, E. R., p 184-190, W. B. Saunders Co.

Dufour, R., (1993), Osmometry - The Rational Basis for Use of an Under-appreciated Diagnostic Tool, A booklet published by Advanced Instruments Inc.

Knauer Electronic Semi-micro Osmometer and Advanced Instruments 3MOplus Micro Osmometer Instrument Manuals.

Refractometry.

Willard, H. H., Merritt, L. L., Dean, J. A. and Settle, F. A., (1981), Refractometry, part of Chapter 14 in Instrumental Methods of Analysis, 6th Ed., p 403-407, Wadsworth Publishing Co.

Atago Refractometer Instruction Manuals.

APPENDIX 1

LABORATORY RULES

1. Students are required to wear laboratory coats (or gowns) and protective footwear for all laboratory sessions. You are also advised to wear goggles when handling corrosive or infective material.

2. No smoking, drinking or eating is allowed in the laboratory.

3. No mouth pipetting is allowed in the laboratory. "Pi- pumps", "Pumpetts" or similar pipette fillers should be use.

4. You are responsible for the apparatus in your locker. All breakages should be replaced from the Preparation Room. You will be charged the full (or part) replacement cost.

5. All spillages should be mopped up at once, water being used to dilute strong reagents. Jars of $NaHCO_3$ are available, on the benches, to neutralise acid spills.

6. Solid waste should be placed in the bins provided and not washed down the sink. (There is a separate bin for broken glass).

7. Any damage to laboratory equipment, including the spillage of reagents into equipment, should be reported to a demonstrator immediately.

8. All instruments must be treated with care and used according to the makers' instructions. If in doubt, ask a demonstrator.

9. Special care must be taken with spectrophotometer and fluorimeter cuvettes:

   Quartz cuvettes can be used for the wavelength range 200 - 1,000 nm. Special ultra-violet grade silica cuvettes have to be used below 200 nm. Glass cuvettes can be used for wavelengths down to 320 nm.

   When using wavelengths above 320 nm, use glass cuvettes where possible, as they are much cheaper to replace, if damaged or broken.

   Pasteur pipettes must not be used to fill or empty cuvettes as these can scratch the optical surfaces.

   You should ensure that cuvettes are filled to about 15 mm from the top. Further filling can lead to spillages into the spectrophotometer or fluorimeter. If spillages do occur, please wipe them up immediately and report the spill to a demonstrator. Underfilling can lead to incomplete cover of the light beam.

   The contents of a cuvette can be mixed with a plastic cuvette stirrer (never a Pasteur pipette) or by placing parafilm on the top and inverting the cuvette a few times.

   Do not stand cuvettes, or any other apparatus, on the spectrophotometers.

Spectrophotometer cuvettes should only be handled by the ground surfaces and not by the optical surfaces. The optical surfaces should be wiped clean and dry with paper tissues before each measurement.

Cuvettes should be cleaned by thoroughly rinsing them with tap water followed by repeated rinsing in distilled water. They should then be blotted dry with paper tissues. For work below 220 nm, the cuvettes should have an additional wash in ethanol (outside as well as inside).

Cuvettes can be drained by inverting them (after decanting the contents) and standing them on a wad of paper tissues. This wad of tissues should be placed on the bench beside the spectrophotometer.

APPENDIX 2

RADIOISOTOPE LABORATORY RULES

1. Laboratory coats must be worn at all times.

2. Eating, drinking, smoking and the application of cosmetics are forbidden.

3. Disposable plastic gloves must be worn when handling radioactive material. These should be monitored (if applicable) before removal. They must be removed before counting the samples or leaving the laboratory. Use paper tissues to turn on taps or switches and to handle the monitor while you are manipulating radioactive material.

4. All work, other than counting, must be carried out on the trays provided which are lined with absorbent paper.

5. The bench surface in the working area should be covered with a detachable non-absorbent material.

6. When possible, a vessel containing a radioactive solution should be placed inside a larger vessel so that any spillage will be retained.

7. No mouth operations are allowed. Pipetting must be carried out with Oxford type pipettors, Propipettes, Pi-pumps, syringes or automatic pipettes, etc. Self-adhesive labels should be used.

8. Work should be carried out in a fume hood where radioactive material is likely to enter the atmosphere.

9. Where a radioactive sample is to be kept, the container should be labelled with "radioactive" label tape as well as the date, isotope and quantity of activity. These labels must be removed from the containers before disposal after the containers have been decontaminated.

10. If a liquid spill occurs:

    (a) Drop a handful of paper tissues on the site of the spill.
    (b) Report the spill to a member of staff.
    (c) Put on plastic gloves, if not already worn.
    (d) Mop up the spill with paper tissues and wash the area with an alkali detergent.
    (e) Monitor the surface, if applicable.

11. If a solid spill occurs:

    (a) Cover the spill with damp paper tissues.
    (b) Report the spill to a member of staff.
    (c) Put on plastic gloves, if not already worn.
    (d) Mop up the spill with damp paper tissues and wash the area with an alkali detergent.
    (e) Monitor the surface, if applicable.

12. Contaminated material must be placed in the appropriate container: combustible waste, non-combustible waste, pipettes and glassware other than pipettes. Liquid waste can be washed down the sink but must be accompanied by copious volumes of water. Washing up will be carried out later by a member of staff. Disposable plastic waste should be washed before being incinerated.

13. Should you become contaminated, ask someone to turn on the tap or use a wad of paper tissues to turn on the tap. Thoroughly wash the contaminated area with an alkali detergent and water; monitor (if applicable) after washing. A shower is provided for gross contamination.

14. Do not handle radioactive material if you have any uncovered cuts.

15. Unnecessary materials such as textbooks, handbags, etc., should not be brought into the radioisotope area.

16. Wash your hands and monitor them (if applicable) before leaving the radioisotope laboratory.

17. Radioactive material must **not**, under any circumstances, be removed from the radioisotope laboratory.

18*. Be aware of the nature and intensity of the radiation emitted by the isotope you are using.

*Radiation Properties of Some Commonly Used Isotopes:

^{57}Co.

Half-life	:	270 days.
Main γ energy	:	122 keV (+ x-rays following electron capture to give ^{57}Fe).
Total decay energy	:	837 keV.

^{125}I.

Half-life	:	60 days.
γ energy	:	35.5 keV (+ x-rays following electron capture to give ^{125}Te).
Total decay energy	:	149 keV.

^{131}I.

Half-life	:	8 days.
Maximum major β energy:		606 keV.
Major γ energy	:	365 keV.
Total decay energy	:	970 keV.

^{3}H (Tritium).

Half-life	:	12.3 years.
Maximum β energy	:	18.6 keV.

^{14}C

Half-life	:	5,730 years.
Maximum β energy	:	156 keV.

^{32}P.

Half-life	:	14.3 days.
Maximum β energy	:	1.71 MeV.

APPENDIX 3

LABORATORY REPORTS

Raw Laboratory Data.

Before you leave the laboratory, you must hand in your raw laboratory data (ie. emission or absorbance readings, titration volumes, etc.) to your demonstrator. This should include your name, that of your partner(s), the date and the title of the practical exercise.

Laboratory Reports.

1. Each written report should be headed with:

a. Name of student;   b. Name of partner(s);
c. Exercise number;   d. Exercise title;
e. Date.

2. The report should consist of an introduction, methods, results, discussion and references.

    The introduction should be short, setting out the aims of the exercise.

    The methods section should be a brief synopsis of the methods used. Flow diagrams are appropriate.

    The results section should include all calculations, tables, graphs, recorder (or integrator) traces and absorption spectra etc. Tables and graphs (on graph paper) should be fully labelled so that it is not necessary to read through the text to discover what the table or graph represents.

Where appropriate, plot the equivalent plasma (or serum) concentration and include the reference range as a bar on the graph. Mark individual readings of the determinations of the unknown on the graph. Eg (From exercise 4-1):

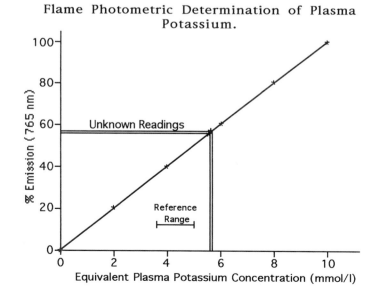

Flame Photometric Determination of Plasma Potassium.

A-6

Most standard curves (response plotted against concentration) will give a straight line. If it is stated that "There is deviation from Beer's law", a plot similar to that shown on page A-12 is the likely result. Use a curve of best fit when drawing a curved graph line. "Flexicurves", which can be bent to give the curve of best fit, can be obtained from scientific bookshops etc.

The curves that you will encounter will be simple. With the exception of some immunoassay and acid-base plots, you are not be likely to encounter sigmoidal curves or any other complex functions. Hence, the curve on the right, below, is the correct way to plot the series of results shown:

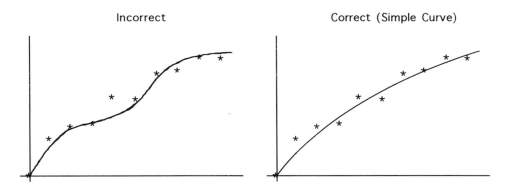

Incorrect   Correct (Simple Curve)

Remember that the blank is the zero standard. It should be treated like any other standard. Hence, the graph line may not go through the origin.

With clinical assays, such as the plasma potassium determination, duplicate values should be averaged and these should be clearly stated, together with the reference ranges. Eg:

Result   :   Unknown Plasma [K]   =   5.6 mmol/l.

Reference Range   :   Plasma [K]   =   3.6-5.0 mmol/l.

The discussion section should include a segment on the physical principles involved in the instrumentation and the chemical principles of the techniques used in the exercise. The clinical applications of the technique, together with a segment on the pathological conditions that produce values outside the reference ranges should be discussed (where appropriate). A brief conclusion may also be appropriate with some exercises.

Any references used in the report should be fully listed in the references section at the end of the report.

Practical reports must be handed in at the following laboratory session. It is difficult to assess your results against the rest of the group, when you hand in a report weeks after the rest of the group's reports have been corrected. It is equally difficult for you to report on an exercise that you did weeks ago.

A loose-leaf binder is recommended for your reports. Unmarked reports should be handed in, one at a time, in a manilla folder. The advantages are that:

1. You retain all of your other reports for study, etc.

2. An accidental loss of a report is not as bad as loss of many reports.

3. It is more convenient for the demonstrator take a stack of them home for marking, when the folder only contains one report.

Write your name on the front of the folder; please indicate which group you attend (if there are a number of different groups doing the exercises).

It is convenient to paste a mark sheet to the inside of the folder so that it is easy to check your marks, if necessary. Make sure that the exercise has been listed on the mark sheet in the front of your folder before you hand it in.

When your corrected reports are returned, please read the comments carefully. If you do not understand any of them, discuss the matter with your demonstrator.

APPENDIX 4

LABORATORY AND REPORT TIPS

1. Before the practical session, read through the exercise and plan your work in view of incubation periods, waiting to use instruments, etc.

2. In addition to the equipment supplied in your locker, you should obtain a laboratory coat (or gown), a Pi-pump and a marking pen (eg. textacolour) which must be waterproof. Acetone is provided for removal of textacolour from glassware.

3. Where a selection of instruments is available (eg. spectrophotometers), alternate the instruments you use, so that you familiarise yourself with all of them.

4. Always cap tubes to exclude condensation dripping from the lid of the water bath, when incubating.

5. With bench centrifuges, make sure that the rubber pads are in the centrifuge buckets and that the bucket trunnions are properly seated in the bucket rotor.

6. Do not leave centrifuges until they have attained enough speed to indicate that they are properly balanced.

7. Do not take bench centrifuges above 3,000 rpm, when using glass centrifuge tubes. 2,000 rpm is adequate for most purposes.

8. Do not stand cuvettes or tubes etc. on spectrophotometers or other instruments.

9. Do not exhaust the air from rubber pipette bulbs in the direction of your face, as these can become contaminated with corrosive reagents.

10. Always check that waterbaths or heating blocks are at the correct temperature.

11. Use a cement mat under burners to protect the bench surface.

12. Label recorder charts fully with time, start, finish, absorbance etc., as soon as you remove them from the recorder.

13. Do not leave the benches untidy.

14. Use the following method for the quick preparation of dilutions:

    Dilute solution A (x mol/l) to produce solution B (y mol/l).

    Take y ml of solution A and make to x ml with H_2O.

    Eg.   The preparation of 6 M HCl from concentrated (11.6 M) HCl.

    Dilute 6.0 ml of conc HCl to 11.6 ml. (Add the acid to H_2O.)

    6.0 ml of 11.6 mol/l HCl contains $6.0 \times 11.6 = 69.6$ mmol HCl.

The diluted solution is, therefore, 69.6 mmol in 11.6 ml.

ie. $\dfrac{69.6}{11.6}$ = 6.0 mmol/ml (ie. 6 M).

Diluting 60 ml of concentrated HCl to 116 ml would probably give a more useful volume.

15. Understand the use of significant figures when expressing results. This is illustrated below:

Reported Figure	Number of Significant Figures	Implied Limits	Implied Error (±%)
1	1	0.5 - 1.5	50
1.1	2	1.05 - 1.15	5
1.11	3	1.105 - 1.115	0.5

As a rough guide, if the method has an error (2 x the coefficient of variation) of about 10%, use 2 significant figures. If it has an error of about 2%, use 3 significant figures.

A more precise guide states: "The implied limit of the last significant figure retained should exceed $\frac{1}{4}$ of the standard deviation (σ) for the measurement."

Eg. For a plasma [sodium] of 140 mmol/l, σ has been found to be 2. Then an observed value of 140.8 should be reported as 141 mmol/l. The implied limit on 0.8 is 0.1. This does not exceed 0.5 ($\frac{1}{4}$ of 2). Hence, one less significant figure is used.

16. Learn to use words like precision, sensitivity, specificity and accuracy correctly. For an assay, these are:

Precision.

The degree of spread of a series of results, on the same sample, around the mean value of those results.

Precision is usually expressed in terms of a range from the result:

Eg. 2.5 ± 0.3 mmol/l (95% confidence limits; ie. ± 2σ).

This means that you are 95% sure that the correct value of that result lies between 2.2 and 2.8 mmol/l.

Where a plot of the measured response, (eg. Absorbance) against the concentration (or amount) of the substance is a curve, or tails off at high concentration, maximum precision will be obtained where there is a maximum change of response for a minimum change in the amount of substance (or concentration), as illustrated below:

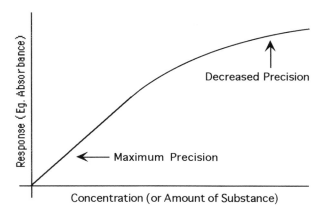

Sensitivity.

The minimum significantly detectable amount. It is the smallest amount that can be distinguished from zero (usually with 95% confidence).

Specificity.

Freedom from interference or cross-reactivity by other substances present.

Accuracy.

The overall closeness of a result to the true concentration of the substance. It depends on the above three factors plus the purity of the standards.

17. Be sure you understand the difference between standards, calibrators and controls (reference samples).

Standards are solutions of known concentration of the pure substance being assayed, against which the unknown samples are compared. They may enter the assay at any suitable stage and are often prepared fresh for each assay. Standards can be divided into primary, secondary and clinical standards.

Primary standards are pure chemical substances used to produce a solution of known concentration. Secondary standards can not be obtained in a pure dry form. However, they can be standardised by titration with a primary standard, or by some other analytical method. Clinical standards can not be obtained pure or standardised against a primary standard. Some examples are glycoproteins such as some hormones, immunoglobulins and some enzymes. These are purified to acceptable limits and assigned a value by an international body (Eg. International Units etc.).

Calibrators are provided by some instrument manufacturers. They are designed to replace standards for specific instruments. Rather than being pure solutions, they usually have a matrix similar to the samples to be analysed (Eg. serum or plasma).

A control (reference sample) is a sample prepared in a large batch, dispensed into vials and frozen or freeze dried. Vials can be thawed out, or reconstituted, and analysed over a period of time. A control is meant to be identical to a typical unknown sample. ie. it has a similar chemical composition and physical characteristics to the sample being analysed. They contain interfering compounds (etc.) not present in the standards.

Pooled serum and bovine serum (no hepatitis or HIV risk) are commonly used. Alternatively, commercial products can be used. These are freeze dried and can be obtained already assayed or unassayed. Assayed controls should not be used instead of primary standards.

Controls are used for:

1. Estimating precision.
2. Detecting analytical variations due to reagent or instrumental defects.

In a hospital laboratory, one or more controls should be run with every assay, entering the assay at the beginning of the assay procedure, with the other unknown samples, and preferably assayed blind (not known to be a control by the analyst). All the assay results should be accepted or rejected on the result for the control. Under no circumstances should all of the results be altered by a factor that would make the control result be within its acceptable limits.

18. Be sure you understand the difference between an analytical instrument and a machine.

An analytical instrument is a piece of scientific apparatus for measuring specific parameters. (Eg. a spectrophotometer.)

A machine is an apparatus for applying mechanical power. (Eg. a centrifuge.)

APPENDIX 5

THE GREEK ALPHABET

α β γ δ ε ζ η θ ι κ λ μ	Α Β Γ Δ Ε Ζ Η Θ Ι Κ Λ Μ	Alpha Beta Gamma Delta Epsilon Zeta Eta Theta Iota Kappa Lambda Mu	ν ξ ο π ρ σ τ υ φ χ ψ ω	Ν Ξ Ο Π Ρ Σ Τ Υ Φ Χ Ψ Ω	Nu Xi Omicron Pi Rho Sigma Tau Upsilon Phi Chi Psi Omega

Greek Letters Used in This Book.

α : α globulins (α_1 and α_2 bands) : Proteins that migrate behind albumin on electrophoresis.
α band : Haemoglobin derivative absorption band nearest the red.
α chain : Heavy chain of IgA.
α particle : Helium nucleus.
[α] : Specific rotation.
Solubility of CO_2 in plasma, at $37\,^{\circ}C$.
Relative retention ratio - Separation factor (HPLC).

β : β globulins (β_1 and β_2 bands) : Proteins that migrate behind the α globulins on electrophoresis.
β band : Haemoglobin derivative absorption band second nearest the red.
β particle : An electron.

γ : γ globulins (γ band) : Proteins that migrate behind the β globulins on electrophoresis. (They remain around the origin.)
γ chain : Heavy chain of IgG.
γ ray : A high energy photon.
Activity coefficient.

Δ : Deletion of nucleotide(s) (molecular diagnostics).

ε : Molar absorption coefficient.

η : Refractive index.
Viscosity.

κ : An immunoglobulin light chain.

λ : An immunoglobulin light chain.
      Wavelength.

μ : μ chain : The heavy chain of IgM.
      Linear flow rate in GC and HPLC.
      Micro (10^{-6}).
      Ionic Strength.

ν : Neutrino.
      Frequency.
      $\overline{\nu}$: Antineutrino.

π : 3.1416.
      Type of electron or bond.
      π^* : Electron anti-bonding orbital.

σ : Standard deviation.

Σ : The sum of.

φ : Osmotic coefficient.
      ϕf : The quantum efficiency of fluorescence.

Ω : Ohms.

APPENDIX 6

SI UNITS

The Conference Generale des Poids et Mesures used the term "Systeme International d'Unites" (SI) in 1960 for a system of suitable and preferred units within the metric system. This particular set of units had been agreed upon by the conference in 1954.

In 1973 the Royal College of Pathologists of Australia recommended the introduction of this set of units into hospital laboratories. Most other countries have also adopted this system.

The basic SI Units are:

Physical Quantity	Name	Symbol
Length	metre	m
Mass	kilogram	kg
Time	second	s
Electric current	Ampere	A
Thermodynamic temperature	Kelvin	K
Luminous intensity	candela	cd
Amount of substance	mole	mol

Derived Units.

These are derived from the basic units above. Eg:

Area	:	m^2
Volume	:	m^3
Speed	:	ms^{-1} or m/s
Concentration	:	$mol\ m^{-3}$ or mol/m^3

Named Derived Units.

These are similar to those above but have been given a name. Eg:

Physical Quantity	Name	Symbol	Derivation
Energy	Joule	J	$kg\ m^2\ s^{-2}$
Force	Newton	N	$Jm^{-1} = kg\ m\ s^{-2}$
Power	Watt	W	$Js^{-1} = kg\ m^2\ s^{-3}$
Frequency	Hertz	Hz	s^{-1}
Pressure	Pascal	Pa	$Nm^{-2} = kg\ m^{-1}\ s^2$
Enzyme activity	katal	kat	$mol\ s^{-1}$

The Use of SI Units in Clinical Biochemistry.

Not all SI units have been introduced as yet. The use of Celsius for temperature is still widely used and has been used in this book.

The retention of the litre (l or L) for the unit of volume is still widely used.

The litre was redefined in 1964 as $10^{-3}m^3$. Concentrations currently expressed in mmol/l are identical with mol/m^3.

Concentration units used in this book also refer to Molar (M) or milliMolar (mM) solutions. These units are mol/l and mmol/l respectively.

% solutions are also used in the book. Unless stated otherwise, this refers to weight/volume (w/v), ie. grams per 100 ml of solution. This notation is retained because of the convenience for the preparation of these solutions.

Hydrogen ion concentration is still often expressed as pH.

$$pH \quad = \quad Log_{10} \frac{1}{[H]}$$

mm of mercury (Torr) is still commonly used as a unit of pressure.

$$
\begin{aligned}
1 \text{ mm Hg} \quad &= \quad 133.3 \text{ Pascals.} \\
1 \text{ kPa} \quad &= \quad 7.502 \text{ mm Hg.}
\end{aligned}
$$

Subdivision of Units.

Units should only be subdivided or multiplied in terms of multiples of 1,000 and not in the 100's of 10's. The prefixes to use are:

Fraction	Prefix	Symbol
10^{-21}	zepto	z
10^{-18}	atto	a
10^{-15}	femto	f
10^{-12}	pico	p
10^{-9}	nano	n
10^{-6}	micro	μ
10^{-3}	milli	m
10^{3}	kilo	k
10^{6}	mega	M
10^{9}	giga	G
10^{12}	tera	T

This system means that the correct subdivision of the m^3 is the mm^3. As there are 10^9 mm^3 in a m^3, it will mean the use of very large numbers when the litre is replaced by the m^3.

Expression of Numbers.

A number should never commence with a decimal point, ie:

0.24 not .24

Numbers should, where feasible, be expressed between 0.1 and 1,000. Eg:

3.29 mmol/l not 0.00329 mol/l

It is recommended that numbers consisting of many digits should be arranged in groups of three with a full space, not a comma, separating them. Eg:

4 265.610 42 not 4,265.61024

However, the comma has been used in this book, as the space can be somewhat confusing.

APPENDIX 7

UNITS OF ENZYME ACTIVITY

This appendix is based on Chapter 4 (Units of Enzymic Activity), p 26-27 of Enzyme Nomenclature - Recommendations (1972) of the International Union of Pure and Applied Chemistry and the International Union of Biochemistry, Elsevier Scientific Publishing Company.

The presence of an enzyme is generally recognised by the occurrence of the chemical reaction that it catalyses, and the amount of the enzyme present may be determined by measuring the rate of the reaction, which, under suitable conditions, is proportional to this quantity.

In the older literature, the rate of reaction was expressed in a variety of ways, often directly related to measured quantities such as change in absorbance, volume of gas absorbed or evolved, or volume of a titration reagent used. In the field of Clinical Biochemistry, enzyme activity was often expressed in units named after the authors of the method used. For example the King-Armstrong Unit was used to express the activity of alkaline phosphatase.

The Enzyme Commission, in 1961, proposed that the reaction rate should be expressed as micromoles of substrate converted per minute, and the enzyme unit was defined as "that amount (of enzyme) which will catalyse the transformation of one micromole of the substrate per minute under standard conditions".

In the above definition, the precise meaning of the term 'amount (of enzyme)' is not specified. Since the definition is meant to apply to enzymes that have never been isolated and weighed, 'amount' cannot be identified with mass. It certainly does not correspond to the chemical concept of 'amount of substance', which can only be defined if the chemical identity and the molecular weight have been established. 'Amount' is then, in this context, a completely abstract concept that is useful for certain computational purposes; it is **not** a determinable physico-chemical quantity.

On the other hand, many enzymes have now been isolated, weighed and characterised chemically. It has, therefore, become desirable to regard enzymic activity in the same manner as other physico-chemical properties. It should be defined operationally and bear a definite relationship to the mass and the amount of substance.

Accordingly, the Enzyme Commission made the following proposals in 1972:

a.	the concept of 'enzyme unit' as a physically undefined 'amount of enzyme' be abandoned;

b.	'enzymic activity' be defined as the rate of reaction of substrate that may be attributed to catalysis by an enzyme;

c.	the phenomenological coefficient that relates the activity under specified conditions to the mass, be called 'specific activity';

d.	the phenomenological coefficient that relates the activity under specified conditions to the amount of enzyme substance (usually expressed in moles) be called the 'molar activity'.

The activity of enzymes that have not been isolated can be measured, but not resolved into its constituent factors, mass and specific activity.

The Katal as the Unit of Enzymic Activity.

The Commission proposed that the unit in which enzymic activity is expressed by the **amount of activity that converts one mole* of substrate per second.** The new unit of enzymic activity is named the **'katal'** (symbol **kat**).

*   The mole has been defined as the amount of substance of a system that contains as many elementary entities as there are carbon atoms in 0.012 kilograms of the nuclide ^{12}C.

The commission realised that an activity of one katal, corresponding to the conversion of one mole of substrate per second, will very often be too great for practical use. In most cases, activities will be expressed in microkatals (μkat), nanokatals (nkat) or picokatals (pkat) corresponding to reaction rates of micromoles, nanomoles or picomoles per second, respectively.

Enzymic activities expressed in the former (1961) international enzyme units (IU) and in katals may be interconverted by one of the following relationships:

$$1 \text{ kat} = 1 \text{ mol/s} = 60 \text{ mol/min} = 60 \times 10^6 \text{ } \mu\text{mol/min} = 6 \times 10^7 \text{ IU}$$

or

$$1 \text{ IU} = 1 \text{ } \mu\text{mol/min} = \frac{1}{60} \text{ } \mu\text{mol/s} = \frac{1}{60} \text{ } \mu\text{kat} = 16.67 \text{ nkat}$$

The katal, being based on the second as unit of time, fits much better with the rate constants used in chemical kinetics than does the old enzyme unit, since reaction rates are commonly expressed in moles (or appropriate submultiples of the mole) per second.

Ambiguities can arise in some cases from the expression "conversion of one mole of substrate". Examples include reactions of the type $2A \rightarrow B + C$, or with polymers (such as starch) from which monomers are removed consecutively.

In such cases, the rate of reaction may be expressed as the number of cycles of a stated reaction per second. In other words, in the case of a reaction $2A \rightarrow B + C$, two moles of A converted per second represent an activity of 1 katal.

In the second example, one mole of reducing group formed from starch, corresponding to the number of glycosidic bonds broken, forms the logical basis for calculating activity.

It has been recommended that enzyme assays be based (wherever possible) on measurements of initial rates and not on amounts of substrate converted by the end of a given period of time, unless it is known that the reaction rate remains constant throughout this period.

If the reaction rate falls off appreciably during the assay because of the formation of inhibitory products, or, in reversible reactions, because the reverse reaction is no longer negligible, the amount of substrate converted is not proportional to the activity of the catalyst.

Derived Quantities.

A number of other quantities are related to the enzymic activity, as follows:

The 'specific activity' of an enzyme preparation that has been defined above, will normally be expressed as katals per kilogram of protein, or suitable multiples thereof, eg. microkatals per kilogram (μkat/kg).

The 'molar activity' of an enzyme is expressed as katals per mole of enzyme.

The 'concentration of enzymic activity' in a solution is defined as activity divided by volume of solution and is expressed as katals per litre, or suitable multiples thereof, eg. microkatals per litre (μkat/l). This is the unit used for the enzyme activity of body fluids such as plasma and urine.

Assay Conditions.

Note that the conditions, such as temperature and pH, are not specified in the definition of the katal. Hence, they should always be stated (Eg. 2.6 μkat/l at 25°C, pH 7.4).

INDEX

ع

ف

$\mathcal{J}$

$\mathcal{K}$

$\mathcal{L}$

$\mathcal{M}$

$\mathcal{N}$